Building Science for Building Enclosures

Second Edition

John F. Straube
Eric F.P. Burnett

www.buildingscienceforbuildingenclosures.com

ISBN: 978-1-0693381-0-5

Cover photo: John Straube

Contents

Detailed Contents

5. Heat transfer 167

6. Water Vapor Transport 241

7. Air Transport 287

Foreword

That a textbook for building science is needed is an understatement. Authors John Straube and Eric Burnett came to building science from structural engineering. In fact, most people come to building science from somewhere else: civil engineering, mechanical engineering, occasionally from architecture. That this is so highlights a problem. Building science is not a mature discipline, and it rarely exists in university departments; indeed, most universities don't quite know what to do with it. Is it the technical side of architecture? Is it the architectural side of engineering? There is clearly a need *to educate current and future practitioners, especially building enclosure professionals, in the relevant building science*. But who is responsible for doing so?

Structural engineering was once an immature profession. It was raucous, undisciplined, filled with great questions—there were many unknowns and few consistent approaches to solving problems. But it was first and foremost engineering. It was not science, although science was used. Structural engineering was a world of coefficients, safety factors, experience and judgment: imprecise and messy rather than precise and neat like physics. Out of this came limit states design, the concept of loads and load resistance, analysis methods and tools—and maturity.

Building enclosure engineering is destined to undergo a similar evolution. Instead of a limiting state such as deflection, we have decay or corrosion. Instead of a wind load, we have a hygrothermal load. We have analysis methods such as hygrothermal models, and we have safety factors. To make sense of all of this, we need a sound foundation. This textbook fills that need. The structural engineering background and insights of Professor Straube and Professor Burnett provide a rational and welcome underpinning to this textbook on building science. Enjoy and mature.

Joseph W. Lstiburek, Ph. D., P. Eng.

President, Building Science Corporation
Westford, MA
August, 2005

Preface to the 2nd Edition

This second edition is long overdue. As soon as the first was in the hands of students and practioners, many small typographical and numerical errors became obvious. Of course new topics, and changing interests influence a textbook and the intervening 15 to 20 years since most of the first edition was written has resulted in some need for changes. Twenty years of extensive experience and research by the author has added tremendously to the knowledge and refinement of the information.

The second edition has been reviewed several times to fix dozens of typographical and numerical errors, many highlighted by users of the text. This second edition also includes signifcant additions and revisions. Almost 200 new pages of additional and enhanced information has been provided in areas of growing interest and relevance to practise, with significant additions to topics such as thermal bridges, air barriers and airtightness, material selection, modeling, continuous insulation, wind washing, and convection loops. The coverage of control layers has been increased, and the "perfect wall" and "pen test" added. Dozens of drawings, hundreds of new references, and some tabular data have been added, updated, and somewhat enhanced.

Numerous editors and reviewers have been involved: Melissa Pound of RDH took on the onerous task of reading every line closely, making sure formatting was consistent, and the author was intelligble. Former colleagues at Building Science Corporation (Joe Lstiburek, Kohta Ueno, and Peter Baker), life-long co-workers Chris Schumacher, Aaron Grin and Jon Smegal and colleagues at RDH (especially Graham Finch, Lorne Ricketts, and Robert Lepage) have an influence on this edition that range from technical review to expert input on specialist areas. I am truly indebted to all of these contributions.

John Straube, Ph.D., P.Eng.

Waterloo, Ontario
Canada
March 2025

Preface to the 1st Edition

In hindsight, this textbook became a project when I was a teaching assistant for Eric Burnett's CE407 Building Science and Technology course at the University of Waterloo in the early 1990s. I produced handouts from his notes, added drawings, and expanded examples. Research, thinking, and debates that we shared during my graduate degrees helped shape the structure of our thinking about building enclosures, and resulted in a broader view than presented in the course itself. In our teaching we identified the need to update, consolidate, and enhance the standard Canadian building science textbook, Neil Hutcheon and Gus Handegord's *Building Science for a Cold Climate*. In our consulting work we both also saw the need for a professional reference textbook that covered all climate zones and focused on quantitative means of analysis and design. The result is this textbook.

Many students and colleagues provided inspiration, critiques, and information. Over a period of almost ten years, students at the University of Waterloo have been the test bed for this material and have shaped it. Many individuals have contributed to the content and its review, including Chris Schumacher, Joe Lstiburek, Alex Lukachko, Achilles Karagiozis, Dave DeRose, Erik Niit, Xing Shiand, and Pierre-Michel Busque.

Finally, I must extend a sincere personal thank you to Angela Burnett, and Michelle McMullen for their significant editing input, Alex Lukachko for redrawing and greatly improving all of the drawings and undertaking the final prepress formatting, and Stephanie Finnegan and Betsy Pettit at Building Science Press for shepherding us through the process of book publishing.

John Straube

Waterloo, Ontario
Canada
August 2005

First Edition acknowledgment from Eric Burnett

This text has been some 24 years in the making—ever since I started teaching a senior undergraduate /graduate course in Building Science and Building Enclosures at the University of Waterloo. That it is finished is largely due to John Straube, my co-author. Without his energy and drive, it would not have been completed nor would it have been as good. I have been associated with John as a teacher (both undergraduate and graduate), as his graduate supervisor (Masters and Ph.D.), as a mentor and, for the last five years, as a colleague and fellow building engineer.

Others also had a hand in this text: some more directly than others. Ted Happold, for whom I worked at Ove Arup in London was probably the first person to impress upon me the fact that *structure* is not everything and that in the case of buildings it is the people who are really important. Other significant influences were Neil Hutcheon, Gus Handegord, Gint Mitalas, and others who worked in the Division of Building Research (DBR) at the National Research Council of Canada in the 1960s and 1970s. Building professionals in North America owe much to the courses, texts, and publications on Building Science of the old DBR. Working for and consulting to organizations such as the Canada Mortgage and Housing Corporation (CMHC), the U.S. National Bureau of Standards (now NIST), and the Portland Cement Association in Skokie, Illinois, all contributed to my evolving interest in building enclosures and the related building science.

Working for and with Morden Yolles and Roly Bergman at M. S. Yolles and Associates in Toronto in the 1960s as well as designing for the Canadian climate were formative experiences. Later I was retained as a senior consultant to Trow Consulting Engineers for more than 10 years; I was exposed to a multitude of problem buildings, which greatly expanded and consolidated my knowledge and experience with building, especially building enclosure, problems.

For some 30 years I was associated with the University of Waterloo and for the last 15 of those years, I initiated and ran the Building Engineering Group (BEG). Recently, as the Hankin Chair, I have been associated with the Pennsylvania State University and have been responsible for the Pennsylvania Housing Research/ Resource Center for nearly ten years. Promoting building enclosure education and the related science within a civil or an architectural engineering department would be a daunting task were it not for the students. I have supervised—and have worked with and learned from—numerous graduate students, undergraduate students, co-op interns and others. Those individuals who have indirectly or directly contributed to my knowledge and hence this text are:

At the University of Waterloo:
Ahmed Anis
Marcus Dell
Cole Howson
Michael Lubgans
Mark Postma
Amanda Reynolds
Chris Schumacher
Peter Sloof
John Straube
Chantal Wegner
Alex Wielhower

At Penn State:
Ed Bosack, III
Rob Bombino
Tony Chamra
Danko Davidovic
John deGraauw
Brennan Glantz
Jim Mendygral
Joseph Pinon
Xing Shi

For their passion, knowledge and commitment to building science, the influence of many professional colleagues—especially the contributions of Joe Lstiburek, Achilles Karagiozis, and Vlad Stritesky—needs to be acknowledged.

The competent and good-humored assistance of my administrative assistant, Michelle McMullen, has been invaluable. My wife, Angela Burnett, lived with the writing of this book and helped to make it more readable. I am grateful.

Eric Burnett

State College
Pennsylvania
August 2005

Chapter 1

Introduction

1.1 Building science and its significance

Buildings are products that provide spaces for human use and, in many instances, occupancy. Whatever else they do, buildings provide sheltered and, to a varying degree, conditioned spaces. For the occupants, the quality of life is directly dependent on the quality of the environment within the building. Since we North Americans almost 90% of our lives indoors [Klepeis et al 2001], the buildings that we use—as homes or for work, for recreation or as institutions—represent the most basic and the most important of all physical products.

The building stock constitutes one of the largest components of any country's capital wealth, and a significant proportion of a country's total productive effort is spent on producing, operating, maintaining, and modifying these built facilities. It is seldom recognized that over the life of any built facility, the cost of operating and maintaining the facility is usually much more than the initial cost of constructing the building.

The impact of buildings on the environment is also massive. For example, the operation of buildings accounts for one-third of the global energy consumption and 26% of energy-related emissions [UNEP 2022]. The energy and carbon released to mine, harvest, manufacture and transport building materials, defined as embodied energy and carbon, is also significant: it accounts for 10% of all energy-related greenhouse gas emissions. As another example of buildings' environmental impact, approximately 100 billion tons of construction renovation and demolition waste is produced annually.

A large proportion of buildings—both new and old—are deficient or inefficient due to faulty design, faulty construction and inappropriate operation and maintenance. The results on the health, safety, comfort, and productivity of the occupants can be measured along with the financial costs to owners, operators and society as a whole.

Constructing a building is only part of the story. All buildings suffer wear and tear and deterioration. The resulting poor performance may take the form of adversely affected appearance, a loss in durability, or excessive consumption of space-conditioning energy,

The annual cost of building deterioration is estimated to be in the billions of dollars; for instance, the Association of Higher Education Facilities Officers has stated that the burden of operating, maintaining, and repairing their stock of buildings is a threat to the quality of university education [Kaiser 1996]. The continuing and cost of operating, maintaining, and repairing poorly designed and poorly built facilities is an increasing burden that has social and other implications. An appreciation of the scale of these consequences has been growing over the last few decades.

The part of the building that separates the interior from the exterior environment is called the *building enclosure* or building envelope. A large proportion of premature building deterioration results from the failure of the building enclosure to perform as desired. Roofing, wall (above grade and below grade), and glazing system failures account for the majority of building defect claims submitted to insurance companies in North America. That moisture is the largest common factor in these failures underlies the focus of this text.

That buildings perform poorly is, unfortunately, a truism. The reasons for the poor performance of buildings in general, and building enclosures in particular, are complex, interdependent and only partly technical. Nonetheless, the technical aspects of poor performance must be addressed. Providing the best building for any given situation requires both a systematic and a comprehensive approach. Because of the complexities of modern facilities, their design, construction, and operation are the domain of many different specialists, such as architects, structural, geotechnical, mechanical, and electrical engineers, energy consultants, and interior designers. Generally missing from these specialists is someone with the appropriate knowledge to be responsible for the enclosure of the building.

Although a fast-growing field, there are still relatively few building enclosure specialists and the discipline of building science remains poorly defined and is not as widely accepted as it should be. A science-based building enclosure education is beginning to develop in North American schools of engineering. The science or the mechanics of enclosure performance must be understood and documented if building enclosures are to be properly analyzed and designed and the many attendant problems diagnosed and resolved. To meet these needs, especially the pressing need for the assessment and solution of building-enclosure-related problems, a new professional is emerging in North America, the so-called building scientist, and the specific subset, the building enclosure specialist.

To avoid or solve many building problems -- especially enclosure problems – a unique mix of engineering analysis and design, physics, material science, meteorology, construction technology, and even human physiology must be applied. This mix is often called building science. Although initially developed and promoted in Europe (especially Scandinavia) and in countries such as Israel and Australia, where it is called building physics, building science as a professional discipline is becoming better known in North America, especially in Canada. Although no definitive definition is currently possible, one attempt is provided here:

> *Building science* is the interdisciplinary collection of knowledge and experience required to understand and predict the behavior (performance) of buildings and their systems, specifically including durability, comfort, energy, environmental separation, indoor air quality, acoustics, lighting, economics, and constructability.
>
> Building science draws heavily on the allied fields of structural engineering, mechanical engineering, and architecture. It relies on the areas of thermodynamics, the transport and storage of heat, air, and moisture, physics, human physiology, material science, meteorology, chemistry, and solid mechanics, among others.

Most of what is called building science in the building industry is some combination of good building practice, rules of thumb (heuristics), and common sense. But building science has a hard core, based on physics and hard-won, evidence-based experience (including actual measurements), and this science is useful for extending current experience to new building practices, materials, and technologies.

As early as 1971, Neil Hutcheon, the pre-eminent building scientist who led the Canadian Division of Building Research for many years, made the argument for building science [Hutcheon 1989]:

> Trial-by-use, although it was the basis of much of the tradition in building, is by no means outmoded, since satisfactory service is still the real and final proof of adequate performance. There is a vast difference, however, between trial-by-use as the primary way of arriving at prediction and use as a confirmation of prediction based on evidence. .... Tradition places the emphasis on how things should be done; science sets out to explain why so that the experience can be carried over to different materials and circumstances.

The objective of this book is to provide an introduction to the science needed for the analysis and design of building enclosures and the performance prediction of "different materials and circumstances." This objective requires:

- the presentation of the relevant science that governs performance,

- an explanation of the basic principles that should be employed in the design of building enclosures, and
- some coverage of the factors that influence the performance of the buildings and their enclosure elements.

Numerous books [Latta 1973, Baker 1980, Givoni 1976] most notably *Building Science for a Cold Climate* [Hutcheon & Handegord 1983], have tried to fill the same need in the past, each with a different focus or approach. More recently, some books have focused on specific aspects [Harriman & Lstiburek 2009, Rose 2005] while others focused on analytics and physics [Hagentoft 2012, Hens 2016, 2017]. Numerous excellent building technology textbooks exist [Allen 2019, Boswell 2013, Mehta 2017], but unfortunately most, with some exceptions [Brand 1990, Lstiburek & Carmody 1993, Straube 2012 Lstiburek 2020], describe building technology and/or practice without science or quantification.

This text is intended for engineers and technical specialists involved in the design, construction, operation, analysis, repair, rehabilitation, and renovation of buildings. The focus is on the building enclosure, i.e., walls, windows, roofs, basements, slabs, etc. The control of heat, air, and moisture is emphasized because of their critical importance. Other important enclosure considerations such as acoustics, fire, and lightings are part of building science, but they are not the focus of this textbook and are addressed in other texts [Vigran 2014, Stollard 2014, IES 2011].

The context in which the building professional works is also relevant. Understanding the roles of the various parties involved in buildings (e.g., architect, structural, mechanical, geotechnical, and electrical engineer, developer, owner, operator, occupant, lawyer, building official, banker) is important. In addition to their individual expertise, each professional is a member of a team and needs to be aware of the bigger picture. For this reason, the nature of built facilities and their design, delivery, and operation are briefly discussed.

1.2 The built facility

A building— any building —is itself a facility or is part of a larger facility. A *built facility* can be broadly defined as one or more buildings built to serve one or more specific functions. On the other hand, a *civil facility* is a constructed facility, such as a bridge or a hydroelectric power plant, that may or may not involve buildings. The built environment comprises both types of facility; indeed the one, civil facilities, provides the infrastructure for the other, and built facilities provide the infrastructure for human activities.

Consider, for example, a sports complex, a convention center, a hospital, a hotel, or a housing development. Each is a built facility, and each may consist of more than one building. Some of the buildings may be unique, such as a hall or stadium, but they are still part of the built facility. A detached house is also a built facility, but it usually consists of only one building.

Clearly there are many different types of built facility. A town may, for example, have one or more hospitals or health care facilities, several schools and other educational facilities, one or more fire and police stations, various types of residential facilities, a range of commercial and manufacturing facilities, and so on. These built facilities can operate only if an appropriate infrastructure is in place, consisting of civil facilities such as a transportation system, utilities, stormwater control, potable water reticulation systems, and other services. The sum of all the civil and built facilities constitutes the built environment for that location.

This text focuses on built as opposed to civil facilities. A built facility always has at least the following components:

- occupants/users/owners, etc.,
- a building or buildings and the parts thereof,
- a service infrastructure/utilities, etc., and
- a distribution/transportation/traffic handling system.

Some 2,000 years ago the Roman architectural commentator Vitruvius stated that buildings must provide "durability, convenience, and beauty". Later writers such as Alberti [n.d.] and Palladio [n.d.] identified the functions and requirements of a building from a historical and evolutionary standpoint. In 1624 Sir Henry Wotton [n.d.] wrote, in his *Elements of Architecture*, that all buildings need to exhibit "firmness, commodity, and delight": firmness implied strength, stability and durability; commodity implied value in terms of utility and economics; and delight meant that it had to please the eye of the beholder. Other writers have provided longer and more comprehensive lists of functions. However, built facilities and buildings are too complex to be defined by a simple list, especially lists that are a mixture of needs, functions, and attributes.

Most descriptions of buildings focus on aesthetics, culture, materiality, technology, or similar attributes. However, to consider the built facility in a systematic and comprehensive fashion it is necessary to distinguish between needs, functions, and attributes. A built facility must serve **human needs**. But it is designed and built primarily to do something or, put another way, to fulfill a specific purpose. This may be referred to as its overall **function**. Moreover, there are a variety of **attributes** or qualitative characteristics that can be expected of any built facility or building. Each of these is considered in turn.

1.2.1 Human needs

If a built facility or a building is considered from the standpoint of serving human needs in general, it is possible to develop a representational hierarchy of these needs. Irrespective of culture, status, wealth, or education, human beings have basic needs for built facilities.

The most fundamental of needs is for a secure and fixed location and thus a site or place, and hence land and property. This most basic need is the first level in the incremental hierarchy of human needs for built facilities that is proposed here. For a modern facility, the term **location** encompasses the developed, serviced, and landscaped site and incorporates most subgrade constructions and amenities.

The second or next most fundamental level of need is for **shelter** from the climatic and other features (people, animals, insects, etc.) of the exterior environment. In contemporary terms, the word shelter connotes security (physical and mental) and health for persons, activities, and materials. In physical terms, shelter is provided by the combination of the roof, exterior walls (above and below ground), and base floor of the building, i.e., by the collection of assemblies that constitute the separator between the exterior and interior environments.

The third level of need is for some degree of practical separation of the internal spaces as well as the servicing of some of these spaces within the facility—for instance, for purposes of cooking, sleeping, working, etc. A collective term for this level of human need is **utility.** Utility includes not only the internal separators (horizontal, vertical, and otherwise) of internal spaces but also the provision of services (such as power, communications, fresh air, and waste removal), equipment, and material furnishings.

The fourth level of need is for **comfort** or some degree thereof, ranging from the control of smoke or smell to the satisfactory conditioning of all interior spaces. The term comfort also encompasses those furnishings, finishes, and whatever else is needed to provide an appropriate degree of physiological and psychological well-being.

The ultimate level of need is for what may be referred to as **delight**. This level of need is not restricted to what pleases the eye. Delight encompasses esthetic, decorative, cultural, spiritual, and even some psychological considerations (such as pride and ego).

This five-level hierarchy, which borrows heavily from Abraham Maslow, is illustrated in Figure 1.1. Maslow was a sociologist who developed the well-known five-level hierarchy of human needs in general [Maslow 1943]. The hierarchy of human needs for built facilities outlined above illustrates a number of points. There is clearly a hierarchy, and each level is interrelated. For example, delight applies to the site, all relevant internal spaces, and the built facility as a whole.

Furthermore, the hierarchy provides a useful way of viewing the overall building process. The overall building process involves identifying and meeting as many as possible, if not all, of the relevant human needs at each level. In North America we are generally wealthy enough to be able take three, perhaps four, levels for granted; provision of the fifth, delight, is a bit more problematic.

Expectations however vary significantly: the performance required or expected of a farmer's home in 17th century Britain, an auto factory workers home in early 20th century Detroit, and a medical imaging suite of a 21st century hospital are vastly different. The different levels of indoor air quality, thermal comfort, structural safety, and durability for these examples are obvious, even though these buildings met the primary needs.

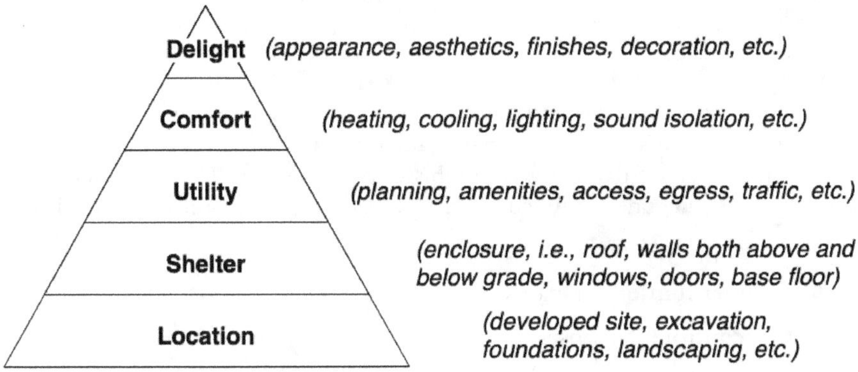

Figure 1.1: The five-step hierarchy of human needs for a built facility

Note: Inherent in all five fundamental levels of human need is the necessity for the appropriate structure to support the building enclosure and meet the need for internal structural separation (horizontal and vertical).

To any student, practitioner, or educator, it should be evident that some knowledge of each and every level of human need is essential, even if only one level (or a subset thereof) is chosen as a specialist professional area. But also note that none of these layers of human need is the sole responsibility of any single profession. How much an individual building professional needs to know is an interesting question, but an even more pertinent issue is the importance of shelter and thus the building enclosure. The educational preparation of most building professionals, whether architects or structural, mechanical, or even architectural engineers is not usually aligned with the hierarchical logic inherent in Figure 1.1.

1.2.2 Functions of built facilities

Clearly, a built facility must meet some or all of the human needs categorized above. Conversely, the built facility has to satisfy all the functional requirements that derive from meeting all the relevant human needs. First and foremost, the built facility has to satisfy one overarching or transcendent function: the provision of a collection of space(s) that make it possible to achieve the purpose for which the facility was built. In short, **the facility's main function is to provide space with a desired environment, or a collection of desired environments, for human use and occupancy.**

The main issue, then, is this: does the facility, i.e., the collection of spaces, do what it is supposed to do? The assembled spaces, i.e., the built product, must make it possible to achieve the overall purpose of the built facility — whether the facility is intended to serve a residential, institutional, commercial, manufacturing, or some other overall purpose or mix of such purposes. Furthermore, each space must not only contribute to the facility's overall purpose but must also serve one or more related but subordinate purposes such as eating, sleeping, working, worshipping, producing, processing, storing, etc.

In summary, the facility and each building must satisfy one overarching or transcendent function. Each building therefore also has to satisfy all of the specific functions that derive from this singular purpose. There are four subsidiary categories of functional consideration for each building, namely:

- **support** functions,
- **distribution** functions,
- **control** functions, and
- **finish** functions.

These four categories are described in some more detail in Figure 1.2. The functions are largely physical in that a specific load must be identified, quantified, and accommodated.

Figure 1.3 illustrates the interrelationship between:

- the built product, i.e., the overall facility, the building(s), and the spaces provided,
- the human needs to be met, and
- all the functions (both the primary function and the subsidiary physical functions) that have to be satisfied.

1. Support:
Accommodate, resist, distribute, and transfer all physical loads, normal, abnormal or exceptional, such as:
• air pressure differentials (wind, HVAC, tornado, hurricane, explosion, etc.)
• ground motion (seismic, settlement, etc.)
• fire
• gravitational effects (dead and live, both static and dynamic)
• impact (vehicles, projectiles, flying debris, etc.)
• volume change (thermal effects, moisture, creep, shrinkage, etc.)
2. Distribute:
Transport or flow into, within, and/or out of the building:
• people (collection and dispersion, e.g., elevators, escalators, etc.)
• goods (collection and distribution, e.g., loading docks, conveyors, elevators, etc.)
• vehicles (distribution, storage, etc.)
• data (telephone and computer links)
• utilities:
o water (hot, cold, potable, gray, etc.)
o air (conditioned, compressed, makeup, etc.)
o gas (natural, propane, refrigerant, etc.)
o electricity (high and low voltage, etc.)
3. Control:
Manage movement into, within, and/or out of the building:
• people (access and egress)
• animals, birds, and insects (access, etc.)
• vehicles (access and egress)
• environmental loads such as:
o precipitation (exclusion of rain, snow, hail, etc.)
o solar radiation (light, UV, etc.)
o heat (temperature)
o air (conditioning loads, water vapor, odor, etc.)
o moisture (water, water vapor)
o sound (quality, quantity)
o light (quality, quantity)
o fire
o gaseous contaminants and particulates (e.g., fire and smoke)
4. Finish:
Provide acceptable appearance and/or suitability of all relevant faces both within and on the outside of the building, considering:
• shape, proportion, color, speculance, reflectance, texture, pattern, relief etc.
• UV exposure, wetting, impact, fire, graphitti, staining

Figure 1.2: Physical functions that a built facility must provide

As shown, the most basic component of any built facility is each individual space. Each space can be identified by means of its:

- location, absolute and relative to other spaces and outside,
- configuration and shape (height, width, breath, area, volume),
- environment (temperature, light, view, indoor air quality, visual, aural, and nasal quality, power supply, telecommunications, etc.), and
- internal finishes, etc.

Having established a hierarchy of human needs and developed various levels and categories of functions, it remains to identify each of the attributes that a built facility needs to demonstrate over its life.

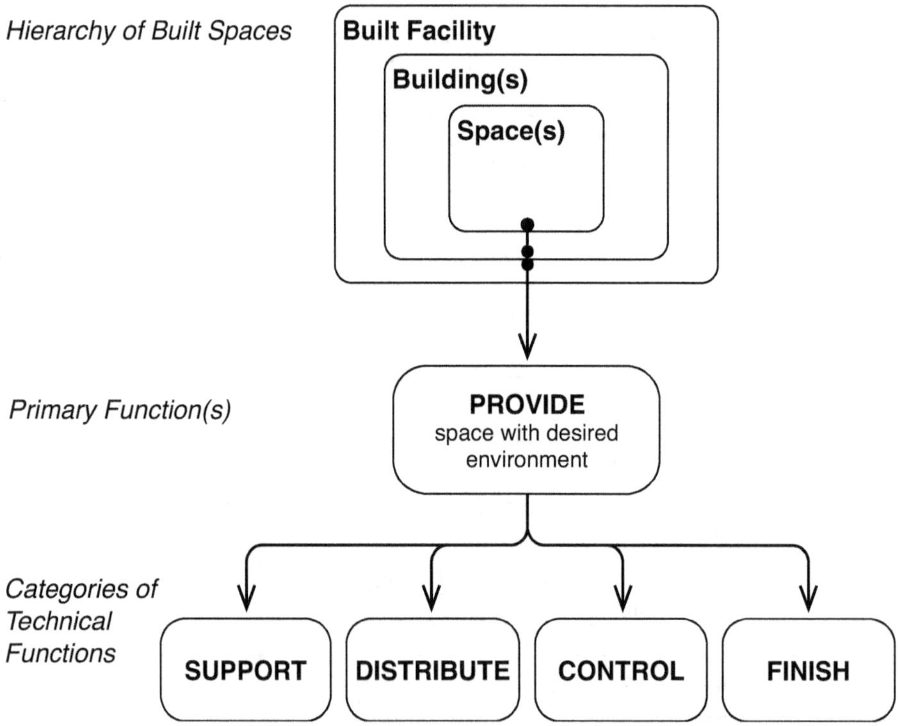

Figure 1.3: Buildings, built facilities, and the functions

1.2.3 Building attributes

The term **attribute** is used here to define a qualitative characteristic that could be expected of a building. A list of common attributes includes:

1. buildability or constructability,
2. affordability, i.e., economic viability or cost effectiveness over both the short and the longer term,
3. safety in relation to life, health, injury, property, and economic enterprise,
4. viewability, including esthetic, cultural, and other visual expectations,
5. utility, including suitability and flexibility,
6. sustainability, including operational environmental impacts, initial embodied energy and carbon, resource use, etc.
7. serviceability (structural and otherwise),
8. operability,
9. maintainability,
10. repairability, including minor and major repairs,
11. durability, including physical, programmatic, and esthetic durability,
12. adaptability in use and performance by modification, extension, or conversion, and
13. disposability, including recyclability.

All of these attributes have to be considered during the stages of conception and design. The first and the last attributes apply to the process of construction and deconstruction respectively. The second and third are always critical, while the second last attribute influences the future. Attributes 4 through 12 all characterize the service life, both the design life and the real service life, of the built facility.

More specific lists can be developed but may have little value without understanding the context: many special buildings and special climates have special needs. However, high-level general lists of considerations for design can be useful.

The most often quoted list of "considerations", expanded from a seminal Neil Hutcheon [1953] paper, includes [Hutcheon 1963]:

1. Control heat flow,
2. Control air flow,
3. Control water vapor flow,
4. Control rain penetration,
5. Control light, solar and other radiation,

6. Control noise,
7. Control fire,
8. Provide strength and rigidity,
9. Be durable,
10. Be aesthetically pleasing, and
11. Be economical.

The major flaw with this list is that it mixes functional requirements, such as "control", with characteristics such as "be durable" or economical, and, oddly, the provision of "strength and rigidity". Although the building science industry has developed many lists with better precision than this simple list, it still encompasses many of the practical aspects of enclosure design.

1.3 Buildings

A building, any building, can be said to comprise four different physical but interdependent parts (Figure 1.4).

- The **structure**, comprised of the superstructure and the substructure below grade, i.e., the part of the building that is primarily involved in accommodating, transferring, or supporting all the so-called structural loads such as gravity loads, wind load, seismic load, etc.
- The **enclosure**, i.e., the part that separates the interior environments from the exterior environments, both above and below ground.

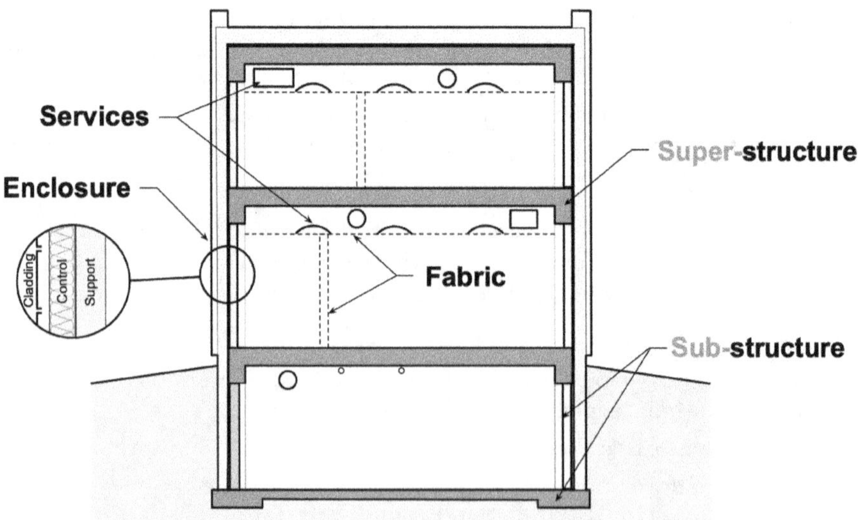

Figure 1.4: The major categories of physical components of a building

- The **service systems**, i.e., the services and related systems (both active and passive) that ensure comfort and utility – for example, space conditioning, water conditioning, waste removal, security, the movement of people and goods, etc.
- The **fabric**, i.e., all those physical components of the basic building, such as finishes, furnishings, fittings, partitions, etc., that cannot be included in the other three parts.

These physical subdivisions are largely reflected in professional areas of focus (i.e., the structural engineer, the services or mechanical, electrical, and plumbing engineer, the interior designer, etc.) and often trade classifications (plumbers, formwork carpenters, millwork, etc.).

One or more of these physical components may appear to be relatively more important than the others. In reality each is an essential and integral part of an operating whole, and the quality of the performance of any building is dependent upon the combined effectiveness of each of the four physical parts. Clearly it does not make much sense for the superstructure to be designed and built without some knowledge and consideration of the enclosure, the service systems, and the building fabric. Similarly, the building enclosure should not be considered in isolation from the structure or space-conditioning equipment.

Moreover, it is worth reiterating here that a building is a part of a larger whole, namely the built facility. Which encompasses the actual building or buildings and the supporting infrastructure, the occupants, and their activities. Figure 1.5 illustrates the linkage between the components of any built facility and the buildings involved, highlighting the position of the building enclosure in the built environment.

It should also be appreciated that the built facility, and thus the building or buildings involved, is the product of at least two different processes. The first is the overall socioeconomic process of supply and demand: the fact that buildings are consumer products and market conditions apply (Section 1.4). The second is the process that encompasses the overall life or life cycle (Section 1.6) of any building or built facility. Both require some discussion.

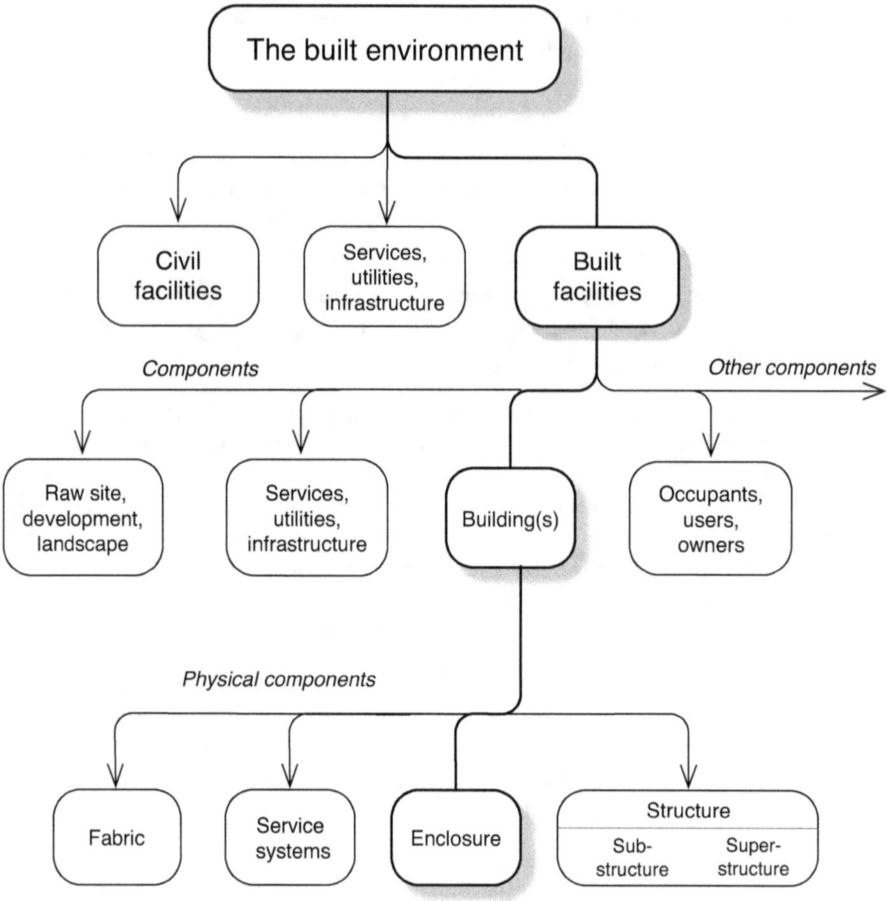

Figure 1.5: Components of built facilities and their buildings

1.4 The socioeconomic market process

Like a car or a box of cereal, a building is a consumer product. However, there are numerous good reasons why the design and construction of buildings do not use exactly the same approaches as those used, for example, for cars, airplanes, or offshore oil platforms.

Buildings are:

- the largest and most expensive of common consumer products,
- individual, not mass-produced, because of variations with regard to location, site, and orientation,

- large and heavy products fixed to land, and hence difficult to transport from a factory in one or a few pieces (thus requiring assembly of many smaller components on the site)
- a large, visible, and intentionally immovable part of the built environment,
- probably the most durable of durable goods and generally exist for long periods of time, and
- not easily disposed of.

Although buildings are too individual and too large to be easily mass-produced, industrial processes have long been used in the production of buildings. The trend has been an increase in the proportion of off-site and mass-produced components. The change has been from the off-site manufacture of hand-hewn stones millennia ago, to pre-assembled cabinetry and heating systems of the past decades, to increasing degrees of prefabrication (components, panels, and modules) today. The assembly-line process applied to other products has been applied by large home builders (by moving people along the line rather than moving the product) since at least the famous Levittown developments in New York State in the 1940s.

Like any consumer product, the production of buildings is generally subject to market forces, such as supply and demand and the relevant socioeconomic conditions linking supplier (or developer) and user (or occupant). This context is shown, somewhat simplistically, in Figure 1.6.

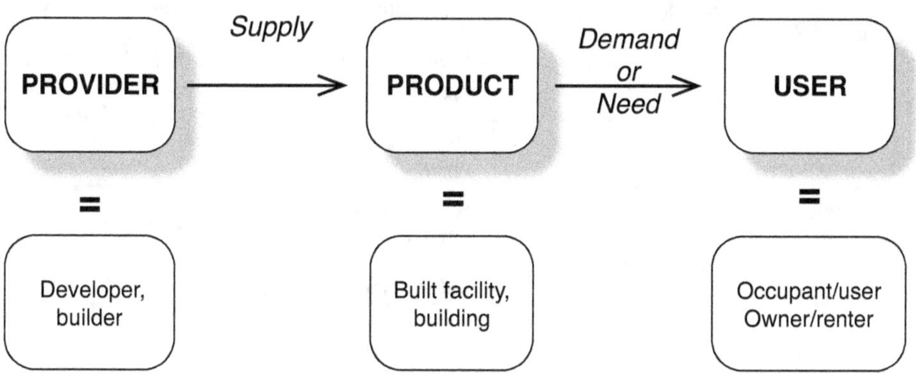

Figure 1.6: The overall market process

Figure 1.6 reinforces the point that most building professionals are in fact members of a team of facilitators in the delivery process for a building. While they might act as developers, their professional careers usually focus on designing, contracting, and constructing buildings that they rarely operate or occupy. Obviously, some knowledge of the social, economic, and political context is necessary in order to design, build, or operate any built facility satisfactorily, but this topic is not addressed any further in this text.

1.5 The product

In this text, we are generally concerned with buildings and built facilities, and in particular with their enclosures. The building as a product varies significantly.

Table 1.1 categorizes the various types of built facility and identifies the associated buildings. Note the following points:

Table 1.1: Types of built facility

Categorization by:		**Nature of Building**	**Comment**
General use	**Size, form, use**		
Residential	Single unit (detached)	Houses*	
	Single unit (attached)	Semi-detached* (2 units), duplex*(2 units), triplex*(3 units), quadraplex*(4 units) row housing*, town homes*, etc.	<4 stories = low-rise ≥4 but ≥12 = mid-rise > 12 = high-rise
	Multiple units	Apartment Dormitory Motel Hotel	
Service Industrial	Storage	Warehouse + Cool storage Freezer storage	Often single story (high bay) structures with perhaps a small mezzanine for offices and light storage.
	Manufacturing	Light industrial +	
	Mixed-use/multipurpose	Production, storage, office +	
Commercial	Retail Office Mixed-use	Shop, plaza, mall Low-, medium- and high-rise Office/retail/residential	
Health	Hospitals, clinics, etc.		
Educational	Schools, colleges, universities, etc.		
Institutional	Prisons, justice facilities, civic buildings, etc.		
Monumental	Museums, galleries, monuments, etc.		
Industrial	Medium to heavy industrial facilities, factories, etc.		

Notes: * In most instances, these buildings may be designed and built without direct professional engineering or architectural involvement.

+ Smaller and/or simpler examples of these buildings can be built without direct professional engineering or architectural involvement.

- Most buildings (those listed with an asterisk) do not necessarily involve any direct input from or participation by members of the engineering or architectural professions. These buildings include most single-family houses as well as relatively small multi-unit residential, commercial, and light industrial buildings. The amount of time and money spent on their design is often minimal. Also note that the smaller the building, the greater the ratio of enclosure area to usable square footage. In addition, the various types of non-engineered buildings are less subject to codification, quality control, regulation, and inspection.
- Small buildings such as sheds, cabins, or temporary lightweight structures are excluded from the categorization in Table 1.2.
- Many buildings are intentionally designed to be mixed-use facilities.

1.6 The life cycle of a built facility

Just like any other product, a building has to be delivered, used, and eventually disposed of. The overall life or the life cycle, from conception to disposal, of most engineered buildings involves at least eight well-defined stages, usually sequenced as follows (Figure 1.7):

1. conception,
2. design, including conceptual, preliminary, development, and final design stages,
3. bidding or tendering or the main contractual stage,
4. building or construction, i.e., fabrication, assembly, and finishing,
5. acceptance or handover stage, the final step of the building commissioning process,
6. operation, repair, and maintenance,
7. renewal, conversion, and/or adaptation, often involving reconception and design, and eventually
8. disposal, including demolition, reuse, and recycling of the parts, i.e., deconstruction.

The order of the stages can be varied, and activities within stages can overlap (consider, for example, a design-build project). Also, the significance of a stage can be reduced or minimized, or a stage such as the bid or commissioning stage may be omitted. The figure also illustrates the effect of wear and tear and the need for repair, as well as the distinct possibility of conversion or rehabilitation. In addition, the possibility of premature failure of the facility—due to business, structural or operational conditions—is recognized.

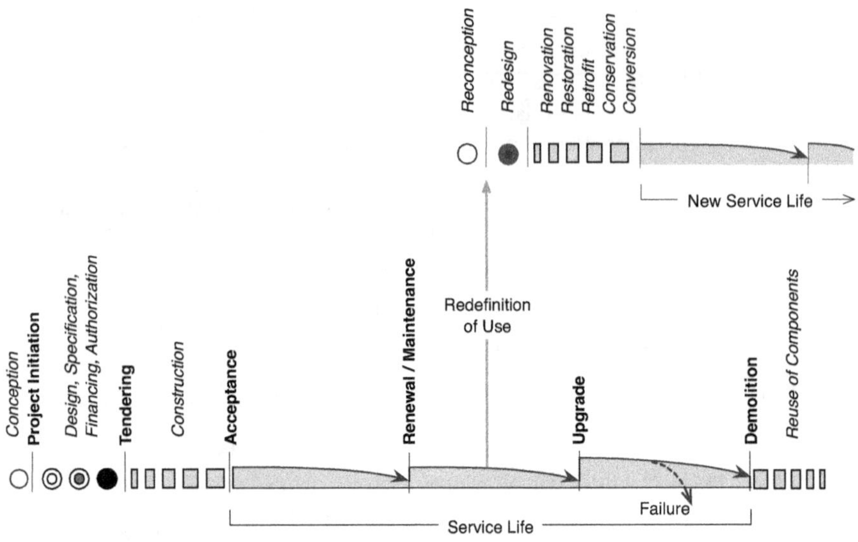

Figure 1.7: Phases in the life of a building (not to scale)

These eight basic stages can be rearranged to better demonstrate the nature of the overall life-cycle process, as shown in Figure 1.8. For instance, it is significant that the process is an ordered sequence of stages, four of which are largely directed at making a major decision, and each of which is followed by an execution stage. Moreover, the process can involve a number of iterative loops brought about by conversion or rehabilitation or a major remodeling.

The element of time, specifically the duration of each phase, is very important. In a generalized representation such as Figure 1.7, time is impossible to quantify. However, the service life of a building, i.e., the period of operation, can range from a period of less than one year (e.g., an exhibition building) to well over 100 years (e.g., a cathedral). Similarly, the amount of time required for conception, design, construction, and even demolition can vary significantly.

Figure 1.9 to emphasize the importance of process in the pursuit and achievement of good performance. In this figure, the five relevant teams of participants are identified, and their possible involvement at each of the eight stages during the process is shown. While it is possible for one individual (e.g., the developer, architect, or designer) to be responsible for the conception of a building or built facility, buildings are developed, designed, built, run, and eventually deconstructed by teams. Hence, it can be seen that management, planning, and communication within and between teams are very important considerations.

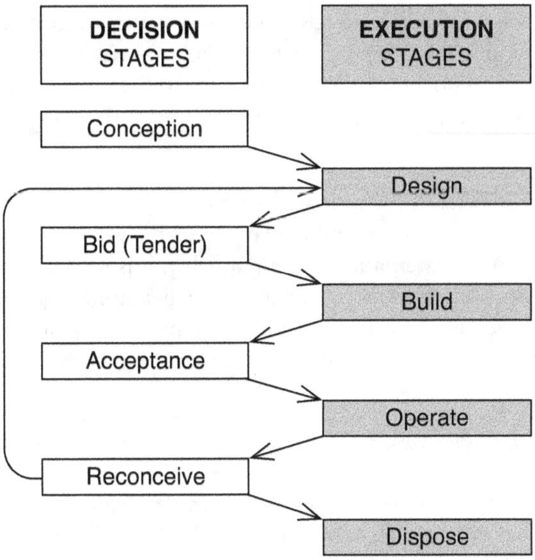

Figure 1.8: Stages in the building life

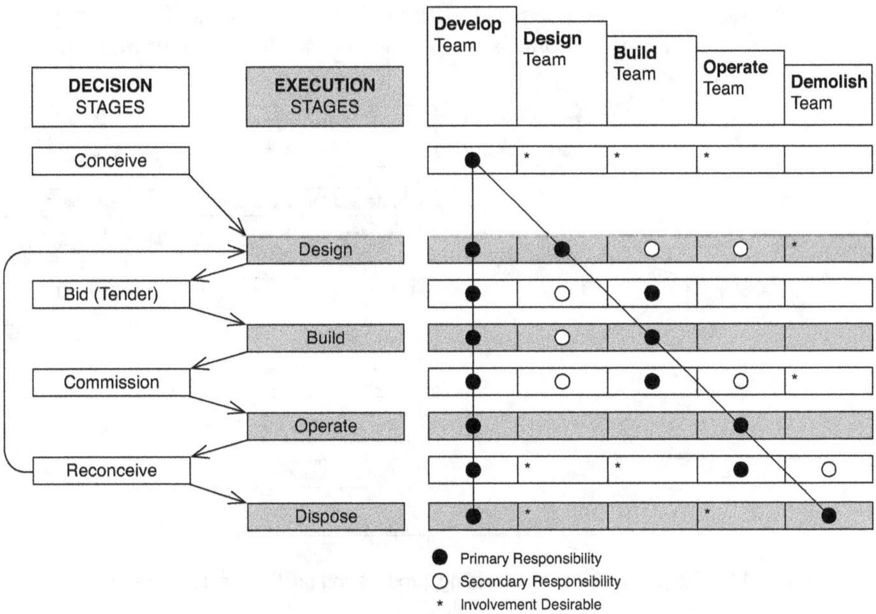

Figure 1.9: Stages in the life of a built facility, the teams at each stage, and their involvement

An additional issue that affects the performance of buildings is **when** decisions are made. Decisions made very early in the conceptual design can have the most profound influence on performance (e.g., a different site or a different type of building system could be chosen, etc.). Conversely, changes made to the design will usually cost much more to implement during construction or after the building has been completed. The importance of timing to building projects have long been studied [Paulson 1976] but remains a major challenge in practice (Figure 1.11). Timing may not be everything but, for built facilities, timing is critical.

Quality, or the level of performance delivered, is an important concern in the practise of building science and hence this textbook. Unless the necessary steps are taken to systematically assure quality, it is unlikely that the common causes of poor building performance—namely ignorance, greed, graft, and process—can be effectively addressed. A popular quotation on quality, attributed to John Ruskin (1819–1900), sums up its importance:

> *…There is hardly anything in the world that some men cannot make a little worse and sell a little cheaper, and the people who consider price only are this man's lawful prey. It's unwise to pay too much, but it's more unwise to pay too little. When you pay too much you lose a little money, that is all. When you pay too little, you sometimes lose everything, because the thing you bought was incapable of doing the thing you bought it to do…*

Figure 1.10 identifies that, at each stage, a conscious decision must be made to assure quality. It is quite possible that this simple figure illustrates the single most problematic issue contributing to the poor performance of built facilities.

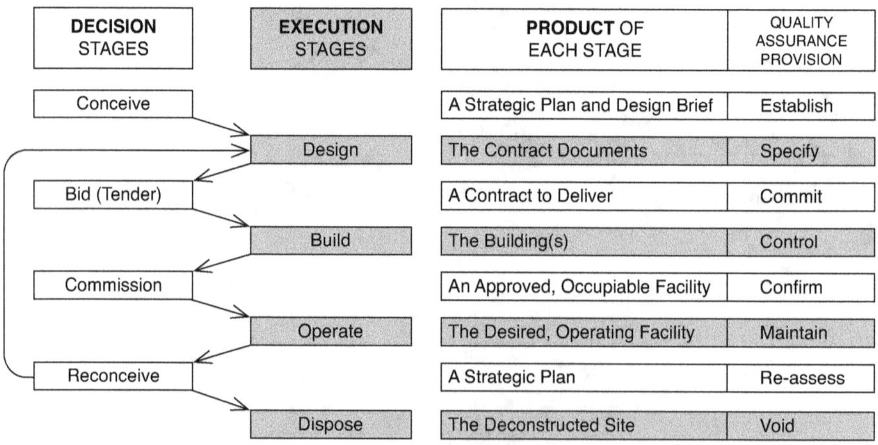

Figure 1.10: Stages in the life of a building, product and quality assurance provision

Figure 1.11 attempts to summarize the important idea that early intervention during design has a disproportionate impact on performance and cost. Good design would make early choices that have a positive impact on performance: making these choices requires an understanding of how different choices impact performance, and this knowledge is fundamentally building science.

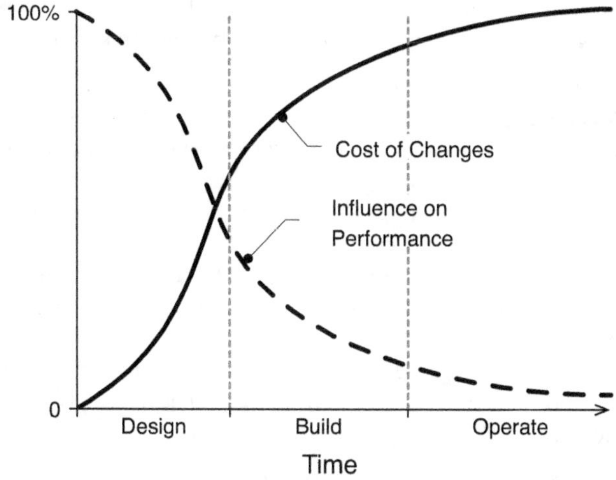

Figure 1.11: Opportunity to influence building performance

1.7 The costs associated with built facilities

Because cost is one of the most important considerations, a few words must be devoted to this topic. In general, over the life of a facility, four sets of costs are associated with any building:

1. The cost of **delivery,** which includes all the costs associated with conception, design, bidding, building, and commissioning. This cost also includes the purchase and development of the site, all taxes, fees, bonding, insurance, etc., as well as relevant costs for initial occupation.
2. The continuing cost of the initial financing, i.e., the **financing cost**, which includes principal, interest, and mortgage insurance. Anyone who has ever bought a house is aware of the relative scale of initial or delivery cost and the real cost of ownership. For example, a down payment of 20% on a $250,000 house means that a mortgage of $200,000 must be arranged. At a fixed interest rate of 5% over 30 years, the monthly carrying charge will only be $1,067. However, the cumulative total of monthly principal and interest payments will be $384,258. The contribution to interest of $184,258 (the cost to buy the money to buy the facility) approaches the initial cost of the facility. If the delivery cost of the facility is fully paid for at occupation, then the continuing cost of the initial financing is zero. In some situations, this cost may, for bookkeeping purposes, be included in costs 3 or 4 below.
3. The cost of **operating the facility** itself, which includes not only the costs for staffing, utilities, and cleaning but also the costs for maintenance and repair, plus some fixed annual expenses related to operation such as taxes, fees, insurance, etc. For the modest single-family home example above,

operating costs will be on the same scale as the mortgage payment, and for a condominium complex or office building, will often be much greater.

Consider the Building Owners and Managers Association (BOMA) data [BOMA 2015] reproduced in Table 1.7. For a data set of thousands of private-sector office buildings, the average total annual operating cost is about \$13 per square foot (\$143/m^2) of rentable space. Other sources such as RSMeans [2015] indicate that, excluding the cost of land and site development, the construction cost of new office space ranges from about \$140 to \$240 per square foot (\$1500 to 2500/m^2) depending on location. It follows that, without any provision for inflation and other time-dependent factors, it might take, on average, about ten years for the sum of all annual operating costs to equal the initial delivery cost. But such buildings are very likely to have a service life of 50 to 75 years. Clearly the operation costs are significant.

4. The **occupant-related costs** associated with the facility, which includes the salaries and all the benefits and other costs entailed in achieving the primary purpose for which the facility was built. Thus, the full costs of all staff housed in a office building are part of the cost of the built facility. Occupant or user costs usually dwarf facility costs but are not often included in high-level cost-benefit analyses.

According to a variety of sources, the net present worth (which depends on financing and inflation rates) of a high-rise office building over its real service life may be quantified as follows:

- item 1 above, delivery (design and construction cost): 1 to 2%
- item 3 above, operating the facility: 3 to 6%
- item 4 above, occupant-related costs (payroll and benefits): 90 to 96%

The proportions shown in Table 1.2 are relatively crude and are not representative of all built facilities. Nonetheless, they demonstrate that all of the costs that occur over the life of the building (life-cycle costs) are very important. They each deserve serious attention, especially if long-term ownership is contemplated.

Two very important points need to be made. First, on an individual basis, the sum of the cost of items 2, 3, and 4 is far greater than the delivery cost (item 1). Second, by a wide margin, item 4 is the largest of the four cost categories for non-residential facilities. This relative cost relationship is not necessarily true for civil engineering construction such as a bridge or a sewage facility and is one of the major differences between civil facilities and built facilities.

Table 1.2: Average annual operating costs for private-sector office buildings [BOMA 2015]

Cost category	Cost ($\$/ft^2$)	Cost ($\$/m^2$)	Percentage of total
Utilities	2.35	25.29	18%
Maintenance and repair	2.11	22.70	16%
Cleaning	1.53	16.46	11%
Administration	1.47	15.82	11%
Grounds and roads	0.27	2.91	2%
Security	0.70	7.53	5%
Fixed operating expenses	4.89	52.62	37%
Annual operating cost	**13.32**	**143.32**	**100**

Understanding these simple economic relationships for built facilities is important. The economics of providing a built facility based on initial cost alone are very different from those based on life-cycle cost. As shown in Figure 1.6, the decisions that are made during the design stage are critical, and it follows that a building designed based on delivery cost is likely to be very different from one designed on the basis of life-cycle costs.

Unfortunately, the decision-makers for the design of many buildings are not the same group of people who will pay over the long term. However, some long-term ownership groups (such as governments, pension funds, and real estate investment trusts) are increasingly focusing on life-cycle costs, and hence insisting on the design of buildings that may cost marginally more to construct but have substantially lower life-cycle costs.

That many facilities were built without consideration of life-cycle costs has had, and continues to have, serious financial implications for many existing government, educational, medical, and institutional facilities. While all of these are facilities that have long-term, usually public-sector, ownership, the real pressures of the moment tend to obscure some of the likely future consequences. We are all victims of the short term.

1.8 References

Alberti, Leon Battista, *The Ten Books of Architecture*, Leoni Edition, reprinted Dover Publications, New York, 1986, Book 1, Chapter 2.

Allen, E. & Iano, J., 2019. *Fundamentals of Building Construction: Materials and Methods*, 7th ed. Wiley.

Baker, M. 1980. *Roofs*. Multi-Science Publications Ltd., Montreal.

BOMA, 2015. *2015 BOMA Experience Exchange Report*. Building Owners and Managers (BOMA) International.

Boswell, C. K., 2013. *Exterior Building Enclosures: Design Process and Composition for Innovative Facades*. Wiley.

Brand, R., 1990. *Architectural Details for Insulated Buildings.* Van Nostrand Reinhold, New York.

Givoni, B. 1976. *Man Climate and Architec*ture, 2nd ed., Applied Science Publishers, London.

Hagentoft, C., 2011. *Introduction to Building Physics.* Studentliteratur, Lund, Sweden.

Hens, H. 2017. *Building Physics – Heat, Air and Moisture: Fundamentals and Engineering Methods*, 3rd Edition; Ernst and Sohn.

Hens, H., 2016. *Applied Building Physics: Ambient Conditions, Building Performance and Material Properties*, 2nd Edition, Ernst and Sohn.

Hutcheon, N. B., Handegord, G. O. P., 1983. *Building Science for a Cold Climate.* NRC/IRC.

Hutcheon, N.B., 1953. "Fundamental Considerations in the Design of Exterior Walls for Buildings", *Proc. 47th Annual Meeting of the Engineering Institute of Canada*, Halifax, May.

Hutcheon, N.B., 1963. *Requirements for Exterior Walls.* Canadian Building Digest 48. Division of Building Research, National Research Council, Ottawa.

Hutcheon, N.B., 1998. "The utility of building science", reprint of the lecture delivered in 1971, *J. Thermal Env. and Bldg. Sci.*, Vol.22, p4-9.

IES, 2011. *The Lighting handbook*, 10th Ed., Illuminating Engineering Society of North America.

Kaiser, H., 1996. *A Foundation to Uphold—a Study of Facilities Conditions at US Colleges and Universities.* Association of Higher Education Facilities Officers.

Klepeis, N.E., Nelson, W.C., Ott, W.R., Robinson, J.P., Tsang, A.M., Switzer, P., Behar, J.V., Hern, S.C. and Engelmann, W.H., 2001. "The National Human Activity Pattern Survey (NHAPS): a resource for assessing exposure to environmental pollutants". *Journal of exposure science & environmental epidemiology*, *11*(3), pp.231-252.

Latta, J.K. 1973. *Walls, Windows and Roofs for the Canadian Climate.* National Research Council of Canada, Ottawa.

Lstiburek, J., Carmody, J. 1993. *Moisture Control Handbook.* Van Nostrand Reinhold, New York.

Lstiburek, J., Harriman, L., 2009. *The ASHRAE Guide for Buildings in Hot and Humid Climates*, 2nd ed., ASHRAE, Atlanta.

Lstiburek, J., 2020. *Moisture Control for Residential Buildings: Principles and Practices*, Building Science Press.

Maslow, A.H., 1943. "A theory of human motivation". *Psychological Review.* 50 (4), pp. 370–96.

Mehta, M.L., 2017. *Building Construction: Principles, Materials, and Systems.* 3rd Ed, Pearson.

Palladio, Andrea, 1965. *The Four Books of Architecture,* translated by Isaac Ware, reprinted Dover Publications, New York, Book 1, Chapter 1.

Paulson Jr, B.C., 1976. "Designing to reduce construction costs". ASCE *J of the Construction Division*, 102(4), pp.587-592.

R.S. Means, 2015. *Building Construction Cost Data.* RS Means; 907 pp. 73rd ed.

Rose, W., 2005. *Water in Buildings: An Architect's Guide to Moisture and Mold.* Wiley.

Stollard, P., 2014. *Fire from first principles: a design guide to international building fire safety.* Routledge.

UNEP, 2022. *2022 Global Status Report for Buildings and Construction: Towards a Zero - emission, Efficient and Resilient Buildings and Construction Sector.* United Nations Environment Programme Nairobi.

Vigran, T.E., 2014. *Building acoustics.* CRC Press.

Vitruvius. 1960. *The Ten Books on Architecture.* translated by Morris Morgan, Dover Publications, New York, Book 2, Chapter 2.

Wotton, Henry, 1968. *The Elements of Architecture*, (London 1624.) University of Virginia Press, Charlottesville.

1.9 Notes

Chapter 2

The Building Enclosure

2.1 The building enclosure

That part of any building, above or below grade, that physically separates the outside, or exterior environment, from the inside, or interior environment(s), is called the *building enclosure* or building envelope. *Environmental separator* is another term used to describe the enclosure, but note that this generic term also applies to separators of two different interior environments (e.g., demising walls between different occupancies). The term building enclosure is preferred to the term building envelope, largely because it is considered both more general and more precise.

This chapter addresses the nature and functions of the building enclosure and the loads it must be designed to accommodate and manage.

2.1.1 The nature of the building enclosure

Both the above-grade and the below-grade portions of the building enclosure are part of a physical system involving three interactive components: the exterior environment(s), the enclosure system, and the interior environment(s). The exterior environment above grade is very different from that below grade, and within any building there can be numerous interior environments. The nature of the building enclosure and its spatial relationship with the other parts of the building are shown in Figure 2.1. Note the variable components and dimensions of the enclosure.

As suggested by Figure 2.1, a building in general consists of a collection of spaces bounded by a set of spatial separators. There are separators between interior environments as well as separators between an interior environment and the exterior environment; collectively, the latter constitute the building enclosure. In many respects, the functions of separators within a building (i.e., floors and interior walls) are very similar to those of the building enclosure. Usually, however, the performance requirements of internal separators are fewer and, in most cases, much less onerous. The performance of internal separators is not directly dealt with in this text, although the fundamental physical principles obviously apply.

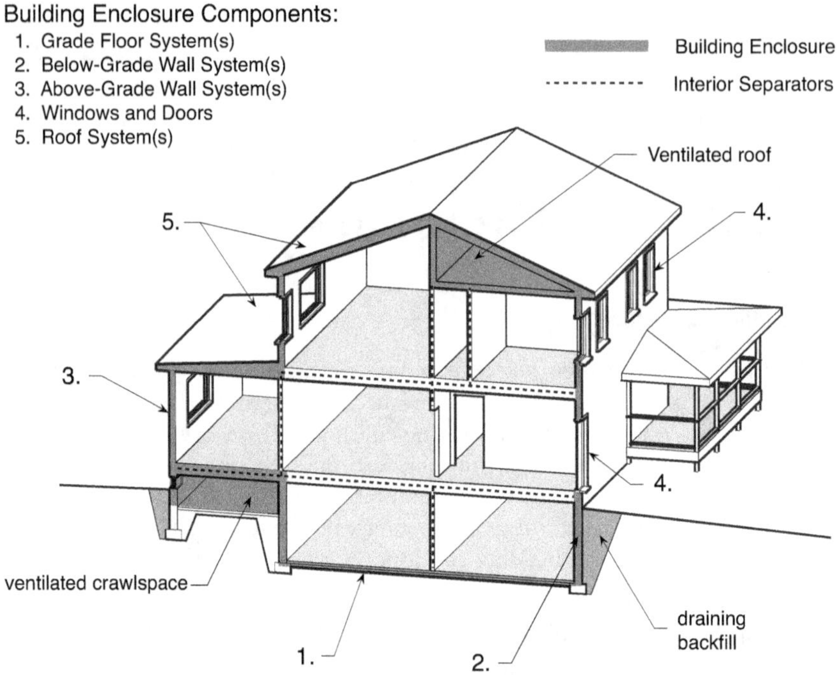

Figure 2.1: The building as a set of separated spaces and the as-built separators

2.1.2 Enclosure components

The typical building enclosure usually consists of the following components:

- the roof system(s) including skylights,
- the above-grade wall system(s) including windows (fenestration) and doors,
- the below-grade wall system(s), and
- the base floor system(s).

Figure 2.2 shows the relationship between these enclosure components and the built facility.

The building enclosure should not be viewed as a combination of numerous one-dimensional or even two-dimensional planar components. Rather, each enclosure component is usually a three-dimensional, multi-layered, multi-material assembly that extends from the inside face of the innermost interior layer (e.g., the paint or wallpaper) to the outside face of the outermost layer (e.g., cladding or roof shingles). Although many assemblies, and their transitions, can be represented adequately as comprised of two-dimensional layers, occasional penetrations (such as by fasteners, flashing, etc.) may play an important role in performance.

Each enclosure component (e.g. window, wall) is an assemblage of layers of specified products (e.g. sheet glass, insulation) or materials (e.g. aluminum, wood). Some products are themselves formed of layers, such as glass-mat faced exterior gypsum, plywood, or foil-faced insulation. A deliberate air space or cavity should also be considered a layer.

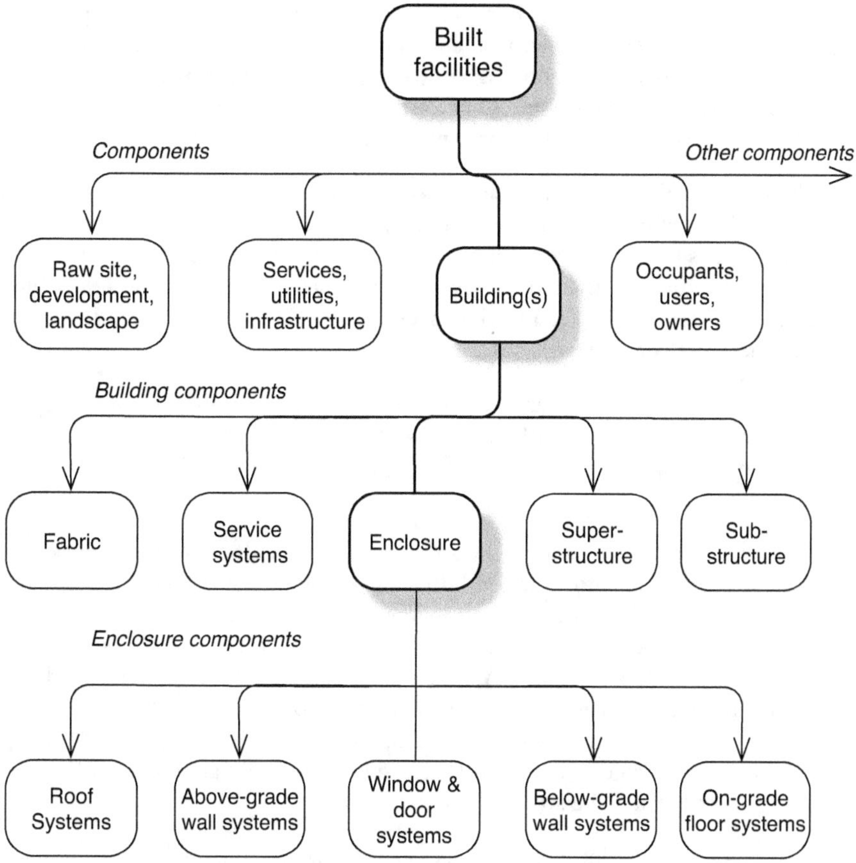

Figure 2.2: The relationship between built facilities and the physical components of the building enclosure

Consider, for example, the sloped roof system shown in Figure 2.1; this assembly consists of all the contiguous layers between the finish on the ceiling (paint) and the outer face of the roof shingles (the granules). The compacted gravel under the basement floor (or below-grade floor) system is a capillary break and part of the enclosure (as, possibly, is the backfill on the outside of the basement wall system, if granular) and hence are also specified layers that are part the as-built enclosure assembly. By common definition each enclosure incorporates all the contiguous (in-contact) layers.

In practice, the importance of clearly understanding the layers that comprise the enclosure and the transitions between different components cannot be overstated[1].

The above definition of the enclosure has some limitations. First, the physical difference between site microclimate modifiers and the enclosure is not that straightforward. Second, although one can define the enclosure as the building component that separates interior from exterior environments, it is not always easy to precisely define "interior" and "exterior."

For example, the backfill around a basement wall or ivy growing on a masonry wall are, by definition, considered to be part of the enclosure. They are also part of the site. What cannot be ignored is the significant impact of this soil or ivy on moisture and thermal loadings. The influence of climate modifiers such as trees, attached unconditioned garages or carports, overhangs, and the soil on the enclosure is not always clear.

The definition of interior and exterior environments can also be confusing, especially in the case of buffer spaces such as garages, screened porches, or storage attics, and for different interior conditions, such as swimming pools, freezer rooms, etc.

Consider an attached garage. The space in the garage does not experience outdoor conditions, nor is it conditioned like the interior of the building. The enclosure could be considered to incorporate both the exterior wall of the house and the garage wall, with the space between (containing the car) being an integral part of the enclosure. Alternatively, the enclosure could be considered to be the garage wall, and the intermediate wall could be viewed as an interior separator with significantly different conditions on each side. The latter definition is adopted in most modern design to ensure that pollutants from the garage do not enter the attached building.

Similarly, two adjacent internal spaces can also have quite different environmental conditions. For the not uncommon scenario of a hotel with a swimming pool next to an air-conditioned exercise room, the differences in temperature and humidity across the separating wall are substantial. In such cases, the loads and performance of the interior separator are more like those of a building enclosure component.

---

[1] Appendix A describes a methodology for assessing enclosure performance continuity that relies on this understanding.

2.2 The separated environments and their modification

The nature or states of the two environments separated by the building enclosure are usually very different. For instance, internal environments are usually specified in relation to the physical needs of people and are defined in terms of temperature, relative humidity, air velocity, and air quality. However, sensitive equipment, materials, or processes (e.g., computers, archived paper, electronics factories) may require different, often more stringent, conditioning requirements than those required by (adaptable) human occupants. Because the interior spaces are usually conditioned, the state of the interior environment is often assumed to be constant but human beings generally require different interior climates depending upon age, gender, clothing, activity level, and physical condition.

The climate-related loads that a building and its enclosure experience are modified versions of the local climate. These microclimates are modified by adjacent buildings, the landscaping (especially trees), and other parts of the enclosure (see Chapter 3). Roof overhangs, for instance, are climate modifiers to the walls below as well as integral parts of the roof system.

The service systems (e.g., lights, motors, stoves) and the contents of a space often have a strong and time-dependent influence on the interior environment because of their energy impacts. The interior environment tends to vary less than either the changing conditions across the layers of the enclosure or the exterior environment. This stability is largely due to a combination of the enclosure's moderating effect on variations in the exterior environment and the conditioning of the interior environment by some energy inputs. For a building to achieve its desired performance, the state specified for each interior environment must be met. Each specific state may be constant, but the specified state will probably vary on a daily, weekly, and/or seasonal schedule.

The exterior or outside environment could be considered to be a three-dimensional space with stochastically varying mass and energy properties. The local climate and weather provide the major, but not the only, exterior loads. Local climatic records, such as average and extreme values, are commonly available, but these data are usually based on open-field weather station records. The building may be some distance from the weather station, and the terrain (hills, other buildings, etc.), the landscaping (trees, shrubs, etc.), and the building itself (overhangs, protrusions, etc.) can moderate the weather that each enclosure component is actually subjected to. As a matter of convention, that portion of the exterior environment that is close to, and affected by, the building is called the *external microclimate*. In fact, different parts of the enclosure are subjected to different exterior microclimates.

The properties of exterior environments and interior environments are separate and major fields of study. Each environment is discussed in some detail in the Chapter 3.

It is not sufficient to have some knowledge of the exterior and interior environments alone. Both the building itself and each component of the building enclosure can have a significant influence on the nature and magnitude of the environmental loads that may have to be accommodated. It follows that some knowledge is needed not only of the building enclosure and the separated environmental microclimates but also of their interaction.

Establishing the characteristics of the local microclimates can be difficult. Within the building, the ambient conditioned state is usually specified in terms of human perception, and the innermost enclosure interface will experience only a modified version of this state. For example, the air temperature experienced by a person in a conditioned room may be different from the temperature at the innermost face of the outside wall. The external microclimate is dependent on the shape and orientation of the building, the influence of the portion of enclosure involved, and the topography, landscaping, windbreaks, overhangs, and other features that modify the ambient exterior environment. Many parameters are needed to characterize an environment, and most parameters could be considered to be time dependent and either a mass or an energy flow. The conceptual nature of these moderating or modifying influences is depicted in Figure 2.3.

Figure 2.3 indicates that, while the building enclosure separates the interior and exterior environments, it really experiences several microclimates. The enclosure interacts with both environments and, in turn, affects both environments. This interaction is usually dependent on the time of day, the day of the week, and the season. There is thus a cyclical (diurnal, workday/non-workday, weekly, and seasonal) aspect to the time-dependent response of the enclosure as it separates and, to some degree, modifies the influence of both environments.

While the enclosure may be the largest modifier, the difference or the modification between the ambient and interface conditions can also be significant. Environmental loads are cyclical, and their impact is not necessarily consistent, instantaneous, or simultaneous. Note that Figure 2.3 will be different for different parameters. Moreover, this simple representation of variation over one cycle of time for one parameter does not accurately or fully represent the complex behavior and interactions that occur.

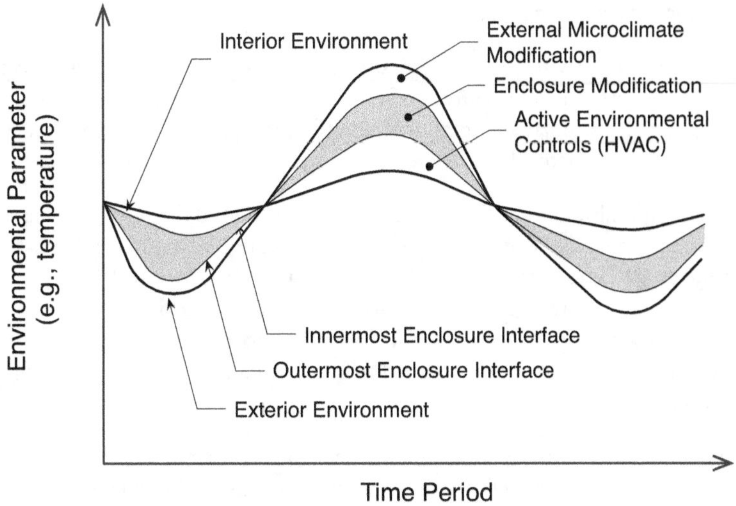

Figure 2.3: Contributors to environmental modification

Especially in the case of low-rise buildings, the effect on the enclosure from the exterior environment can be intentionally reduced or moderated by the use of microclimate modifiers such as plantings, fountains, building overhangs, and windbreaks. For example, a roof overhang (an integral part of the roofing system) will moderate the amount of rain on the wall below and will therefore be a microclimate moderator for the vertical portion of the building enclosure. With taller buildings, it is more difficult to modify the external microclimate.

2.3 Enclosure loads

The performance of any portion of the enclosure should be considered in relation to the various loads generated by:

i) the exterior environment,

ii) the interior environments, and

iii) the enclosure itself.

The generic term *load* refers to any event, phenomenon, or characteristic that can affect the enclosure. Each and every load that impacts the building enclosure can usually be categorized into one of just five types, namely:

- energy related (thermal, light, sound, etc.),
- moisture related,
- air (or gas) related,
- ground related.

2.3.1 Exterior environmental loads

The exterior environment, both above and below grade, is responsible for many different loads. Sources of these loads are the natural climate or the weather (both normal and abnormal effects), human or human-made effects, and nature or natural phenomena.

A considerable body of empirical and experimental data exists on the interaction of exterior climate, site, and building [Watson and Labs 1983, Olgyay 1963, Geiger 1965, DeKay & Brown 2014, Moore 1993]. Research, notably by Fanger [1970] and Givoni [1981], has identified and defined the physiological aspects of human response to both interior and exterior climates, and these data have been distilled into convenient forms for building designers [ASHRAE 2017].

Table 2.1 identifies, by type and source, some of the loads due to the exterior environment that could affect the enclosure. It should be evident from this list of loads that the design or analysis of the enclosure involves consideration of not only the usual structural loads but also mass (air, moisture, etc.) and energy (heat, light, sound, etc.) loads. Moreover, for design purposes, we need to have some knowledge of both average and extreme conditions for each relevant load. The performance of the enclosure is more often dominated by longer-term events (a 48-hour cold snap, 50 years of exposure to UV) than the superstructure (a peak wind gust of 3 seconds, a two-minute duration earthquake).

2.3.2 Interior environment loads

The interior environment consists of occupied, used, and, typically, conditioned spaces. As shown in Table 2.2, the main sources of loading are human induced, the result of natural phenomena, or due to conditioning. Variations in the state of the interior "climate" and the related extreme values tend to be smaller or less than those occurring within either the enclosure or the exterior. This occurs not only because the building enclosure modifies the effect of exterior variations but also because the interior environment is generally augmented by the addition or removal of mass and energy, e.g., the building or parts of it may be mechanically conditioned.

For a building to successfully meet its desired performance, the required state of each of the interior environments must be ensured. The state of an interior space may either be maintained constant or may vary in time as a function of the nature and degree of space conditioning, internal gains (equipment, lighting, solar effects, and human activities), and the modifying influence of the enclosure.

Table 2.1: Loads from the exterior environment

	Type				
	Energy related	**Moisture related**	**Air related**	**Ground related**	**Gravity related**
Weather or natural climate	Ambient conditions, solar	Relative humidity, fog, rain, ice, snow	Barometric pressure, wind		Water, snow, hail
Abnormal climatic effects	Reflected solar, lightning	Tornado, hurricane, flooding	Tornado, hurricane	Frost heave, landslide	Wind-borne missile
Natural phenomena	Fire, groundwater	Adfreezing, freezing	Radon, methane, soil gas	Seismic, landslide, settlement, termites, plants, etc.	Hydrostatic pressure, soil pressure
Human-made weather	Climate change, city effect	Smog, acid rain	Wind-related vortex/swirl	Mining subsidence	
Human-induced events	Fire	Fire (hoses, sprinklers, etc.), water main break	Smoke, sonic boom, sound		Impact, blast, wear and tear

Table 2.2: Loads from the interior environment

	Type				
	Energy related	**Moisture related**	**Airflow related**	**Ground related**	**Gravity related**
Interior space	Ambient conditions, solar	Relative humidity, water (sprinklers, etc.)	Barometric pressure, wind, stack, fan-induced		Water
Natural phenomena	Fire	Fungal growth, mold	Radon, methane	Settlement, termites, plants, etc.	
Human-induced events	Fire, people	People, flooding, combustion equipment	Smoke, sound, explosion		Impact, wear and tear, dead and live loads

The enclosure is only one part of the larger system that moderates or controls the environments on both sides (especially the inside) of the enclosure at some location on the building. The building itself and the enclosure both influence the exterior microclimate that interfaces with the exterior surface of the building enclosure at a specific location. For example, the impact of wind, rain, and solar radiation on the enclosure is influenced by the buildings shape. The mass and energy inputs provided by heating, ventilation, and air-conditioning (HVAC) equipment or internal gains of course clearly modify the interior environment.

Table 2.3: Loads from the enclosure

	Type				
	Energy related	**Moisture related**	**Airflow related**	**Ground related**	**Gravity related**
Element or component being considered	Volume change, shape change, fire	Relative humidity, built-in moisture, volume change, fungal growth, mold, creep, shrinkage, etc.	Off-gassing, airflow, air pressure differentials		Self weight, live loads
Adjacent elements	Fire	Fungal growth, mold	Smoke		Dead loads, live loads

2.4 The functions of the enclosure

At the most basic level, the primary function of the building enclosure is to separate the interior and exterior environments. In reality, the building enclosure has to provide the "skin" to the building, i.e., not just separation but also the visible facade. Unlike the superstructure or the service systems, the enclosure is seen and is therefore of critical importance to owners, the architect, and the public. The users or occupants are concerned with both sides of the building enclosure. The appearance and the operation of the enclosure have an influence on the interior environment and on factors such as productivity and satisfaction. The enclosure is a major contributor to providing delight.

In general, the physical functions of the building enclosure may be grouped into three categories like the building categories introduced in Chapter 1. Namely, all enclosure functions can be categorized as:

1. **Support**, i.e., to support, resist, transfer, and otherwise accommodate all the structural loads imposed by the interior and exterior environments (including occupany), by the enclosure, and by the building itself. The enclosure or portions of it can be an integral part of the building superstructure—usually by design but sometimes not.

2. **Control**, i.e., to control, regulate, and/or moderate all the loads due to the separation of the interior and exterior environments — largely the flow of mass (air, moisture, etc.) and energy (heat, fire, sound, light, etc.).

3. **Finish**, i.e., to finish the enclosure surfaces — the interfaces of the enclosure with the interior and exterior environments. Each of the two interfaces must meet the relevant visual, esthetic, required by occupants.

However, some parts of an enclosure (e.g. above suspended ceilings) do not require a finish and products or layers labelled with the term finish also often have structural (e.g., impact) and control functions (e.g. fire, acoustics, vapor).

A fourth building-related category of functions could also be imposed on the enclosure, namely:

4. **Distribution** i.e., to distribute services or utilities such as gas, power, communication, water in its various forms, and conditioned air to, from, and within the enclosure itself.

The enclosure may also serve other, usually non-physical, purposes such as advertising or as a symbol or image, e.g., as a statement about power or security or wealth, etc. However, these overall, non-physical functions are not directly considered in this text.

Almost all enclosures must satisfy support, control, finish, and distribution functions. However, only the support and control functions must be provided over the entire surface of the enclosure: **control and support functions must continue across every penetration, every interface, and every assembly** (see also Appendix A). The lack of this required continuity is the cause of the vast majority of enclosure performance problems. The need for finish and distribution varies over the extent of the enclosure.

The **support function** is of primary importance. Without structural integrity, the remaining functions are useless. However, the industry has reached a high level of understanding and accomplishment in this area. Support systems have evolved from massive elements pierced at a few locations to efficient primary structural systems (such as steel or concrete frames) with lightweight framed infill and sheathing. The trend to lightweight enclosures is likely to continue as the demand for more resource-efficient buildings grows.

For the enclosure to fulfill its function of separating the indoor environment from the outdoor environment, many **control functions** are required. Like support, these control functions are required everywhere for acceptable performance, and hence continuity of these control functions, including at penetrations and connections, is critical to explicitly consider because of the difficulty of achieving continuity. The most important control function with respect to durability is often rain (water) control followed by airflow control, thermal control, and vapor control. The level of fire and sound control required varies with code requirements and the owner's desires.

Unlike the control and support functions, which rely on continuity to achieve performance, the **finish function** is optional, and may not be needed in some areas. For example, finishes are often not provided above suspended ceilings, in service spaces, or in industrial buildings, where the finish is often deemed unimportant. A structural material such as concrete, or masonry, that is left unfinished is just that: unfinished. However, even a coat of paint or the specification of a texture or color is explicitly defining a finish function. Confusion arises because finish systems may, and often do, have other functions: paint may provide moisture protection to metal and a white roof may be selected for thermal control. Hence it is important to recognize that cladding and interior finishes routinely have both required finish and control functions.

The exterior finish layer of an enclosure is often termed *cladding*, but cladding almost always does more than provide finish functions, it also often part of the rain control, UV resistance, and even may be integrated into the enclosure support function.

The **distribution function**, a building function, is often imposed on the enclosure. Large proportions of most enclosures do not need to fulfill this building function, but it is highlighted because it usually impacts the control-related functions. For example, service entrances penetrate the entire enclosure, and pipes, ducts, and wires that run through insulated stud spaces or window frames can significantly influence their performance.

For the building enclosure to perform satisfactorily, each and every load must be managed to the level expected and all the functions must be met. A list of control, support, and finish loads is provided Table 2.4 along with the functional category that manages each load.

This table is important because it shows that the number of loads is considerable and that each load affects at least one functional category for the enclosure and often more than one.

The responsibilities or interests of the various design professionals are also shown in Table 2.4. Each profession has a stake in the performance of the enclosure, and the relative importance of this stake can be identified. The table highlights the need for better professional education, much better interaction and communication between the various professions, and a clearer understanding of their responsibilities.

In practice it is useful to divide the functions of the enclosure into various subcategories to which one can assign actual products, subsystem assemblies, trades, performance requirements, and test standards. With respect to the control functions, these are termed **control layers**.

Table 2.4: Enclosure loads and building professions

		Category of functions				Professional responsibility			
	Specific loads	Interior finish	Support	Control	Exterior finish	Architect	Structural	Fire, Sound	Energy
Causal phenomenon or load – Primarily structural	Gravity – dead (assembly, etc.)		●				●		
	Gravity – live (people, snow, etc.)		●				●		
	Wind		●	○	•	•	●		
	Ground movement (seismic, settlement, etc.)		●				●		
	Explosion		●				●		
	Rheological (creep, shrinkage, etc.)		●		○		●	○	
	Impact (vehicles, missiles, people, etc.)	•	●		•	•	●		
Causal phenomenon or load – Primarily environmental	Fire		●			●	○	○	
	Heat (thermal, etc.)		○	●			•	●	●
	Air (pressure, movement, leakage, etc.)		○	●		●	○	○	○
	Moisture (built-in, precipitation, etc.)		○	●	○	●	•	○	•
	Smoke	•		●		○		●	
	Solar radiation (incident, reflected, etc.)		•	●	○	○		○	●
	Chemical attack/atmospheric (acid rain, etc.)			●	○	○	•		
	Particulate matter (dust, VOCs, etc.)	•		●	•	○		●	
	People (wear and tear, etc.)	○		●	•	●	○	○	
	Insects, birds, animals, termites, rodents, etc.	•		●		•	●		
Causal phenomenon or load – Primarily perceptual	Light (natural, incandescent, fluorescent, etc.)	•		●		●		●	○
	Sound	○		●	●	•		●	
	Visual – local	●			●	●	•		
	Visual – contextual	●			●	●	•		
Professional Responsibility	Architectural (Architect)	●	○	○	●				
	Superstructure (Structural Engineer)		●	○	○				
	Interior Environment	○		●	○				
	Energy Specialist	○	•	●	○				

Primary significance ●
Secondary significance ○
Tertiary significance •

The most important and commonly defined enclosure control layers are, in approximate order of importance:

1. water[1] control layer (Chapters 8 and 12),

[1] The term water is intended to denote the liquid form. For the above-grade enclosure this primarily means rainwater and meltwater control. For below-grade application, ground water and surface water need to be controlled.

2. air control layer (Chapter 7 and 11),
3. thermal control layer (Chapter 5), and
4. vapor control layer (Chapter 6).

The chapter references in the above list indicate where the physics of the associated control function are covered in this text.

Each control "layer" may be a single material or a subassembly of materials that together provide the control desired. In many cases, two or more of the control functions are provided by a single material, layer, or sub-assembly.

Confusion about the classification of the functional roles of enclosure components and materials is far too common, and this confusion can cause serious performance and durability problems. Consider the follow examples:

- Vinyl wallpaper is often applied as a finish, but in fact functionally acts as a low-permeance vapor control layer. This unintentional control of diffusion can, and too often does, create serious mold problems in air-conditioned buildings, especially in hot-humid climates.
- Gypsum wallboard (drywall) is often seen as fulfilling a finish function, but in fact the paint is more often the finish and the drywall often serves as a control layer for fire, sound, and/or airflow. If a designer or builder stops the drywall above a suspended ceiling because a finish function is not needed here, the required fire, sound, and airflow control will also be missing.
- Although the primary function of foam plastic insulation is to limit the flow of heat, depending on its material properties and installation details it may also be used to control vapor, air, and/or rainwater.

Figure 2.4 provides a concept sketch of an ideal enclosure assembly developed from the principles of building science described in early publications of the Canadian Division of Building Research [Hutcheon 1964, Latta 1967, 1973]. The term "perfect wall", first coined by Dr. Joe Lstiburek [Lstiburek 2007], is widely used to describe this approach. The concept is considered ideal because:

a) the functional layers are clearly identified, thereby avoiding confusion and misunderstanding by the many stakeholders involved in enclosures,

b) the specific arrangement of the layers has very wide application for most climates, occupancies, and common building material choices, and

c) the arrangement often provides high-performance (limiting thermal bridging for example) and durability (by protecting the primary control layers).

The benefits of this arrangement will become clear as the physics of enclosure behavior are described later in this text.

The arrangement or order of the layers is important. To be the "perfect wall" the thermal control layer must be located outside of the air, water and vapor control layers, which themselves should be located outboard of the enclosure support structure (and building superstructure). The order of the air water and vapor control layers themselves are not important, and in practice they are often combined into one product.

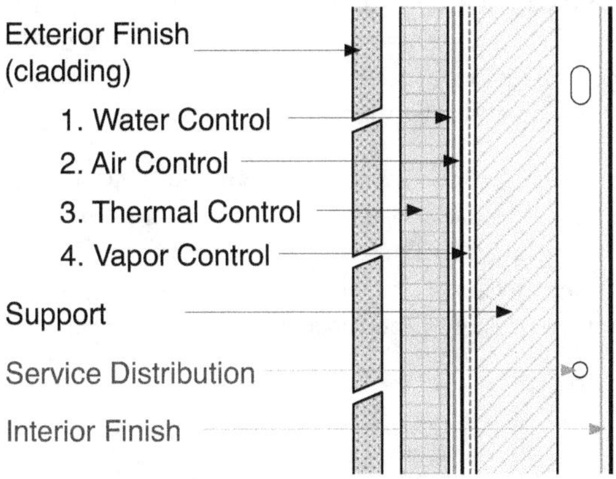

Figure 2.4: Ideal enclosure assembly and functional layers

Many successful enclosure systems do not follow the "perfect" principles (e.g., insulated glazing units, most architectural precast, exposed membrane roofs). Their success therefore relies on specific material choices, and layer properties that vary with climate zone and occupancy. The understanding and analysis to make the correct choices for these "non perfect" enclosures is a major part of building science and the information in this textbook.

Only four control functions, and control layers, are discussed above. These have widespread practical significance. However, the fire control layer is another very important control layer that routinely needs to be defined (sometimes both an exterior and an interior fire control layer are needed). Similarly, a sound control layer can often be necessary or desirable. Although sound and fire are not a focus of this text, identifying sound and fire control layers, and ensuring their continuity in design and construction is as beneficial for performance as it is for water, air, and heat.

2.5 Design

Design, always at the core of engineering, has taken on increased interest in today's academy and industry, as well as capturing the public imagination.

Although design is a process, the term is also used to describe a product and the result of the process.

Design can be defined as a constrained decision-making process during problem solving. Engineers use the results of analysis, often quantitative, to guide their decisions, but judgment and intuition (both gained by experience) are equally critical. Design of physical objects, such as enclosures, involves the selection of materials, and then forming and arranging them to meet a particular purpose. For all engineering design, clear definitions of the problem(s) and constraint(s) are necessary to assess whether a design solution ("design") is successful. This text attempts to provide the framework for design of the building enclosure by identifying needs and constraints as well as providing methods and information for analysis.

The three primary design dimensions of any building have been identified and discussed, i.e., the possible loads, the functions, and the attributes of performance (Figure 2.5). These apply not only to the building enclosure but also to the other physical components of the building. is a schematic that shows these three design dimensions for each of the physical components of a building.

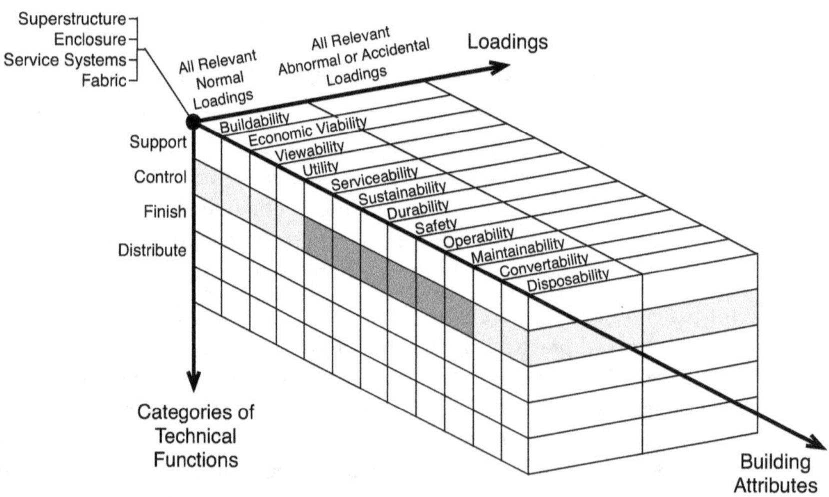

Figure 2.5: The primary design dimensions for any enclosure

Function, attribute, and load each constitute a dimension for design in the sense that the design team needs to systematically and explicitly consider each function, each attribute, and each load, as well as all relevant combinations of these loads. In each step, a technical design criterion should be set up in such a manner that the modeled, estimated, or computed response is equal to or better than the permissible performance threshold. In mathematical terms this process can be written as follows:

$$\sum_{c}^{n} \sum_{w}^{m} \{R \leq R^{s}\}_{\min \$}^{\max \beta} \quad \text{(Eq. 2.1)}$$

For all (i.e., the sum of) n relevant combinations of loads (c) and for all m relevant loads (w), the modeled or computed response (R) must be demonstrably better (larger or smaller as the case may be) than a specific limit or threshold value (R^s) for this particular response. This conditional criterion needs to be resolved for all relevant situations, and this has to be accomplished at minimum cost ($) and for maximum benefit (β).

In structural design, it was once commonplace to provide a "safety factor" to increase the resistance of, for example, a beam and thereby reduce the probability of failure in service. This concept has become more sophisticated with the development of limit states design (LSD). Although it has been proposed that LSD be applied to building enclosure design [Bomberg & Allen 1996], the approach has not yet been widely applied. The concept of safety factors has less direct meaning or application to many moisture and thermal control aspects of building enclosures than to structural design. In most practical cases, achieving a design that, on average, resists the environmental loads likely to be imposed is sufficient for success. In other cases, one may design an enclosure to perform adequately for 95% or 99% of the hours per year and accept degraded performance targets for the remaining hours.

Note that in this text we are largely, but not exclusively, concerned with only one physical component, i.e., the building enclosure; the control of only three loads, namely heat, air, and moisture; and all the performance attributes but in particular serviceability, durability, safety, and maintenance. A more complete coverage would also investigate fire, sound, and light as deeply. However, the focus of this text is in response to the most widespread failures the authors have observed in industry, as well as the largest shortfalls in academic literature and textbooks.

In this and the previous chapter, a concerted attempt has been made to develop a systematic, logical, and consistent framework or context to study, analyze, and design both buildings and the building enclosure. Numerous authors [Hutcheon 1953, AJ Handbook 1974, Allen 1995, Reid 1984] have addressed this issue. Although not as technical, James Marston Fitch's writings, particularly *Architecture and the Esthetics of Plenty* [Fitch 1961] and *American Building* [1975] deserve to be much better known. Interested readers should review these sources.

2.6 Historical development of the building enclosure

Building probably began with simple forms of construction being used for shelter from the wind, sun, and rain. Gradually, as the desire for better shelter grew, suitable materials were identified, and construction skills were developed. Vernacular architecture throughout the world is usually characterized by the judicious and advantageous use of readily available local materials and an experiential understanding of climate and site [Fitch & Branch 1960]. These forms of building evolved over generations and, since the requirements were relatively simple and change was usually very slow, the design, the building materials, and the construction techniques evolved at a pace dictated by matching needs with available resources.

As society flourished, construction materials and techniques developed from reeds and mud into manufactured baked mud and burnt clay brick [Sandstroem 1970]. For several thousand years, walls in Europe and elsewhere were built of masonry, wood, or clay material. Because of their size or massiveness, these walls were exceptionally strong and durable and provided a modicum of thermal protection through their natural heat storage and thermal insulation capabilities. Vernacular adobe buildings of New Mexico [Knowles 1974] and brick and concrete Roman buildings such as the Trajan Baths were built to absorb, store, and release ambient heat and capture solar energy so that the net energy flow over several days was balanced. Where the climate was colder and masonry not as readily available, houses of logs or half-timber with clay infill or earth toppings were built for their greater insulating properties.

The Industrial Revolution dramatically changed the situation, leading to the rapid development of new materials, products, and techniques. New forms of energy generation and equipment facilitated space conditioning and extended humankind's environment to include less hospitable climates: the population of Alaska and Las Vegas exploded as space conditioning became available. The building structure, its form, assembly techniques, and materials underwent radical change in the relatively short period between the 19th century and the present time. Specialization and mass production, the hallmarks of the Industrial Revolution, were slowly introduced into the building industry.

The superstructure, and to a much lesser degree the enclosure, began to be considered separately as a specialized component. Many buildings evolved as a structural endoskeleton with an enclosing skin. In the West, traditional massive wall systems gave way to skeletal structural systems, often with non-load-bearing enclosures. The influential architect Corbusier formally proposed the "Dom-Ino" system in 1914, which explicitly separated the enclosure, structure, and interior separations to allow for more rational construction and freedom in design.

The shift from log-and-plank buildings to the balloon wood-frame construction system (possibly invented by either Augustine Deodat Taylor in 1833 [Condit 1968] or George Washington Snow in the 1830s [Randall 1972]) was slow. Balloon-frame construction was extensively used for reconstruction after the Great Fire of Chicago in 1871. In Europe, cast-iron structural frameworks with load-bearing infill were being used in English mills and warehouses by the turn of the 19th century. In the mid-1800s, truly flexural frameworks of cast iron and, later, the much stronger and more ductile wrought iron, first appeared in the form of train sheds in England.

While they were atypical or temporary structures, the Palm House at Kew Gardens (1845) and the Crystal Palace (1851) were the first examples of mass-produced, prefabricated buildings with thin enclosures distinct from the structure [Fitch 1961]. Hippolyte Fontaine's St. Ouen Docks warehouse in Paris (completed 1865) was the first multistory building to carry all structural loads, including that of the exterior masonry walls, on an iron frame [Condit 1968, p. 124].

Buildings with framed structural systems and specialized enclosures were developed to their modern state in commercial high-rise construction in the US, primarily in Chicago and New York City [Friedman 2010]. The first iron-framed skyscraper (the Home Life Insurance Building in 1884) and the first all-steel frame building (the Rand McNally Building by Burnham and Root Architects in 1890) were both built in Chicago [Randall, pp. 105–107 and pp. 118–120].

The birth of the modern skyscraper is often defined as 1884, when the Home Life Insurance Building designed by William Le Baron Jenney began construction. In this building there were no load-bearing walls, since the metal skeleton carried the entire building weight. The thin masonry facade was hung from the frame like a curtain, on shelf angles fastened to spandrel beams [Condit 1964 pp. 83–85]. By 1895, the D.H. Burnham and Co. design for the Reliance Building in Chicago contained many of the technical components of the modern high-rise office tower: curtain walls, elevators, and a wind-braced frame of steel. Louis Sullivan, who has been credited with the statement that "form follows function," brought the esthetic statement of the skyscraper to maturity with the 1899 Carson Pirie Scott Department Store [Schueller 1990].

The adoption of self-supporting building endoskeletons of steel and, later, reinforced concrete led to the development of thin building "skins" consisting of lightweight facade panels (usually spandrel panels) and large expanses of window. The enclosure most often associated with skyscrapers, the metal and glass curtainwall, had its start in the Oriel Chambers building, built in Liverpool in 1864. The technological improvements and the wealth that followed the Second World War reinforced the trend towards thin metal and glass enclosures. This trend spread from high-rise commercial construction to almost all industrialized buildings. The combination of low-energy-efficiency enclosures, low-cost fluorescent lighting, suspended ceilings, and fully air-conditioned mechanical conditioning systems became common after 1945 [Banham 1969].

This form of building rapidly became the universal standard promoted by the architectural vision of architects such as Le Corbusier, Mies Van de Rohe, and the Bauhaus School. The limited regard for site, climate, and locality were exemplified by this movement's "International Style" label. The Equitable Savings and Loan Building, completed in 1948 in Portland, Oregon, was the first fully sealed and air-conditioned office building clad with curtain wall (at least in the US)[Moore 1993, p.5].

By the early 1950s, the vertical enclosures of New York's influential Lever House (now known to be a highly energy-inefficient building) and the United Nations Building were both dominated by large areas of heat-absorbing, tinted, double-pane curtain wall glazing made habitable by large air-conditioning systems. Both of these buildings have since undergone significant enclosure replacement programs to address durability and performance issues.

These post-war changes in building structure and form, in materials, and in methods of assembly did not always result in an improvement in durability or better control of the interior environment. The epitome of these developmental changes was and, to some extent still is, the skyscraper. Tall buildings are also exposed to greater environmental extremes because of their exposure; the ability of the enclosure to contribute to the control of the interior environment has often been inadequate and durability has become a very serious concern.

If unprotected from the sun, the large expanses of glass on modern skyscrapers produce solar gains that result in large and rapid temperature fluctuations and glare. These lightweight, smooth, exterior wall components were often inadequate at controlling the flow of heat, air, noise, and water vapor. The facades of these buildings were subject to driving rain which not only induced water leakage to the interior but also accelerated deterioration hidden within the systems. Deficiencies in lighting, solar control, energy use, and human comfort had to be compensated for by providing ever more powerful mechanical conditioning [Banham 1969]. The heating, ventilating, and air-conditioning equipment has often failed to meet the prescribed standards of human comfort while also consuming significant amounts of energy.

Henry Cowan has provided a good overview of the development of the technology of buildings in his book *An Historical Outline of Architectural Science* [Cowan 1977]. First issued in 1966, a second edition was published in 1977. What is especially prescient is the fact that one chapter is entitled *"Environmental Design Replaces Structure as the Principal Problem of Building Science."* Cowan was clearly aware that the building enclosure and the service systems for conditioning and servicing the usable spaces within the building were becoming much more important. News of these changes, however, has only recently begun to penetrate most of the schools of architecture, architectural engineering, and structural engineering in the United States and Canada.

In recent years, the field of building science and a growing awareness of the interrelationship and interaction between the building and both the interior and exterior environments has led to improvements in building performance. Expectations for comfort, durability, indoor air quality, and utility have risen and continue to rise. Because of rapidly changing materials, building techniques, and equipment, the ability to predict the performance of buildings has become much more important. An increase in the number of books with a building science core has also increased rapidly in the last decades [Boswell 2013, DeKay & Brown 2014, Harriman and Lstiburek 2009, Straube 2012].

The need to conserve global material and energy resources, reduce carbon emissions, and provide better resilience all require buildings that perform better. Spurred on by the oil crisis of 1973, buildings in the industrialized world have been constructed that are better insulated, more airtight, and more responsive to the sun, the climate, and the local microclimate—essentially a return to some of the principles of vernacular buildings. While some of the changes in building materials, designs, and systems may have led to new problems, such as interstitial condensation and poor indoor air quality, the problems have prompted a greater awareness and understanding of buildings and their enclosures. It is becoming clear that building designers must have some knowledge of building science and the performance of the building enclosure in order to design better building enclosures and better buildings.

2.7 References

AJ Handbook of Building Enclosure,1974. Ed. A.J. Elder, The Architectural Press, London.

Allen, Edward, *How Buildings Work*. Oxford University Press, New York, 1995.

ASHRAE Handbook of Fundamentals, American Society of Heating, Refrigeration and Air-Conditioning Engineers Inc., Atlanta, Georgia, USA, 2017.

Banham, R.,1969. *The Architecture of the Well-tempered Environment*, The Architectural Press, London.

Bomberg, M., Allen, D. ,1996. "Use of generalized limit states method for design of building envelopes for durability." *Journal of Thermal Insulation and Building Envelopes* 20, no. 1, pp. 18-39.

Boswell, C. K. 2013. *Exterior Building Enclosures: Design Process and Composition for Innovative Facades*. Wiley, Hoboken, New Jersey.

Condit, C., 1968. *American Building*. The University of Chicago Press, Chicago.

Condit, C., 1964. *The Chicago School of Architecture.* The University of Chicago Press, Chicago.

DeKay, M., Brown. G.Z., 2014. *Sun, Wind, and Light: Architectural Design Strategies.* Wiley, 3rd Ed.

Fanger, F.O., 1970. *Thermal Comfort, Analysis, and Applications in Environmental Engineering.* McGraw-Hill Inc., New York, 1970.

Fitch, J.M., Branch, D.P., 1960. "Primitive Architecture and Climate," *Scientific American*, Vol. 203, pp. 134-145.

Fitch, James M., 1961. *Architecture and the Esthetics of Plenty.* Columbia University Press, New York.

Fitch, James M., 1975. *American Building: The Environmental Forces That Shape It.* Schocken Books, New York.

Friedman, Donald, 2010. *Historical Building Construction: Design, Materials, & Technology*, 2ed, W.W. Norton, New York.

Givoni, B., 1981. *Man, Climate, and Architecture.* Van Nostrand Rheinhold, New York.

Harriman, L. and Lstiburek, J. 2009. *The ASHRAE Guide for Buildings in Hot & Humid Climates.* ASHRAE, Atlanta, Ga..

Hutcheon, N.B., 1953. "Fundamental Considerations in the Design of Exterior Walls for Buildings", *Proc. 47th Annual Meeting of the Engineering Institute of Canada*, Halifax, May.

Hutcheon, N.B., 1964. *Principles Applied to a Masonry Wall.* Canadian Building Digest 50. Division of Building Research, National Research Council, Ottawa.

Knowles, R., 1974. *Energy and Form.* MIT Press, Cambridge, Massachusetts.

Latta, J.K., 1967. *Precast Concrete Walls – A New Basis for Design.* Canadian Building Digest 94. Division of Building Research, National Research Council, Ottawa.

Latta, J.K. 1973. *Walls, Windows and Roofs for the Canadian Climate.* National Research Council of Canada, Ottawa.

Lstiburek, J., 2007. "The Perfect Wall", *ASHRAE Journal.* ASHRAE, Atlanta, Georgia, May.

Moore, F., 1993. *Environmental Control Systems.* McGraw-Hill, New York.

Olgyay,V., 1963. *Design with Climate.* Princeton University Press, Princeton, New Jersey.

Randall, F., 1972. *History of the Development of Building Construction in Chicago.* Arno Press, New York.

Reid, E., 1984. *Understanding Buildings.* MIT Press, Cambridge, Mass.

Sandström, G. E. 1970. *Man the Builder.* McGraw-Hill Inc., New York, esp. p.8-18.

Schueller, Wolfgang, 1990. *The Vertical Building Structure.* Van Nostrand Reinhold.

Straube, J.F., 2012. *High Performance Enclosures: Design Guide for Institutional Commercial and Industrial Buildings in Cold Climates.* Building Science Press, Massachusetts.

Watson, D., and Labs, K., 1983. *Climatic Building Design.* McGraw-Hill Inc., New York.

Chapter 3

Environmental Conditions

3.1 Background

The fundamental function of the building enclosure is to separate the interior and exterior environments; hence, an understanding of both of these environments is important to understanding the loads on the building enclosure. As shown in Figure 3.1 and Figure 3.2, these loads are the result of both differences between interior and exterior environmental conditions (e.g., temperature differences) and absolute exposure (e.g., ultraviolet radiation from solar exposure on the exterior cladding). This chapter investigates environmental conditions and some of the loads related to enclosure design.

With respect to building loads, mechanical means (heating, cooling, ventilating, lighting, etc.) are used to provide the desired interior conditions when the enclosure and site modifiers cannot moderate the entire load. Since the enclosure is the focus of this text, the environmental conditions that result in an enclosure load are given more coverage than human comfort or the needs of mechanical space conditioning equipment selection.

Modern enclosure systems employing complex assemblies and an every-changing market of products has led to an increased awareness of the effects of material and system hygrothermal properties. This has in turn led to an increase in the use of hygrothermal enclosure modeling (Chapter 14) which has driven more interest, and more research, into defining environmental boundary conditions relevant for modeling.

The demands of building codes have driven the need for hourly energy modeling which has also created the need for research and documentation of climatic elements specifically for energy analysis [ASHRAE 2021]. The need for hourly weather data is key for both types of modeling, as the response time of most enclosures is generally in that order (shorter time steps are of limited value for most modeling applications).

Basic weather data has been collected and archived for well over a century in most developed countries of the world. The most important environmental measures for purposes of hygrothermal enclosure design can be listed as:

- temperature (both absolute and differences),
- humidity (both absolute and relative humidity),
- solar insolation (infrared, visible light, and ultraviolet radiation),
- wind (speed, direction, and the resulting pressure differences), and
- precipitation (in the form of rain, snow, hail, sleet, etc.).

Many other building-related measures are important, such as the number of freeze-thaw cycles, the amount and direction of wind-driven rain, time of wetness, and ultraviolet radiation exposure, but these can often be estimated from the other five primary measures. In fact, even the five primary determinants listed above are highly correlated—for instance, extensive rainfall typically implies little sun, high humidity, and above-freezing temperatures.

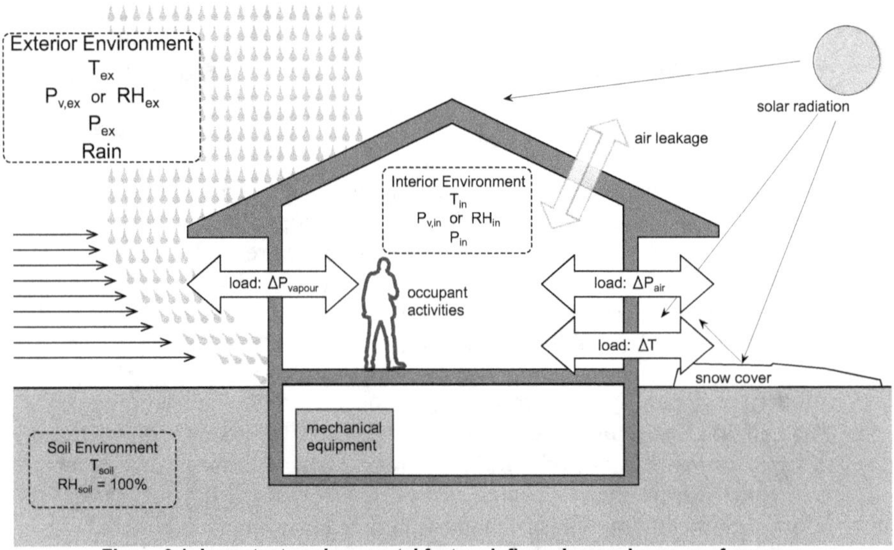

Figure 3.1: Important environmental factors influencing enclosure performance

Interior and exterior environmental conditions can conveniently be studied as separate entities, but it must be borne in mind that interior conditions are influenced by the exterior environment as well as the activities and processes within a building (e.g., stack-effect pressures, interior humidity, and mechanically-induced building air pressures).

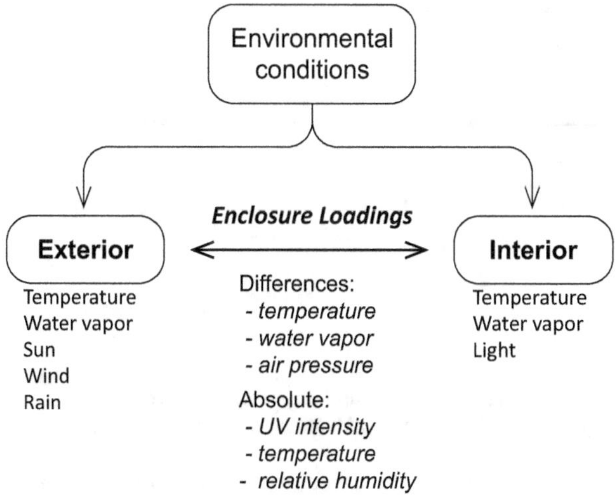

Figure 3.2: Environmental conditions and enclosure loads

Several attempts have been made by researchers to relate the durability and performance of enclosures to specific interior and exterior environmental factors [Sanders 1996], but all of these attempts presume a certain type of enclosure and/or ignore potentially important environmental elements. The approach taken in this chapter is to define the primary environmental loads independent of enclosure type. The designer or analyst can then use such information to assess, for example, the exposure and susceptibility of a specific enclosure to certain types of performance problems, such as driving rain penetration, heat loss, or exfiltration condensation.

The environmental conditions around a building can vary significantly in both space and time (for example, the outdoor temperature is usually quite different in, say, Miami compared to Toronto and usually also differs at any given location between winter and summer). Subsequent sections deal with the interior and exterior building environment from both a temporal and a spatial perspective.

3.1.1 The impact of climate change

Climate change has become a pervasive public policy issue, attracting more attention each year. Climate is a long-term measure of weather, and although warm days can be followed by cold days, climate change is measured by long-term trends. Measurements of temperature at hundreds of sites around the world over the last century and more have shown small (numerically) but steady increases in temperature (Figure 3.3).

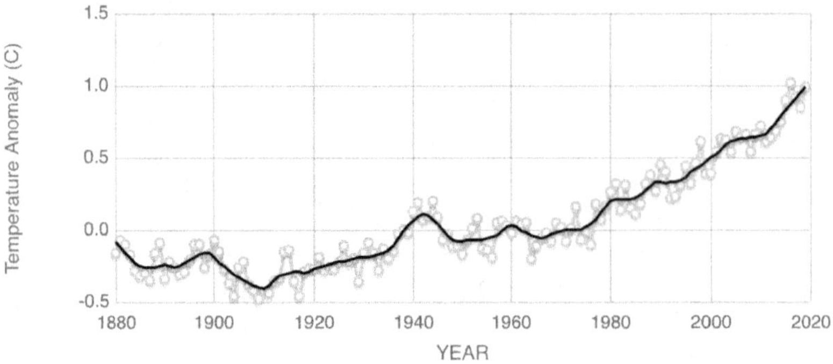

Figure 3.3: Global land-ocean temperature index
[Source: NASA's Goddard Institute for Space Studies, 2019]

The impacts of climate change on environmental conditions are many and complex. It is simple to predict that the air temperature around most buildings will be warmer 30 years from now; however, other aspects of climate change, such as the amount of rain, the frequency of storms such as hurricanes, the intensity of flooding, and the change in solar radiation exposure are more difficult to predict. All of these examples will have some influence on how buildings perform and hence how their enclosures should be designed.

The biggest impact of climate change on building enclosure design and analysis, however, is the demand for better performance from owners, specifiers, and regulatory agencies. The drive to lower energy use in buildings encourages designers to produce better insulated, more airtight enclosures. The imperative to reduce carbon in building materials is forcing alternative building materials to be considered for use, materials with which the building industry has little or no long-term experience. The ability to recover from and adapt to climate impacts is termed *resilience.* Designing buildings that are more resilient in the face of a changing climate includes building enclosure design issues in addition to resilient infrastructure.

3.2 Spatial climate considerations

3.2.1 Exterior climate

Exterior environmental conditions can vary dramatically and often unpredictably. Various terms are used to define different temporal scales (e.g., climate vs. weather) and different spatial scales (e.g., climate zone vs. microclimate). Climatologists and meteorologists have developed most of these terms, but the focus of these professions is somewhat different from that of building or enclosure designers. Hence, a review and some building-specific definitions are necessary.

Weather refers to the collective meteorological conditions at a specific location at a specific time. On the other hand, **climate** is a descriptor of average weather conditions at a specific location over the long term [Barry 1992]. Both climate and weather typically refer to conditions over an extended area, at least several square kilometers in normal usage. The term **microclimate** is used for relatively small-scale spatial situations and the related variations in meteorological conditions. Microclimate may be used to describe the climate conditions within, for example, a small valley, or on the south slope of a hill, or on a wall locally sheltered by an overhang.

Any precise definition of environmental conditions must incorporate both a temporal and a spatial scale (Table 3.1).

Table 3.1: Spatial climate variations

Prefix	Synonym	Approximate dimension	Example	Modifiers
Macro	Zone	500 to >1000 km	Gulf Coast, Prairies	Ocean currents, vegetation (type, coverage)
Meso	Region	1 to 500 km	New York City, Great Lakes	Topography, vegetation, elevation, development intensity
Micro	Site	10 m to 10 km	Leeside of south-facing hill	Slope, terrain, vegetation water bodies, orientation
Micro	Building	1 to >100 m	East face, behind row of trees	Massing, shape, solar absorption, vegetation, overhangs, orientation
Micro	Enclosure surface	0.1 to 10 m	Below a window sill	Orientation, surface features, color
Nano	Detail	1 to 100 mm	At a screw hole or drip edge	Texture, cracks, grooves, shadows

The **macroscale** is analogous to a climatic zone, e.g., the Prairies, hot-dry areas, etc., and has a scale of several hundreds to thousands of kilometers [Whittow 2000, Mayhew 1997]. In *Design with Climate*, Olgyay [1963] recommended a total of only four regions to cover all of North America, based primarily on the Koeppen classification. Koeppen's classification was first proposed in 1900, and it is based on the response of flora to a range of variables.

Lstiburek [2001] used modified definitions of these regions to define seven North American climate zones for building design: (see Figure 3.4).

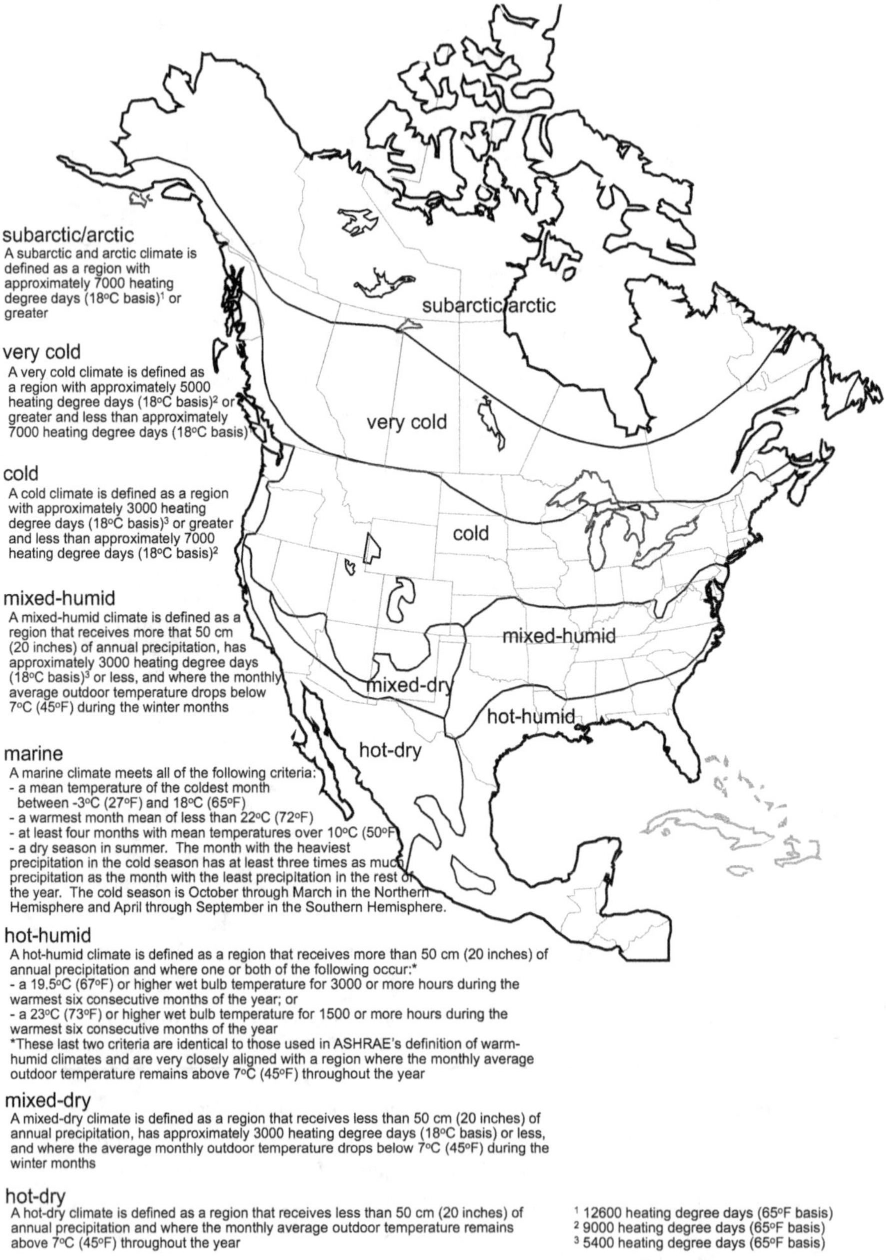

Figure 3.4: Lstiburek's North American climate zone classification [Lstiburek 2005]

Variations of this classification, one of the more widely used in practice by designers, have now been incorporated into US and Canadian building codes (Figure 3.5).

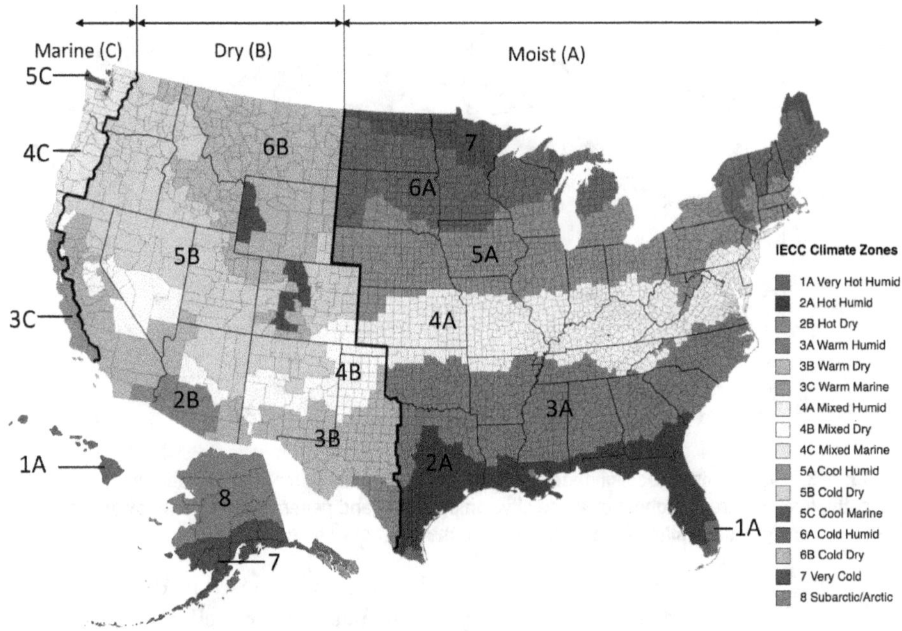

Figure 3.5: ASHRAE US climate zones [PNNL 2022]. Note the marine, dry, and moist lines

The **mesoclimate** or local climate, e.g., the climate of a city or county, may be defined as the subset of the macroclimate that is commonly affected by local topography, vegetation, and human activity [Whittow 2000, Mayhew 1997]. For building design purposes, mesoclimates tend to have a range of about 1 km to as much as 100 km. The spatial scale of the mesoclimate varies dramatically with topographical features—local climates can vary significantly within 2-5 km in a valley, along a coast, etc., but may be quite similar at locations 50 km apart as is the case on the Prairies and in the desert. The mesoclimate is rarely modified by a single built facility, but it can be modified by intense urban development, deforestation, etc. For example, the so-called "heat island effect" (the increased temperatures in large cities caused by solar absorption and heat storage in large unshaded areas of pavement, concrete, and roofing) commonly causes average summertime temperature to increase by 3 to 5 °C (Figure 3.6) in large cities.

The site-specific **microclimate**, with a scale from a few tens to several hundreds of meters, is often very important for building design. Specific site features such as wind breaks, earth berms, ground cover, etc., modify the climate at each building site differently. Together, the site-specific and building-specific microclimates may be considered to represent the building's "exposure." Geiger's *The Climate Near the Ground* [1961] provides a wide range of examples of specific site conditions as well as very detailed measurements (e.g., see Figure 3.7).

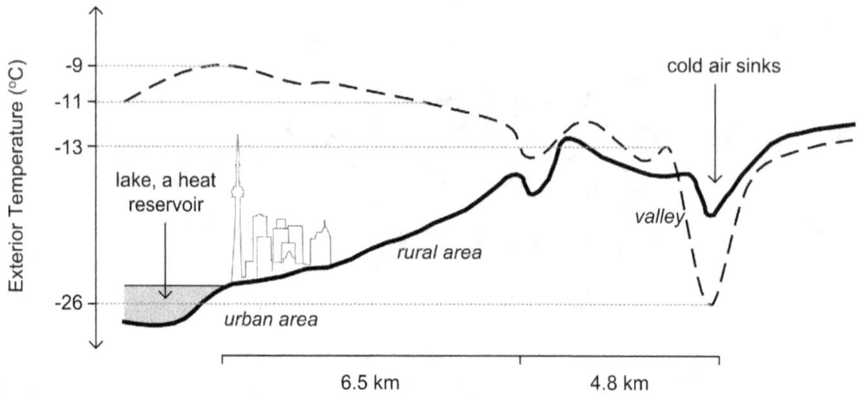

Figure 3.6: Variations in local climate around Toronto, Canada [after Fitch & Bobenhausen 1999] Note: the temperature is highest over the city, which stores and generates heat, and lowest at bottom of a grassed valley, which loses heat to the night sky and collects falling cold air

The **building-specific microclimate** is that climate caused by direct interaction of the site microclimate with the built facility itself, i.e., the size, shape, massing, color, reflectivity, etc. For example, the brick cladding on a wall may store solar energy and warm the air near the building; the two parts of a built facility may channel snow or wind between them; or a courtyard shaded by a building can be protected from sun and wind (Figure 3.9).

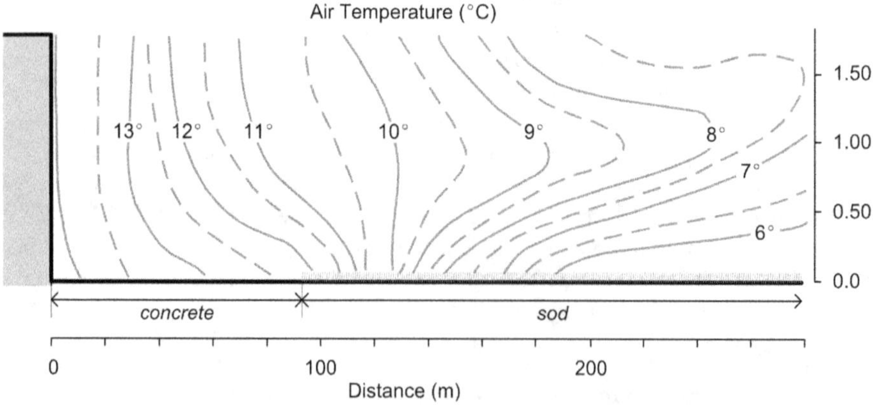

Figure 3.7: Measured site microclimatic variations [after Fitch & Bobenhausen 1999] Note: the temperature in degrees Celsius over concrete and grass surface next to building

The **enclosure-specific microclimate** is that climate on or very near the surface of the enclosure caused by direct interactions with the building enclosure itself and has a scale of a tenth of a meter to as much as ten meters. For example, overhangs influence the deposition of rain and solar radiation on the wall below, and roof parapets influence the local wind pressures both in front and behind them (Figure 3.10).

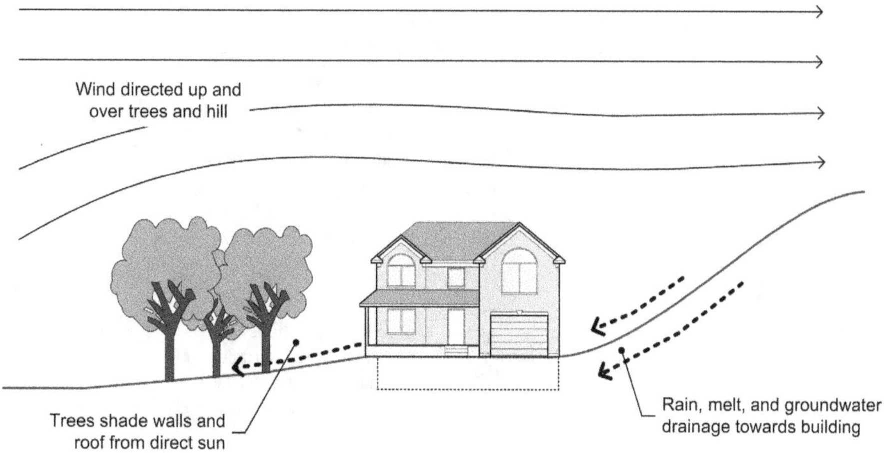

Figure 3.8: Examples of site-specific microclimate modifiers

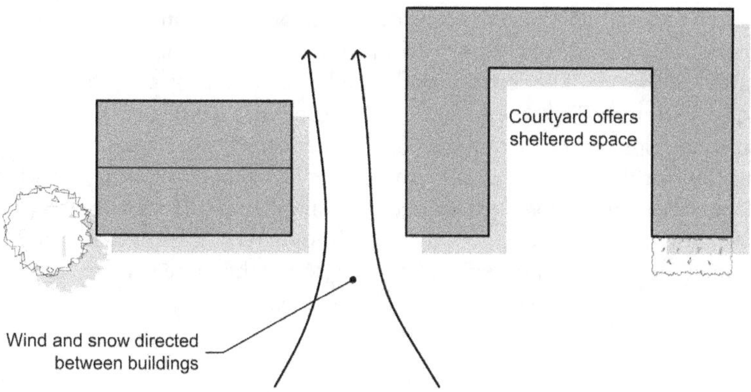

Figure 3.9: Building-specific microclimate

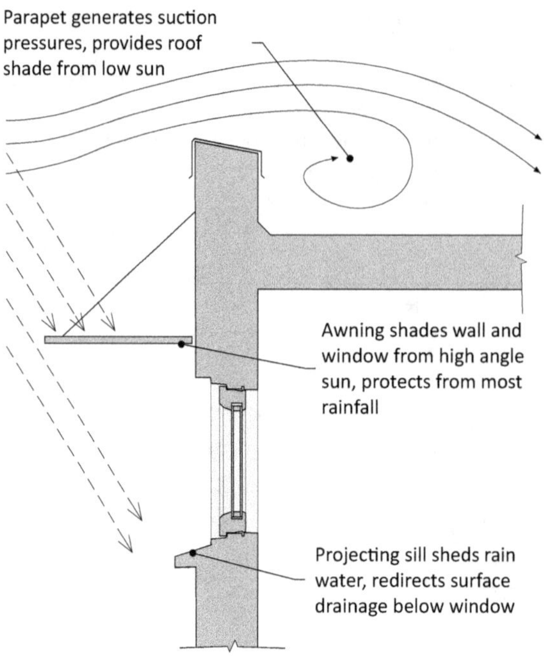

Figure 3.10: Enclosure-specific microclimate

The **nanoclimate** has a spatial scale ranging from less than a millimeter to nearly a meter. This environmental scale is primarily of interest for assessing and understanding local material deterioration. For example, the environmental conditions experienced by the area just below a lap in hardboard siding, the area around a nail or screw, and the underside of a projecting fascia board are quite different. They apply to areas of from less than one to several square meters (Figure 3.11). An understanding of nanoclimatic conditions is often critical for the assessment of numerous practical building deterioration phenomena including corrosion, differential metal corrosion, freeze-thaw deterioration, wood decay, and surface mold growth. Plant biologists sometimes use the term epiclimate to describe the same climate scale near the surface features of plants [Stoutjesdijk and Barkman 1992].

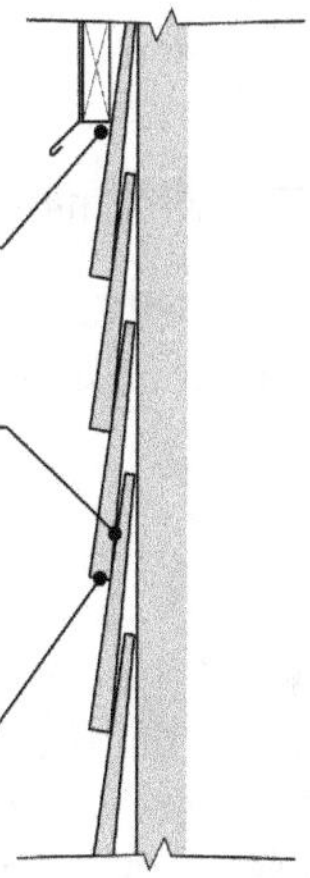

Figure 3.11: Example of nanoclimate at lapped siding

3.2.2 Interior environment

Interior conditions usually vary less than exterior conditions, unless a building serves distinctly different and specialized functions (e.g., freezer storage next to a retail space). However, small environmental changes can be important for human comfort, and relatively small differences in air pressure between one part of a building and another can have significant effects on the performance of a building.

In most cases, interior spatial variations can be predicted and controlled by the designer, although this is not often done. In practice, spatial variations are important if air is supplied or exhausted to one part of a building (e.g., a bathroom exhaust), possibly causing a local negative pressure relative to the exterior and hence generating an air pressure differential across both the enclosure and internal separators. Temperatures within an interior space may vary slightly because of large heat sources (e.g., some processes such as cooking or welding) or large heat sinks (e.g., poorly insulated parts of the enclosure).

Air moisture content within a building often does not vary appreciably over small distances because of the ease of water vapor transport, but differences can exist across interior partitions, for short durations near large sources of moisture like showers or cooktops, and around large continuous sources in large spaces, like a hot tub. Note that a change in temperature when the absolute air moisture content is relatively constant will cause the relative humidity to vary.

Thermal bridging, natural convection, and radiation can combine to generate important spatial variations of temperature (and hence RH) over the interior face of enclosures. These nanoscale variations can explain mould, moisture damage, and fading patterns in internal recesses, reveal, window jambs, etc. These variations are discussed in more depth in Chapter 5.

3.3 Temporal scale of environmental conditions

Environmental conditions can change quickly and sometimes unpredictably with time. Hence, various statistics and measures have been developed to assess the time scale.

3.3.1 Exterior climate

The temporal scale of the exterior climate has already been used to distinguish climate from weather, but there are other aspects to consider as well. Typical climate data sets (usually based on a 30-year set of data), from the longest to the shortest time scale, include:

- annual average, annual average extreme, and extreme values,
- seasonal mean (e.g., summer or winter),
- mean and average daily maximum and minimum of the coldest and hottest months, or of January and July,
- daily mean, annual or extreme maximum, minimum, and range (the difference between the maximum and minimum),
- hourly average, 15-minute samples, and peak values.

Different temporal scales apply to different building and enclosure analysis problems. For example, average annual values may be useful for choosing deep ground exterior temperatures, whereas the design of cladding to resist wind gusts will often conduct analysis using a time duration of about one second. Hence, no single temporal scale is the most appropriate for an enclosure analysis.

Figure 3.12 is a representative annual plot of temperatures averaged over a very long time period. The long-term variation in average temperature over a year is often sinusoidal in continental climates. Daily variations (typically induced by solar effects) are imposed onto the long-term average, and weather systems moving through an area have an impact for two to five days.

The design of space conditioning equipment was traditionally based on 1% or 2.5% design conditions. That is, the space conditioning equipment was sized to provide the energy necessary to maintain indoor conditions for all but 1% or 2.5% of the hours of the hottest or coldest month (typically July and January) or season (typically the three summer months or three winter months) (Figure 3.12).

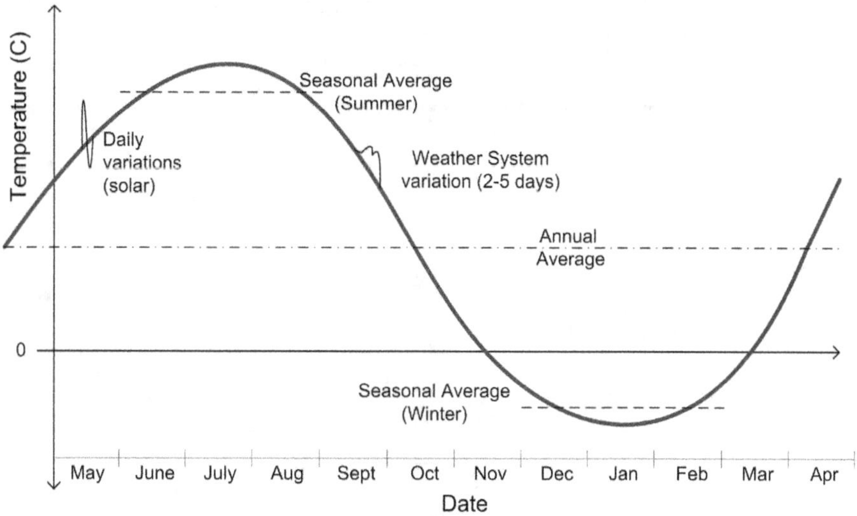

Figure 3.12: Temporal variations in weather

ASHRAE has recently moved to defining design conditions as being based on the percentage of hours exceeding a specific value on an annual basis. Hence, the 0.4% temperature used for cooling design is that temperature exceeded for 0.4% of the 8760 hours in a year (35 hours). Similarly, the 99% temperature used for heating design is that temperature below which only 1% of hours fall per year (88 hours). Since these values are climatic data (averaged over 15 to 30 years), a given year may have more or fewer hours above and below a certain threshold. The *ASHRAE Handbook–Fundamentals* provides such data for hundreds of locations throughout North America and the world.

Building enclosure design requires more information than the extreme conditions typically used in HVAC design. In fact, for enclosure analysis, the average monthly temperature may be one of the most useful basic data sets. Because of the emphasis on equipment sizing in design, average temperatures are not as readily available to design professionals. Online resources such as www.weatherbase.com and www.worldclimate.com provide annual average temperature and rainfall data. The National Centers for Environmental Information (NCEI) of the National Oceanic and Atmospheric Administration (NOAA) are sources of detailed climate data for thousands of locations around the world.

In most of North America, the year is divided into four seasons. These four seasons have little meaning in some parts of the world (winter, for example, is not a term used in India, whereas the monsoon season is very important there). However, seasons are important to North American building design. It is convenient and relatively accurate to divide the seasons into the months listed in Table 3.2. Although the summer and winter solstices usually are close to the

21st of June and December, this is not a good measure of seasons if temperature is used as the primary metric.

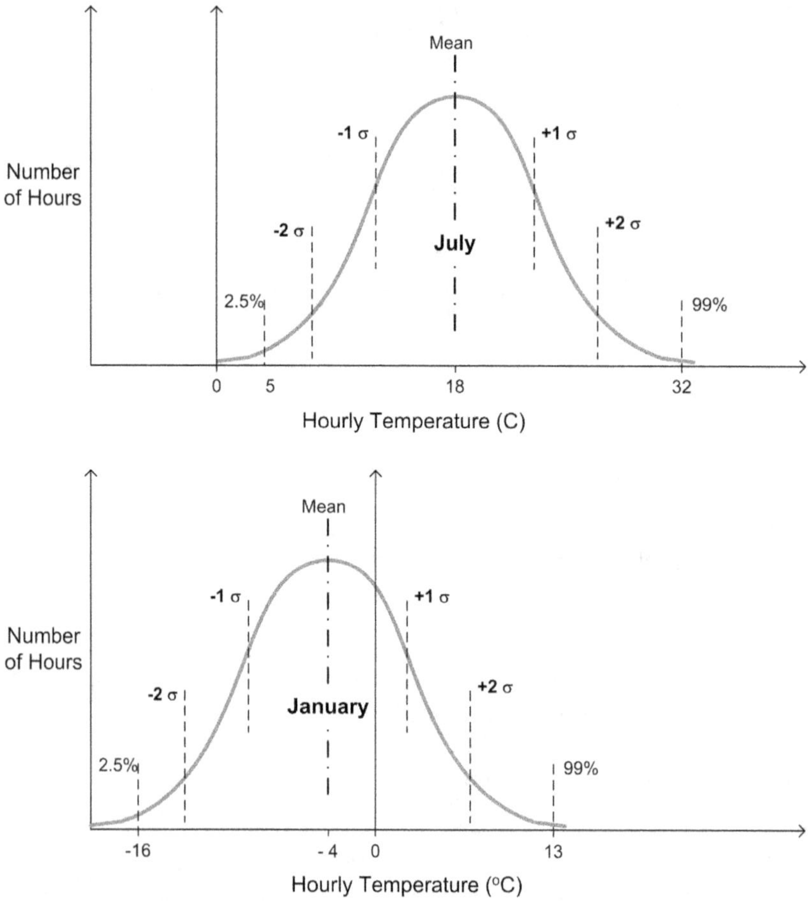

Figure 3.13: Statistical measures of the exterior environment in a northern cold climate

Table 3.2: Seasons and months in North America

Season	Months
Winter	December, January, February
Spring	March, April, May
Summer	June, July, August
Fall (Autumn)	September, October, November

3.3.2 Interior environmental conditions

Interior conditions vary much less than exterior conditions and have rhythms that reflect human activities. In contrast to exterior conditions, the designer has direct control over the interior conditions. For example, a typical schedule for an office space cycles through a week and includes a workday schedule and a weekend/holiday schedule. Each day the conditions will vary during the day, with slightly different conditions during workday evenings and during night or non-working hours.

The schedule for a school might be similar in some respects to that of an office, but the annual summer vacation period needs to be accounted for. In fact, many performance problems have been reported in schools because of humid interior conditions during summer vacation time. As further examples, a church has a worship space of large volume which typically requires heating/cooling for only a few hours each day (and yet may have significant thermal mass slowing the response rate), whereas the basement or office areas of the church may require a small amount of heating and cooling for many hours each day. Such differences in the time scale can create real challenges for designers, and a reason for the enclosure designer to work closely with the mechanical systems designers, occupant representatives, and architectural teams.

3.4 Exterior climate elements

Meteorologists define a climate element as a property or condition of the atmosphere that helps specify and describe the weather. Many climate elements are measured at weather stations around the world at hourly or daily intervals, including temperature, humidity, wind, rain, snow, sunshine, water balance, etc. The elements most important for building enclosure design and assessment are considered below.

3.4.1 Temperature

The most quoted climate measure is temperature. As shown later, temperature is in fact not as important as some other climate elements, but it is easy to measure and has been recorded for well over a century. Figure 3.13 maps the average July temperatures in North America. Average summer (often July daily average) and winter (January) temperatures are widely used to assess the climate.

The average temperature for a month provides no information about the variability of the temperature over a day. The average daily maximum less the average daily minimum for a month or year is defined as the average daily temperature range. The daily temperature range is important for assessing the effect of thermal mass on space conditioning loads and the effect of temperature cycling on material durability.

Average daily temperature ranges can be found tabulated in the *ASHRAE Handbook–Fundamentals* and other sources. High-altitude arid climates tend to have very wide ranges of up to 20 °C (36 °F), while tropical climates with intensive vegetation and surface water have much smaller ranges, perhaps only 5 to 7 °C (9 to 13 °F). This is due to a combination of surface thermal mass and the role that altitude and water vapor content from plants and rain play in night sky radiation.

Figure 3.14: Mean July temperatures over North America (°C)

Temperature over the year is aggregated in two important building-related climatic measures: heating degree days (HDD) and cooling degree days (CDD). HDD are often used directly in simple calculations of the space-heating energy consumption of buildings. The HDD value for a location is defined as the number of days that the temperature is below a certain base temperature, usually 18.3 °C (65 °F), multiplied by the difference in average daily temperature below the chosen base point. For a specific weather station, HDD18 is defined and evaluated as:

$$\text{HDD18} = \sum_{i=1}^{365} (18 - \overline{t}_{daily}) \text{ for } t_i < 18°\text{C} \qquad \text{(Eq. 3.1)}$$

The definition is based on daily average temperatures because these data were more widely available when the concept of HDD was developed.

An initial base temperature of 18 °C (65 °F) was chosen almost 100 years ago since this was the temperature below which it was assumed that supplemental heating would be required to maintain comfortable interior temperatures. This temperature is termed the *balance temperature*. The balance temperature can be defined as the exterior ambient temperature at which the internal gains are equal to the heat losses to the exterior, so that the heating system is not needed, i.e., the heat loss is balanced with gains.

The value most widely available is 18 °C (65 °F). For well-insulated, airtight modern buildings or buildings with large internal heat sources (such as some factories, many offices, etc.) a balance temperature of 10 °C (50 °F) is often more appropriate, and data for HDD10 (or HDD50) is available in ASHRAE's climate data sheets.

Figure 3.15 maps the HDD18 for North America. Multiply values by 9/5 (1.8) for HDD65.

Cooling degree days are similarly defined as the number of days that the outdoor average temperature exceeds the base temperature, times the difference between the outdoor average temperature and the base temperature. CDD are sometimes based on the same 18 °C (65 °F) base temperature as HDD, but this base temperature is too high for most commercial buildings (because such buildings have more internal heat gains). Hence, CDD10 (CDD50) data are more commonly used.

Climate change and city heat island effects have combined to generate a slow rise in temperatures in many weather stations that have collected data for long periods of time (i.e., over 50 years). As a result, building design climate data is changing. Many sources of data use a recent 30-year record such as 1980–2010 or 1990–2020; differences in temperature and humidity are routinely observed when such data are compared to earlier periods such as 1950–1980.

Figure 3.15: Mean heating degree days (18 °C base)

3.4.2 Ground temperature

Although the temperature of the soil and ground around and below buildings has been measured in many locations around the world, the documentation of these temperatures is far less detailed than for air temperature. The average annual ground temperature is close to the average annual air temperature. However, the temperature varies significantly with depth and with time of year because of the very significant thermal mass of soil, water, and rock that comprises ground. The moisture content, soil/rock type, ground cover, and exposure all influence the ground temperature.

A plot of measured soil temperature vs. depth over the year (Figure 3.16) for Toronto, Canada provides an example of a typical site. Soil near the surface interacts the most with the air, and hence shows a variation that is close to the air temperature; however, the January temperature is warmer than air temperature and the July temperature is cooler. As one measures deeper, the thermal mass of the soil reduces correlation with the air temperature and delays the response. Hence, at 3.0 m (10 ft) below grade, the warmest temperature occurs in October and the coldest in April, about a three-month delay. The variation in ground temperature at 3.0 m (10 ft) depth is only about 3 °C (5 °F) above and below the annual average.

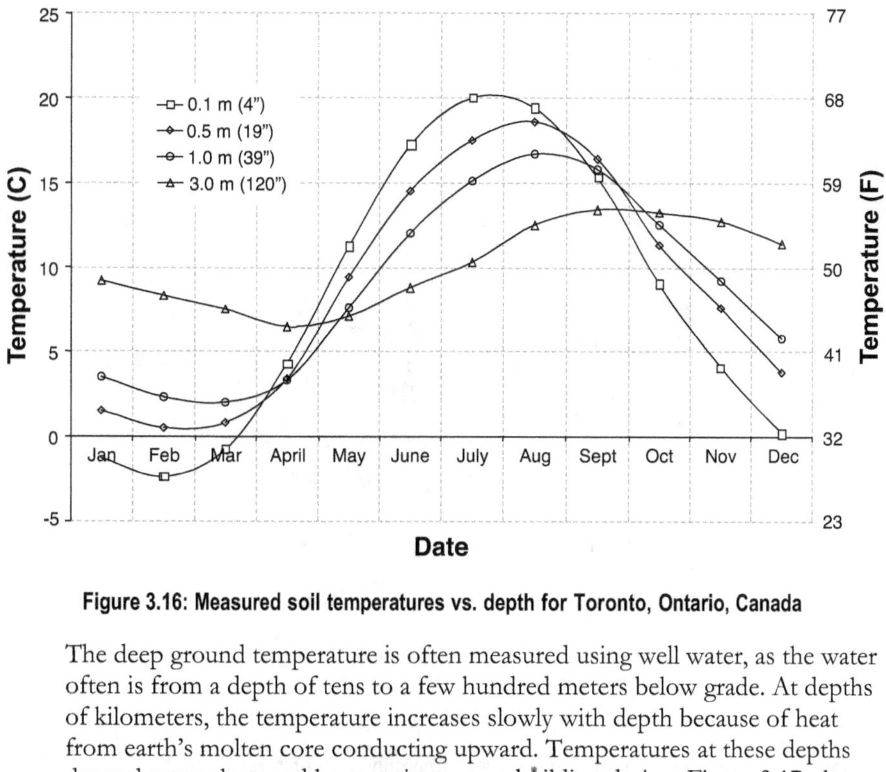

Figure 3.16: Measured soil temperatures vs. depth for Toronto, Ontario, Canada

The deep ground temperature is often measured using well water, as the water often is from a depth of tens to a few hundred meters below grade. At depths of kilometers, the temperature increases slowly with depth because of heat from earth's molten core conducting upward. Temperatures at these depths depend on geology and have no impact on building design. Figure 3.17 plots isotherms and several local station temperatures on a map [from Kasuda 1965] to provide a sense of the variations. Note that average annual ground temperatures below freezing define the permafrost zone.

The construction of a building usually influences the ground temperatures, sometimes significantly. This must be considered when calculating building-ground heat exchange (Chapter 5).

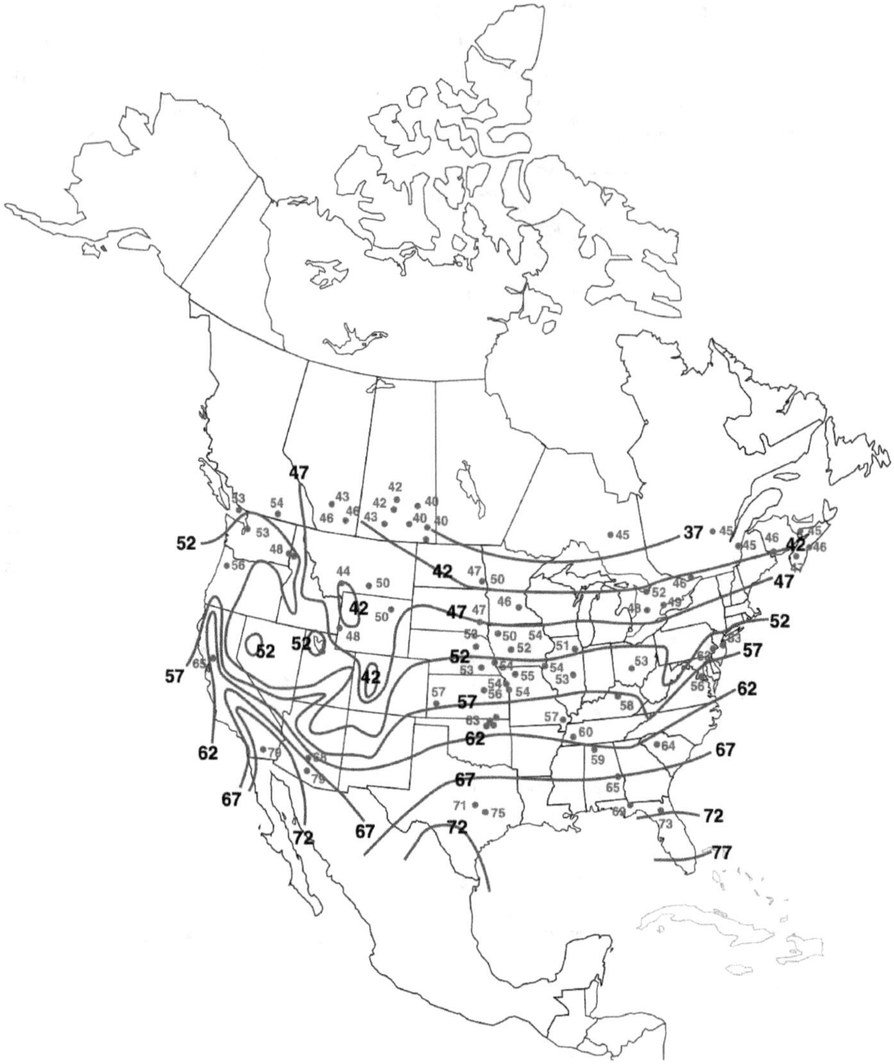

Figure 3.17: Stable deep (more than 10 m = 33 ft) ground temperatures (°F) [after Kasuda 1965]

3.4.3 Moisture content of the air

The moisture content of the outdoor air is important to the performance of building enclosures because it can be either a source of moisture or a sink for both the enclosure and the building interior.

The moisture content of the outdoor air can be calculated from measurements of dewpoint temperature, vapor pressure, or a combination of temperature and relative humidity. Moisture in the outdoor air is generated from precipitation and stored in plants, soil, and surface bodies of water. Hence, locations with high rainfall and vegetation cover tend to have high relative humidity. Since the capacity of air to hold moisture varies dramatically with temperature (see also Chapter 4), locations with high temperatures and rainfall (one description for a tropical climate) also tend to have very high air moisture contents. Thus warm, humid, tropical southern Florida has more moisture in the air than the wet but cool Pacific Northwest.

The summertime dewpoint temperature is an important value for the design of building dehumidification (and air-conditioning) systems. Plotting summer dewpoint temperature over the US (Figure 3.18) makes clear how very different the eastern side of the continent is from the west[1].

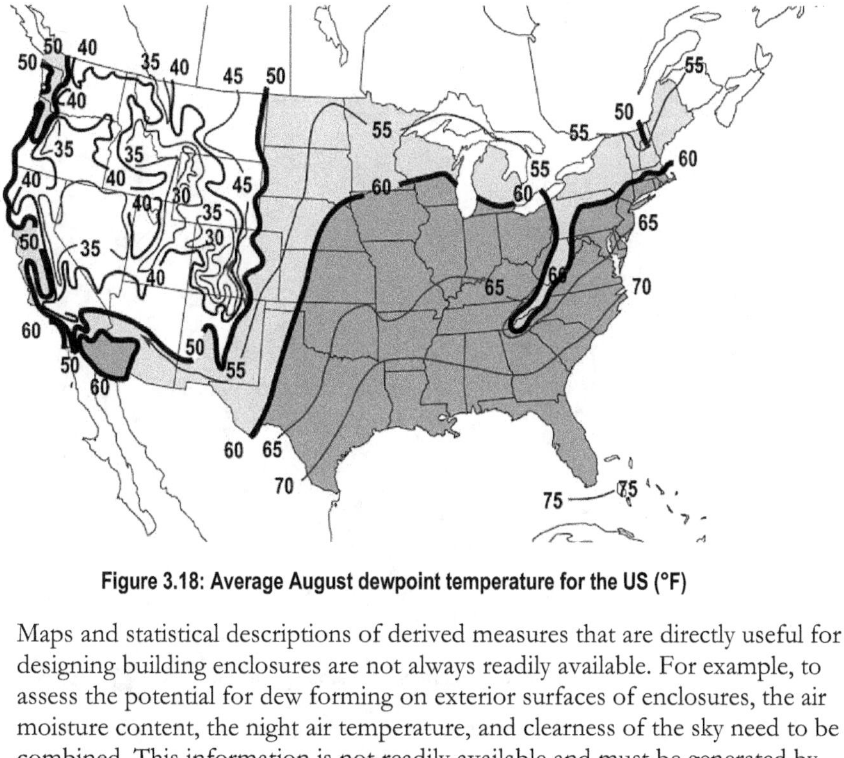

Figure 3.18: Average August dewpoint temperature for the US (°F)

Maps and statistical descriptions of derived measures that are directly useful for designing building enclosures are not always readily available. For example, to assess the potential for dew forming on exterior surfaces of enclosures, the air moisture content, the night air temperature, and clearness of the sky need to be combined. This information is not readily available and must be generated by the analyst if important for the problem at hand.

3.4.4 Sun

The sun is one of the two primary sources of energy on earth (stored energy from the molten core being the other). Understanding its behavior is therefore important with respect to the performance of enclosures and the energy consumption of buildings.

---

[1] This map plots temperature in °F. It assumed readers of this textbook can easily convert values to °C if desired.

The earth rotates about its axis every 24 hours. The earth also revolves around the sun once every 365.2425 days in an elliptical orbit, at an average distance of 154.4 million km (94.45 million miles). The earth–sun distance is 154.4 million kilometers on June 21 and 144.6 million kilometers (91.34 million miles) on December 21 (Figure 3.19). The earth is tilted at an angle of 23 degrees and 27 minutes with respect to its orbital plane (Figure 3.20). This tilt is termed the earth's declination. At noon on June 21, the sun is directly perpendicular to the earth's surface at 23.5 °N—hence the Tropic of Cancer. On that date, all parts of the earth south of the Antarctic Circle (66.5°S, or 90° - 23.5°) experience 24 hours of darkness, while all areas north of the Arctic Circle (66.5 °N) experience 24 hours of light. At noon on December 21, the sun is perpendicular to the Tropic of Capricorn.

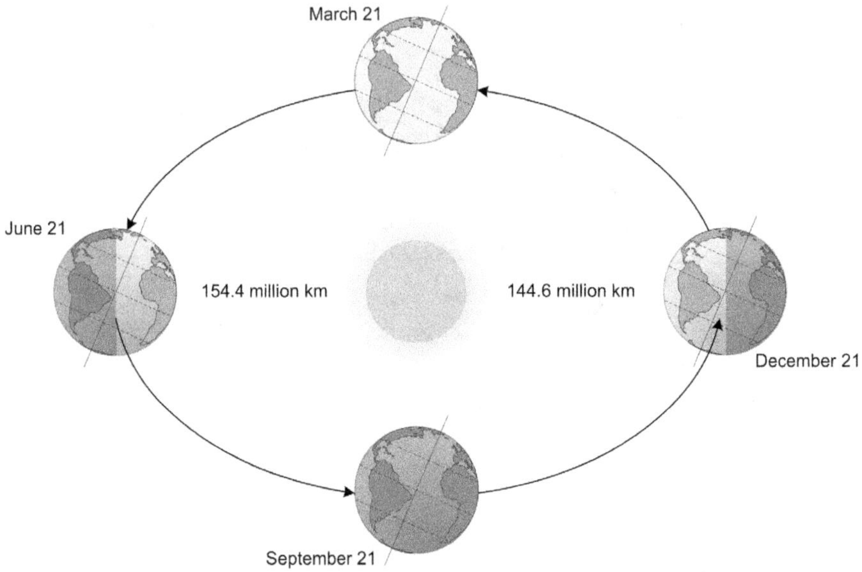

Figure 3.19: Earth's orbit around the sun

These facts, combined with the thermal storage mass of the earth and oceans, helps explain the seasons, prevailing winds, the length of the day, monsoons and hurricanes, and practically all other mesoscale weather phenomena.

On June 21, usually the longest day of the year, the angle of the sun's rays with respect to the earth are also the steepest in the northern hemisphere. The northern hemisphere therefore receives the most intense and greatest amount of solar radiation on this date. However, the thermal lag of the earth delays the hottest days of the year for another month or more. Similarly, the least solar energy is delivered in December 21, and the coldest days of the winter tend to be a month or more later.

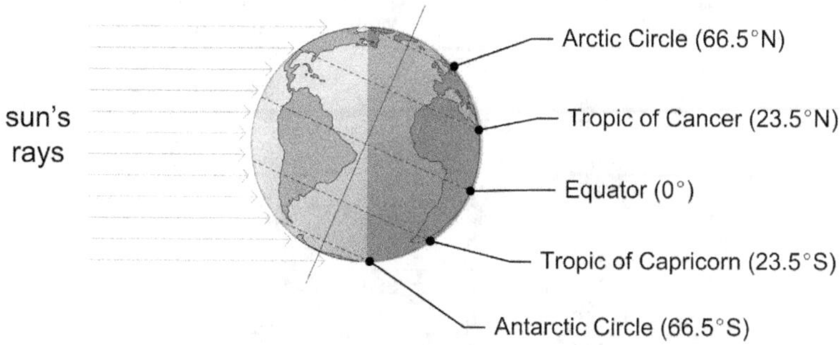

Figure 3.20: Earth's tilt on December 21

It is therefore easy to see how overhangs on a building can help control solar radiation. For example, at 40 °N, the angle of the sun above the horizon at solar noon on June 21 will be 90 - 40 + 23.5 = 73.5°, whereas on December 21 it will be only 90 - 40 - 23.5 = 26.5°. This knowledge of the sun is critical for the proper design of buildings (heating, cooling, daylighting) as well as for building enclosures (since the sun affects enclosure temperatures and the deterioration of some materials.

Figure 3.21 and Figure 3.22Figure 3.21 provide examples of clear-day solar information at 45 °N on July 21 and January 21 respectively. The angle of incidence of the sun on any arbitrary plane surface at any time of the year can be accurately calculated. Only the occurrence of clouds and mist, and variable ground reflectance add a degree of randomness to the solar radiation striking a building enclosure. Much more detailed information is available from a host of solar radiation resources [Duffie & Beckman 2013].

Hour	North Wall	East Wall	South Wall	West Wall	Roof
1 - 4	0	0	0	0	0
5	32	71	5	5	9
6	**117**	472	38	38	119
7	83	651	68	63	286
8	87	**679**	107	82	454
9	97	606	209	97	595
10	107	457	318	107	704
11	114	252	394	114	772
12	116	126	**420**	126	**795**
13	114	114	394	252	772
14	107	107	318	457	704
15	97	97	209	606	595
16	87	82	107	**679**	454
17	83	63	68	651	286
18	**117**	38	38	472	119
19	32	5	5	71	9
20-24	0	0	0	0	0
TOTAL	1390	3820	2698	3820	6673

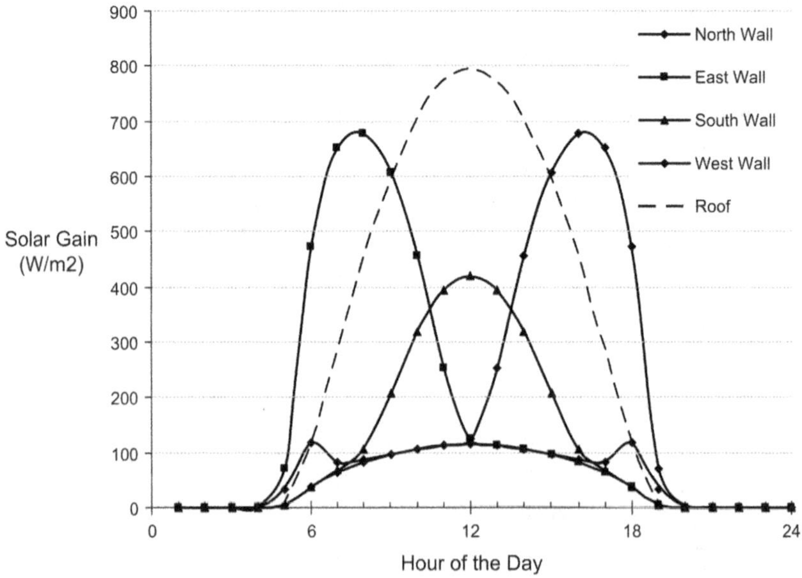

Figure 3.21: Solar gains - clear day values on July 21 at 45ºN (W/m²)

Hour	North Wall	East Wall	South Wall	West Wall	Roof
1 - 7	0	0	0	0	0
8	8	199	137	8	17
9	29	**422**	454	29	111
10	42	358	646	42	222
11	50	177	756	50	299
12	**53**	57	**792**	57	**326**
13	50	50	756	177	299
14	42	42	646	358	222
15	29	29	454	**422**	111
16	8	8	137	199	17
17 - 24	0	0	0	0	0
TOTAL	311	1342	4778	1342	1624

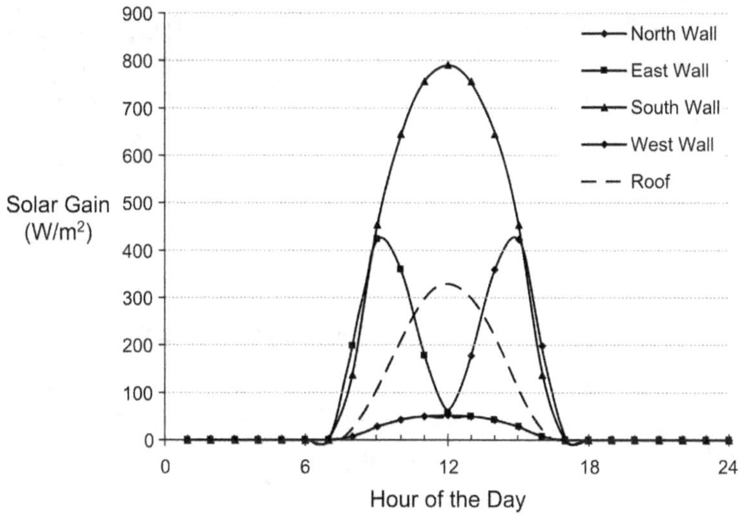

Figure 3.22: Solar gains - clear day values on January 21 at 45°N (W/m²)

3.4.5 Wind

From both a structural and a hygrothermal design point of view, wind is a major climatic element to be considered. Because of its importance to building structural design, wind behavior has been studied extensively, especially for structural design purposes. Many wind engineering resources are available; this section reviews wind pressures on buildings primarily as they relate to the hygrothermal aspects of enclosure design.

Wind is a stochastic phenomenon, so it is usually necessary to describe it with the aid of statistics, averages, extremes, and probability distributions. Figure 3.23 presents a representative plot of wind speed measured in the free unobstructed wind. The fluctuating component has a Gaussian (normal) probability distribution in the free wind, but this may be modified once the wind interacts with a building. It has been found that the average wind speed from one averaging period to the next does not change appreciably in constant weather conditions if the averaging period is about 15 minutes to one hour long. However, the speed changes often and quickly enough that maximum (i.e., gusts) and minimum velocities can only be captured if measurements are taken at about one-second intervals.

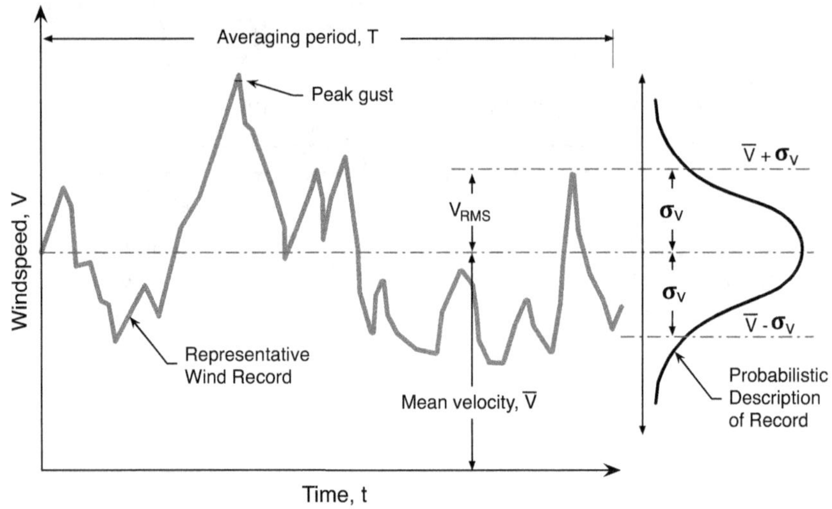

Figure 3.23: Temporal and statistical representation of wind velocity

Wind speed can be easily and accurately transformed into a pressure in the center of a large square plate placed perpendicular to the airflow direction by Bernoulli's equation:

$$p = \frac{1}{2}\rho V^2 \qquad \text{Eq. 3.2}$$

where p is the stagnation pressure (Pa),

ρ is the air density (kg/m^3), and

V is the air velocity or speed (m/s).

To normalize wind pressures, the actual wind-induced pressures acting normal to a building face are often expressed as a fraction of a reference stagnation pressure in the free wind at a particular height. The free wind speed reference height is often chosen to be at the top of the building (eaves height for buildings with pitched roofs is the convention) or 10 m (33 feet) above the ground (the height at which most weather stations measure wind speed).

The pressure coefficient, C_P, is defined as:

$$C_P = \frac{P_{building}}{P_{stagnation}} \qquad \text{Eq. 3.3}$$

For example, if the pressure recorded at the point of interest is 80 Pa and the stagnation pressure calculated from measured speeds upwind at the same height as the top of the building is 100 Pa, C_p will be given as 0.8 or 80%.

The pressure coefficient can be plotted over the surface of a building for different wind conditions to provide an understanding of how the wind interacts with building shapes independent of speed. Figure 3.24 and Figure 3.25 plot pressure coefficients for typical blunt rectangular buildings.

Code approaches to wind design

Different national codes and standards prescribe slightly different approaches to calculating the peak wind loads required to design structural systems and cladding. Some of these approaches can easily be adapted to calculate wind pressures on enclosures that occur regularly and thus impact their hygrothermal performance. The methodology of the National Building Code of Canada (NBCC) will be described here, but the US Standard ASCE-7 will produce similar results[2].

The NBCC tabulates the design pressure, q, for many locations based on long-term records of mean hourly wind velocities measured at 10 meters above grade with a probability of occurring in any year of 10%, 3%, and 1%. To convert velocities to pressure, the NBCC assumes an air density of 1.2929 kg/m^3 and standard temperature and pressure conditions. The design pressure given in the NBCC is simply the stagnation pressure for the assumed density:

$$q = \frac{1}{2}\,(1.2929)\,V_{10}^{\;2} \qquad \text{Eq. 3.4}$$

where q is the design pressure (Pa),

V_{10} is the design hourly mean wind speed at 10 m above grade (m/s).

The pressure acting at any point on a building surface (whether interior or exterior) is drastically affected by the building's location, height, and shape, as well as the wind direction and local topography. The NBCC recognizes these variables by modifying the design pressure with several factors.

The external pressure or suction on the building surface in the NBCC is given by the equation:

$$P_e = q \cdot I_w \cdot C_t \cdot C_e \cdot C_g \cdot C_{pe} \qquad \text{Eq. 3.5}$$

where P_e is the external static design pressure (positive or negative, i.e., suction) acting normal to the surface,

q is the specified design pressure (Pa),

I_w accounts for the importance of the building based on occupancy/use,

---

[2] ASCE7 uses three-second duration wind speeds, not hourly average pressures, for design but the equations to convert to pressure, adjust for height, gusts, and building shape all take the same form as they are based on the same science.

C_t accounts for site topography that is different than a flat plane,

C_e is the exposure factor which accounts for spatial changes in wind speed with height and terrain (see later section),

C_g is the gust factor which accounts for temporal variations in wind speed and is given a value of 2.0 for the primary structure (equivalent to a 3 s long gust) and 2.5 for cladding design (equivalent to a 1 s gust),

C_{pe} is the building- and cladding-specific external pressure coefficient averaged over the area considered.

Recommended values of C_{pe} are documented for simple building shapes in design standards and technical publications. Illustrative summaries are provided in Figure 3.24 and Figure 3.25. Stathopolous [1989] provides plots for intermediate-height buildings. To obtain more accurate values for complicated building geometries and topographies, it is usually necessary to perform wind tunnel testing or sophisticated computational fluid dynamics (CFD) simulations.

The pressure acting across the enclosure is the difference between the exterior and interior. Therefore:

$$P_{(e\text{-}i)} = P_e - P_i = P_e - q \cdot I_w \cdot C_t \cdot C_{ei} \cdot C_{gi} \cdot C_{pi} \qquad \text{Eq. 3.6}$$

where $P_{(e\text{-}i)}$ is the design pressure across the enclosure acting normal to the surface,

C_{ei} is the exposure factor as above but evaluated at mid-height of the building,

C_{pi} is the internal pressure coefficient, and

C_{gi} is the internal gust factor, normally set to 2.0 unless detailed analysis shows otherwise.

The internal pressure coefficient, Cpi, depends on the air leakage of the building enclosure and the distribution of this leakage about each orientation. A normal building enclosure is assumed to have a modest amount of air leakage uniformly distributed over the entire area. For this common scenario values of -0.15 and 0 is used, whichever is less favorable. If there is a dominant opening, either intentionally (a large open door or loading dock) or unintentionally (a broken window during a hurricane), the internal pressure will change drastically, and both gust factors and dynamic effects must be accounted for.

In the Canadian code approach, q is equivalent to a static design pressure for an extreme hourly average wind speed. The 1-in-50-year pressure given in the NBCC is the highest hourly average pressure likely to occur for one hour every 50 years. If the gust factor of 2.5 is applied, the resultant pressure is likely to occur for about one second every 50 years.

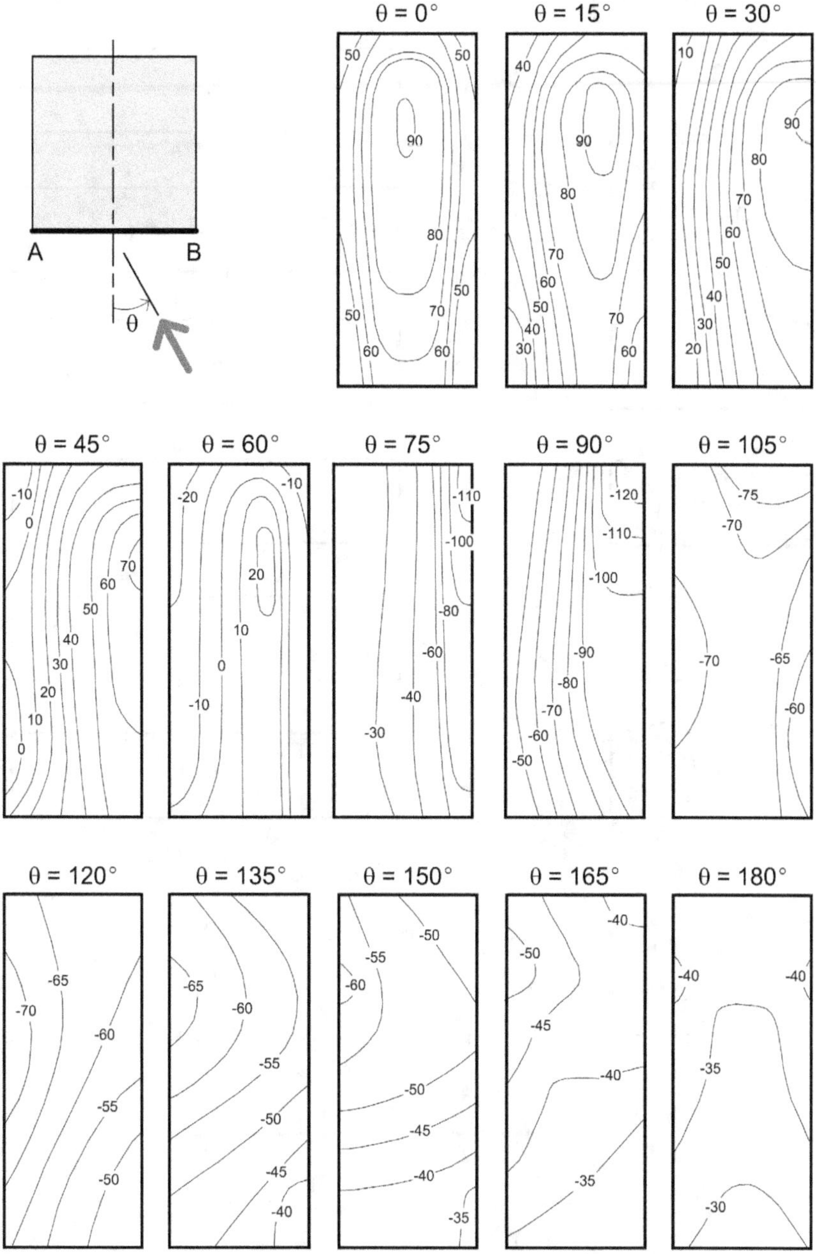

Figure 3.24: Local external pressure coefficients ($C_{pe} \times 100$) for tall building with varying wind direction [Davenport & Hui 1982]

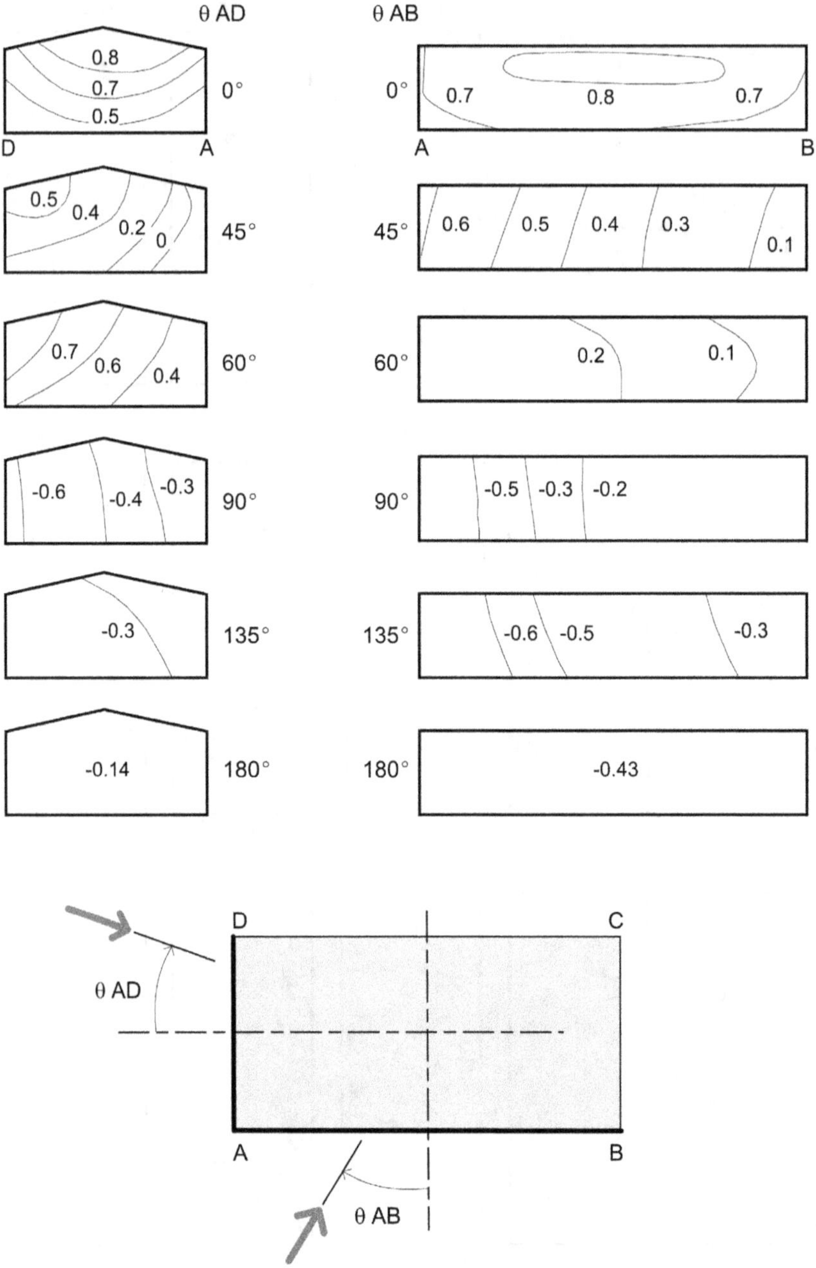

Figure 3.25: Exterior pressure coefficients for low-rise building [ASHRAE 2021]

This approach may be acceptable for defining structural design loads, where safety is a primary concern, but it is not a satisfactory means of modeling wind for in-service conditions. For an understanding of the real behavior of a building during its service life, a more accurate and realistic description of the wind and its effects is necessary. The wind is not a steady, smooth-flowing stream of air. The wind velocity and direction are constantly changing, and this causes temporal and spatial pressure variations on the face of a building. As a result, the building enclosure typically experiences a much lower mean pressure (compared to code-based extreme design values) with rapidly changing variations (as opposed to the static approximation of the NBCC).

The following sections describe the mean pressure gradients, temporal variations, and spatial-temporal variations in the wind and on a building face.

Vertical wind speed gradients

The mean velocity of the wind (and hence the stagnation pressure) varies with height and terrain. The fact that the wind velocity (and pressure) varies with height affects the pressures on a building face and is accounted for in the NBCC by an exposure factor C_e (and by K_z in ASCE-7).

A common approach is to model the velocity increase with height using a power law, (logarithmic laws are also used). The increase of mean velocity with height may be estimated (in consistent units) using:

$$V(z) = \left(\frac{z}{z_{ref}}\right)^{\alpha} V_{ref} \qquad \text{Eq. 3.7}$$

where V(z) is the mean velocity at height z,

z is the height above the ground,

z_{ref} is the reference height above the ground,

α is the gradient exponent, and

V_{ref} is the mean velocity at the reference height.

Note that the reference height might be chosen to be the standard height at which wind records are collected (10 m or 33 ft). It can be seen from Figure 3.26 that the wind speed increases more slowly as the terrain becomes rougher (because of more retarding friction with the ground) and stops changing at a greater height (the edge of the so-called boundary layer or gradient wind height).

The wind velocity gradient exponent is given in both the NBCC and the ASCE-7 standard for a variety of terrains. The exponent varies from less than 0.1 (for flat, smooth terrain or open water upwind) to over 0.33 for highly complex upwind conditions such as those in city centers with many tall buildings.

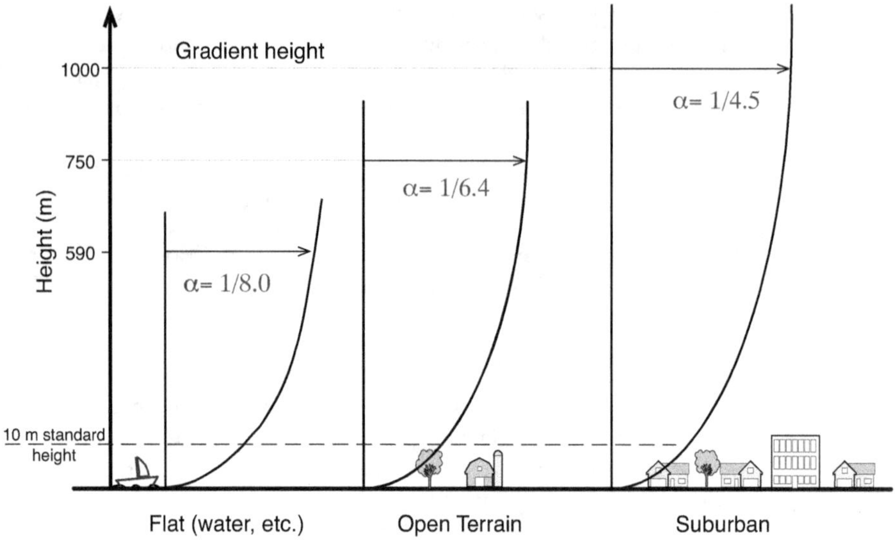

Figure 3.26: Variation in wind velocity with height and terrain

Making use of the power law for velocity, the mean stagnation pressure gradient of the wind can be described by:

$$P(z) = \left(\frac{z}{z_{ref}}\right)^{2\alpha} P_{ref} \qquad \text{Eq. 3.8}$$

where P(z) is the mean pressure at z m above the grade (Pa) and

P_{ref} is the mean measured or calculated pressure at z_{ref} (Pa).

Spatial wind pressure variations

Airflow around the building induces spatial pressure gradients greater than those that exist in the wind. On the windward face of a building under a perpendicular wind, the pressures drop from a maximum along the vertical centerline to a negative pressure at the building edge. This negative zone rarely extends toward the center of the face by more than 1 or 2% of the building width. The variation of mean pressures on the face of a high building is more complicated since the velocity of the wind increases with height.

Complications of building geometry, surface roughness, and upstream obstructions mean that wind tunnel testing is necessary for accurate predictions of mean spatial variations on the face of a building. Results from many tests on tall buildings with square plans has resulted in pressure gradient diagrams as shown in Figure 3.24.

Example: Consider a rectangular, 30 m tall, 12-story hotel in rough terrain. Assume a gradient exponent of 0.15 and an interior pressure coefficient of -0.3). What would typical in-service pressure differences across the enclosure due to wind be predicted to be?

The average hourly wind speed at 10 m above grade in Waterloo, Ontario (and in many other Canadian locations), is between 3 and 4 m/s. If we use an average of 3.5 m/s, the velocity at the top of the building will be:

$$\overline{V(30)} = \left(\frac{30}{10}\right)^{.15} \cdot 3.5 = 4.13 \text{ m/s}$$

This results in a stagnation pressure at 30 m above grade of:

$$P_{stagnation} = 0.647 \cdot 4.13^2 = 11.0 \text{ Pa}$$

Assuming a C_{pe} of 0.8 as representing a large area of the windward façade, and C_{pi} = -0.30 for interior pressures, the hourly average pressure across the windward face of the enclosure would be predicted to be about

$$(C_{pe} - C_{pi}) * P_{stag} = (0.8 - -0.30) \cdot 11.0 = 12.1 \text{ Pa.}$$

Inspection of Figure 3.24 shows that the maximum inward pressure will occur at about 80% of the height up the building when the wind is blowing perpendicular to the face. Maximum hourly pressures will be slightly higher over a small area (where C_{pe} is just over 0.90). Short peak gusts (lasting 1 to 3 seconds) will be between 2 and 2.5 times higher than those calculated.

Figure 3.24 also shows that the average suction pressure along walls parallel to the wind direction and pressure coefficients would often reach C_{pe}= -0.80 to 1.20 are, which, combined with C_{pi} = -0.30, results in a net pressure coefficient of -0.50 to 0.90. Therefore, a suction pressure of 6 to 10 Pa is a reasonable estimate near the upper sides of the building. The leeward side could be expected to have a C_{pe} of around -0.35 and thus a net pressure of only about 1 Pa drawing air outward through the enclosure on this façade.

Naturally, structural design loads would be much higher (100 times higher), because the 1-in-50 year design stagnation pressure represents not average pressures but rare short-term gusts.

Temporal variations: gusts

For dynamic behavior, peak cladding and window loads, and pressure-moderated wall systems, wind gusts of short duration are the primary interest. The wind velocity is quite steady when averaged over periods of 10 minutes to 1 hour, but varies significantly about this mean at any time t. This has led to the analysis of a time-varying wind speed signal V(t), by separating it into its mean hourly velocity V, and its varying component, v(t), also called the gust vector.

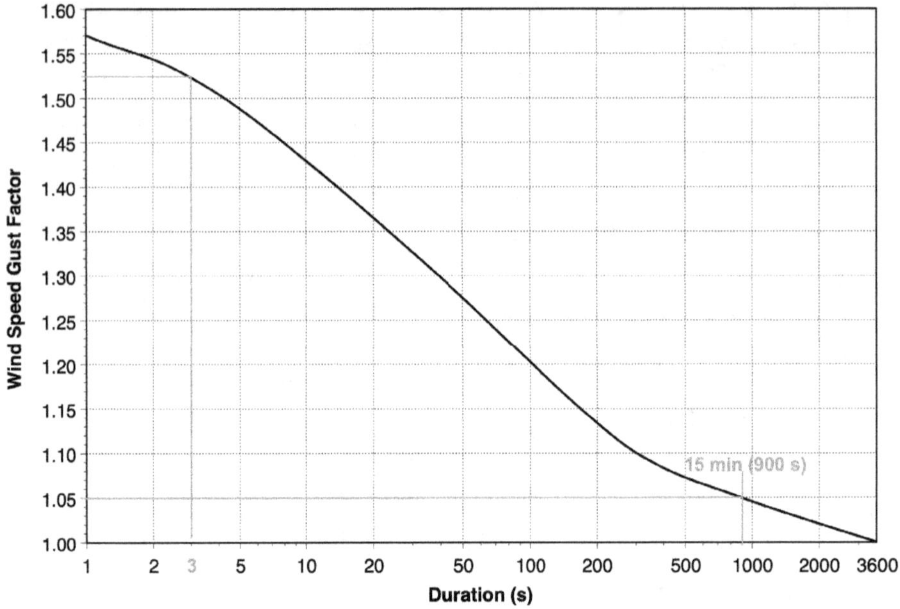

Figure 3.27: Durst gust factor correlation [1960]

To assess extreme pseudo-static wind pressures, most codes apply a gust factor, C_g, to the hourly average pressures. A maximum gust magnitude for a typical building size is $C_g = 2.0$, whereas a higher value would be applied to a small area such as a cladding component ($C_g = 2.5$). This factor attempts to predict appropriate peak pressures that will act on a building or cladding component, but it does not accurately reflect the typical behavior of gusts or their spatial extent.

Numerous correlations exist to adjust the wind speed from the hourly average to the duration of the gust of interest. One of these correlations was developed by Durst [1960] (Figure 3.27). For a gust duration of 3 seconds, a gust factor of 1.525 is predicted. This would result in a pressure increase of 2.3 times (1.525 squared). For a 15-minute (900-second) long gust, the gust factor would be only 1.05.

3.4.6 Rain

Rain is a common source of moisture that causes building problems and has been an important climatic load for as long as buildings have been built. Hence, a close study of rain is justified. The annual average rainfall over North America is shown in Figure 3.28.

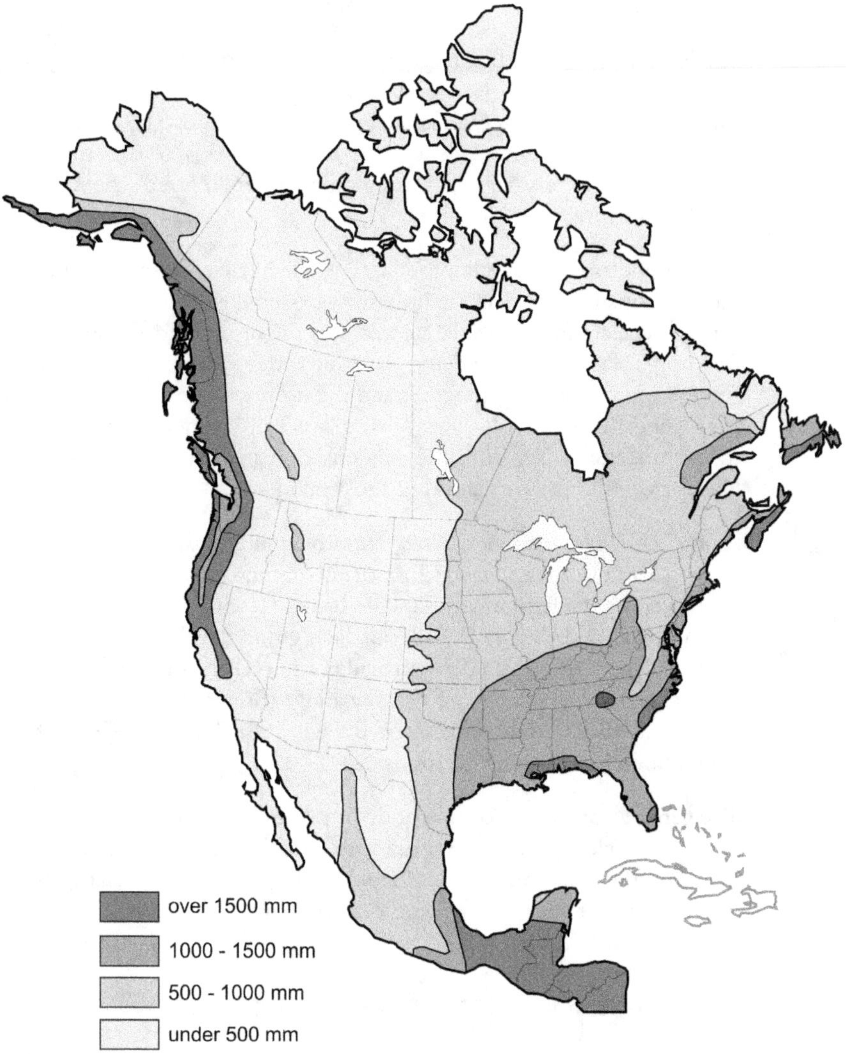

Figure 3.28: Average annual rainfall in North America

Condensation of water vapor onto condensation nuclei creates cloud droplets (aided by cooling as a result of air rising). Cloud droplets are perfectly spherical in shape because surface tension forces are much larger than aerodynamic or gravity forces. They range from 1 to 50 μm in diameter, with an average of 10 to 25 μm [Barry 1992, Grey 1970, Beard 1976]. Such small drops fall so slowly that their movement is governed by the wind currents aloft, and they act as suspended particles which can be experienced as a heavy fog at grade.

The precise mechanism of raindrop formation is not known, but the most widely accepted theory is based on ice accumulation around frozen condensation nuclei. This theory predicts that drops can form only within clouds that are below 0 °C. Because the vapor pressure at the surface of ice is lower than that for water, ice nuclei quickly grow in size by condensation. If the crystal diameter increases to more than about 0.10 to 0.25 mm, it begins falling. As the ice crystal falls through to warmer air layers, the ice melts and a raindrop forms.

If two raindrops in a rainfall touch, they will likely coalesce into a larger drop. Larger drops fall faster and can therefore overtake and absorb smaller drops, further accelerating their fall. However, if the resultant drop becomes too large, the droplet falls quickly and aerodynamic forces overcome the surface tension forces holding the drop together and the drop will break apart. Both Grey [1970] and Beard [1976] suggest that drops with diameters greater than about 6 to 7 mm are unstable. Both the condensation and coalescence mechanisms are likely at work in the formation of most raindrops.

As the rainfall intensity increases, the probability of two raindrops colliding increases and the average raindrop diameter increases. This conclusion is supported by the field measurements of Best [1950], Laws and Parsons [1943], and Marshall and Palmer [1948], but each group developed its own relationship. Markowitz [1976] provides an excellent summary. More recent researchers have made use of radar readings [Rogers 1962, Caton 1966] and sophisticated electronic drop sizers [Dingle 1962, Bradley 1974] to further explore raindrop size distributions.

The general conclusion of this body of research is that the nature of the distribution clearly varies not only with rainfall intensity but also with the type of storm, the cloud height, etc. Figure 3.29 is a plot of the relative probability distribution of raindrop diameter as a function of rainfall intensity given by Best [1950]:

$$F(\varnothing) = 1 - \exp\left\{-\left(\frac{\varnothing}{1.30 \cdot r_h^{0.232}}\right)^{2.245}\right\} \qquad \text{Eq. 3.9}$$

where, $F(\varnothing)$ is the cumulative probability distribution of drop diameters for a given rainfall intensity,

$\varnothing$ is the equivalent spherical raindrop diameter (mm), and

r_h is the rainfall rate on a horizontal plane ($mm/m^2/h$).

While precise knowledge of the distribution of drop sizes may not be available, the physical characteristics of the drops themselves are well understood. The aerodynamics of raindrops are of special interest for building applications, because these aerodynamic properties influence how rain is driven against building facades.

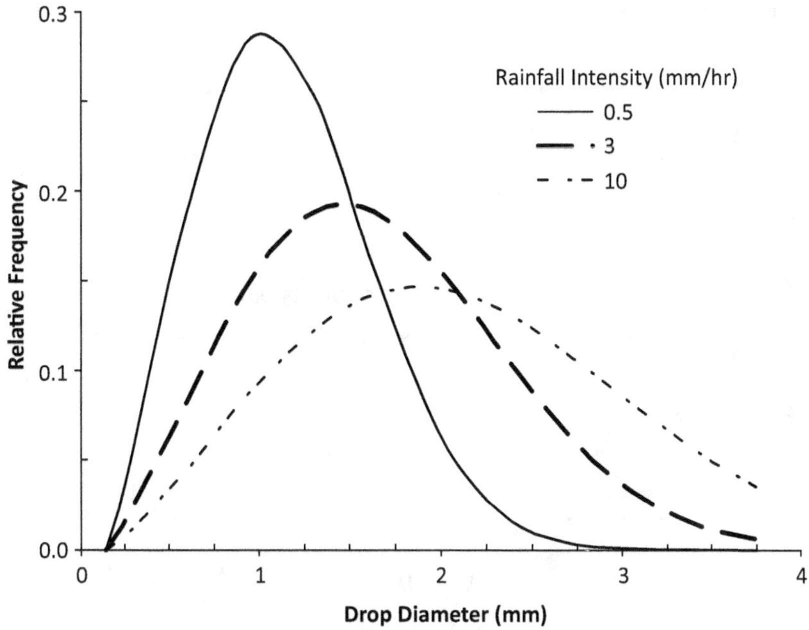

Figure 3.29: Distribution of raindrop sizes with rainfall intensity

Beard [1976] collected the available experimental data for raindrop size, shape, and drag coefficient and, based on fluid dynamics principles, developed general relationships for the terminal velocity of raindrops as a function of air pressure and air temperature. These relationships match most available experimental data, including the comprehensive and widely accepted data of Gunn and Kinzer [1949]. Figure 3.30 is a plot of this data. Dingle and Lee [1972] developed a simplified equation, accurate to ± 2.5%, for raindrop diameters of more than 0.3 mm:

$$V_t(\varnothing) = -.1660 + 4.91840\varnothing - .88802\,\varnothing^2 + .05489\varnothing^3 \leq 9.20 \quad \text{Eq. 3.10}$$

where, $V_t(\varnothing)$ is the terminal velocity of a raindrop in still air (m/s), and $\varnothing$ is the diameter (mm) of the raindrop.

More precise (± 0.5%) and much more unwieldy equations are available; see Beard [1950].

Rainfall intensity is related to the duration of the rain event and the likely occurrence of the event. Short rain events can be more intense, and these rain events that deliver a large amount of rain in a short period of time are rare. These relationships are defined by intensity-duration-frequency curves, often available for specific locations because of their importance for the design of stormwater management systems.

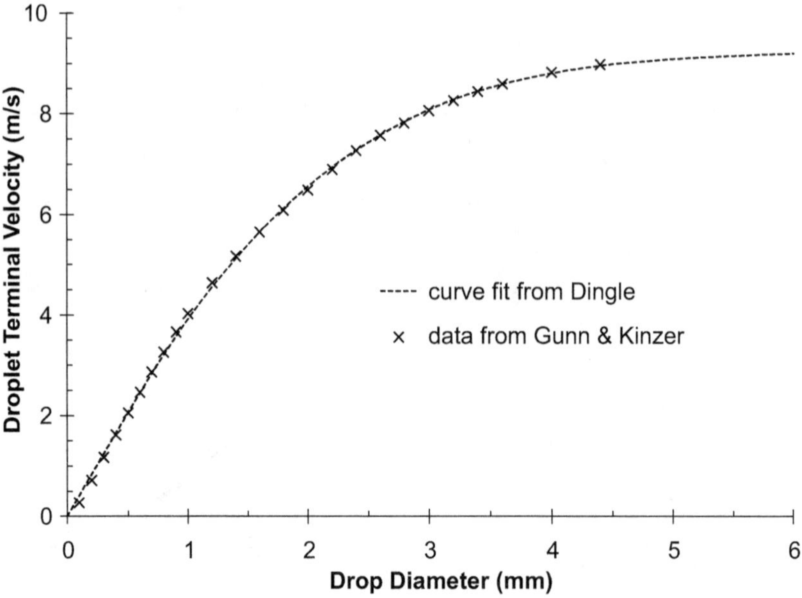

Figure 3.30: Velocity of water droplets in still air

Coincidence of wind and rainfall

Because of the meteorological causes of rain, wind and rain are often coincident, although the most intense rains do not occur at the same time as the worst winds (except, perhaps, in hurricane regions). It has been found that rainfall intensities are less than 5 mm/h during more than 90% of rain events [Kuenzel 1994b, Straube 1998] and both the data in Table 3.4 and that by Underwood and Meentemeyer [1998] indicate average rainfall intensities are in the order of 1 mm/h. In extreme weather events, rates ten times as high can be measured for short periods of time (hours) on an annual or decadal basis. Information on extreme rain intensities and their return periods is widely available for use in stormwater drainage design (e.g., Figure 3.31).

The statistical distribution of rainfall intensity is important for understanding enclosure response. For example, several ASTM rain penetration standards (e.g., E514 and E333) presume very high driving rain deposition rates (134 to 200 mm/h respectively). These extremely high rates are relevant to only a small portion of the world and very special localized enclosure conditions and are applied simply to flood the wall surface with water, not to reflect reality.

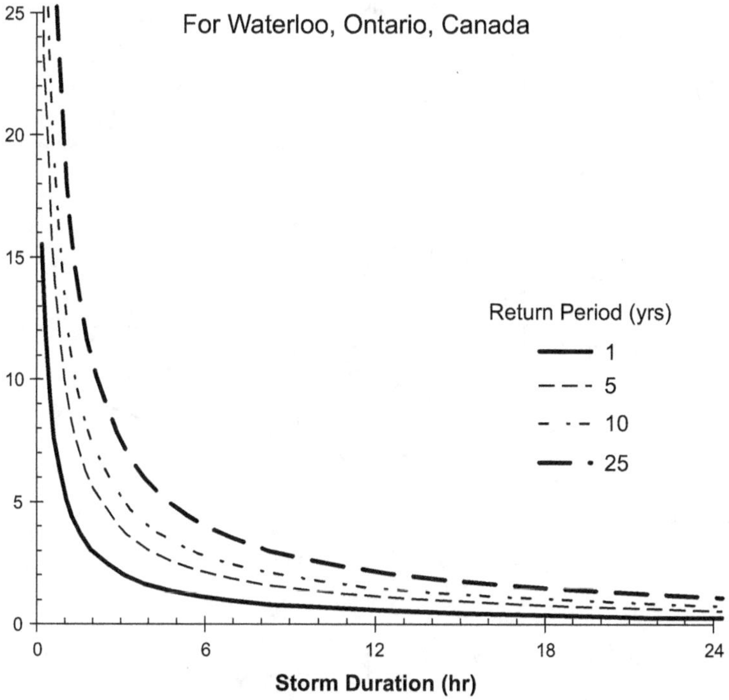

Figure 3.31: Example intensity-duration-frequency (IDP) curve available for stormwater design

Coincidence of wind direction and rainfall

The direction of wind during rainfall is often important as it helps define the rain load on a particular wall orientation. It is well documented that the wind direction distribution is different during rain events (e.g., Skerlj [1999]). Although all climates are not the same, researchers [Straube & Schumacher 2005, Surry et al 1995, Levelton Engineering 2004] did find many climates in which the predominant wind direction was quite different than the wind direction during rain.

In many climates, most of the rain comes from one direction, and the difference in driving rain deposition between the orientation with the most rain and that with the least can be a factor of ten or more. Understanding this difference can be useful in forensic analysis as well as design. Figure 3.32 plots the wind direction frequency rose for all hours of a year in Toronto Canada. In this climate, the wind comes much more from the East during rainfall and much more from the West and North during all hours of the year.

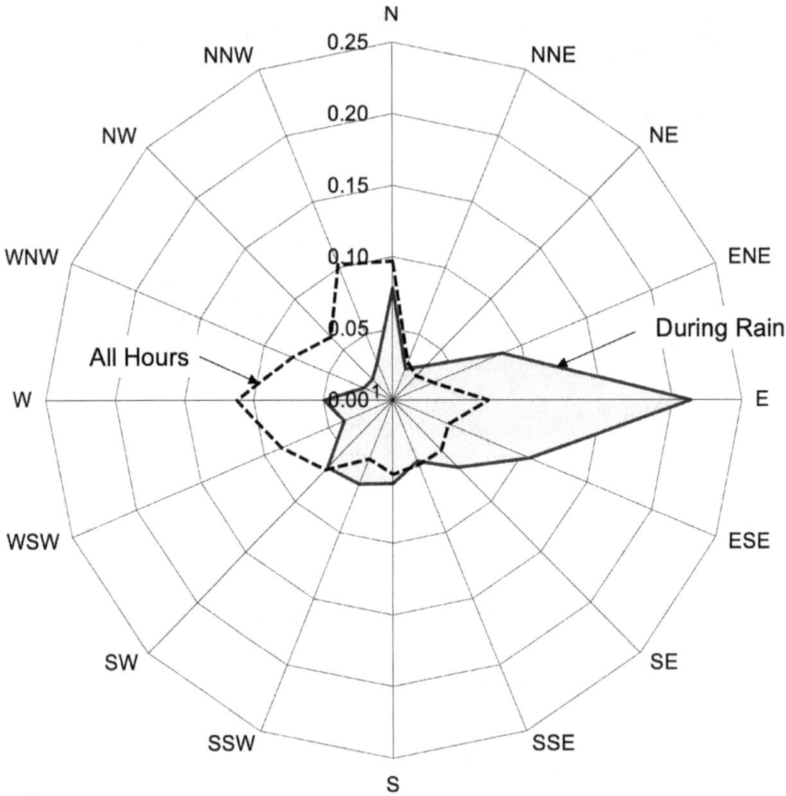

Figure 3.32: Average annual wind frequency during rain for Toronto, Canada

Coincidence of wind speed and rainfall

Wind speed during rainfall is critical for two reasons: the wind directs rainfall against vertical facades and it also generates air pressure differences across enclosures. High stagnation pressures can drive rain into sufficiently large cracks and openings (see Chapter 12). For most enclosure walls, the increase in air pressure with wind speed is unimportant relative to the size of the increase in rain deposition with wind speed. Hence, practitioners often observe an increase in rain penetration control problems in walls exposed to high wind pressures. These rain penetration problems are more likely to be a result of an increase in rain deposition rather than a result of higher air pressure.

Figure 3.33 plots the results of 25 years of hourly wind and rain data (1965–1990) for Seattle, Washington. Wind speed is slightly higher during rainfall (by about 10%) but the wind speed is, on average, relatively low. The plot shows that the hourly average stagnation pressures during more than 99% of all rain events are lower than 20 Pa. Many locations in North America have similar average wind speeds (i.e., 3 to 4 m/s at 10 m above grade) and hence these results are typical of most North American and Northern European locations. Areas subject to hurricanes have a different profile: every few years, very high wind speeds combine with high rainfall rates to generate very demanding loads.

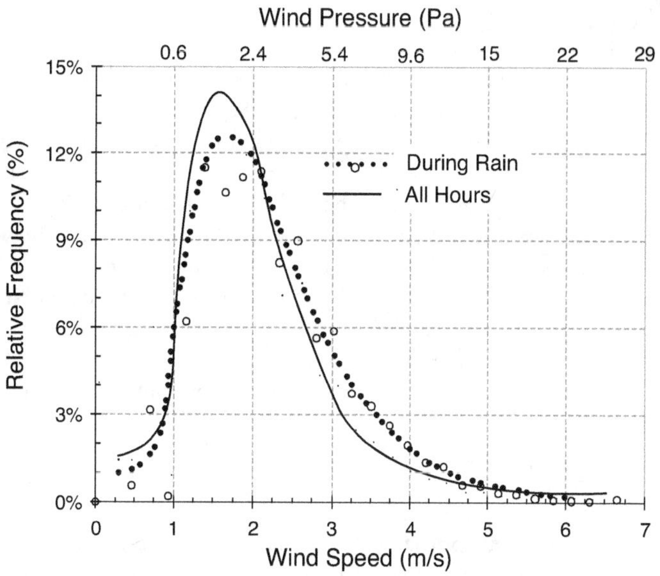

Figure 3.33: Coincidence of rainfall and wind speed (and wind pressure) for Seattle, WA

3.4.7 Driving rain

Wind-driven (driving) rain, the result of the interaction between wind and falling rain, is very important for building enclosure design, as it is the largest liquid water load on most enclosures.

Despite the importance of driving rain to building performance, there is a lack of quantitative data related to the magnitude, duration, frequency, and deposition of rain on buildings. Rainfall itself is only of interest to roofs and sky-facing sloped walls. If rain fell vertically, there would be little difficulty in preventing rainwater from reaching the surface of a wall. However, even a breeze will shift the falling raindrop off its gravity-driven vertical trajectory.

Driving rain has been investigated by building scientists and meteorologists in the field for many years. Driving rain has been measured under real conditions in Europe for almost 100 years (continuously for more than 60 years at one location) although not usually on building systems. Lacy, of the Building Research Establishment in Britain, conducted the seminal English-language study in the early 1960s [Lacy 1965]. He recommended a simple equation to predict driving rain in the free wind (away from obstacles and other buildings) during typical rain events that still provides a reasonable estimate today:

$$r_v = \frac{2}{9} V {r_h}^{8/9}$$ Eq. 3.11

where, r_v is the rate of rain passing through a vertical plane in space,

r_h is the rate of rainfall (rain falling through a horizontal plane), and

V is the wind speed.

Improved scientific knowledge led to more refined analysis. Predicting the actual trajectory taken by a raindrop in the free wind is relatively straightforward: raindrops fall to the ground at their terminal velocity and are blown sideways at the speed of the wind (Figure 3.34). As shown earlier, the speed at which raindrops fall is connected to the size of the drop (essentially, large drops fall faster than light drops in a non-linear fashion). The wind speed carries the drops along horizontally. The combination of the two forces determines the trajectory of the drop, and simple geometry can then be used to assess the amount of rain passing through a vertical plane. This assessment is complicated by the fact that raindrops do not come in a single size, but the correlations between raindrop size distribution and rainfall intensity have been presented above.

Predicting wind-driven rain

A simple and practical approach for estimating rain deposition on vertical building walls is to divide driving rain deposition into the effects of the undisturbed wind (the free wind driving rain) and the effect of the aerodynamics and geometry of the local building. An equation developed using this approach [Straube 1998] is:

$$r_{vb} = RDF \cdot DRF \cdot \cos(\theta) \cdot V(h) \cdot r_h$$ Eq. 3.12

where r_{vb} is the rain deposition rate on a vertical building surface (l/m²/h),

r_h is the average rainfall rate on a horizontal plane, e.g., the ground (mm/h = l/m²/h),

V(h) is the wind speed at the height of interest, h (m/s),

θ is the horizontal angle between the normal to the wall and the wind direction,

DRF is the driving rain factor, which accounts for interaction of the wind and rain in the undisturbed wind,

RDF is the rain deposition factor, which accounts for the shape and aerodynamic features of a building.

The driving rain factor (DRF) is the inverse of the median raindrop terminal fall velocity. For average rainfall conditions, a DRF of about 0.2 is appropriate; for intense downbursts it may be as low as 0.14, and for light drizzle it may be as high as 1.0.

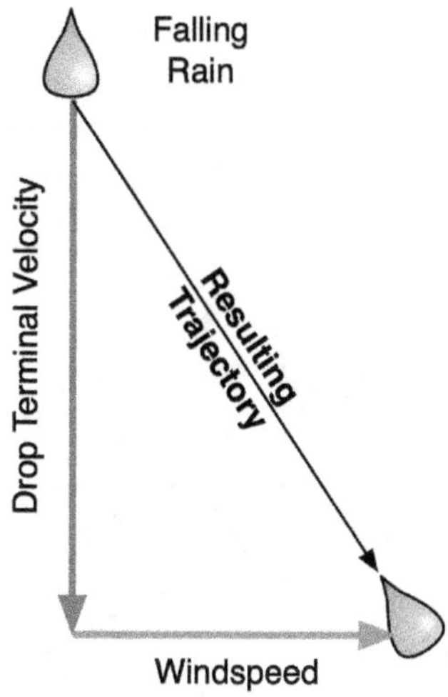

Figure 3.34: Driving rain

A simple and reasonably accurate equation for the DRF is:

$$DRF = 0.22 \cdot r_h^{-0.123} \qquad \text{Eq. 3.13}$$

The fact that the wind velocity, V, is included in Eq. 3.12 should be expected. The trajectory of a raindrop is directly related to windspeed (Figure 3.34). For a given rainfall intensity (and thus a given droplet size distribution) increasing windspeed will relate directly to an increase in driving rain deposition. Hence, calculating the wind speed at the height of interest is important (and is a fundamental limitation of the "catch ratio" concept found in the literature).

Hence, buildings, or parts of buildings, exposed to high windspeed are exposed to high levels of driving rain deposition. Based on the information presented earlier about wind, it can be understood that wind speed (and hence rain deposition) will be much higher close to the ground on hilltops, on the seashore, or in otherwise exposed conditions than is the case for buildings protected by other houses, by trees, or by hills. Sites protected from high wind speeds are also protected from driving rain. The fact that wind speed increases with height (especially once one is higher than low-level obstacles like trees and neighboring housing) means that the driving rain exposure of tall buildings is much higher than for low-rise buildings. The top of an exposed high-rise building can experience five times greater wind velocity than that experienced by a house in a suburban setting. This difference in wind speed relates directly and quantitatively to the quantity of rain deposition and leads to the important conclusion that high-rise and other wind-exposed (e.g., seaside) buildings must be designed to control much larger volumes of driving rain than low-rise buildings sheltered by trees and other neighboring buildings.

Using Lacy's simple equation (which is based on wind speed at 10 m above grade), Underwood and Meentemeyer [1998] developed a contour map of the

US (Figure 3.35) documenting total annual driving rain, driving rain event duration, seasonal driving rain quantities, and more. This data is very useful but does not translate directly to the load of rainwater on a vertical wall without considering the wind direction (θ) during rainfall, the height of the location on the building, the upwind terrain, and the shape and exposure of the building (RDF).

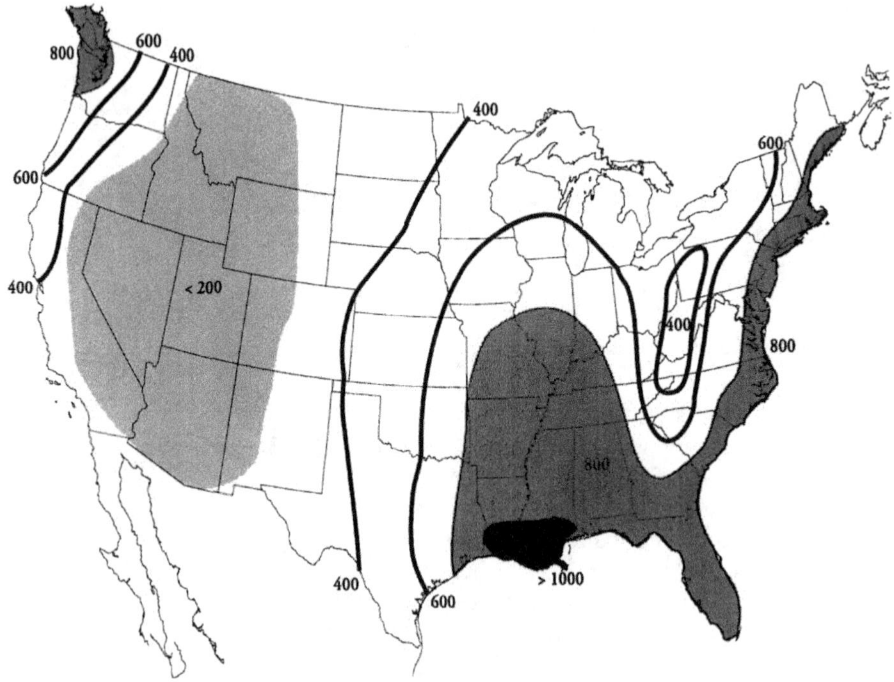

Figure 3.35: Total annual wind-driven rain (mm/yr) for the United States [Underwood and Meentemeyer 1998]

The more refined methods of Eq. 3.12 were used to calculate driving rain statistics for 14 major North American cities representative of different climate zones in Table 3.3 [ASTM 2009].

Several important and useful observations can be made from this data:

- average annual rainfall varies widely as expected, from 211 mm/yr (8.3 inches/yr) in Phoenix to 1487 mm/yr (58.5 inches/yr) in Miami,
- wind speed during rain events is usually slightly higher, by an average of 15%, than when it is not rainy,
- average wind speed at 10 m during rain events ranges from 3.9 to 6.0 m/s (8.7 to 13.4 mph), or a stagnation pressure of 9.8 to 23.2 Pa,

- total driving rain in the free wind is a significant fraction of falling rain, varying from 71 to 120% of rain on a horizontal surface,

Table 3.3: Wind-driven rain climate statistics for 14 representative cities

City	Rainfall	Wind Speed (m/s)		Driving Rain		
	(mm/yr)	All Hrs	During Rain	(hours) Duration	(mm/yr) Total	Worst Direction
Los Angeles CA	334	3.6	4.6	156	283	173
Minneapolis MN	621	4.6	4.9	250	599	226
Toronto ON	685	4.3	4.8	586	646	308
Kansas City MO	914	4.5	4.8	292	729	311
Miami FL	1487	4.3	4.6	339	1057	443
Houston TX	1214	3.8	5.0	418	945	470
Boston MA	972	5.4	6.0	731	1165	531
Atlanta GA	1270	3.9	4.6	515	1092	544
Seattle WA	912	3.9	4.5	1155	948	738
Vancouver BC	1239	3.3	4.0	1300	1156	811
Denver, CO	265	3.8	3.9	267	274	105
San Francisco	511	4.5	4.4	290	540	359
Halifax, NS	1421	4.2	5.3	721	1456	762
Phoenix AZ	211	3.0	4.7	150	182	89
Average	861	4.1	4.7	512	791	419
Max	1487	5.4	6.0	1300	1456	811
Min	211	3.0	3.9	150	182	89
	(Inches/ year)	(mph)	(mph)	(hours)	Inches/ year	
Los Angeles CA	13.1	8.1	10.3	156	11.1	6.8
Minneapolis MN	24.4	10.3	11.0	250	23.6	8.9
Toronto ON	26.9	9.6	10.7	586	25.4	12.1
Kansas City MO	36.0	10.1	10.7	292	28.7	12.2
Miami FL	58.5	9.6	10.3	339	41.6	17.4
Houston TX	47.8	8.5	11.2	418	37.2	18.5
Boston MA	38.3	12.1	13.4	731	45.9	20.9
Atlanta GA	50.0	8.7	10.3	515	43.0	21.4
Seattle WA	35.9	8.7	10.1	1155	37.3	29.1
Vancouver BC	48.8	7.4	8.9	1300	45.5	31.9
Denver, CO	10.4	8.5	8.7	267	10.8	4.1
San Francisco	20.1	10.1	9.8	290	21.3	14.1
Halifax, NS	55.9	9.4	11.9	721	57.3	30.0
Phoenix AZ	8.3	6.7	10.5	150	7.2	3.5
Average	34	9.1	10.6	512	31	17
Max	59	12.1	13.4	1300	46	32
Min	8	6.7	8.7	150	7	4

- annual driving rain on a vertical plane oriented in the worst direction (representative of the load on a vertical enclosure wall) ranges from 30 to 81% of annual rainfall (i.e., of rain on a roof), and

- driving rain on the enclosure wall facing the worst direction ranges from 38 to 78% of the total driving rain, demonstrating that some climates may have twice as much directionality than others.

There are numerous techniques to demonstrate directionality. One is to consider only the 15-degree sector of the wind rose and calculate the driving rain excluding any rain from angles outside this sector. This may be appropriate for climatic assessment. For building enclosure design, it is more useful to consider the cosine-corrected contribution of rain (that is, driving rain from an angle that is not normal to the façade is reduced by multiplying by the cosince of that angle) to a particular orientation (i.e., facade) for the +/-90 degrees from a particular direction on the wind rose. This approach is taken in Figure 3.36 for Seattle, WA (the DRF was calculated every hour based on the rainfall intensity), and the significant variation in direction for this climate can be seen.

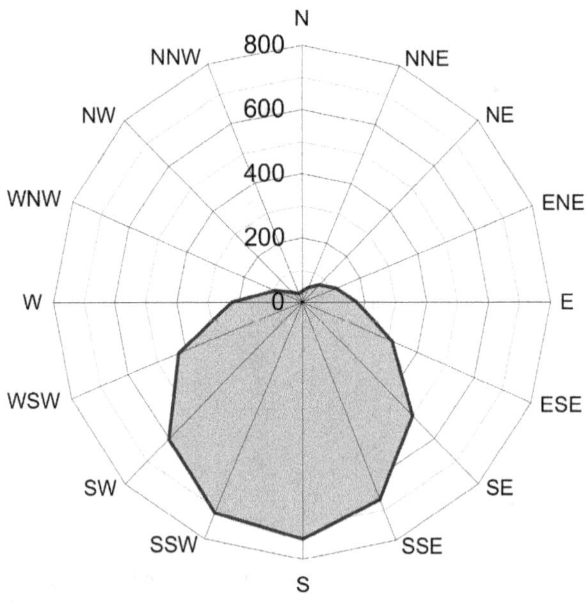

Figure 3.36: Wind-driven rain direction and quantity (mm/yr) for Seattle, WA

Rain deposition on buildings

As summarized in Eq. 3.12 the RDF (rain deposition factor) is the ratio of rain in the free wind to rain deposition on a building, which accounts for the effect of building shape and size on rain deposition. Baker and Robinson, of the National Research Council of Canada, produced an excellent and still valid qualitative study of driving rain wetting patterns and staining [Baker and Robinson 1975]. For example, the top and edges of all buildings, but especially high-rise buildings, tend to experience more rain deposition (Figure 3.37). By separating the building influence from the climate load, the use of RDF allows for more quantitative assessments[3].

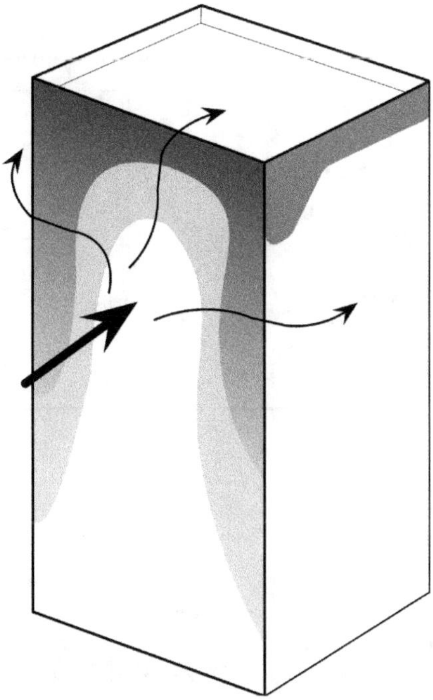

Figure 3.37: Typical rain deposition pattern and airflow on a tall cuboidal building

The value of the RDF theoretically depends on the building geometry and size, wind turbulence intensity, and raindrop size. In general, since most rain events are of low intensity and have a similar distribution of raindrop sizes, average values of RDF measured in the field over a range of driving rain events have shown somewhat consistent results.

The values in Figure 3.38 (compiled from various driving rain measurements, computer modeling, and theoretical considerations) are recommended for initial studies where more detailed information may be difficult to obtain. An RDF of 1.0 is appropriate near the upper corners of blunt-edged, rectangular buildings. Over small areas (a few square meters) with abnormal geometries, much higher RDF values may be more realistic. An RDF of much less than one is likely over most areas of most buildings, especially the lower central portions of the wind-facing facade.

[3] The *catch ratio* is defined as the ratio of the amount of water on the wall to the amount in the rain. This metric, though widely used by researchers, has serious practical limitations as it does not allow an analyst the ability to separate the effects of rainfall intensity, and height. It is an accurate measure of the particular building being studied exposed to the specific rain fall but does not provide general applications to a different building. Hence, the use of DRF and RDF is recommended.

The details of the effect of building shape on RDF, as well as the effect of surface shedding and drainage details, will be explored in more depth in Chapter 12.

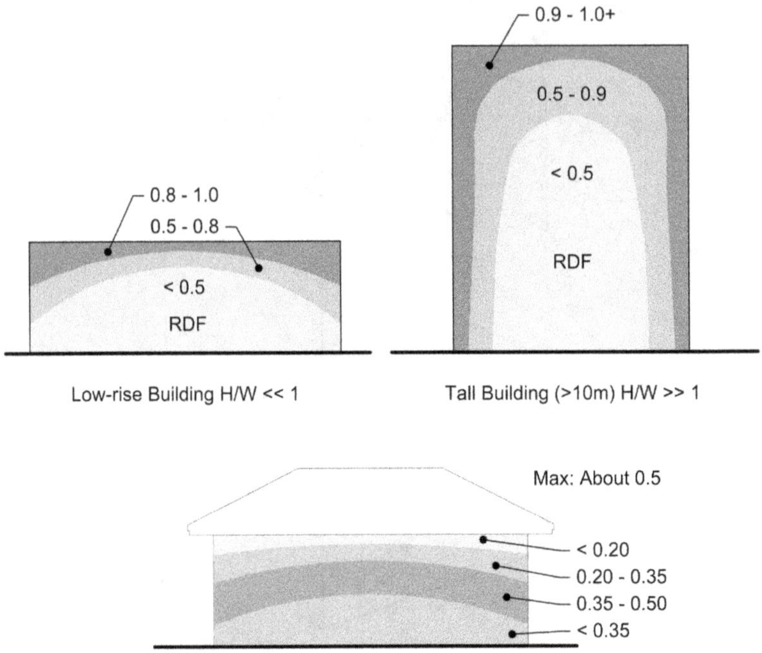

Figure 3.38: Rain deposition factor (RDF)

3.5 Interior environmental elements

The interior environmental conditions within buildings are of great interest to building designers and, of course, mechanical engineers who are responsible for the space conditioning. For the designer of an enclosure, the interior conditions are important because they generate enclosure loads. In fact, the nature of the conditioned space is the primary area where the concerns and interests of the designers of service systems (often mechanical and electrical engineers) and the designer of the enclosure overlap. The need for a holistic view of interior conditions, service system operation, and enclosure performance has been highlighted in recent years, and these relationships are likely to continue to grow in importance.

Buildings can have a wide variety of interior conditions, and these conditions may vary throughout the building. These conditions are defined by use, that is, by occupancy, processes, and contents. Occupancy affects the condition because each group of people (school children vs office workers), their type and level of activity (sleeping vs playing sports), and their clothing (hospital gowns vs heavy work clothes) have different requirements. Many processes within buildings may also require specific environmental conditions (in fact, this is often why they are conducted inside buildings). Some building contents, such as frozen vegetables and rare paintings, also require special conditions.

The required interior conditions may also vary with time because either the use or the intensity of use changes with time. Whereas a frozen-storage building requires constant temperature conditions, a stadium may require some temperature control only during major events. A church typically requires good temperature control and some humidity control only when the full congregation meets to worship, whereas the temperature control required during choir practice would ideally be restricted to specific areas.

The fundamental interior environmental variables that are controlled in practice are temperature, humidity, and air pressure (the velocity of air is also controlled in air-based HVAC systems to ensure comfort for the occupants, but this has no effect on the enclosure design). Based on these three variables Lstiburek [2001] identified three basic types of building interior environments:

I. Temperature controlled (humidity and pressure not directly controlled), e.g., most houses, apartments, warehouses, factories, and simple buildings.

II. Temperature and humidity controlled (pressure is not directly or tightly controlled), e.g., museums, most offices, some factories, some small commercial buildings.

III. Temperature, humidity, and pressure controlled, e.g., many special-use facilities, data processing centers, biological and chemical laboratories, and factories.

Lstiburek's classification can be extended and refined to include other features. Table 3.4 has been developed assuming that there are primary design variables that are explicitly controlled (typically by monitoring and consequential action) and others that are controlled incidentally because of the primary control variable. In almost all cases, control of one or more of the three variables affects the others, but the effect is generally small.

Most older buildings are Type I with rather loose temperature control of heating only, e.g., by a thermostat. If cooling only is provided (with no humidity control) the building is a true Type I building. Despite common belief, most buildings with air-conditioning systems impact the interior humidity only peripherally to their primary function of cooling. The lack of a humidistat is a clear sign that humidity is not being directly controlled by the system. Hence, the typical air-conditioned building or home is a Type Ib building (see Table 3.4). The interrelationship of temperature and humidity control has resulted in numerous building-related performance problems, including excessive interior relative humidity (RH) levels and condensation within ductwork and equipment.

Most buildings with residential occupancy use the Type Ib approach and deciding on the interior Relative Humidity and operating pressure is a challenge if hygrothermal analysis is to be conducted. That RH and air pressure are often critical to enclosure performance highlights the importance of this challenge. There is an increasing volume of research to assist designers and analysts [TenWolde & Walker 2001, Arena & Karagiozis 2010, Bagge et al 2010, Cornick et al 2008, Rousseau et al 2007].

Table 3.4: Types of conditioning for buildings

Type	Temperature Control	Humidity Control	Pressure Control	Examples
I a	◉			Heated house, warehouse
I b	◉	◎		Heating and normal A/C
I c	◉		◎	Heating + exhaust fans
I d	◉	◎	◎	Heating+ A/C + exhaust fans
II a	◉	◉		Museum, fruit storage
II b	◉	◉	◎	Pressurized + controlled
III	◉	◉	◉	Special labs, chip fabrication
IV	◉		◉	Dust-sensitive manufacturing
V		◉	◉	Special food storage
VI			◉	Cement factory

Note: ◉ Directly controlled ◎ Incidental / Implicit

Type II buildings directly control both temperature and humidity. Because of the number of humidity control problems observed in practice with Type Ib buildings, more designers are choosing to specifically separate temperature control from relative humidity control [Mumma 2001]. This approach results in a true Type II building and is appropriate for many building uses and occupancies.

Only since the Second World War have Type II buildings become more common. In cold weather, the exterior air moisture content is lower than is desirable for interior levels, and humidification may be employed. If the building is airtight and humidity is generated by occupants and processes, ventilation can be used to control (by dilution) the interior humidity. Buildings that control temperature and use humidifiers and exhaust ventilation to control humidity can be considered to be Type IIa buildings, although the level of RH control is usually modest. Some houses in cold climates (ventilation controlled by a humidistat during cold weather) use this approach.

Museums and some special manufacturing and storage uses tend to require tighter humidity control over the entire temperature range. Consequently, they use air-conditioning systems that place a higher priority on humidity control. The air moisture content must be directly measured by sensors in these systems to allow for direct control.

In many Type II buildings, the interior is incidentally slightly pressurized or depressurized by air-handling units combined with leaky ducts, exhaust fans, etc. These types of interior environments are classified as Type IIb.

Direct control of pressure differences across the enclosure (Type III) is sometimes applied in special-use buildings to accommodate the fact that the enclosure has some air permeability. Depressurization can be used to avoid exfiltration condensation in cold weather or to avoid the uncontrolled loss of interior gases and particles. Pressurization is useful to avoid infiltration condensation in hot weather or to exclude outdoor pollutants. For example, to control dust entry or the uncontrolled release of pathogens, a microchip fabrication plant and a medical research laboratory could, respectively, be pressurized and depressurized. Similarly, depressurization could be used to control interstitial condensation in the enclosure of a pool building in a cold climate. This latter approach may be an economical retrofit strategy.

Although Type I, II, and III buildings encompass most situations, there are also other types:

- Type IV buildings, which control only pressure and temperature (e.g., for some manufacturing situations).
- Type V buildings, which control only pressure and humidity (e.g., special food storage facilities).
- Type VI buildings, which control only pressure (e.g., to control dust from a manufacturing process).

3.5.1 Requirements of use

Human comfort conditions

The most common requirements defining the interior conditions of buildings are human comfort conditions. A significant body of work is available to aid in the understanding of human thermal comfort, including ASHRAE Standard 55 [ASHRAE 2017] and Chapter 8 of the *ASHRAE Handbook–Fundamentals*. The following provides a brief review.

Human thermal comfort can be considered as an energy balance. Heat is generated internally (as a function of activity and metabolism) and lost or gained from the body to the interior environment (Figure 3.38). The most important modes of heat transfer include convection and evaporation of sweat (both strongly affected by local air velocity, RH and air temperature), radiation to the surroundings, and conduction to the floor. The insulation value (defined by the unit clo) and coverage of clothing and footwear affects all of these transfer modes.

Based on research into this energy balance, recommended temperatures for human comfort range from about 20 to 25 °C (68 to 77 °F) in winter and 23 to 27 °C (73 to 80 °F) in summer for office conditions, assuming normal seasonal dress and activities. The range of temperatures that provide comfort depends on the humidity levels: occupants exerting themselves will cool by sweating, a process encouraged by low ambient relative humidity. High interior relative humidity levels will feel warmer.

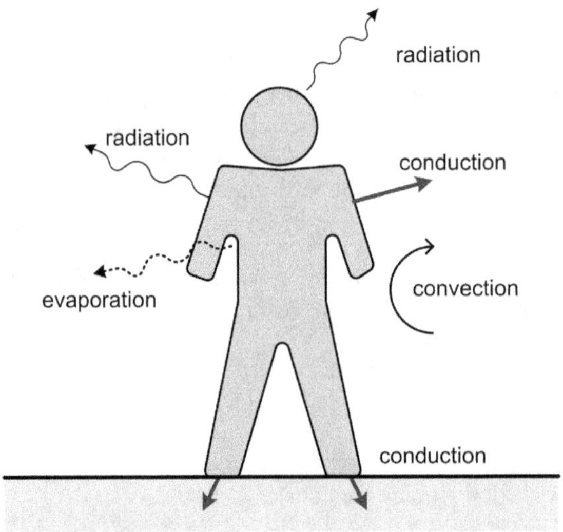

Figure 3.39: Energy balance of a human in a building

Comfort can be strongly affected by radiation from and to a person's surroundings. The radiation effect is measured by the **mean radiant temperature** or MRT. It is defined as the temperature that a radiatively black sphere would need to operate at if it were to exchange no net radiation with the environment being described.

If the surroundings are colder than the air temperature, the person will feel cooler than indicated by the measured air temperature, and vice versa for warm surroundings. Hence, air temperature will not provide an accurate measure of comfort if the surfaces enclosing the space are at significantly different temperatures. This phenomenon is used to great advantage in radiant heating and cooling systems, and is the cause of thermal discomfort experienced by occupants in the perimeter zone of buildings with poor enclosures.

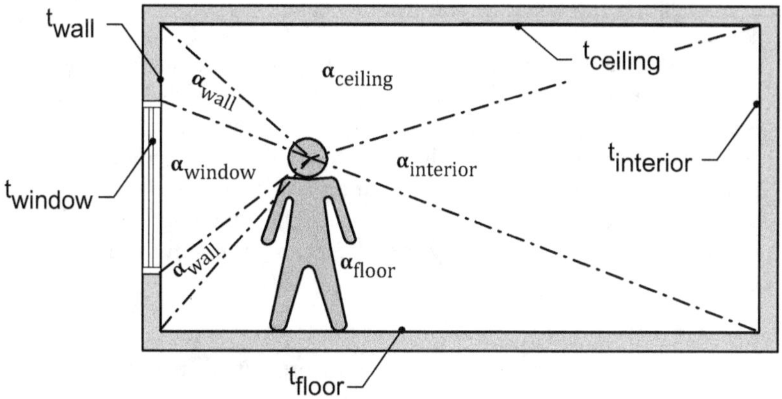

Figure 3.40: Mean radiant temperature for occupant in a perimeter room

The MRT is affected by integrating the product of the temperature of each surface surrounding the occupant area and the spherical angle subtended. In many cases, simpler 2D analysis can be conducted by considering the angle and the temperature of each surface. For example, as shown in Figure 3.40 an occupant near a large window in cold weather will radiate heat much more quickly to the window because it is far colder (t_{window}) than the interior partition ($t_{interior}$). The wall is likely to be warmer, and the subtended view angle is also less. The ceiling and floor represent the largest view angles but are typically close to air temperature. If the ceiling were the underside of a poorly insulated roof enclosure, the large ceiling angle would mean the ceiling temperature would play a large role in the MRT, and hence human comfort.

To capture the important impact of radiation on human comfort indoors, the term operative temperature, t_o, is used. By definition, t_o is the temperature of an isothermal, radiatively black (see Chapter 4) surroundings in which the occupant would exchange the same amount of heat by radiation and convection as in the actual non-uniform environment being described.

For normal interior conditions (air velocity less than 0.4 m/s and radiant temperatures under 50 °C) the operative temperature can be calculated as:

$$t_o = \frac{t_{air} + MRT}{2} \qquad \text{Eq. 3.14}$$

where t_{air} is the air temperature in the space, and

MRT is the mean radiant temperature of the surfaces surrounding the person.

Other important comfort factors include the difference in temperature between a person's head and feet, the floor temperature for bare-footed people (e.g., in swimming pools, bathrooms, and bedrooms), asymmetrical radiation loss (e.g., especially near windows with poor thermal quality), and drafts, especially cold drafts falling from windows and from mechanical air-conditioning equipment.

Ongoing research continues to examine the role of relative humidity with regard to human comfort. It has been proposed that enthalpy, not temperature, is a better predictor of comfort. It has been suggested [Brager and de Dear 2000] that occupants of naturally ventilated buildings are comfortable over a broader range of temperature and humidity variation if these conditions are tied more closely to exterior conditions, i.e., the temperature may rise above the set point without provoking complaints about discomfort if it is very warm outside.

Air velocity is a human comfort factor but is not that important for building enclosure performance. Airflow, driven by pressure differences, is the important factor for enclosure performance. Air velocity increases the heat and moisture transport from the environment to the body. Hence, higher air velocities provide a cooling effect due to evaporation of sweat and direct convection of body heat. However, above a velocity of about 1 m/s it becomes difficult to work, since papers are blown off desks, eyes and noses begin to dry out, and so on.

Other uses

Many buildings accommodate uses that demand interior environmental conditions quite different from, or more stringent than, those required by humans. These special conditions are often obvious and are part of the design process. However, the impact of these special situations is sometimes (far too often) not accounted for in the design of the building enclosure.

Some building uses that create different interior conditions are listed in Table 3.5. The role of humidity is usually a critical factor: high levels of interior moisture in climates with cold weather require additional care in the design and construction of the enclosure. Low interior temperatures can allow condensation to occur near interior surfaces. The largest problem observed in practice is the reliance on standard building materials, assemblies, and practices for these special interior environments.

Table 3.5: Building uses with special interior environments

Building Use	Special Interior Environment
Swimming pools, saunas, printing plants, etc.	Higher temperature and absolute and relative humidity
Frozen storage	Very low temperature, low absolute humidity, and high relative humidity
Skating rinks	Low temperature and high relative humidity
High-tech manufacturing	Very stable temperatures and humidity, low particulate loads
Hospitals	Controlled bio-aerosols and humidity
Museums & Archives	Tightly controlled humidity and temperature (commonly 50%RH regardless of climate or collection)

3.5.2 Temperature

Temperature has been controlled for longer than any other interior variable. In most cases, the desired range of temperatures falls within the narrow range of human comfort. Exceptions are cold storage facilities, some livestock raising, etc.

The air temperature within a building is a balance of the heat produced inside, the heat gained or lost through all means, and the heat energy stored in the materials that constitute a building. This energy balance is often difficult to predict since there are typically many different sources of heat gain and loss, and the movement of heat to and from storage is complex. Table 3.6 lists some of these sources and sinks.

It follows that control of the gain and loss of heat from within and outside the building is needed if temperature is to be maintained at its intended level. For buildings with large internal gains (such as computer-intensive and highly glazed offices or light-intensive retail stores), cooling becomes important. This is especially true for large buildings since their ratio of surface area to volume is small, thus limiting the amount of heat that can be lost via the enclosure. Such buildings are labeled **gain dominated**. For buildings with few internal gains (e.g., detached houses and small warehouses) and a large surface-area-to-volume ratio, heating may be required and the term **enclosure dominated** is used.

Table 3.6: Sources and sinks of heat in buildings

Heat Gain (sources)	Heat Loss (sinks)
Computers, lighting, photocopiers	
People, animals	
Furnace / boiler / heating equipment	Air-conditioning equipment
Solar gain through windows and on the enclosure	Night sky radiation
Conduction through the enclosure from outdoor air or soil	Conduction through the enclosure to outdoor air or soil
Air movement through the enclosure and intentional openings	Air movement through the enclosure and intentional openings
Processes, equipment	

3.5.3 Humidity

Humidity is a generic term for moisture in the air. Humidity is an important component of the interior environment since it is both a major determinant of human health and comfort and the source of moisture for biological growth (i.e., mold, viruses, bacteria, dust mites) in the building.

The moisture content of the interior air is a balance of moisture production and removal—a negligible amount of moisture can be stored in the air. Sources of moisture include the exterior air (when the moisture content of the exterior air is greater than the interior), the activities within the building, evaporation of standing water, and materials releasing (desorbing) moisture. Moisture sinks include the exterior air (when the moisture content of the exterior air is lower than interior), mechanical dehumidification, condensation of water, and material storing (adsorbing) moisture. This balance is clearly dynamic and driven largely by exterior weather conditions and interior activities.

Canadian studies [Hamlin & Gusdorf 1994, Hansen 1984] have found the moisture production rate within detached homes was below 3 kg/day for less than 10% of the time and less than 21 kg/day for 90% of the time. Studies in other countries [Sanders 1996] tend to find that moisture production rates are in the same range, and that typical average values are 8 to 15 kg/day. The more people and the more time they spend in the home, the higher the moisture production rate. Two adults can be assumed to produce 8 kg/day, the first child 4 kg/day, the second child, 2 kg/day, and the third and each additional child 1 kg/day. (Note, the number of people in a dwelling is often assumed to be equal to the number of bedrooms plus one.)

Another source of moisture may be the combustion of hydrocarbons. Although most such combustion is (or should be) directly vented to the exterior, standalone heaters are occasionally used. Natural gas and kerosene respectively produce about 0.15 and 0.10 liters of water per kWh of heat produced.

Table 3.7 lists estimates of the amount of moisture produced by various sources associated with human occupancy.

Table 3.7: Moisture production due to human occupancy (typical family of four) (from various sources)

Source	Strength (liters per day)
Humidifier	2-20+
Hot tub, Whirlpool	2-20+
Firewood, per cord	1-3
Washing floors etc.	0.014 per m^2
Dishwashing	0.5
Clothes Drying (unvented)	2– 3 (per load)
Cooking Dinner for four	0.5 to 0.7 (2+ with gas range)
Defrosting (frost free) fridge	0.5
Typical bathing/washing per person	0.2 to 0.4
Shower (ea)	0.5 -1.0
Bath (ea)	0.06 - 0.1+
Uncovered crawlspace to wet basement	0.04 to 0.1 / m^2
Unvented gas appliance	0.15 /kWh for natural gas, 0.10 /kWh for kerosene
Seasonal desorption (or new materials)	3 - 8 depends on the house construction
Ten medium-sized ferns	2.5
One 2 m tall indoor tree	25-50
Pets	0.2-0.5 (20 kg dog)
Total (Typical)	2 to 40, Average 10-15

The interior humidity level in a building is in constant flux with the interaction of indoor moisture production, air exchange rate, and outdoor air moisture content, as well as the vapor stored and released from building materials. The interior vapor pressure, and hence RH (see the psychrometrics section in Chapter 4), can be calculated from the following equation:

$$P_{v,i} = P_{v,e} + K * G_w / (n * V) \qquad \text{Eq. 3.15}$$

where, $P_{v,i}$ and $P_{v,e}$ are the interior and exterior vapor pressures respectively (Pa or in Hg),
G_w is the moisture production rate inside the space (kg/hr or lb/hr),
n is the number of air changes per hour (1/h),
V is the volume of the space (m^3 or ft^3), and
K is a constant (135800 for SI units, 641 for IP units).

Data from Table 3.7 and Eq. 3.15 above can be used to generate plots of the relationship between indoor relative humidity and building volume, moisture production rate, ventilation rate, and outdoor conditions. Figures 3.29 and 3.30 below provide guidance on how these impacts can be assessed for a house and an apartment of typical size and occupancy.

Note that a modern airtight home has less than 0.25 air changes per hour in normal conditions, and the rate may be as low as 0.1 ACH if proper mechanical ventilation is not provided. An older leaky house, or an over-ventilated space, may have over 1 ACH in normal conditions. Based on typical condensation potentials, "safe" interior RH levels might be considered below 60% at 5 °C outdoor temperature and below 30% at –10 °C or colder.

It can be seen that the addition of a 10 kg/day source (e.g., a clothes drier vented to interior) can increase humidity in an airtight house from about 30% RH to almost 50% at a very plausible air exchange rate (due to ventilation and natural air leakage) of 0.25 ACH. Hence, a single large source of moisture can cause the interior humidity to rise from relatively dry to dangerously high (in cold weather) conditions. However, this also demonstrates that the addition of ventilation can be used to manage high interior moisture sources. The energy penalty of large ventilation rates can be reduced by the use of some form of energy recovery ventilation.

During warm and humid outdoor conditions, ventilating an interior space using air that has the same or a higher absolute humidity level will not be able to lower the interior air moisture content. The only means of controlling interior humidity under these conditions is to remove it (using air-conditioning / dehumidification equipment) or store it in highly adsorbent materials until conditions are more favorable.

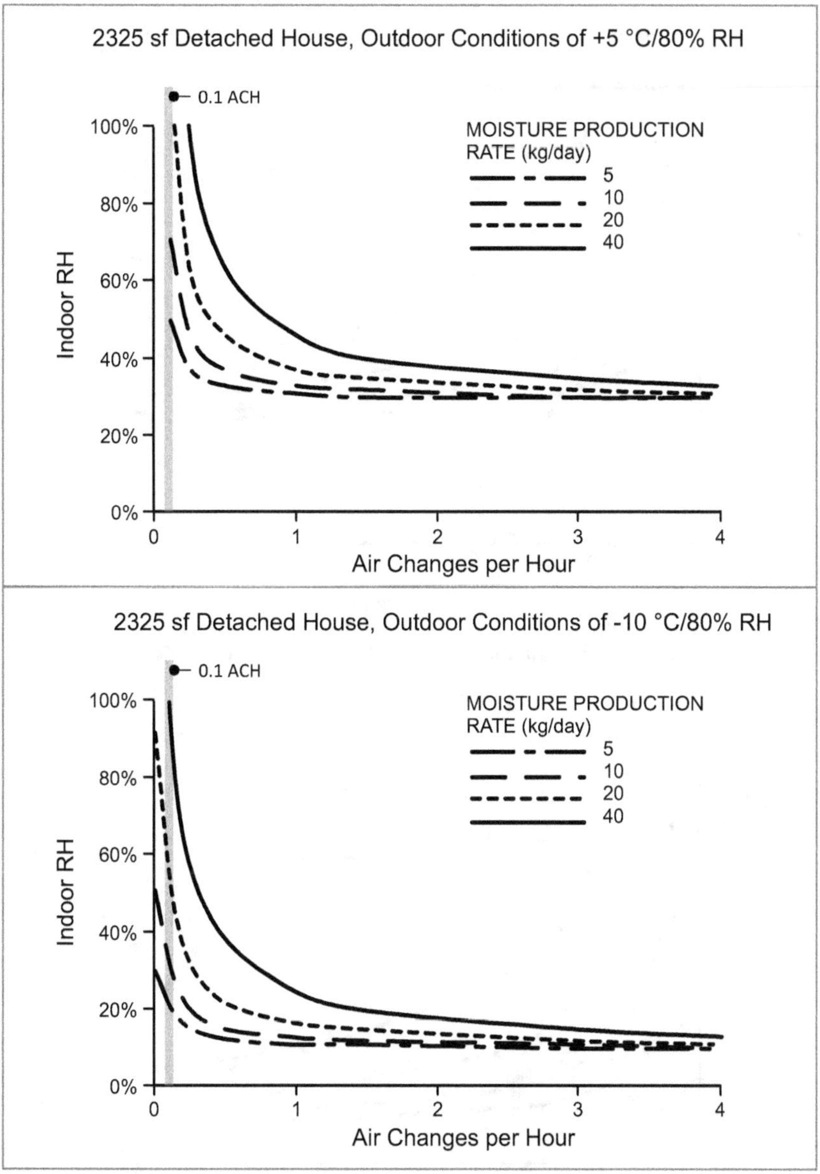

Figure 3.41: Moisture production in a (215 m^2) 2325 ft^2 detached house

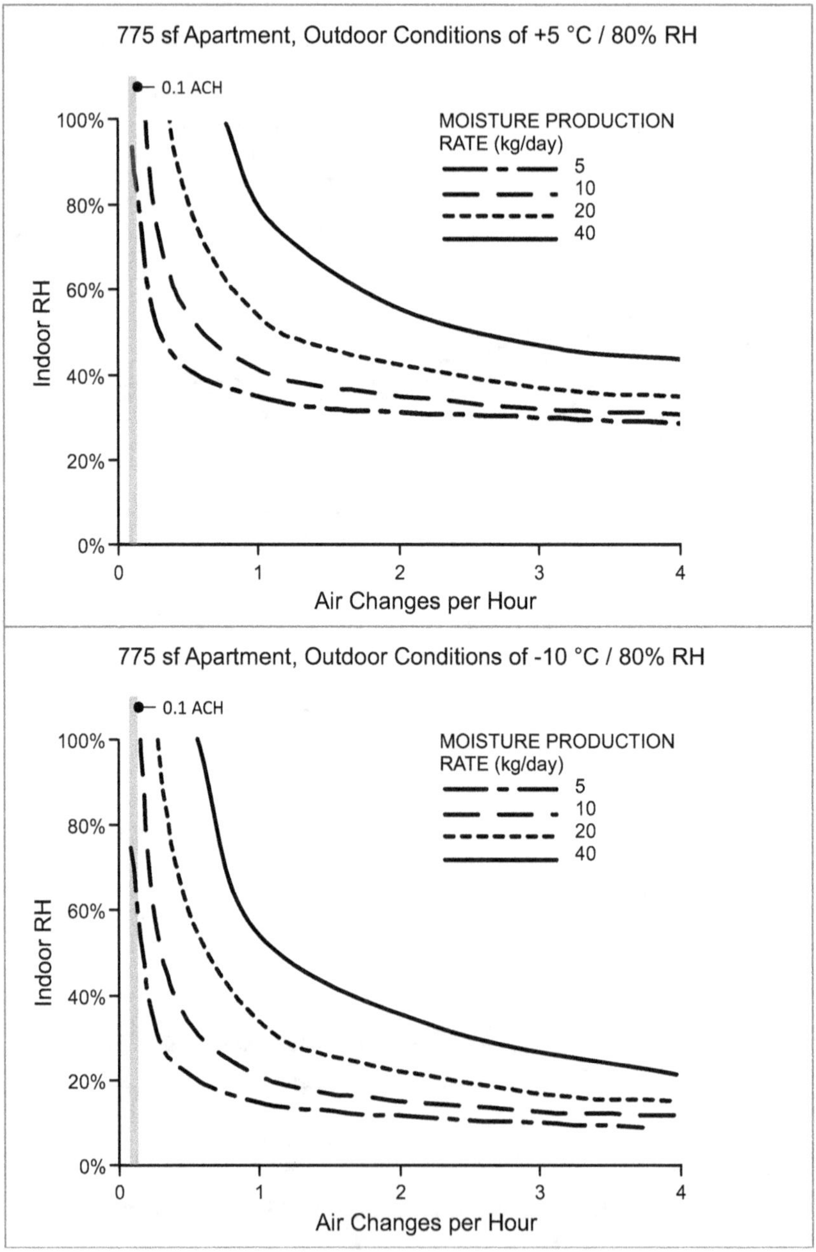

Figure 3.42: Moisture production in a (72 m²) 775 ft² apartment

3.5.4 Pressure differences

Although not a climate element, the air pressure difference across the enclosure is an important environmental condition. It should be measured in any complete building investigation and considered in enclosure design since it is the force that drives airflow through the enclosure. Controlling airflow is one of the fundamental functions of the building enclosure.

Pressure within a building is generated by stack effect (buoyancy), wind effects, and fans or blowers in mechanical equipment. Pressure differences across the enclosure are therefore the result of indoor and outdoor temperatures (buoyancy), weather, site, and building shape (wind), and the nature and operation of mechanical equipment.

Wind pressures act on all sides of a building, creating suction on some sides of a building and pressures on the other as described in detail above. The hourly average magnitude of these wind-induced pressure differences (which can act in either direction) is typically in the range of 4 to 10 Pa for sheltered buildings close to the ground and from 10 to 50 Pa higher above grade.

Stack-effect pressures are generated by differences in air density with temperature, i.e., hot air rises and cold air sinks. The air within a building during cold weather acts like a bubble of light hot air in a sea of dense cold air. This is studied in Chapter 7. In warm weather the situation is reversed, although differences in air temperature, and hence density, are usually not as great. Stack-effect pressures increase with height and with temperature difference between the inside and the outside. Actual pressures (which act in both directions) due to stack effect range from only 2 Pa in housing during hot weather to well over 40 Pa in high-rise buildings during cold weather.

Mechanical equipment in the form of appliances that supply or exhaust air to the building modify the pressure within it. Exhaust-only ventilation (including clothes driers, combustion appliances, range hoods, bath fans, and HVAC systems) results in a net negative pressure acting across the building enclosure. Supply-only ventilation pressurizes a building space. Pressures due to mechanical equipment should be maintained to less than a few pascals, although malfunctioning building systems can have pressure differentials of 10 Pa and more. As building pressures are not often measured (except in special-use buildings), control of pressure is practically difficult[4].

These pressures are arithmetically added together at every location in a building to generate a net pressure. If the net pressure is an inward-driving pressure, the air flowing inward through the enclosure will draw possibly polluted outdoor air and within-wall air pollutants into the building. Pressurizing a space will force air outward through the enclosure, which could cause serious condensation problems in cold weather.

---

[4] As noted, standard practice in mechanical engineering is to pressurize a commercial building by providing more supply air flow than exhaust (or relief) flow. This is an increasingly poor practice: as building enclosures become tighter and ventilation fans become variable speed, pressure control of the building is lost, and often excessively high.

Pressures can vary significantly within a building (and thereby cause performance problems) due to the effect of internal separators, but vary little within well-connected interior spaces. Pressures vary dramatically over time, often over short periods of time, because of the combined effect of the dynamic wind, operation cycles of mechanical equipment, the use of doors and windows, and changing temperatures.

Chapters 7 and 11 cover aspects of airflow control and pressure management in greater detail, while previous sections of this chapter describe the pressure effects induced by the wind.

3.5.5 Indoor environmental quality

Indoor environmental quality (IEQ) is a term that encompasses much more than temperature and humidity and comfort. IEQ embraces aspects of lighting, sounds, particulate loading, gaseous pollution, mold spores and other bio-aerosols, and volatile organic compounds (VOCs). Indoor air quality (IAQ) is the subset of IEQ that deals with air alone.

There is an increasing body of evidence indicating that IEQ has a direct effect on the health, well-being, motivation, and productivity of people in buildings. For example, daylighting of otherwise similar stores increases the volume of sales, and daylighting office space has been shown to reduce absenteeism. More difficult to quantify, but probably just as important, are other factors that influence human productivity, such as visual and aural acuity and privacy.

Proper fresh air and space humidity control (IAQ) avoids mold growth and dust mites, both of which influence occupant health. Chapter 9 of the *ASHRAE Handbook–Fundamentals* covers these aspects of environmental health in some detail.

Poor IAQ is often identified as the cause of the group of symptoms called Sick Building Syndrome (SBS). It is important to recognize, however, that technical building science factors may not be the most important in explaining air quality complaints. One of the larger, most comprehensive studies of SBS [Marmont et al 2006] succinctly summarized this:

> "… the physical environment of workstations and office buildings appears to have less effect on symptom prevalence than features of the psychosocial environment. SBS symptoms are more about the jobs people do, their psychosocial environment, and their ability to control conditions in their office than about the physical environment of the building or workstation where the job is performed."

Controlling humidity is one important aspect of IAQ, as is the supply of fresh air to dilute interior pollutants and odors. However, these design variables can be in direct opposition: in summer, increasing ventilation increases humidity in many humid climates, and certainly also increases energy consumption

[Lstiburek 2002]. Hence, simply increasing fresh air ventilation is not a reliable means of increasing IAQ. However, reducing indoor sources of pollutants (cleaning supplies, finishing materials, etc.) always improves air quality, and better lighting design (reducing glare and shadows) always improves IEQ, often with a reduction in energy use. Simple prescriptions of increasing window areas and/or fresh air ventilation rates and/or relative humidity as means of improving IEQ are almost always misguided.

The control of relative humidity is the most important intersection between building enclosure performance and indoor air quality concerns: increasing RH in cold weather dramatically increases the risk of both surface and hidden interstitial condensation and their attendant moisture damage. However, there is little solid science to support the contention that interior relative humidity higher than about 25% or 30% is beneficial to comfort or health in cold climates. A recent, tightly controlled study [Grun et al 2012] of occupants of a sealed space (airliner cabins) found that complaints of dry eyes, nose, throats, or skin did not change significantly between environments with a humidity of 10% (very low) and 40% (moderately high for a cold climate): reasons other than humidity must be proposed to explain the comfort complaints.

3.6 Commentary

Understanding environmental conditions is a critical part of understanding and predicting the performance of building enclosures. Builders of vernacular buildings were often well attuned to their exterior environment and intentionally manipulated the microclimate to their benefit. The design of enclosures for modern buildings, especially complex and large buildings, demands a thorough and often detailed understanding of environmental conditions.

The increasing use of hygrothermal enclosure modeling demands the use of sufficiently accurate environmental conditions to provide useful results. Although this chapter has reviewed some of the issues, the references provided at the end of this chapter identify sources containing much more detailed and comprehensive information. In particular, the current ASHRAE Handbook and the Straube [2009].

3.7 References

A Dictionary of Geography. ed. Susan Mayhew, Oxford University Press, Oxford, 1997.

Arena, L., Karagiozis, A. Mantha, P., "Monitoring of Internal Moisture Loads in Residential Buildings— Research Findings in Three Different Climate Zones", *Proc. of Thermal Performance of Exterior Envelopes of Whole Buildings XI*, Clearwater Beach, ASHRAE, Dec. 2010.

ASHRAE/ANSI Standard 169 *Climatic Data for Building Design Standards.* ASHRAE, Atlanta, Georgia, 2013.

ASHRAE/ANSI Standard 55. *Thermal Environmental Conditions for Human Occupancy.* ASHRAE, Atlanta, Georgia, 2017.

Barry, R.G., and Chorley, R.J., *Atmosphere Weather and Climate.* 6th ed., *ASHRAE Handbook–Fundamentals.* American Society of Heating Refrigeration and Air-conditioning Engineers, Georgia, 2021.

Bagge, H. Johansson, D., Lindstrii, L. "Indoor Hygrothermal Conditions in Multifamily Dwellings— Measurements and Analysis", *Proc. of Thermal Performance of Exterior Envelopes of Whole Buildings XI*, Clearwater Beach, ASHRAE, Dec. 2010.

Beard, K.V., "Terminal Velocity and Shape of Cloud and Precipitation Drops Aloft," *J. of the Atmospheric Sciences*, Vol. 33, 1976, pp. 851-864.

Best, A.C., "The Size Distribution of Raindrops," *Quart. J. Royal Meteor. Soc.*, Vol. 76, 1950, pp. 16-36.

Bradley, S.G., and Stow, C.D., "The Measurement of Charge and Size of Raindrops: Part II. Results and Analysis at Ground Level," *J. of Appl. Meteor.* Vol 13, Feb 1974, pp. 131 - 147.

Brager G. and de Dear, R. "A Standard for Natural Ventilation," *ASHRAE Journal*, October 2000, pp. 21 – 27.

Caton, P.G., "A study of the raindrop-size distributions in the free atmosphere," *Quart. J. Royal Meteor. Soc.*, Vol 92, 1966, pp. 15-30.

Cornick, S., and Kumaran, M.K., "A Comparison of Empirical Indoor Relative Humidity Models with Measured Data", *J. of Bldg. Phys.*, Vol. 31, No. 3, pp. 243-268, Jan. 2008.

Davenport, A. G., and H. Y. L. Hui, *External and Internal Wind Pressures on Cladding of Buildings*, Boundary Layer Wind Tunnel Laboratory, The University of Western Ontario, London, Ontario, 1982.

Dingle, A.N. and Hardy, K.R., "The description of rain by means of sequential raindrop-size distributions," *Quart. J. Royal Meteor. Soc.*, Vol. 88, 1962, pp. 301-314.

Dingle, A.N., and Lee, Y., "Terminal Fall Speeds of Raindrops," *J. of Appl. Meteor.*, Vol 11, August 1972, pp. 877 - 879.

Duffie, J.A., and Beckman W.A., *Solar Engineering of Thermal Processes.* 4th Edition. John Wiley & Sons, New York, 2013.

Durst, C. S., 1960. "Wind speeds over short periods of time". Meteor. Mag, 89(1056), pp. 181-187.

Fitch, J. M. and Bobenhausen, W. *American Building.* Oxford University Press, New York, 1999.

Geiger, R., *The Climate Near the Ground.* Harvard University Press, Cambridge, MA, 1961.

Grün, G., Trimmel, M., Holm, A., "Low humidity in the aircraft cabin environment and its impact on well-being: Results from a laboratory study", *Building & Environment*, Vol 47, pp. 23-31, 2012.

Gunn, R, and Kinzer, G.D., "The terminal velocity of fall for water drops in stagnant air," *J. Meteor.*, Vol 6, pp. 243-248, 1949.

Hamlin, T. and Gusdorf, J., *Moisture Production in Canadian Single Family Houses.* CMHC, Ottawa, 1994.

Handbook on the Principles of Hydrology, ed. Donald M. Grey, National Research Council of Canada, Ottawa, 1970, pp. 2.2-2.3.

Hansen, A.T., *Moisture Problems in Houses.* CBD-231, National Research Council of Canada, May 1984.

Henriques, F.M.A., "Quantification of wind-driven rain - an experimental approach", *Building Research and Information*, Vol. 20, No. 5, 1992, pp. 295-297.

Inculet, D. and Surry, D., *Simulation of Wind Driven Rain and Wetting Patterns on Buildings.* A report published by the CMHC, February 1995.

Karagiozis, A., and Hadjisophocieous, G., "Wind-Driven Rain on High-Rise Buildings", *Proc. of the BETEC/DOE/ASHRAE Thermal Performance of Building Envelopes VI*, Dec. 1995, pp. 399-406.

Karagiozis, A., Hadjisophocieous, G., and Cao, S., "Wind-Driven Rain Distributions on Two Buildings", *J. of Wind Engineering and Industrial Aerodynamics*, Vol 67 and 68, 1997, pp. 559-572.

Künzel, H.M., 1994. *Regendaten für Berechnung des Feuchtetransports*, Fraunhofer Institut für Bauphysik, Mitteilung 265.

Kusuda, T., Achenbach, P.R., 1965. *Earth Temperature and Thermal Diffusivity at Selected Stations in the United States.* National Bureau of Standards NBS Report 8972, Washington, 1965. and *ASHRAE Transactions*, Vol. 71 (1), pp. 61-75.

Lacy, R.E., "Driving-Rain Maps and the Onslaught of Rain on Buildings", *Proc. of RILEM/CIB Symposium on Moisture Problems in Buildings,* Helsinki, (Building Research Station Current Paper 54, HMSO Garston, U.K) 1965.

Laws, J.O., and Parsons, D.A., "Relation of raindrop size to intensity," *American Geophys. Union Trans.*, No. 24, pt. 2, pp. 453-460, 1943.

Lstiburek, J., 2002. "Residential Ventilation and Latent Loads", *ASHRAE Journal*, April.

Lstiburek, J. 2005. *Builders Guide to Cold Climates.* Building Science Press, Westford, MA.

Markowitz, A.M., "Raindrop Size Distribution Expressions," *Journal of Applied Meteorology*, Vol. 15, 1976, pp. 1029-1031.

Marmot, A.F, Eley, J., Stafford, M., Stansfeld, S.A., Warwick, E., Marmot, M.G., "Building health: an epidemiological study of "sick building syndrome" in the Whitehall II study", Occup Environ Med, Vol. 63, pp. 283–289, 2006.

Marshall, J. S., and Palmer, W. M., "The distribution of raindrops with size", *J of Meteor.*, Vol. 5, Aug. 1948, pp.165-166.

McQuiston, Parker, Spitler,1999. *Heating Ventilation and Air Conditioning.* Fourth Edition. John Wiley & Sons, Inc., New York.

Mumma, S.A.,2001. "Designing Dedicated Outdoor Air Systems," *ASHRAE Journal*, May, pp. 28-31.

Olgyay, Victor, 1992. *Design with Climate.* Van Nostrand Reinhold, New York.

PNNL 2022. Antonopoulos, C.A., Gilbride, T.L., Margiotta, E.R. and Kaltreider, C.E.. *Guide to Determining Climate Zone by County: Building America and IECC 2021 Updates* (PNNL-33270). Pacific Northwest National Laboratory, Richland, WA.

Penguin Dictionary of Physical Geography. Ed. John Whittow, Penguin Books, London, 2000.

Rogers, R.R., and Pilie, R.J., "Radar Measurements of Drop-Size Distribution," *J. of the Atmospheric Sciences*, Vol. 10, Nov. 1962, pp. 503-506.

Rousseau, M., Manning, M., Said, M.N., Cornick, S.M., Swinton, M.C. and Kumaran, M.K., "Characterization of Indoor Hygrothermal Conditions in Houses in Different Northern Climates", *Proc. of Thermal Performance of the Exterior Envelopes of Whole Buildings X*, Clearwater Beach, FL., Dec., 2007.

Sanders, C. *Environmental Conditions: Final Report Task 2*. International Energy Agency, Energy Conservation in Buildings and Community Systems Program, Annex 24 Heat, Air and Moisture Transfer in Insulated Envelope Parts (HAMTIE). Laboratorium Bouwfysica, K.U.-Leuven, Belgium, 1996.

Stathopoulos, T. Dumitrescu-Brulotte, M.,1989. "Design recommendations for wind loading on buildings of intermediate height," *Can. J. of Civil Eng.,* Vol. 16, No. 6, pp. 910-916.

Straube, J.F., 2009. Chapter 6 "Exterior Climate Data for Hygrothermal Analysis of Building Enclosures" in *ASTM Manual 18 Moisture Control The Key Factor in Mold Prevention* – 2nd Ed., edited Heinz Trechsel, Mark Bomberg, West Conshohocken, Pa.

Straube, J.F., Schumacher, C.J. 2005. "Driving Rain Data for Canadian Building Design", *Proc of 10th Canadian Building Science and Technology Conference*, Ottawa May.

Stoutjesdijk, P. and Barkman, J.J.,1992. *Microclimate, Vegetation and Fauna.* Opulus Press AB, Knivsta, Sweden.

TenWolde, A., Walker, I. 2001. "Interior Moisture Design Loads for Residences", *Proc. of Thermal Performance of Exterior Envelopes of Whole Buildings VIII*, Clearwater Beach, FL, ASHRAE, Dec.

Underwood, S.J., Meentemeyer, V., 1998. "Wind-Driven Rain for the Contiguous United States for the Period 1971-1995", *Physical Geography*, Vol 19, No. 6, pp. 445-446.

Chapter 4

Materials, Energy and Moisture

4.1 Introduction

Analyzing and understanding building enclosure performance requires some knowledge of a number of the physical sciences, including physical chemistry, thermodynamics, material science, and fluid dynamics. This text cannot present all the information necessary but a selective review, from a building science perspective, is presented in this chapter. The scope and emphasis is on those basic science topics with which engineers and architects in the building industry may be unfamiliar. Various specialized texts are available if more depth is needed.

4.2 Matter and Building Materials

The use of materials is fundamental to the practice of architecture, engineering, and of course, to building enclosure design. The physical components of civil engineering and building systems are built from materials, whether they are cladding, beams, roofing, pipe, wires, backfill, or roads. In the selection and design of building enclosures, understanding materials takes on a special significance because, unlike a bridge or a pipe, the environmental conditions that a material is exposed to varies significantly with the arrangement of the layers of an assembly. Building materials used in enclosures also often fulfill multiple functions which relate in more diverse, less specialized requirements.

From the perspective of performance, the job of the designer is to select materials and arrange them in such a manner as to fulfill the function of that component for its desired service life. Durability can be defined as the ability of a material, component, or system to perform its intended function for the specified or expected service life without unplanned repair and maintenance.

A commonly shared phrase is that there are no non-durable materials, only materials used in a non-durable manner. Selecting the correct material in the correct arrangement for the environmental loadings expected will result in the desired performance and durability. Failure to do so explains many failures.

For example, although paper is highly susceptible to water, locating paper-faced gypsum board on the interior face of an office's exterior wall is acceptable because that wall can be designed to ensure the paper is not exposed to water. Using paper-faced gypsum board in a shower stall or directly behind cladding would be a bad design choice and would result in poor durability because it is likely that small leaks will occur and will wet the paper.

To make the appropriate choice and optimize the use of materials, designers and specifiers require some understanding of materials, their properties, and behaviour.

To produce designs with the desired durability, requires an understanding of:

- The characteristics of the material,
- The environmental conditions the materials will be exposed to, and
- The mechanisms that may damage or change the material.

The following sections provide a very brief and general introduction to categories of material characteristics and damage mechanisms. Chapter 3 reviewed environmental conditions acting on the interior and exterior of the building enclosure, but much of Chapters 5 and 6 is intended to provide the understanding and analytical tools to allow the environmental conditions within the enclosure assembly (often termed *interstitial*) to be predicted.

4.2.1 Matter and its forms

Matter is what makes up the physical universe. All natural matter is composed of some combination of the 92 natural elements. They are usually further organized into compounds: substances made of the union of two or more elements in definite proportion by weight. Two or more atoms of the same or different elements, form molecules when they are bonded together. Some molecules are relatively small, such as water molecules, while others are very large. Many polymers, for instance, are composed of tens of thousands of atoms in large molecular chains.

Atoms form molecules and compounds because of the forces of attraction between them. These forces may result in the formation of strong bonds, such as ionic, covalent, or metallic bonds. These bonding forces act over small distances; for example, many atomic bonds act over a distance of 0.1 to 0.3 nm (a nanometer is one billionth of a meter). Van der Waal's forces are weaker forces that act over larger distances, perhaps as much as 100 nm, but are most significant at distances of less than 20 nm. Atoms, compounds, and very small particles may be held together by electrostatic forces—relatively weak forces that can act over larger distances (several hundred nm), but play an important role in some materials, such as gels, sols, and dispersions.

4.2.2 States of matter

Matter is often classified by its state: gas, liquid or solid. Plasma and adsorbed compounds are also sometimes considered a state. Plasma is a compound with its atoms stripped of their electrons, typically by high temperatures. Adsorbed matter is gases or liquids that are loosely attached to the molecules of liquids or solids.

A gas consists of molecules with a high level of kinetic energy, i.e., the molecules are moving about randomly at a high speed (hundreds of meters per second typically). In fact, the root mean square velocity (V_{RMS}) of the molecules is a direct measure of the gas temperature and the partial pressure that the gas exerts.

Liquid matter does not strongly resist forces that act to change its shape because the molecules are free to move with respect to each other. Liquids still have sufficient molecular repulsion to resist forces acting to change their volume. Hence, a liquid retains more shape and offers greater resistance to disturbance than a gas does—this is clearly the case with the water molecule since liquid water is much more viscous and difficult to compress than steam (water vapor) is.

Solids exhibit an atomic or molecular structure that impart rigidity and a fixed shape. The atoms and molecules vibrate about fixed positions. Many stiff and strong solids are formed into crystals, but amorphous (unstructured solids) are also common.

4.2.3 Material Classifications

Classifications of matter, based on their properties or their structure, are used to group the many millions of material variations into manageable classes.

The most common classes of building materials can be categorized as minerals, metals, polymers, and composites.

Minerals or ceramics are compounds formed from the combination of metallic and non-metallic elements. Common ceramic materials include glass, cement, and clay, all critically important materials for the building industry. Ceramics tend to be non-reactive, hard, moderately to very strong, stiff, and brittle. Processed mineral-based materials, such as gravel and stone, are widely used. Glasses are often sufficiently important that they may get their own category.

Metals are mixtures (alloys) of metallic elements, usually in crystalline form. They have free electrons that can move readily through the crystals. This explains the thermal and electrical conductivity of metals, as well as their relatively high degree of reactivity (which explains corrosion, or the reaction with oxygen). Metals are strong due to their bonding, and ductile due to their micro-structure: these are ideal properties for many structural applications. Different types of metals may be alloyed to create phases within the solid body, or unique metallic compounds that have special properties.

Polymers are molecules based on carbon, hydrogen and other non-metallic elements. They are characterized by the large size of their molecules, and tend to have low density, low stiffness and strength, and high sensitivity to temperature and UV. A polymer may be composed of ten to tens of thousands of molecules in chains of repeating mer units. For example, polyethylene is a polymer made from chains of ethylene CH_2-CH_2-and so on. Many polymers contain hydrogen bonds with a carbon backbone chain, but replacing hydrogen with fluorine produces a more durable polymer (Teflon®). Most polymers are called "organic" (and are often based on hydrocarbons) because their chemistry is carbon based. Silicone is a very important inorganic polymer (which acts in many ways like carbon-based polymers, but the bonds are stronger) in construction practice. Important sub-classifications of polymers include *elastomers* (highly elastic, flexible), *thermosets* (irreversibly cured, stiff), *thermoplastics* (can be softened, melted and reformed).

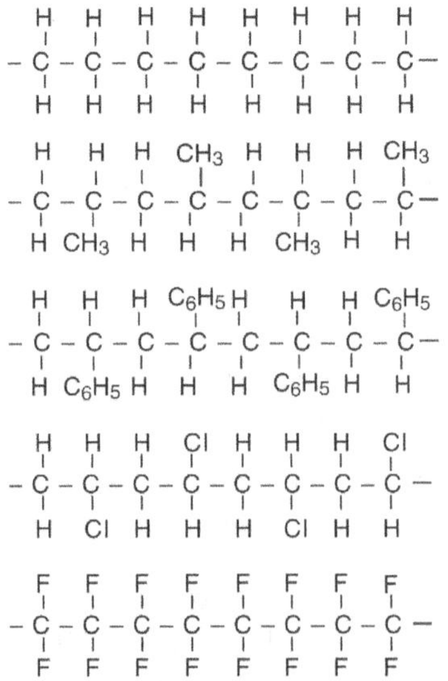

Polyethylene, **PE**

Polypropylene, **PP**

Polystyrene, **PS**

Polyvinyl chloride, **PVC**

Polytetrafluoroethylene, **PTFE**

Figure 4.1: Some Common Carbon-based Polymers

Natural polymers (usually carbohydrates), in the form of wood, natural fibres (e.g., cotton), bone, leather etc. tend to be very complex and have traditionally been very important, often being awarded their own category: plant- and animal-based materials.

Composites or hybrids are materials made of more than one material type. These are increasingly important, as they allow materials engineers to design specific properties that take optimal advantage of different materials.

Examples include reinforced concrete (steel and concrete), fiberglass (glass fibers in a polyester resin matrix), and wood-fiber reinforced cement. Concrete is a mixture of different ceramic elements and steel is a mixture of carbon and iron, but these are not considered composites because the mixture is occurring at the molecular level. Glass-fiber reinforced cement composites are on the other hand considered a composite because the combination occurs at a macro level.

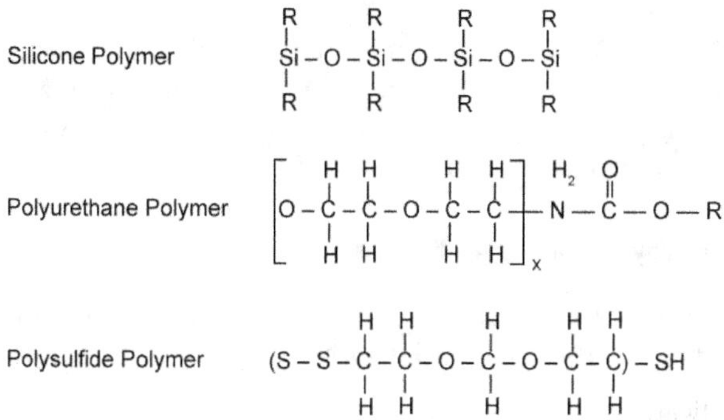

Figure 4.2: Common Polymers used as building sealants

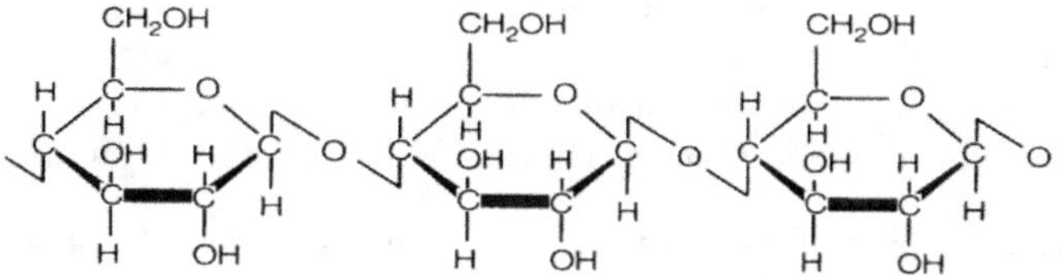

Figure 4.3: Cellulose, a natural carbohydrate polymer $(C_6H_{10}O_5)_n$

Other Classifications

Organic and **inorganic** is another useful classification of materials. Organic materials contain carbon and are often created by or from life forms on Earth. For example, wood, bone, and cotton are all organic carbon-based materials. Oil and natural gas are carbon-rich liquid and gaseous compounds joined with hydrogen to form hydrocarbons[1]. These compounds are used for man-made organic materials such as most modern polymer plastics. In the past, and increasingly again today, natural renewable sources of carbon, predominately

---

[1] A hydrocarbon is an organic molecule consisting entirely of hydrogen and carbon.

carbohydrate[2] were used to produce natural-sourced plastics such as Celluloid, gutta-percha, Bakelite and Cellophane.

Plant or **mineral** is another traditional distinction between materials. Minerals are defined as substances that 1. occur naturally, 2. are made of substances that were never alive (i.e., hence coal is not a mineral), 3. have an invariant chemical make-up (hence sand is not a mineral, but quartz is), and 4. exhibit atoms are arranged in a regular crystalline pattern (the calcium in milk is not a mineral, since it is in solution, but is a mineral if it forms solid crystalline deposits).

Other classifications of convenience, such as semi-conductors, advanced materials, biomaterials, natural material, etc. are frequently made but these are usually not rigorously defined from a scientific point of view, nor are they very useful for guiding building material selection.

4.2.4 Material Durability for Building Applications

Three common and powerful environmental agents of deterioration explain most building enclosure material damage:

1. heat,
2. ultra-violet radiation (UV), and
3. moisture.

Oxygen is a fourth common and important agent, as it allows oxidation, but is so ubiquitous and hard to avoid that it has little impact on enclosure design. As materials, especially polymers, become more sophisticated and widely used, special chemical interactions, unique to the situation, have become more common as deterioration agents. These defy simple categorization.

The general classes of materials described earlier have rather specific and common possible failure mechanisms that can guide material selection.

Minerals / ceramics are immune to breakdown from UV exposure or high heat—if a material is expected to be exposed to high levels of UV for a long time (e.g. cladding in the Middle East, roofing), materials from these two categories should be selected. Concrete masonry, cement, stucco, glass, and mortar are typically considered durable materials in wet environments and can tolerate high temperature and low temperatures with no ill effects. However, two common deterioration mechanisms attack porous mineral materials, both moisture-based: freeze-thaw action and salt expansion (subfluorescence). Both require significant exposure to moisture (the material must be nearly saturated). Glass is a non-porous mineral that can be attacked by specific chemicals and alkalis (e.g., cement- and lime-based materials).

---

[2] A carbohydrate is commonly defined as a large biological molecule, or macromolecule, consisting of carbon (C), hydrogen (H), and oxygen (O) atoms, usually with a hydrogen:oxygen atom ratio of 2:1 (as it is in water) giving rise to the hydrate label; the empirical formula is $C_x(H_2O)_n$.

Polymers, including natural polymers, are susceptible to UV degradation, especially the hydrogen-carbon and carbon-carbon bond: the energy level and wavelength of UV can penetrate polymers and cause the long polymer chains to be broken (weakening the material) and then cross-link (laterally connecting the shortened chains together), causing brittleness. Sunburn, greying of wood, and brittle and weak plastic lawn furniture are all common experiences of this degradation. Some polymers are much less susceptible to UV than others (those that replace carbon and hydrogen bonds with silicone and fluorine are almost immune), and many use additives of some form to absorb, block, or reflect UV.

Oxidation attacks polymers as well, especially at higher temperature. Oxidation mechanisms can be triggered, enhanced, and accelerated by the presence of UV radiation (termed photo-oxidation) and moisture. Hydrolysis, a chemical reaction where water breaks down chemical bonds, can affect some polymers as well, although usually only over long periods of time and at high temperatures.

If the temperatures are high or low for long enough, polymers will often degrade. At high temperatures oxidation accelerates (polymer chains are split and oxygen atoms are inserted), softening or flow begins (e.g. asphalt), and off gassing (volatile organic compounds are released) increases. Some polymers are much more resistant than others. Wood has excellent performance (no softening or degradation) up to well above 80°C and polymers without hydrogen-carbon bonds can perform well at elevated temperatures. Common temperature resistant polymers include silicone (up to 200°C), fluropolymers like Teflon®/PTFE (150°C), and NOMEX® fabrics (over 250°C).

Low temperatures often result in a significant increase of stiffness and reduction in ductility: plastics become brittle when it is very cold. Low temperatures cause some polymer chains to become rigid and interlock. Temperatures of -20 to -40°C can cause some plastics to shatter with even a small impact. High-performance polymers like Teflon®, NOMEX®, and many silicones can remain flexible down to below -80°C, as does wood.

Many man-made plastics can be very durable if they are protected from UV radiation and maintained at close to room temperature (hence their predicted lifespan of centuries if discarded in landfills), but they must contain anti-oxidant additives to slow oxidation if exposed to high temperatures, UV-absorbing/blocking additives if exposed to UV, and often require plasticizers if exposed to cold temperatures. In all cases, high temperatures reduce stiffness and strength, and low temperatures increase stiffness and brittleness.

Natural polymers (cellulose, starch) are affected by all of the above mechanisms and can also, in the presence of moisture, be eaten by decay fungi[3]. This is a critical concern with paper and other processed wood products, and the eventual demise of almost all natural polymers such as wood and bamboo. However, natural polymers are usually relatively insensitive to temperature variations (the mechanical properties of wood are surprisingly stable within the range of -40 to +60°C) and hence wood is often used for its structural properties in the liquified natural gas industry (where wood performs structurally down to -160 °C).

Metals, like minerals, are unaffected by UV and temperature in building enclosure applications. They do, however, corrode, or, technically speaking, react with oxygen via an oxidation process. Metallic corrosion is an electrochemical process and hence requires moisture and is accelerated by the electrical conductivity of salt solutions. As sufficient oxygen is available in essentially all practical situations, it is moisture and the electrochemistry that must be managed to ensure durability. Alloys of steel (notably stainless steel) and pure aluminum form strong corrosion products that adhere to and protect themselves, which provides long life in most cases. However, mechanical abrasion and some atmospheric impurities can cause rapid corrosion of even these two relatively durable classes.

Composites have the most complex responses because they combine different materials which may be of different categories: concrete reinforced with epoxy-coated reinforcing steel may be susceptible to attack on any one, or all three, of polymers, metals, and mineral.

Durability Dose-Response Influences

The intensity, duration, and frequency of exposure to a particular agent of deterioration will generally increase the rate of deterioration. A material susceptible to UV degradation can be exposed to hundreds of hours of outdoor light before deterioration is observed, but exposure to thousands of hours of bright sunshine will cause total failure. Masonry may be exposed to many hundreds of rain events, followed by freezing, with no damage, but if the masonry is saturated by long and intense exposure to rain, a single freezing event may cause a visible crack.

Hence the design of building materials, and their selection, is filled with nuance, as some materials of the same class (i.e., plastic, metals, minerals) can be much more resistant to deterioration than others: despite this, it is important to recognize that all carbon-based polymers will be damaged by UV (only the rate of damage can be slowed), all masonry can be damaged by freeze-thaw (only the degree of saturation required can be increased), and all metals will corrode (only the rate can be reduced).

---

[3] Even hydrocarbon-based polymers can be attacked by some fungi as well as by hydrolysis, but the process is generally much slower than natural polymers.

Applications to Building

The designer of a building, especially of a building enclosure, has tight control over the environmental conditions that will be experienced by materials over the life of the product by selecting the materials and their relative arrangement. Whether the material that forms the layer of an air barrier, a seal between water barriers, or the head of a cladding fastener, the expected long-term performance may be judged. Small volumes of unique conditions – nanoclimates – can form and must be foreseen, for example, a pocket that traps water, a joint that allows UV penetration, a screw hole in a coated metal, etc. Galvanic corrosion, an important mechanism, usually only acts over small areas of contact that must be expected and guarded against.

Many building products are not made of a single material, but complex composites and hybrids of polymers, metals, and/or ceramics, with a vast array of additives and coatings. This allows manufacturers to supply the most economical, safe, and environmentally benign product for a specific application but massively complicates product selection. A bituminous air-barrier may have excellent performance based on the SBS-bitumen used, but plasticizers added to enhance flexibility at low temperature may migrate out of the product over time and cause premature failure. Or the polyethylene facer used as structural reinforcement for the product may break down quickly with short UV exposure during construction, rendering the product weak.

The interior layers of an enclosure or partition are usually not exposed to water, UV, or a wide range of temperatures, and are therefore easier to specify. Building superstructure and enclosure support structures can, by design, be protected from the primary agents of deterioration. The primary risks therefore tend to be from abnormal events, i.e., fire or flooding. Cladding is often one of the most critical choices in terms of material durability as it is exposed to the most challenging environmental conditions.

If a cladding material is expected to be exposed to water, oxygen, and temperatures above freezing, polymers, composites, and mineral-based materials would be preferred over metals. But if there is high UV exposure, and/or high temperatures, mineral-based materials would, as a class, have the highest durability. However, if the material sought is to be used in a freeze-thaw environment, mineral products may be susceptible to damage. Hence, a corrosion-resistant metal alloy, such as stainless steel or zinc, or a metal protected with polymers or mineral-based coatings (i.e, paint) would be an option. Cladding at grade (wet) in a low UV environment (a cloudy city) could be made of a polymer.

Closure

Understanding the fundamental characteristics of the different classes of materials can be a starting point for material selection during enclosure design. Given the wide variations between materials of the same class, and the proliferation of composite and hybrid materials, what has been presented above is just a rough guide as a starting point. There are numerous building material reference books available [Illston & Domone 2010, Ashby et al 2007].

4.3 Heat and energy

Heat is a measure of the vibratory energy stored in matter: the faster the rate of vibration, the greater the heat content. The relationship of heat content to temperature is the most straightforward for ideal gases—the root mean square velocity of the individual gas molecules is a direct measure of the temperature. Of course, the most common method of measuring thermal energy, heat, in a material is to measure its temperature.

As temperature increases, the molecules of a fluid collide more often and more energetically with one another and with the sides of any container. These collisions are what cause pressure on the sides of a container. Since a gas molecule can move freely, gaseous matter offers almost no resistance to change in shape (friction) and very little resistance to a change in volume (compression). If we compress a gas, the number of molecules per unit volume increases but, provided the temperature remains the same, the velocity of the molecules will not change. However, because the same number of molecules are moving at the same speed but in a smaller container, the pressure increases because the rate of molecular collision increases, especially with the walls of their container. Even at high pressures, the space between molecules remains large relative to the size of the molecules.

If a sufficient amount of energy is removed from a gas, the strength of attraction between molecules will eventually become stronger than the kinetic energy of the moving molecules, and the gas will become a liquid. Similarly, if a gas is sufficiently compressed, the molecules are forced into close contact and a liquid forms.

The removal of even more energy from a material slows the movement of molecules, and intra-molecular forces begin to dominate. A solid is formed when the molecules cease to move about freely. Molecules in a solid state remain in a fixed position but do continue to vibrate. When even this vibration ceases, the temperature of the material has reached absolute zero (-273 °C or -460 °F). Solids are characterized by their resistance to change in either shape or volume.

The relative forces of attraction between molecules and the kinetic energy of molecules with a certain velocity clearly vary with different types of molecules and with different mixes of molecules. For example, at one extreme, tungsten becomes a liquid at 3370 °C (6100 °F), and at the other extreme, helium becomes a solid at -269 °C (-450 °F).

The more energy in a solid, the more the molecules vibrate, and the more intensive the collisions with their neighbors. Therefore, solids are expected to expand with temperature. The kinetic theory of gases described above also leads to the expectation that heating a liquid or gas will cause it to expand. The vibrational or kinetic energy causes molecules to move farther apart from each other, and the same number of hot molecules will therefore occupy more space than an equal number of colder molecules.

4.3.1 Sensible Energy and Heat Capacity

The energy used to increase the velocity or vibrations of molecules is called *sensible energy* (because one can sense an increase in temperature in the material). The sensible energy required to raise a unit mass of material one unit of temperature is called the *heat capacity* of the material. In common language, the term "thermal mass" is often used. The heat capacity can be a strong function of temperature and pressure for many gases, but it can be assumed to be constant over limited ranges of temperature (as it is for liquids and solids).

All materials above absolute zero contain some heat energy. The amount of this energy, ignoring state changes and assuming linear behavior, can be found from:

$$E = c \cdot m \cdot T \qquad \text{(Eq. 4.1)}$$

where Q is the amount of heat energy (J),

c is the heat capacity (kJ/ kg · K or J / g · K),

m is the mass (in g or kg), and

T is the absolute temperature (K).

Most materials do have phase changes and their heat capacity is different for each phase. It is much more useful in practise to consider the amount of heat energy required to change a material from one temperature to another. This can be described by:

$$E = c \cdot m \cdot \Delta T \qquad \text{(Eq. 4.2)}$$

This equation, and the invariant value of c, is sufficiently accurate for most building science applications and over narrow temperature ranges. Table 4.1 lists the heat capacity values for common materials.

It is often useful to consider the heat capacity per unit volume rather than per unit mass. Fluids (such as air and water) are widely used in heating and cooling systems to transport thermal energy, and the flow rate is almost always measured in volumetric units.

Table 4.1: Specific heat capacity of some common materials (at 25°C unless noted)

Material	Specific heat capacity (kJ / kg · K)
Air at 25°C	1.006
Air at -20 °C	1.005
Air at 80 °C	1.009
Alcohol, ethyl	2.40
Aluminum	0.90
Argon	0.52
Asphalt	0.92
Brass	0.38
Brick	0.84
Concrete, brick, rock, minerals	0.8 – 0.9
Copper and brass	0.40
Foam plastics, plastics	1.5
Glass	0.67
Granite	0.80
Gypsum	1.1
Hydrogen (H_2 at 1 atm)	14.3
Ice at 0 °C	2.11
Ice at -20 °C	1.95
Marble	0.90
Krypton	0.25
Polycarbonate	1.2 – 1.7
Porcelain	1.10
Sandy clay	0.8 – 0.9
Silicon rubber sealant	1.45
Snow	2.10
Steel, iron	0.45 - 0.50
Water vapor at 0 °C	2.05
Water, liquid at 0 °C	4.218
Water, liquid at 20 °C	4.182
Water, liquid at 80 °C	4.196
Wood / wood products, 0%MC 20 °C	1.6
Wood / wood products, 12%MC 20 °C	2.0
Wood / wood products, -20 °C	1.84

The *volumetric heat capacity* can be calculated as:

$$c_v = c \cdot \rho \qquad \text{(Eq. 4.3)}$$

where ρ is the density of the material in consistent units.

For example, air at room temperature has a density of about 1.2 kg/m^3 and a specific heat capacity of 1.0 kJ/ kg · K. Hence, the volumetric heat capacity is:

$$c_v = c \cdot \rho = (1.0\ \text{kJ/kg} \cdot \text{K}) \cdot (1.2\ \text{kg/m}^3) = 1.2\ \text{kJ/m}^3\ \text{K}$$

This value can be used to conveniently calculate the rate of heat delivery to a space given the volumetric flow rate and temperature drop of supply air.

4.3.2 Latent Energy and State Change

When materials undergo a change of state, additional energy is required to reorganize the molecules into the new state. There is a significant difference in total energy content between a gas and a liquid at the same temperature, for instance. A state change therefore gives off or takes up a material-specific amount of energy called *latent energy.* The term latent is used because the state change occurs with no change in temperature, unlike sensible heat energy, which can be sensed.

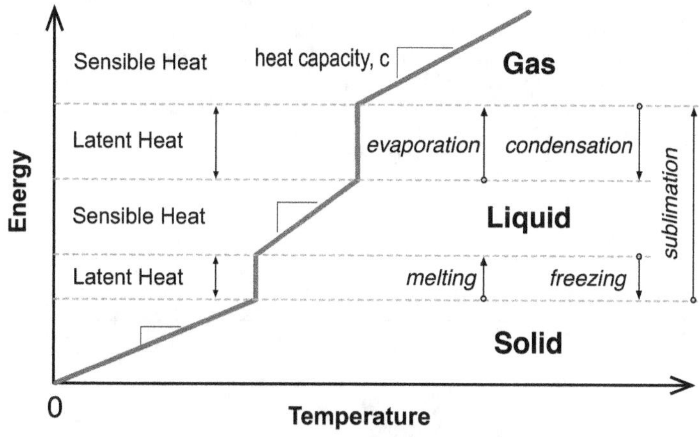

Figure 4.4: Simplified energy-temperature relation for a typical material

Depending on the state change, the latent energy required to change a unit mass of material from one state to another is called the *heat of fusion* (from solid to liquid) or *heat of vaporization* (from liquid to gas).

It should be noted here that chemical energy – the energy stored in the bonds that join atoms into larger units – is very different than latent energy. This energy can be converted to heat energy only by breaking some of the bonds. For example, the combination of oxygen and hydrogen in the presence of a

flame converts chemical energy to heat energy in the reaction
$O_2 + H_2 \Rightarrow H_2O$ + heat.

Thermal (i.e. heat) energy can be transported by three modes: convection, conduction, and radiation. These modes of transport are discussed in some detail in Chapter 5. Radiation, however, deserves additional attention as it has significance beyond heat transmission. The fundamentals of radiation are covered in the next section.

4.3.3 Radiation

Radiation is energy emitted by matter as a result of molecular vibration. The emission can be attributed to changes in the electron configurations of the molecules in the material. Lasers (the acronym of Light Amplification by Stimulation of Emitted Radiation) can produce very intense monochromatic (one frequency) light by stimulating certain molecules to a specific energy level. Neon lights emit a specific frequency when high-voltage electricity stimulates electrons to jump from their orbitals about the atomic nucleus.

In one analogy, radiation is considered to be a form of wave energy, with a specific frequency and amplitude (Figure 4.5). In another analogy, radiation is considered to be a particle (named a photon) with a certain amount of energy. Each analogy is useful for different analytical purposes, and the wave-particle duality of radiation is commonly accepted today. For building science applications, the wave analogy tends to be more useful.

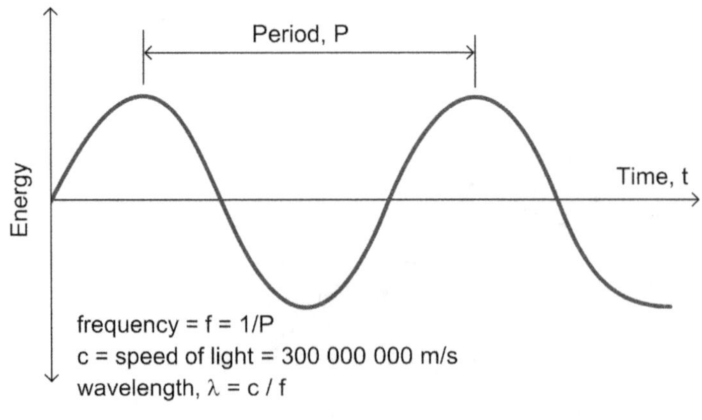

Figure 4.5: Wave representation of electromagnetic radiation

Figure 4.6 shows the range of typical frequencies and wavelengths related to different types of radiation. The visible light frequencies cover the range from 380 to about 780 nm. Each color (the acronym ROYGBIV is a memory aid) has a specific frequency range associated with it. Ranges for each color are listed in Table 4.2. Figure 4.6 also illustrates the reasoning behind the terms infra-red (infra meaning *below*) and ultra-violet (ultra meaning *beyond*).

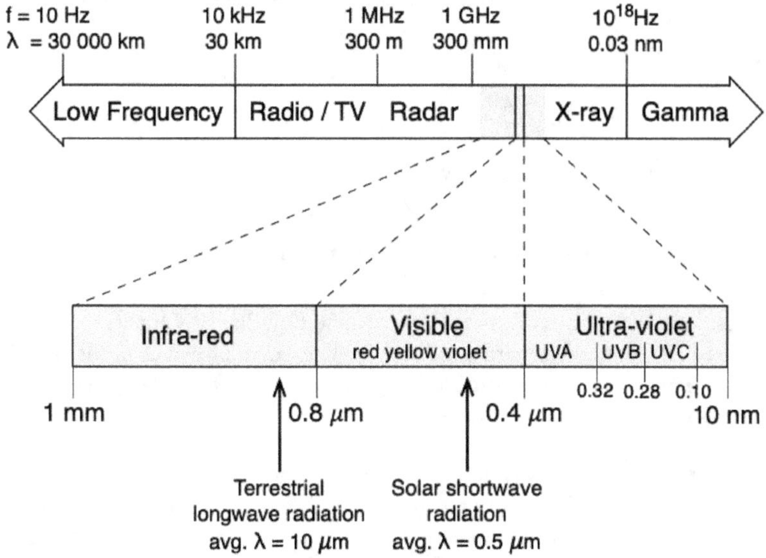

Figure 4.6: Wavelength and frequency ranges for common types of electromagnetic radiation

Table 4.2: Approximate wavelength versus color

Color	Wavelength range (nm)
IR-C "Far Infra-Red"	3 000 – 10 000
IR-B	1 400 – 3 000
IR-A "Near Infra- Red"	780 – 1 490
Red	610 – 800
Orange	590 – 620
Yellow	560 – 590
Green	495 – 560
Blue & Indigo	450 – 495
Violet	380 – 450
UV-A	320 – 400
UV-B	280 – 320
UV-C	100 – 280

Planck's distribution, which gives the emissive power $W_{b\lambda}$ (in W / m^2·μm) for a unit wavelength interval λ (in micrometers) at a temperature of T (in Kelvins), was developed from quantum mechanical considerations:

$$W_{b\lambda} = \frac{C_1}{\lambda^5 \cdot [e^{C_2/\lambda T} - 1]} \qquad \text{(Eq. 4.1)}$$

where $C_1 = 3.742 \times 10^8$ W·μm^4/m^2 and

$C_2 = 14390$ μm·K are radiation constants derived from Planck's and Boltzmann constants and the speed of light.

Consider the distribution of energy predicted by Planck's equation, plotted in Figure 4.7. Note the logarithmic scales. The range of wavelengths presented in this figure is normally considered the range of *thermal radiation*: 0.1 to 100 μm. The following should be noted:

- At any given wavelength there is a large increase in emitted radiation with increasing temperature. In fact, a doubling of the absolute temperature results in a 16-fold increase in emitted radiant energy. The peak wavelength becomes shorter (i.e., the frequency increases) with higher temperatures. The peak emission, and most of the total energy, is concentrated in a narrow frequency band.

- Thermal radiation from surfaces over the range of normal building temperatures i.e., -40 °C (-40 °F) to 100 °C (212 °F) will be infra-red of an average wavelength of 10 μm, and hence far from the visible range. Radiation with wavelengths of this order is termed *long-wave radiation* or, because of the temperature range that emits it, *terrestrial radiation.*

- Only surfaces with temperatures of more than about 800 K (500 °C / 1000 °F) will radiate visible (dark red) frequencies. Surfaces heated further, to about 1300 K (1000 °C / 1900 °F), will glow yellow (a wavelength of about 580 nm). If a surface is heated to more than about 1500 K (1200 °C / 2200 °F), it will emit all visible light frequencies and appear white.

- The surface temperature of the sun is about 5800 K (5500 °C / 10,000 °F). The figure shows that the peak emission of the sun's energy (*solar radiation*) is centered in the visible range with a wavelength of about 0.5 μm. Wavelengths of this order are termed *short-wave radiation* (in contrast to long wave). Human eyes evolved to respond to this powerful radiation source.

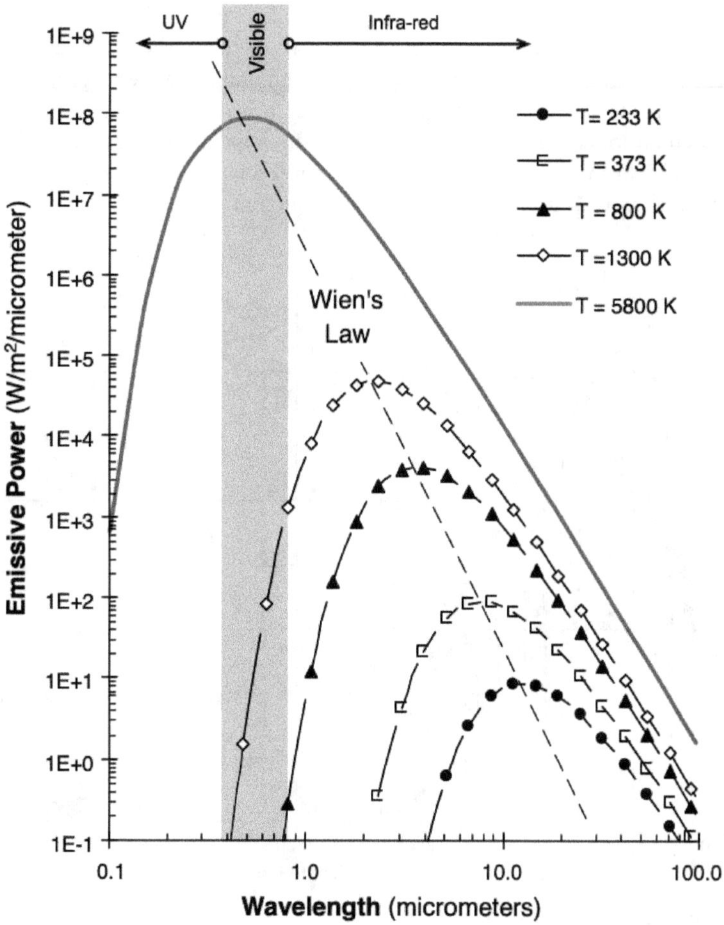

Figure 4.7: Plank's spectral distribution of thermal radiation from a black body

The inefficiency of incandescent light bulbs can now be appreciated. Although the tungsten filament is operated at a temperature of about 2500 to 2900 K (4000 to 4750 °F), near its melting point, Planck's equation shows that much of the energy is emitted as infra-red energy. (The emissivity of tungsten in the visible range is also less than 0.5 at these temperatures). As a result, most of the electrical energy is converted to infra-red heat energy, and only one to two percent is converted to visible light. Fluorescent and LED lights do not increase the temperature of a filament to generate light but stimulate the emission of light radiation directly. Hence, they are much more efficient.

The radiation given off by the sun has been measured for decades. A plot of a reference solar spectrum, ASTM G173, is plotted in Figure 4.8. The similarity to Planck's distribution should be noted. The spectrum outside the earth's atmosphere is much more consistent because components of the atmosphere, such as dust, water vapor, ozone, and CO_2, absorb, reflect, and scatter the incoming radiation. The total energy radiated to the earth over all frequencies is the integration of the extra-terrestrial radiation curve: the generally accepted standard value is 1367 W/m^2. How much radiation reaches the earth depends on many factors, especially the thickness of the atmosphere and the aforementioned components. A bright sunny day at sea level typically contains around 1000 W/m^2, which is a common reference value used in testing (for example, when rating the output of photovoltaic solar panels).

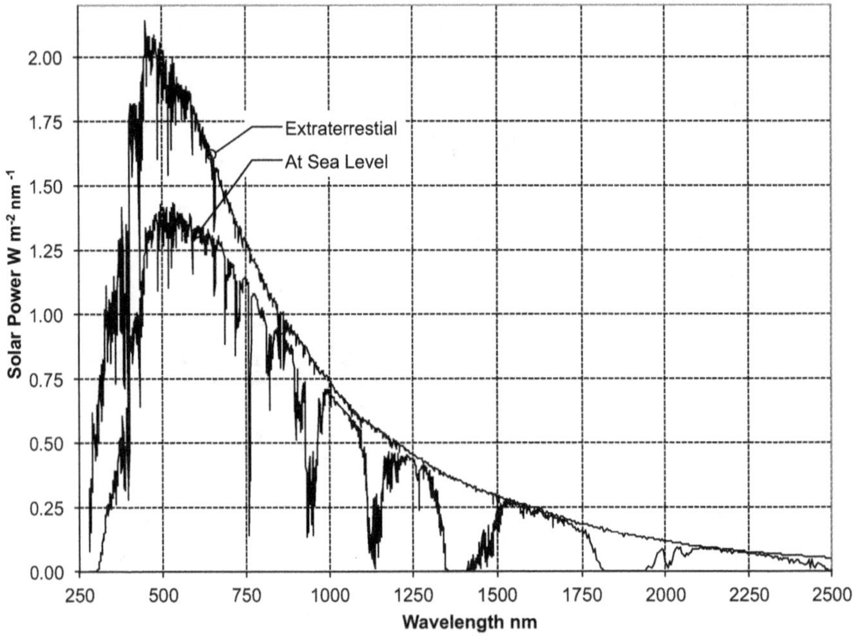

Figure 4.8: ASTM G173 Reference Solar Spectrum (showing absorption due to H_20 etc)

The shape of Planck's equation shows there is a temperature (or wavelength) of peak radiation intensity. **Wien's equation**, named after the scientist who first calculated the wavelength of peak intensity, is very simple:

$$\lambda_{max} = 2898 / T \qquad \text{(Eq. 4.3)}$$

where T is the absolute temperature of the emitting material (K), and

λ_{max} is the wavelength (in micrometers) at which maximum radiation is emitted.

Using Wien's equation, one can estimate the surface temperature of the sun from the measured peak spectral distribution of daylight. The wavelength being given off by many building objects can also be estimated using Wien's equation: in fact, this is how infra-red thermometers function.

The total area under the curve of Planck's distribution, i.e., the total rate of radiation emission from a perfect black body, was calculated by **Stefan and Boltzmann** to be:

$$q = \sigma \cdot T^4 \quad \text{(Eq. 4.4)}$$

where q is the emitted heat flux (W/m^2)

T is the absolute temperature of the surface (K), and

σ is the Stefan-Boltzmann constant, (5.67×10^{-8} W / $m^2 \cdot K^4$).

Most real materials are not perfect radiators of energy. The measure of the ability of a material to emit thermal radiation is called the emissivity, ε. The emissivity of a surface is the ratio of the energy actually emitted at a given temperature and wavelength to what would be predicted by Planck's equation for a perfect radiator. The practical and widely used version of the Stefan-Boltzmann equation has been modified to account for this by the simple addition of the emissivity term:

$$q = \varepsilon \cdot \sigma \cdot T^4 \quad \text{(Eq. 4.4)}$$

The emissivity, ε, ranges from 0 to 1 and does not change very much for a particular material over the range of typical building temperatures but is unique for each wavelength and can change significantly with material temperature over a wide range. The emissivity value normally quoted is for room temperature (IR) wavelengths unless otherwise stated. Many natural building materials have an emissivity of about 0.9 at room temperature.

Thermal radiation that strikes a surface can be absorbed, reflected, or transmitted. Absorption is the transformation of radiant energy into thermal energy stored by the molecules. Reflection is the return of radiation from a surface without a change in frequency. Transmission is the passage of radiation through a material without a change in frequency. All thermal energy falling on a surface must be subject to one of these three actions. Therefore,

$$\alpha + \rho + \tau = 1 \quad \text{(Eq. 4.5)}$$

where α is the **absorptance**, the fraction of incident radiation absorbed;

ρ is the **reflectance** (or albedo), the fraction of incident radiation reflected; and

τ is the **transmittance** or transmissivity, the fraction transmitted.

Some special cases should be considered. For opaque surfaces $\tau = 0$, and hence $\alpha + \rho = 1$. In practice there are no perfectly transparent materials ($\alpha + \rho = 0$ and $\tau = 1$). In glass for example, the transmittance is actually only relatively high, not 1.0, and some energy is absorbed, as is evident from the increase in glass temperature during solar exposure that can easily be felt.

Kirchoff proved that for any given wavelength, $\varepsilon = \alpha$ (**Kirchoff's Law**). However, for some materials the emissivity (and thus absorptance) changes substantially at different temperatures (which is the same as saying different wavelengths).

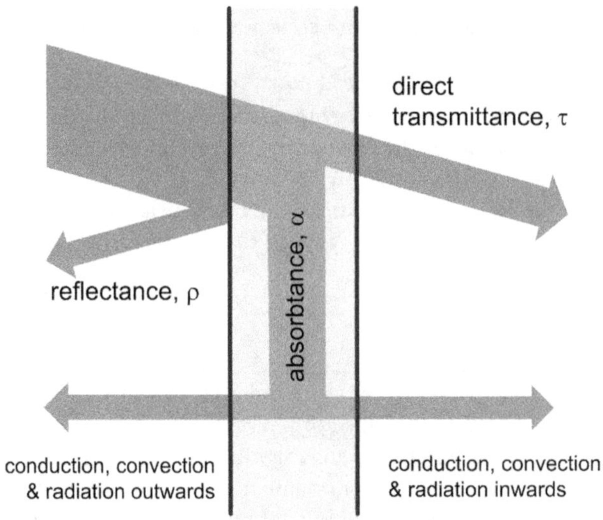

Figure 4.9: Reflectance, transmittance, and absorptance for glass

A material that is both a perfect emitter ($\varepsilon = 1$), and a perfect absorber ($\alpha = 1$) is a very special case; such an ideal body is called a *black body*. This is the simplest case, and often assumed in simple calculations, but materials that approach true black body behaviour are actually very rare in reality.

A material for which the emissivity (and thus absorptance) is the same for all wavelengths is called a gray body. A *gray body* is one which for which $\alpha = \varepsilon < 1$. Many natural building materials are assumed to be gray bodies for radiation calculations.

There are many practical building applications that use the variation in radiation properties of materials. For example, clear UV-protective coatings are nearly opaque at UV frequencies, but nearly transparent to visible frequencies.

Table 4.3 lists the solar absorptance and thermal emittance at room temperature for a range of different materials, taken from a range of sources, but most are from the extensive listings in Henninger [1984] and Parker [2000].

Table 4.3: Absorptance and emittance of common building materials

Material	**Solar absorptance (α)** (Visible light)	**Thermal emittance (ε)** (at room temperature)
Bare human skin	0.6 – 0.8	0.98
Most non-metallic materials	Varies with color and texture	0.90
Red brick	0.60 – 0.80	0.90
Yellow or buff brick	0.50 – 0.70	0.90
White or cream stucco or brick	0.30 – 0.45	0.90
Concrete unpainted	0.75	0.90
Smooth, white painted surface	0.20	0.90
Black, non-metallic surfaces	0.85 – 0.95	0.90 – 0.98
White asphalt shingles	0.79	0.91
Black asphalt shingles	0.95	0.91
Carbon lampblack	0.98	0.95
Spectrally selective absorbers	>0.85	< 0.12
Silver vapor deposited	0.04	0.02
Gold vapor deposted	0.19	0.02
Bright aluminum paint	0.20 – 0.30	0.30 – 0.40
Polished aluminum (foil)	0.10 – 0.30	0.03 – 0.04
Green anodized aluminum	0.66	0.88
Clear anodized aluminum	0.27-0.35	0.76-0.84
Sheet aluminum	0.39	0.25
Galvanized steel sheet (new)	0.39	0.04
Oxidized galvanized steel	0.20	0.29
Machine-rolled stainless steel	0.39	0.11
Steel, polished	0.10	0.07
Window glass	0.04 – 0.40	0.90 – 0.94
Ice	0.03 – 0.15	0.95
Snow – fresh	0.20	0.82
Water	0.84 – 0.93	0.90
Green grass	0.74	0.90

Advanced window glazings typically employ special metallic coatings, only a few molecules thick, that allow only certain wavelengths through while blocking others. Such glazings are said to be spectrally selective. Hence, the properties for this type of glass vary with wavelength (Figure 4.10).

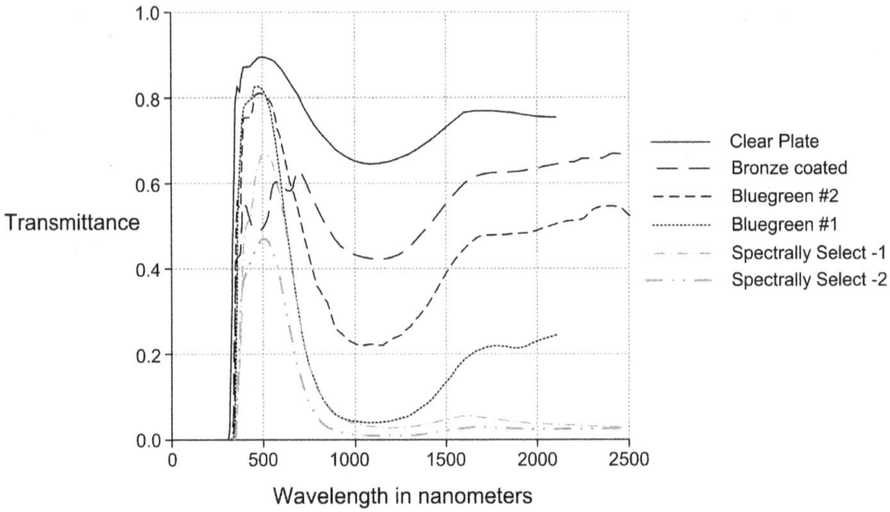

Figure 4.10: Radiation transmittance as a function of wavelength [ASHRAE 2021]

Many materials absorb differing fractions of solar radiation (wavelength about 0.5 μm) and infra-red terrestrial radiation (wavelength about 10 μm). These properties (and Kirchoff's law) can be used to advantage when designing solar collectors to absorb radiation (*selective absorbers*) and when selecting materials to reject solar radiation (termed *cool* surfaces). In the former case, a high solar absorptance and a low infra-red emittance are desirable. In the latter case, a material with a low solar absorptance and a high emittance should be chosen.

Cool surfaces, such as cool roofs, can be used to reduce unwanted solar heating, both to reduce heat gain through the enclosure and to maintain materials at more moderate temperatures and thereby enhance their durability, reduce the heat island effect in cities, and reduce heat gain through roofs.

In the realm of cool surfaces, reflectance, the complement of absorptance, is often used. Another term for solar reflectance is *albedo.* Visually dark, or deeply textured surfaces will have low reflectance (high absorptance) and light, shiny, and smooth or polished surfaces will have high reflectance. It can be seen from Table 4.3 that many materials (other than shiny metallic ones) have high thermal emittance values.

A cool surface will absorb little solar radiation (centered in the 0.5 micrometer range) because of low absorptance and emit as much thermal radiation (centered in the 10 micrometer range) as possible because of high emittance.

Figure 4.11 provides a plot of a number of materials demonstrating the range of possible radiation performance combinations. The upper left corner contains "cool" materials (low solar absorptance and high infra-red emissivity), and the lower right, the hottest materials (high solar absorptance and low infra-red emissivity) when exposed to the sun.

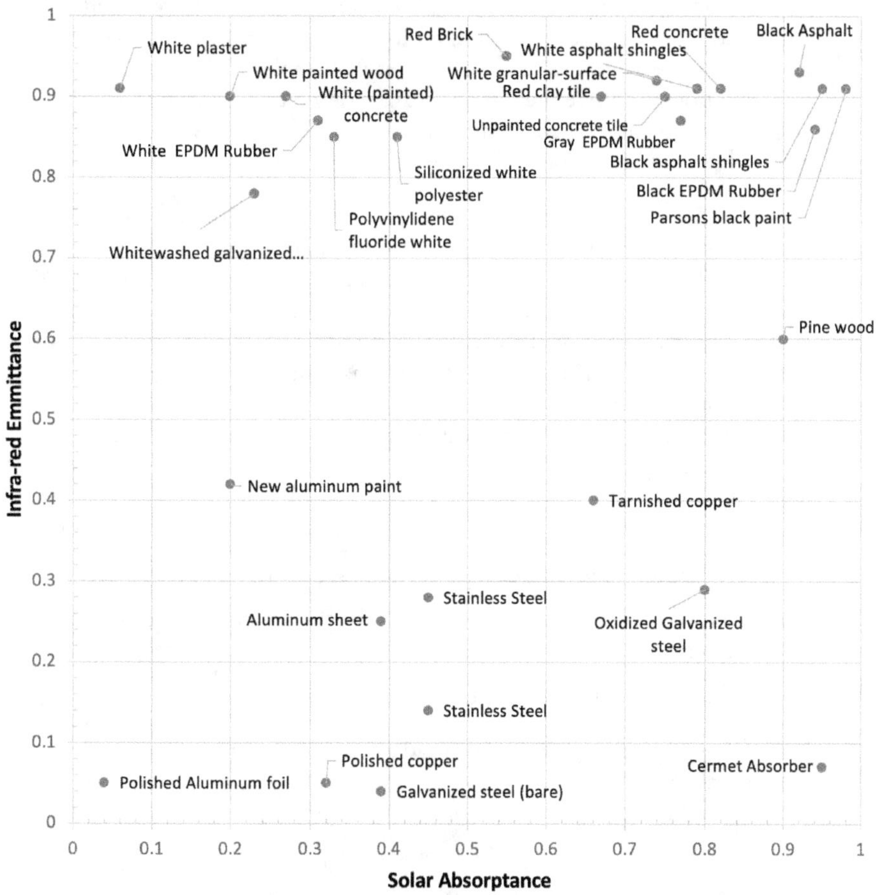

Figure 4.11: Solar and thermal radiation properties of several materials [based on data from Lawrence Berkeley National Labs, Florida Solar Energy Center, NASA]

To measure cool roofs and walls, an industry measure called the Solar Reflectance Index (SRI) has been developed and added to building code requirements in some warmer climates. SRI combines the emittance and reflectance into a single number. It defines standard black (reflectance 0.05, emittance 0.90) as 0 and a standard white (reflectance 0.80, emittance 0.90) as 100. Highly reflective surfaces with high emissivities can have values of SRI over 100. Reference should be made to ASTM Standard 1980 [ASTM 2011] which defines this.

A flat plate exposed to the sun designed to collect solar heat energy (a solar hot water heater for example) will have the opposite properties than a cool roof. These often use coatings that are selective absorbers and occupy the lower right of

Both Planck's and Stefan-Boltzmann's equations are derived for an ideal black body and must be modified through the use of the emissivity ratio to predict real material behavior. The closest physical example of a black body is a small hole in a hollow sphere the interior surface of which is painted with a highly absorbent and emissive material such as lamp black. Outer space also closely approximates a black body.

Unlike conduction or convection, the transport of heat by electromagnetic waves does not require the presence of a material. In fact, radiation occurs most efficiently through a vacuum. Since all materials, even cold ones, radiate, it is the **net** radiation exchange that defines the heat transfer. In general, calculating this value for all potential geometries is complicated. However, for the special and useful case of a relatively small surface exchanging radiation with a large area (e.g., between a wall and the outdoors, a roof and the sky, a human body and the walls and ceiling of a room), an approximation of the radiative heat flux can be found from:

$$q = Q/A = \varepsilon \cdot \sigma \cdot (T_S^4 - T_a^4) \qquad \text{(Eq. 4.6)}$$

where T_s is the surface temperature (K),

T_a is the outdoor temperature (K) usually assumed to be equal to the ambient air temperature, and

A is the surface area (m^2).

Numerous practical applications of radiative heat transfer are covered in Chapter 5.

4.4 Water

Water is the one material that has the most influence on biology and buildings. Water is part of all aspects of building material manufacture, installation, and operation as well as being a critically important deterioration agent. Hence, a deep study is warranted.

4.4.1 The water molecule

Water, in all its states, is a molecule with one oxygen and two hydrogen atoms. Hydrogen has a single positive charge, and oxygen two negative charges. The water molecule is formed from a covalent bond (one in which each atom shares electrons with the other atoms in the resultant molecule). The two hydrogen molecules are separated by an angle of 104.5°. The covalent OH bond has an average separation distance of about 0.096 nm, and the equivalent ionic diameter of the molecule is about 0.282 nm (Figure 4.9) [Franks 1984, Wilson et al 1987].

It can be seen from Figure 4.12 that the centroid of the two positive charges is not coincident with the centroid of the two negative charges. This spatially unbalanced distribution of charges results in a permanent dipole or a polar molecule, i.e., one end is permanently positive, and the other end is permanently negative.

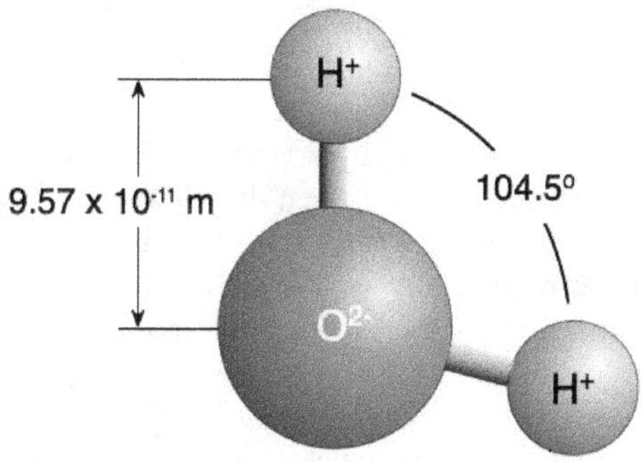

Figure 4.12: The H_20 molecule

The polar nature of the water molecule is pivotal to the way it interacts with other molecules [Glasstone and Lewis 1960, Powers 1968]. The principal force of attraction in water is the orientation effect, i.e., the positive hydrogen of one water molecule attracts the negative oxygen of another and causes the two molecules to link like two magnets.

Another contributor to intermolecular attraction for non-polar molecules is a phenomenon called the dispersion effect. This van der Waal's-type attraction is caused when the fluctuation in the electronic cloud of one molecule causes a similar displacement of electrons in a neighboring molecule. Since the charges are opposite, an electrical attraction is generated.

The induction or polarizability effect, another van der Waal's force, is similar to the dispersion effect but operates between polar and non-polar molecules only. If a (polar) water molecule approaches a molecule with an electron cloud that it can influence, the polar nature of the water molecule ensures that an intermolecular electrostatic bond forms in a manner similar to the dispersion effect.

The orientation and dispersion effect alone cannot explain the properties and behavior of water [Pauling 1988]. Hydrogen bonding occurs when the hydrogen atom of one molecule actually forms a weak chemical bond with the oxygen atom of another water molecule (or another solid molecule). Although largely electrostatic in nature, this bonding is different from the orientation and dispersion effect because a partial sharing of an electron (from the hydrogen) is involved. The hydrogen bond is generally weaker than ionic, covalent and metallic interatomic bonds because these bonds involve the complete sharing of electrons, but it is stronger than van der Waal's-type intermolecular bonds.

Hydrogen bonding between water molecules accounts for the viscosity, boiling point, and strong surface tension of water. Because of hydrogen bonding, water molecules cannot leave the surface of a body of water as readily as they could without this intermolecular attraction. The energy required to break this bond is quite high, and this fact partly explains the large amount of energy necessary for the evaporation of water.

Both the dispersion effect and hydrogen bonding cause water molecules to have a strong cohesive force between themselves. In a mass of water, shown schematically in Figure 4.13 this gives rise to attraction of molecules to those all around it, such as molecule A. However, at a transition of water to another fluid, such as air, there are no forces of attraction to molecules at the surface to the air above. Hence molecule B experiences stronger attraction to the neighbouring molecules at the surface. This additional attraction is called surface tension and is exceptionally strong in water (0.072 N/m).

The dispersion effect and hydrogen bonding also cause water molecules to have strong adhesion with other polar materials. This attraction to other materials results in adsorption and absorption (a result of the important force called capillarity covered in more detail in Chapter 8). Capillary rise in small tubes can be quite substantial, sufficient to draw water to the top of a tall tree, or many meters up a concrete column.

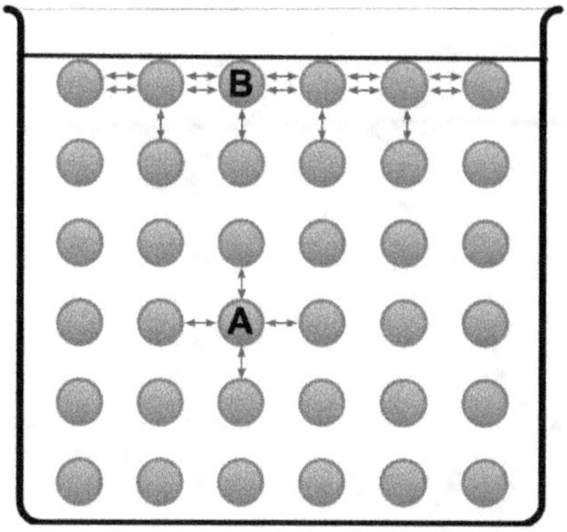

Figure 4.13: Surface tension

Liquid water tends to exist in clumps; at room temperature these clumps are about 80 molecules in size, i.e., water behaves more like $H_{160}O_{80}$ [Cherry 1979]. As the temperature increases, the clumps break up and become smaller; in the vapor state, the molecules have so much energy that they act individually. The difference in size between the liquid water molecule clump and the individual vapor molecule explains how materials such as Gore-Tex™ and Tyvek™ can simultaneously be watertight and vapor permeable[4].

Water has one rare characteristic among matter: as it is cooled close to freezing it expands, contrary to what would be expected. This occurs because water molecules are arranged in a solid six-sided crystal when water freezes from a liquid to a solid (Figure 4.14). Although the energy in the resulting crystal form is lower than for alternative arrangements at the same temperature (and hence stable), the crystal occupies more volume than six liquid molecules.

Water crystals begin forming as the temperature approaches zero. Above zero degrees Celsius, the crystals constantly break apart again or are dispersed. As water is cooled it contracts and its density increases. Water has its highest density at 4 °C (39 °F), i.e., the molecules are closest together at this temperature.

[4] The term "breathable" is sometimes used, but this term should be avoided because it implies there is some movement of air.

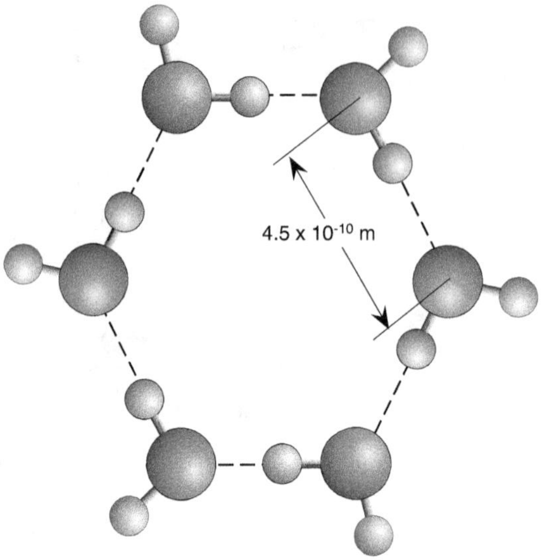

Figure 4.14: Ice crystal

As the temperature drops further, molecules begin to arrange themselves into ice crystals which occupy more volume; therefore, the density decreases, so much that ice floats on water. Between 4 °C and 0 °C (32 °F) the volume of water increases by 9 %, as an increasing proportion of molecules form into crystals. At about 0 °C all of the water molecules are organized into crystals and, as the water freezes, the crystals interconnect and a solid matrix is formed. As the temperature below freezing, further contraction once again takes place as for any other solid, although the density of the solid does not rise above that of liquid water under normal conditions.

The expansion experienced as water is cooled to 0 °C (32 °F) causes frost-heave forces and may result in freeze-thaw damage to some building materials. Note, however, that the mechanism of freeze-thaw damage in porous brittle materials (e.g., brick and concrete) is far more complicated than simple water expansion.

4.4.2 Moisture states and state changes

In the realm of most building science problems, moisture can exist in all three phases or states: solid (ice), liquid (water) and gas (water vapor), as shown in Figure 4.15. A fourth state, the adsorbed state, also exists on a molecular level, with characteristics somewhere between liquid water and vapor.

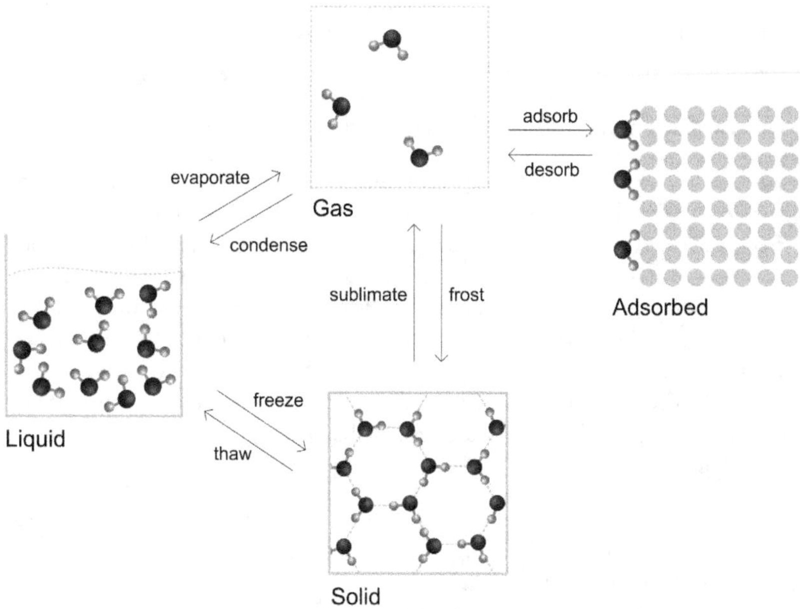

Figure 4.15: Moisture states and state changes

The adsorbed state

At normal temperatures water can exist in the adsorbed state. Water vapor molecules close to solid molecules can be attracted by the complementary polar nature of the solid material molecules, or they can induce polarity in the solid molecules by the dispersion and induction effect. Water vapor molecules may strike a solid surface and rebound, or they may be captured for a time by attractive forces. During the time that a molecule is held, it is said to be adsorbed.

The process of capturing the vapor molecule is called adsorption. Because of the vibrational and translational energy in single water vapor molecules, any single molecule will remain adsorbed for only a short time. On average, this duration is measured in microseconds. The higher the temperature, the higher the energy level and the shorter the time that a vapor molecule will remain on a surface. The more vapor molecules in the air—i.e., the higher the vapor pressure—the greater the number of molecules that will, on average, be adsorbed. The dynamic equilibrium established at a given temperature results in a stable amount of adsorbed moisture on a surface at that temperature.

Adsorbents are materials that have the ability to adsorb gases. Virtually all materials are *desiccants*, i.e., they attract and hold water vapor. Wood, natural fibers, brick, and concrete have a moderate ability to hold water. However, the term desiccant is normally reserved for materials with a very high affinity for water such as those used to remove moisture from the air in HVAC systems and equipment. The term *hygroscopic* is often used to describe those materials that exhibit a moderate affinity. Many porous building materials are hygroscopic, and they can store significant amounts of moisture as adsorbed water vapor on their very large internal surface areas.

Because the mobility of the vapor molecule is reduced by the attraction to the surface on which it is adsorbed, the behavior of adsorbed molecules is different from their behavior in either the pure liquid or vapor states. For example, adsorbed water does not freeze until temperatures are very low because the reorganization of the molecules into an ice crystal form is not the lowest energy state for adsorbed water. Powers, in his study of cement paste and concrete, calculated that adsorbed water would not form ice crystals (i.e., freeze) above -78 °C [Powers and Brownyard 1946]. However, after several layers of adsorbed water molecules accumulate, subsequent layers begin to behave more like liquid water.

Water molecules that are hydrogen-bonded to other solid molecules are said to be chemically bound. These *bound molecules* are not part of the adsorbed state, and typically require much more energy (i.e., heat) to remove. Such chemically bound water molecules can form *hydrates*, as in gypsum (which incidentally is about 20% chemically bonded water by weight), and the considerable energy required to break these bonds explains why gypsum has such high fire resistance.

Moisture state changes

Moisture frequently changes from one state to another, even within a material layer, and these state changes can greatly affect building enclosure behavior. Each state-change process is given a name (see Figure 4.15 above).

As noted earlier, the energy used to increase the velocity or vibrations of molecules is called sensible energy or sensible heat (because one can sense an increase in temperature in the material). The sensible energy required to raise a unit mass of material one unit of temperature is called *the heat capacity* of the material. The heat capacity is a function of temperature and pressure for many gases, but it can be assumed to be constant over limited temperature ranges.

Vaporization is the general term that describes the process by which liquid water changes to vapor. Vaporization can occur at the air-water surface, where it is called *evaporation*. Or it can occur within the mass of the liquid water and move to the surface in bubbles, where it is called *boiling*. Boiling occurs when the saturation vapor pressure is greater than the atmospheric pressure (at 100 °C for standard atmospheric pressure). Evaporation, in contrast, can occur at any temperature. Boiling occurs at a lower temperature on a mountaintop because the atmospheric pressure acting on the surface of the water is lower than at sea level. When vaporization occurs at temperatures below zero (or more precisely, from ice), it is called *sublimation.*

Condensation is the process by which water vapor changes to liquid. If water vapor changes directly to the solid state without passing through the liquid phase, the process is called *frosting* (or deposition if not dealing with water).

When materials change their state, additional energy is required to reorganize the molecules into the new state. For example, there is a significant difference in total energy content between a gas and a liquid. A state change therefore releases or absorbs a material-specific amount of energy called latent energy or *latent heat.* Depending on the state change, the latent energy required to change a unit mass of material from one state to another is called the *heat of fusion* (from solid to liquid) or the *heat of vaporization* (from liquid to gas).

Water's heat of vaporization, the energy to evaporate one gram of water, is 2250 J at 100 °C (212 °F) and 2500 J at 0 °C (32 °F). Condensation of one gram of water vapor releases the same amount of energy. If water's heat of fusion (the energy required to change ice to water (333 J/g)) and the heat of vaporization, (2250 J/g at 0 °C (32 °F)) are provided, the ice will change state directly to vapor at 0 °C (32 °F). Figure 4.16 plots a simplified and linearized energy-temperature relationship for water (the numerical values are approximate).

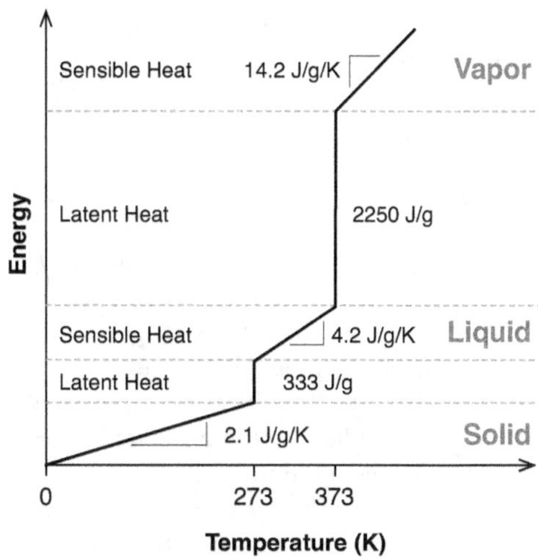

Figure 4.16: Simple linearized energy-temperature relation for water

The heat associated with adsorption or desorption is less than that for fusion or vaporization and varies with the material on which adsorption is occurring and somewhat with temperature.

4.5 Gases and the Atmosphere

In building science, an understanding of two gases is very important: air, the mixture of gases that makes up our atmosphere, and water vapor, the gaseous phase of water. To enable the importance of gases to buildings to be understood and to allow their behavior to be predicted, the basic gas laws are reviewed here.

4.5.1 Ideal gas laws

Ideal gases are defined as those gases whose behavior follows a set of simple laws, described below, over a range of temperature and pressures. Air and water vapor are ideal gases under almost all conditions found in building enclosure application.

As described earlier, the kinetic theory of matter is a useful model for understanding gases. For example, the temperature of a gas is directly related to the root mean square velocity of the molecules. The pressure that a certain volume of gas will exert on the sides of a container can be found directly from its kinetic energy ($^1/_2$ · mass· velocity^2) and from recognizing that the pressure can act on the six faces of a cube:

$$P = \frac{1}{3} \cdot \frac{N \cdot m \cdot v_{RMS}^{2}}{V} \qquad \text{(Eq. 4.7)}$$

where P is pressure,

V is volume,

N is the number of molecules,

m is the mass of each molecule of gas, and

v_{RMS} is the root mean square velocity of the molecules.

Gases are characterized by their ability to change volume and shape relatively easily. Anyone who has pumped up a bicycle tire with a hand pump realizes that a large amount of air can be compressed into a small space. As more air is pumped into the tire, the reading on the pressure gauge increases. Robert Boyle (1627-1691) conducted a number of experiments and found that, at a constant temperature, the volume of a given mass of gas is inversely proportional to pressure. **Boyle's Law** can be written as:

PV = constant or, most usefully,

$$P_1 \cdot V_1 = P_2 \cdot V_2 \qquad \text{(Eq. 4.8)}$$

Therefore, it can be seen that a gas at atmospheric pressure will occupy twice the volume as at two atmospheres[5]. The tire of a racing bicycle may be pumped to 90 psi (or 75 psi above atmospheric pressure). Since one standard atmosphere is approximately 15 psi, such a tire contains approximately six times the volume of air at atmospheric pressure.

Several scientists have investigated the effect of temperature on gases. **Charles's Law** indicates that for gases at the same pressure:

$$\frac{V_1}{T_1} = \frac{V_2}{T_2} \qquad \text{(Eq. 4.9)}$$

The **ideal gas equation** combines the above ideal laws, in a commonly used form:

$$\frac{P_1 \cdot V_1}{T_1} = \frac{P_2 \cdot V_2}{T_2} \qquad \text{(Eq. 4.10)}$$

or

$$P_2 = (P_1\ V_1\ T_2)\ /\ (T_1\ V_2)$$

Avogadro concluded that a given mass of an ideal gas contains a specific number of molecules. **Avogadro's number** is 6.022×10^{23} molecules per mole of gas. (A mole of any material is the number of grams equal to the total of the atomic mass of each of the elements in the molecule). Carbon dioxide, for instance, has the molecular formula, CO_2. The atomic number of carbon is 12, and that of oxygen 16. Therefore, there are 12+(2x16)=44 g per mole of CO_2 and 6.022×10^{23} molecules of CO_2 weigh 44 g.

The constant in the ideal gas equation can be based on the number of moles of gas, in which case the equation becomes the familiar:

$$P \cdot V = n \cdot R \cdot T \qquad \text{(Eq. 4.13)}$$

For an ideal gas, in SI units, the ideal gas constant R = 8.314 J/mol·K.

Through experiment it has been found that one mole of an ideal gas fills 22.414 liters at a temperature of 273.15 K and 101 325 Pa pressure. These conditions are usually referred to as Standard Temperature and Pressure (STP).

In gram weight units, the most useful form of the ideal gas equation for building science, the ideal gas equation can be written:

$$P \cdot V = w \cdot R_g \cdot T \qquad \text{(Eq. 4.14)}$$

where w is the mass of the gas and

R_g is a gas-specific constant.

---

5 "Atmosphere" is commonly used to decribe pressures in mulitples of sea level atmospheric pressure.

This gas-specific constant can be found from:

$$R_g = R/M \qquad \text{(Eq. 4.15)}$$

where M is the molar mass of the gas or the gas mixture of interest, and

R is the universal ideal gas constant (8.314 J/mol·K.)

The molar mass of air is about 28.96[6] g, and that of water vapor is 18.02 g, so:

$$R_{air} = 8.314 / 0.02896 = 287.1 \text{ J/(kg·K)}$$

$$R_{wv} = 8.314 / 0.01802 = 461.5 \text{ J/(kg·K)}$$

Dalton's Law of partial pressure states that each constituent of a gas mixture exerts pressure according to the ideal gas laws. For example, atmospheric pressure is the sum of the individual pressures (calculated according to the ideal gas equation) of each of the constituents. This law is very important for understanding how mixtures of gases, like water vapor in air, or carbon monoxide in exhaust plumes act. Depending on the water vapor content, water vapor may constitute less than 100 Pa or more than 3000 Pa of the total atmospheric pressure of 101 325 Pa.

4.5.2 The Atmosphere

By mass, Earth's atmosphere is composed of (Figure 4.17):

- 78.084 % nitrogen,
- 20.947 % oxygen,
- 0.934 % argon,
- 0.04 % carbon dioxide (and climbing), plus
- a variety of trace gases (and an increasing number of pollutants).

Water vapor is not included in these numbers since it varies dramatically with space and time and is typically a modest component of the atmosphere. At room temperature, water vapor accounts for about 1000 Pa of normal atmospheric pressure, 101 325 Pa (14.7 psi). However, vapor pressure can, if unrelieved, generate significant forces. This property can be demonstrated by placing a tightly sealed can of water over a hot fire. As the water turns to vapor, enough pressure can be generated to burst the can. With few practical exceptions (sealed glazing units being one), the pressure within a building

---

[6] These values are based on the mixture of gases that form air, and the accurate atomic mass considering the mix of isotopes present.

enclosure and the building can usually be equalized with the atmospheric pressure, thereby avoiding large differential forces.

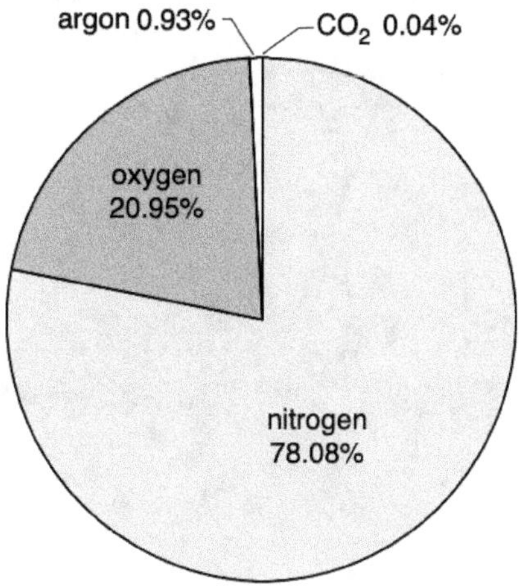

Figure 4.17: Major constituents of air

The global average annual temperature of the air at sea level is about 15 °C (60 °F), and the average temperature drops at a rate of approximately 6.5 °C per km of height (3.5 °F / 1000 ft). This drop with height is termed the *lapse rate.*

The *troposphere* extends from the lowest level of air on Earth to 11 000 m (36,000 ft) above sea level and is that part of the atmosphere that affects buildings directly (Figure 4.18). The temperature at 11 000 m (a common cruising altitude for an airliner) is therefore 15 °C - 11· 6.5 °C = -56.5 °C (-70 °F). This is the average annual temperature of the air outside an airliner. The *stratosphere* is the atmosphere between 11 and 20 km (36 000 to 65 000 ft). The temperature remains relatively constant in this region.

Atmospheric pressure is often expressed as 101,325 Pa, 14.7 psi, or 760 mm of mercury (Hg). One mm of Hg is defined as one torr. This is a long-term average at sea level. Pressures can change significantly with temperature, altitude, and weather systems. For example, a low-pressure storm (e.g., a hurricane) may have a pressure of 90 kPa, and the pressure a mile above sea level, e.g. Denver, will be around 85 kPa on average. Lower air pressure means fewer molecules per unit volume, and this has practical significance when providing oxygen for combustion in furnaces, automobiles, and aircraft.

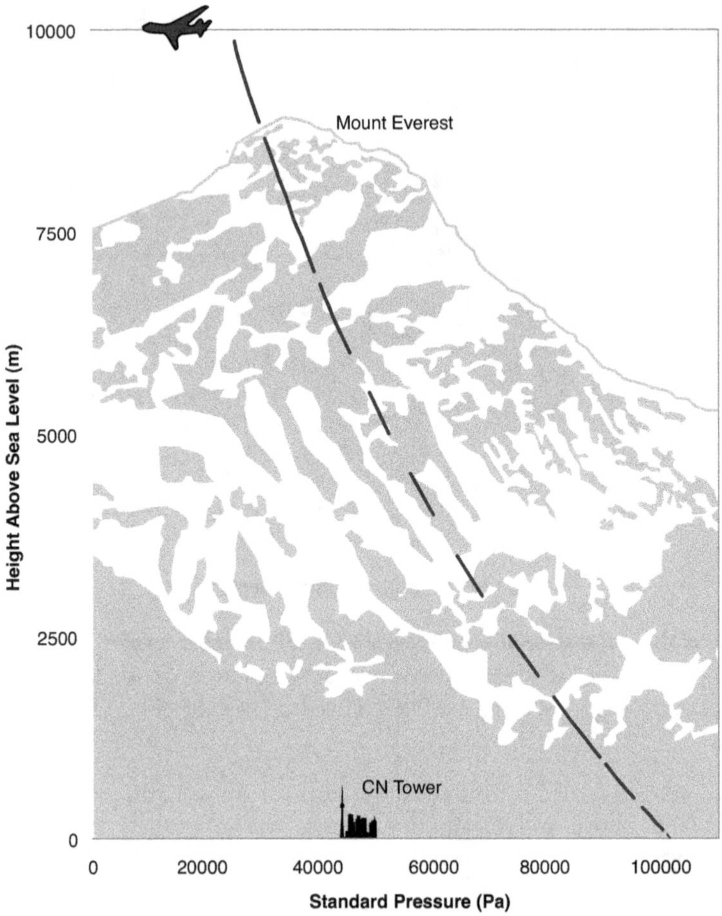

Figure 4.18: Atmospheric pressure change with height

The density of a column of air is expected to decrease with height because for each increase in height there is less atmosphere pressing down from above. However, air density increases with decreasing temperature, and the temperature decreases with height. Combining both effects, the standard pressure of the atmosphere within the troposphere can be described by:

$$P(h) = 101\,325 \cdot (1 - 22.6 \times 10^{-6} \cdot h)^{5.526} \text{ Pa.} \qquad \text{(Eq. 4.16)}$$

The change in pressure with height is sometimes important. For example, lower pressures at high altitudes result in a lower boiling temperature for all liquids (such as water and refrigerants) and can generate considerable forces if sealed containers are subjected to large altitude changes. For example, airtight canisters filled with air can burst open in the luggage bay of an airliner (which typically cruises at 10 to 12 km above sea level). In building applications, hermetically sealed insulated glazing units can bulge, bow, or fail if they are transported to an altitude significantly different from that at which they were sealed. The atmospheric pressure changes over the height of a building are relatively small: weather-system-related deviations from the standard barometric pressure of 101,325 Pa can often be larger (e.g., ±2000 Pa) than changes over the height of even very tall buildings.

4.6 Water Vapor & Psychrometrics

The moisture in air is one of the most important sources of wetting in buildings, and hence can cause human comfort and enclosure performance problems. The design professional needs to know how to design and construct buildings that are safe from damage caused by water vapor. To do this, it is essential to understand the physics of water vapor explained below. The transport of water vapor by diffusion and air movement, its deposition and removal by condensation and evaporation are covered in later chapters. Some terms are explained in the next section, and a list of definitions follows.

4.6.1 Terms, definitions, and units

Moisture: the H_2O molecule in liquid, solid or gaseous states.

Saturated air: air containing its maximum vapor content.

Condensation: the process by which water vapor changes to liquid.

Evaporation: the process by which water changes from liquid to gas at the water-air surface.

Boiling: the process by which water changes from liquid to gas within the mass of the liquid. Boiling occurs when the saturation vapor pressure at the temperature of the liquid water is greater than the total atmospheric pressure.

Sublimation: the process by which water changes its state directly from solid to gas without passing through the liquid state.

Humidity: a general, imprecise term for moisture in the air.

Absolute humidity or **Humidity ratio** (W or W_w): a fundamental quantity defined as the mass fraction of water vapor and dry air:

Units: Metric: dimensionless (g/g) or grams/kg dry air

Imperial: dimensionless (lb/lb) or grains/lb dry air

Vapor pressure (p_w): the partial pressure of water vapor in a mixture of gases.

Units: Metric: Pa or kPa

Imperial: psi or inches Hg

Saturation vapor pressure (p_{ws}): the maximum partial pressure of water vapor that can exist, i.e., the vapor pressure at saturation at a specific air temperature.

Relative humidity (Ø or RH): is the ratio of the amount of water vapor in the air to the amount of water vapor in saturated air at the same temperature.

Units: Metric: dimensionless or %

Imperial: same as metric

Dewpoint temperature (td): is the temperature at which condensation can begin to occur on a surface

Units: Metric: °C Celsius

Imperial: °F Fahrenheit

Some of these terms are further explained below.

4.6.2 Psychrometrics

The fundamental study of relationships between temperature, water vapor, and energy content of air is called *psychrometrics*. Relevant aspects of psychrometrics for building science applications include calculating dewpoint temperatures, relative humidity, and water vapor pressure.

When moisture is in its gaseous state it is variously called water vapor, humidity, or steam. Water vapor behaves as an ideal gas and acts as part of the mixture of gases (primarily nitrogen and oxygen) that constitute air. The terms *absolute humidity, humidity ratio, relative humidity, and vapor pressure* all reflect the same fact: air contains varying amounts of moisture in vapor form.

At any given temperature, there is a maximum amount of vapor that the air can hold. No more vapor can be added to air by evaporation or sublimation. The warmer the air, the more water vapor that can be held. When air contains the maximum amount of water vapor for its temperature condition, it is said to be *saturated.* The vapor pressure of an air mixture at saturation is called the *saturation vapor pressure*, p_{ws}. A useful approximate equation for saturation vapor pressure (in Pa) over water at a temperature T (in Kelvin) is:

$$p_{ws} = 1000 \cdot e^{\left(52.58 - \frac{6790.5}{T} - 5.028 \ln T\right)} \quad \text{(Eq. 4.17)}$$

where p_{ws} is the saturation water vapor pressure [Pa]

T is the absolute temperature [K].

The ratio of the actual amount of water vapor in the air to the maximum possible amount of water vapor in the air at that temperature (i.e., the amount of water vapor in saturated air) is called the *relative humidity*.

$$\phi = \frac{W_w}{W_{ws}} = \frac{p_w}{p_{ws}} \qquad \text{(Eq. 4.18)}$$

Saturated air has a relative humidity of 100%. If air contains half as much vapor as the amount it would contain at saturation, it has a relative humidity of 50%.

Absolute humidity is just another name for the *humidity ratio.* It is defined as the ratio of the mass of water vapor to the mass of (dry) air.

$$W_w = \frac{\text{mass of water vapor}}{\text{mass of dry air}} = \frac{m_v}{m_a} \qquad \text{(Eq. 4.19)}$$

The humidity ratio is the most widely used measure of air moisture content in HVAC design. It is normally given in units of grams of water per kg of dry air or grains per pound.

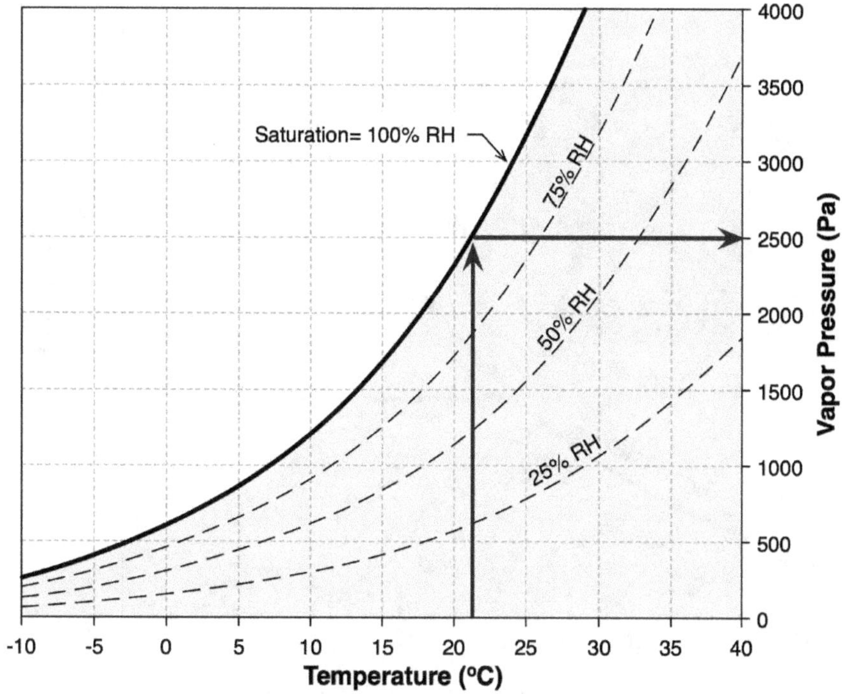

Figure 4.19: Saturation vapor pressure vs. temperature—the psychrometric chart

A *psychrometric chart* is a plot of air, it moisture content, and some other physical properties. Versions of the *psychrometric chart* that can be found in many heating, ventilating, and air-conditioning textbooks. There are many versions of psychrometric charts: some variations use different units and many add additional information such as density and enthalpy.

For building science applications it is useful to have a chart that measures air moisture content in the partial pressure of water vapor. Figure 4.19 plots the moisture content of the air in terms of vapor pressure versus air temperature and relative humidity. For example, the chart can be used to find the saturation vapor pressure of air at 21 C: around 2500 Pa.

Figure 4.20 is a simplified psychrometric chart in moisture units of humidity ratio, that is, the ratio of grams of water vapor per kilogram of dry air. One use of a psychrometric chart is to quantitatively describe the processes of heating, cooling, humidifying and dehumidifying air. For building science applications, a primary use of this chart is to predict, without calculations, at what temperature condensation can occur and even how much.

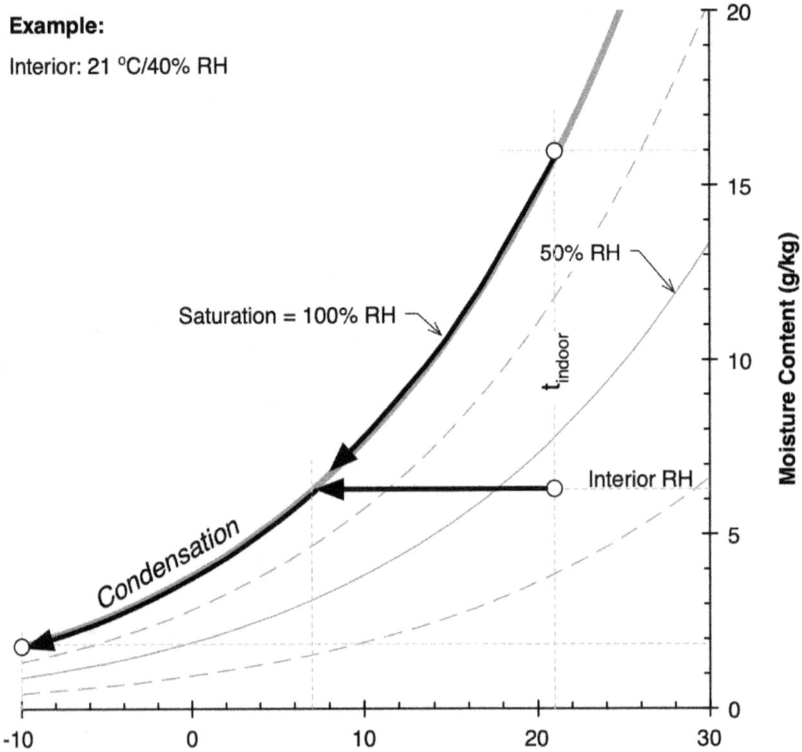

Figure 4.20: Cooling and condensation

The chart can be used to show that one kilogram of saturated (100 % RH) air at room temperature (21 °C) holds 16 g of water vapor and that saturated air at -10 °C contains just less than 2 g/kg. Hence, the chart can show that cooling one kg of saturated air from 21 °C to –10 °C will result in condensation and (16 – 2) = 14 grams of condensate.

Room temperature air at 21 °C and 40 % RH holds exactly 40% as much water vapor as saturated air at this temperature (i.e., 6.25 g/kg). Cooling this air will not cause condensation immediately. Following a horizontal line on the psychrometric chart toward the left (cooler temperatures) shows that as the temperature of this parcel of air drops, its relative humidity will climb. Eventually, when the horizontal line intersects the saturation line, condensation will occur. This important temperature point is called the *dewpoint temperature* and is found by dropping a vertical line to the temperature axis (Figure 4.20). For air with 40 % RH at 21 °C (70 °F), the dewpoint temperature is about 7 °C (45 °F).

The dewpoint temperature of a sample of air is an important measure of the absolute moisture content Two different samples of air with different temperatures and relative humidities but with the same absolute moisture content will have the same dewpoint temperature. These other samples can lie anywhere along the same horizontal line on the psychrometric chart.

Condensation will occur on any surface that is colder than the dewpoint temperature of the air around it. This important fact has a great many practical building science applications. Rather than using a chart, the dewpoint temperature can be estimated from:

$$t_d = \frac{4030}{18.689 - \ln(p_w)} - 235 \quad [°C] \qquad \text{(Eq. 4.20)}$$

It is often more practical to use the mass form of the ideal gas law (Eq. 4.13) instead of the mole form widely used in chemistry:

$p_w \cdot V = w \cdot R_{wv} \cdot T$

where p_w is the partial water vapor pressure

w is the mass of water vapor (g),

R_{wv} is the water vapor specific gas constant (0.4615 J / g·K),

T is *absolute* temperature (K), and

V is the volume of the sample (m³).

Re-arranging this equation results in a relationship for calculating the water vapor pressure of a sample:

$$p_w = \frac{w \cdot R_{wv} \cdot T}{V} \qquad \text{(Eq. 4.21)}$$

Combining equations gives (p_t is total atmospheric pressure, about 101 300 Pa): the ideal gas law for air and the one for water vapor

$$W_w = \frac{0.622 \cdot p_w}{p_t - p_w} \qquad \text{(Eq. 4.22)}$$

The density of air is often an important variable, and can be calculated from the ideal gas law or from simple correlations.

$$\rho = \frac{352}{T} + \frac{344.84}{T^2} \qquad \text{(Eq. 4.23)}$$

where ρ is the density of air (kg/m^3), and

T is the temperature (K).

Finally the heat capacity of dry air and the enthalpy of moist air can be found from

$$c_p = 1030 - 0.2 \cdot T + 0.0004 \cdot T^2 \text{ in J/kg·K} \qquad \text{(Eq. 4.24)}$$

$$h = 1.006\,t + W \cdot (2501 + 1.805\,t) \text{ in [J/g]} \qquad \text{(Eq. 4.25)}$$

The equations required to calculate most useful moist air properties are summarized in Table 4.4.

Table 4.4: Summary of approximate psychrometric equations

Property	Equation	Eqn #
Partial water vapor pressure	$p_w = \frac{w \cdot R_{wv} \cdot T}{V}$ [consistent units]	4.21
Density of dry air	$\rho = \frac{352}{T} + \frac{344.84}{T^2}$ [kg/m^3 ,K]	4.23
Dewpoint temperature	$t_d = \frac{4030}{18.689 - \ln(p_w)} - 235$ [°C, Pa]	4.20
Saturation vapor pressure	$p_{ws} = 1000 \cdot e^{(52.58 - \frac{6790.5}{T} - 5.028 \ln T)}$ [Pa, K]	4.17
Relative humidity	$\phi = \frac{W_w}{W_{ws}} = \frac{p_w}{p_{ws}}$ [consistent units]	4.18
Humidity ratio	$W_w = \frac{0.622 \cdot p_w}{p_t - p_w}$ [kg/kg, Pa]	4.22
Heat capacity (dry air)	$c_p = 1030 - 0.2 \cdot T + 0.0004 \cdot T^2$ [J/kg·K, K]	4.24
Enthalpy	$h = 1.006\,t + W \cdot (2501 + 1.805\,t)$ [J/g]	4.25

Some of the listed equations are not necessarily exact, but they provide sufficient accuracy for most building science applications. These equations provide all the information needed to generate one's own psychrometric chart.

Doing so is a useful exercise, since the spreadsheet can be used for a wide range of different calculations.

Error! Not a valid bookmark self-reference. and Table 4.6 reproduce precise moist air properties from the ASHRAE *Handbook of Fundamentals* for several data points.

Table 4.5: Saturated (100%RH) air properties from -40 °C to 60 °C[1]

Temperature [°C]	Dry Density [kg/m³]	Saturation Vapor Pressure [Pa]	Saturation Humidity Ratio [g/kg]	Saturation Humidity Ratio [g/m³]
-40	1.52	13	0.08	0.12
-35	1.48	22	0.14	0.20
-30	1.45	38	0.23	0.34
-25	1.42	63	0.39	0.55
-20	1.40	103	0.63	0.88
-15	1.37	193	1.18	1.62
-10	1.34	288	1.78	2.37
-5	1.32	424	2.62	3.43
0	1.29	615	3.80	4.88
5	1.27	877	5.43	6.83
10	1.25	1235	7.67	9.45
15	1.23	1714	10.70	12.89
20	1.20	2350	14.77	17.37
25	1.18	3182	20.17	23.13
30	1.16	4262	27.31	30.46
35	1.15	5645	36.70	39.70
40	1.13	7402	49.02	51.21
45	1.11	9610	65.18	65.45
50	1.09	12362	86.43	82.89
55	1.08	15760	114.6	104.1
60	1.06	19930	152.3	129.6

[1] Source: ASHRAE *Handbook of Fundamentals*, 2021

Table 4.6: Dewpoint temperatures of moist air from -20 °C to 40 °C

Temperature [°C]	80%RH	60%RH	40%RH	20%RH
-20	-22.3	-25.1	-29.0	-35.4
-15	-17.4	-20.4	-24.5	-31.1
-10	-12.5	-15.6	-19.9	-26.8
-5	-7.6	-10.8	-15.3	-22.5
0	-2.7	-6.1	-10.7	-18.2
5	1.9	-1.8	-6.6	-14.3
10	6.8	2.7	-2.6	-10.6
15	11.7	7.4	1.6	-6.8
20	16.5	12.1	6.1	-3.1
25	21.4	16.8	10.6	0.5
30	26.2	21.5	15.0	4.7
35	31.1	26.1	19.5	8.8
40	35.9	30.8	23.9	12.9

4.7 Worked psychrometric examples

Example: Calculate the density of air at 20 °C and 75 % RH and compare the value with that for dry air at the same temperature. Will air at 21 °C and 25 % RH rise in a room filled with 20 °C and 80 % RH air?

Solution: Air will rise if it is less dense than the surrounding fluid. Hence, one needs to calculate the density of the air at the two different conditions. The mass of the water vapor and the mass of the dry air both needed to be calculated.

- Water vapor saturation vapor pressure at 20 °C (Equation 4.17):

  $$p_{ws} = 1000 \cdot e^{\left(52.58 - \frac{6790.5}{T} - 5.028 \ln T\right)}$$

  $$p_{ws} = 1000 \cdot e^{\left(52.58 - \frac{6790.5}{293.15} - 5.028 \ln 293.15\right)}$$

  $$p_{ws} = 2348.1 \text{ Pa}$$

- Partial pressure of water vapor at 75 % RH

  $$p_w = \emptyset \, p_{ws}$$

  $$p_w = 75\% \cdot 2348.1 \text{ Pa}$$

  $$p_w = 1761.1 \text{ Pa}$$

- Density of water vapor, using ideal gas law with R_{wv} = 461.5 J/(kg·K)

$$\rho_w = \frac{p_w}{R_{wv} \cdot T}$$

$$\rho_w = \frac{1761.08 Pa}{461.5 \cdot \frac{J}{kg \cdot K} 293.15 K}$$

$$\rho_w = 0.01302 \text{ kg/m}^3$$

The partial pressure of the dry air in the mix equals total air pressure less water vapor pressure. Assuming standard atmospheric pressure of 101 325 Pa:

- Density of dry air, using ideal gas law, R_{air} = 287.1 J/(kg·K), is

$$\rho_{air} = \frac{p_{air}}{R_{air} \cdot T}$$

$$\rho_{air} = \frac{p_{total} - p_w}{R_{air} \cdot T}$$

$$\rho_{air} = \frac{101325 - 1761}{287.1 \cdot 293.15}$$

$$\rho_{air} = 1.1830 \text{ kg/m}^3$$

- The total density is the sum of the dry air and the water vapor:

$$\rho_{mixture} = \rho_{air} + \rho_w$$

$$\rho_{mixture} = 1.1830 \text{ kg/m}^3 + 0.01302 \text{ kg/m}^3$$

$$\rho_{mixture} = 1.1960 \text{ kg/m}^3$$

Dry air at the same temperature has a density of:

$$\rho_{air} = \frac{101325}{287.1 \cdot 293.15}$$

$$\rho_{dry\text{-}air} = 1.2039 \text{ kg/m}^3$$

Thus, dry air at a given total pressure and temperature has a higher density than moist air at the same pressure and temperature. The water molecules are lighter than diatomic oxygen and nitrogen (the major constituents of air), so the addition of water molecules to a mixture at a constant pressure somewhat counter intuitively reduces the density.

- Now, as before, find the densities of the 21 °C and 25 % RH sample.

$$p_{ws} = 2497.4 \text{ Pa}$$

$$p_w = 25\% \cdot 2497.4 \text{ Pa}$$

$$p_w = 624.3 \text{ Pa}$$

$$\rho_{mixture} = \frac{624.3}{461.5 \cdot 294.15} + \frac{101325 - 624.3}{287.1 \cdot 294.15}$$

$\rho_{mixture} = 1.1970 \text{ kg/m}^3$

- For air at 20 °C and 80% RH, use p_{ws} from before,

$p_w = 80\% \cdot 2348.1 \text{ Pa}$

$p_w = 1878.5 \text{ Pa}$

$$\rho_{mixture} = \frac{1878.5}{461.5 \cdot 293.15} + \frac{101325 - 1878.5}{287.1 \cdot 293.15}$$

$\rho_{mixture} = 1.1955 \text{ kg/m}^3$

As air at 21 °C and 25 % RH is more dense than air at 20 °C and 80 % RH, the air will sink.

Example: A 2 g balloon contains 1.75 litres of air at 25 °C and 5 % RH at an air pressure difference of 3000 Pa across its rubber skin. Will it float in a room with air at 25 °C and 80 % RH?

Solution: Using the same approach as before the vapour pressure, and ambient air density at 25°C and 80 % RH can be calculated

$p_w = 80\% \cdot 3180.5 = 2544.4 \text{ Pa}$

$$\rho_{mixture} = \frac{2544.4}{461.5 \cdot 298.15} + \frac{101325 - 2544.4}{287.1 \cdot 298.15}$$

$\rho_{mixture} = 1.1725 \text{ kg/m}^3$

Assuming that the pressure inside the room is 101,325 Pa, the pressure in the balloon will be 101,325 + 3000 = 104,325 Pa. The density of the air inside the balloon can then be calculated as:

$p_w = 5\% \cdot 3180.5 \text{ Pa}$

$p_w = 159.0 \text{ Pa}$

$$\rho_{mixture} = \frac{159.0}{461.5 \cdot 298.15} + \frac{104325 - 159.0}{287.1 \cdot 298.15}$$

$\rho_{mixture} = 1.2181 \text{ kg/m}^3$

- The force on the balloon is calculated using Archimedes' principle, the difference between the mass of the air in the balloon plus the mass of the ballon and the mass of the room air displaced generates buoyancy.

$F_{net} = (m_{balloon} + m_{air\ in\ ballon} - m_{room\ air\ displaced}) \cdot g$

$F_{net} = [0.002 \text{ kg} + (1.2181 \text{ kg/m}^3 - 1.1725 \text{ kg/m}^3) \cdot 0.00175 \text{ m}^3] \cdot g$

$F_{net} = 0.0204 \text{ N}$ [downwards] with $g = 9.81 \text{ m/s}^2$

Thus, the gravity force on the balloon exceeds the buoyant forces and the balloon will sink.

Example: Calculate the mass flow rate of 20 lps of air at 25 °C and -10 °C.

Solution: Since no information is given regarding the relative humidity, we will assume the air is dry. (This assumption simplifies the calculation, and as can be seen above, moisture adds little to the mass in most cases. As with most engineering calculations, one needs to check assumptions for each case.)

The density of air at 25 °C air is:

$$\rho = \frac{352}{T} + \frac{344.84}{T^2}$$

$$\rho = \frac{352}{298.15} + \frac{344.84}{298.15^2}$$

$$\rho_{air} = 1.1845 \text{ kg/m}^3$$

At a volumetric flow rate of 20 lps, the *mass* flow per unit time (θ) would be:

$$\frac{dm}{d\theta} = 1.1845 \text{ kg/m}^3 \cdot 20 \text{ l/s} = 1.1845 \text{ kg/m}^3 \cdot 0.020 \text{ m}^3\text{/s}$$

$$\frac{dm}{d\theta} = 0.0237 \text{ kg/s}$$

Similarly, for a mass flow rate for air at -10 °C and 20 lps

$$\rho_{air} = 1.3426 \text{ kg/m}^3$$

$$\frac{dm}{d\theta} = 1.3426 \text{ kg/m}^3 \cdot 20 \text{ l/s} = 0.0269 \text{ kg/s}$$

These calculations show that, as expected, for equal volumetric flow rates the one at a cold temperature will have a higher mass flow rate than that for the same volumetric flow rate at higher temperatures, all other factors being equal.

Example: Calculate the volumetric flow rate of 1 kg/s air at 25 °C and -10 °C.

Solution: Volumetric flow rate for air at 25 °C and 1 kg/s

$$\rho_{air} = 1.1845 \text{ kg/m}^3 \text{ at } 25\text{ °C}$$

$$\frac{dV}{d\theta} = (1.1845 \text{ kg/m}^3)^{-1} \cdot 1 \text{ kg/s}$$

$$\frac{dV}{d\theta} = 0.844 \text{ m}^3\text{/s} = 844 \text{ liters per second.}$$

Volumetric flow rate for air at -10 °C and 1 kg/s

$$\rho_{air} = = 1.3426 \text{ kg/m}^3 \text{ at } -10\text{ °C}$$

$$\frac{dV}{d\theta} = [1.3426 \text{ kg/m}^3]^{-1} \cdot 1 \text{ kg/s}$$

$$\frac{dV}{d\theta} = 0.745 \text{ m}^3\text{/s} = 745 \text{ l/s.}$$

Example: Calculate the mass flow rate of moisture from a room at 21 °C and 40 % RH to outdoors at -5 °C and 80 % RH if 10 liters per second leave the room. Repeat for inward airflow, indoor conditions of 25 °C and 60 % RH and outdoor conditions of 32 °C and 80 % RH.

Solution: This can be solved by multiplying the difference in moisture content between inside and outside by the mass flow rate.

Vapor pressure at 21 °C and 40 % RH:

$$p_{ws} = 1000 \cdot e^{(52.58 - \frac{6790.5}{T} - 5.028 \ln T)}$$

$$p_{ws} = 2497.4 \text{ Pa}$$

$$p_w = 40\% \cdot 2497.4 \text{ Pa}$$

$$p_w = 999 \text{ Pa}$$

The moisture content of this air is:

$$W_w = \frac{0.622 \cdot p_w}{p_t - p_w}$$

$$W_{in} = \frac{0.622 \cdot 999}{101\,325 - 999}$$

$$W_{in} = 0.00619 \text{ kg/kg or } 0.62 \text{ g vapor /kg air}$$

The mass flow rate of 10 liters/second of indoor air is:

$$P_w = 999 \text{ Pa}$$

$$\rho_{mixture} = \frac{999}{461.5 \cdot 293.15} + \frac{101325 - 999}{287.1 \cdot 293.15}$$

$$\rho_{mixture} = 1.199 \text{ kg/m}^3$$

In practice we usually ignore the contribution of water vapor to density and would be more likely to calculate the density of dry air. The impact of this assumption is small.

$$\frac{dm}{d\theta} = 1.199 \text{ kg/m}^3 \cdot 10 \text{ l/s} \cdot 1 \text{ m}^3/1000 \text{ l}$$

$$= 0.012 \text{ kg/s}$$

Vapor pressure at -5 °C and 80% RH

$$p_{ws} = 1000 \cdot e^{(52.58 - \frac{6790.5}{T} - 5.028 \ln T)}$$

$$p_{ws} = 424 \text{ Pa}$$

$$p_w = 80\% \cdot 424 \text{ Pa}$$

$$p_w = 339 \text{ Pa}$$

The moisture content of this air is:

$$W_w = \frac{0.622 \cdot p_w}{p_t - p_w}$$

$$W_{w,out} = \frac{0.622 \cdot 339}{101325 - 339}$$

$$W_{out} = 0.00209 \text{ kg/kg}$$

Net moisture transfer to the outdoors is

$$\text{Moisture transferred} = (W_{w,in} - W_{w.out}) \cdot \frac{dm}{d\theta}$$

$$= (0.006193\text{- } 0.00209)\ (0.0120)$$

$$= 4.921.10^{-5} \text{ kg/s or } 177 \text{ g/hr}$$

The moisture transfer is about 177 g/hr (almost a full cup per hour). Clearly a significant accumulation of water can leave the building as a result of air leakage. In some cases this moisture could condense within the enclosure and become a significant source of water and its attendant problems.

Example: Repeat the previous calculation for 25 °C / 60 % RH indoors (summer conditions) and 32 °C / 80 % RH outdoor conditions (hot, humid weather).

Solution: Follow the procedure above

Vapor pressure at 25 °C and 60 % RH:

$$p_{ws} = 1000 \cdot e^{(52.58 - \frac{6790.5}{T} - 5.028 \ln T)}$$

$$p_{ws} = 1000 \cdot e^{(52.58 - \frac{6790.5}{298.15} - 5.028 \ln 298.15)} = 3180 \text{ Pa}$$

$$p_w = 60\% \cdot 3180 \text{ Pa} = 1908 \text{ Pa}$$

The moisture content of this air is:

$$W_w = \frac{0.622 \cdot p_w}{p_t - p_w}$$

$$W_{in} = \frac{0.622 \cdot 1908}{101325 - 1908} = 0.0119 \text{ kg/kg}$$

Vapor pressure at 32 °C and 80% RH:

$$p_{ws} = 1000 \cdot e^{(52.58 - \frac{6790.5}{305.15} - 5.028 \ln 305.15)} = 4772 \text{ Pa}$$

$$p_w = 80\% \cdot 4772 \text{ Pa} = 3818 \text{ Pa}$$

The moisture content of this air is:

$$W_{in} = \frac{0.622 \cdot 3818}{101325 - 3818} = 0.0244 \text{ kg/kg}$$

The density of air at 32 °C air is:

$$\rho = \frac{352}{T} + \frac{344.84}{T^2}$$

$$\rho = \frac{352}{305.15} + \frac{344.84}{305.15^2} = 1.157 \text{ kg/m}^3$$

At a volumetric flow rate of 10 lps, the mass flow rate would be:

$$\frac{dm}{d\theta} = 1.157 \text{ kg/m}^3 \cdot 0.010 \text{ m}^3\text{/s} = 0.0116 \text{ kg/s}$$

Net moisture transfer to the outdoors is

$$\text{Moisture transferred} = (W_{w,in} - W_{w.out}) \cdot \frac{dm}{d\theta}$$

$$\text{Moisture transferred} = (0.0244 - 0.0119) \cdot 0.0116 \text{ kg/s}$$

$$= 1.45 \cdot 10^{-4} \text{ kg/s or } 522 \text{ g/hr}$$

Thus, the moisture transfer in hot, humid summer conditions is about 522 g/hr, almost three times more than the cold weather condition examined. Therefore, similar or larger amounts of moisture can be transferred into a building or building enclosure in hot, humid weather than is the case in cold, dry weather.

4.8 References

ASHRAE, 2021. *Handbook of Fundamentals.* American Society of Heating Refrigeration and Air-conditioning Engineers, Atlanta.

Ashton, H.E., 1970. *Irradiation Effects on Organic Materials* . Canadian Building Digest 121, National Research Council of Canada, Ottawa.

Ashton, H.E., 1969. *Weathering of Organic Building Materials.* Canadian Building Digest 117, National Research Council of Canada, Ottawa.

Cherry, John A., Freeze, Allan R., 1979. *Groundwater.* Prentice-Hall, Englewood Cliffs, N.J.

Franks, Felix. 1984. *Water.* Royal Chemistry Paperbacks.

Glasstone, S., 1960. Lewis, D., *Elements of Physical Chemistry.* van Nostrand & Co., Princeton.

Henninger, J. H., 1984. *Solar absorptance and thermal emittance of some common spacecraft thermal-control coatings.* NASA Reference Report 11, Goddard Space Flight Center, Greenbelt, MD.

Illston, J.M., and Domone, P.L. 2010. *Construction Materials: Their Nature and Behaviour.* 4th ed., CRC Press.

Ashby, M.F. Shercliff, H., Cebon, D., 2007. *Materials: Engineering, Science, Processing and Design.* Butterworth-Heinemann.

Parker, D S, J E R McIlvaine, S F Barkaszi, D J Beal and M T Anello, 2000. *Laboratory Testing of the Reflectance Properties of Roofing Material.* FSEC-CR-670-00. Florida Solar Energy Center, Cocoa, FL.

Pauling, L., 1988. *General Chemistry.* Dover Publications, New York.

Powers, T., Brownyard, T., 1947. "Studies of the Physical Properties of Hardened Portland Cement Paste- Part 8. The Freezing of Water in Hardened Portland Cement Paste," *J. Of Amer. Conc. Inst.*, Vol. 43, April.

Powers, T., 1968. *The Properties of Fresh Concrete.* J. Wiley & Sons, New York.

Steinhagen, H.P., 1977. *Thermal conductive properties of wood, green or dry, from -40° to 100°C : A literature Review.* Forest Products Lab, Madison, WI.

ASTM, 2011. Standard 1980-11. *Standard Practice for Calculating Solar Reflectance Index of Horizontal and Low- Sloped Opaque Surfaces.* ASTM International, West Conshohocken, PA.

Wilson, Alan D., Nicholson, John W., Prosser, Harvard J. editors, 1987. *Surface Coatings 3: Waterborne Coatings.* ed. Elsevier Science Publishers, Essex, England.

4.9 Notes

Chapter 5

Heat Transfer

5.1 Introduction

Understanding heat transfer and the temperature distribution through building materials and assemblies is important for assessing energy use, thermal movements, durability, comfort, and the potential for moisture problems.

Heat flow can be a transient or a steady process. In the transient state, temperature and/or heat flow vary with time. Steady-state heat flow occurs when the temperature and heat flow reach a stable equilibrium condition that does not vary with time. Depending on the problem, the assumption of steady-state conditions may provide sufficiently accurate predictions of actual heat flow and temperature conditions. However, for some problems the assumption of steady flow can result in significant errors that may have practical implications.

Heat flow occurs in one, two, or three dimensions. In almost all real situations, heat flow occurs in three dimensions but, from a practical point of view, it is often acceptable to simplify considerations to only one-dimensional or two-dimensional approximations.

Heat transfer occurs by three primary mechanisms, acting alone or in some combination:

i. conduction,

ii. convection, and

iii. radiation.

Changes in state (especially in refrigerants and moisture), although not strictly a heat transport mechanism, must also be considered since these state changes absorb and release heat energy, i.e., latent heat.

Conduction is the flow of heat through a material by direct molecular contact. This contact occurs within a material or between two materials in contact. It is the most important heat transport mode for solids; it is sometimes important for liquids, and it is occasionally important for gases.

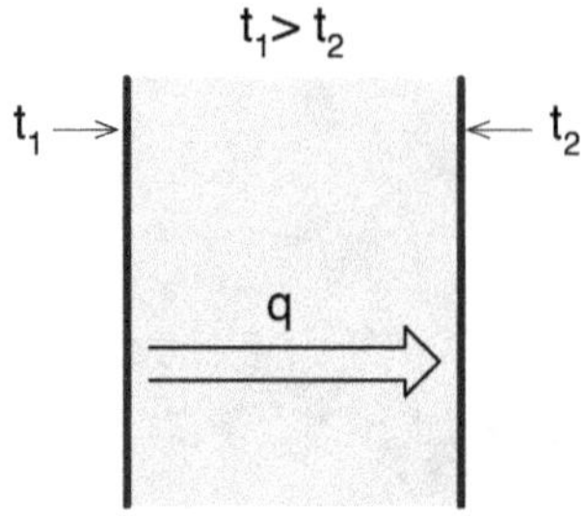

Figure 5.1: Conduction of heat through a solid

Convection is the transfer of heat by the movement or flow of molecules (liquid or gas) with a change in their heat energy content. This is an important heat transfer mode between fluids and solids, or within fluids. (Figure 5.2).

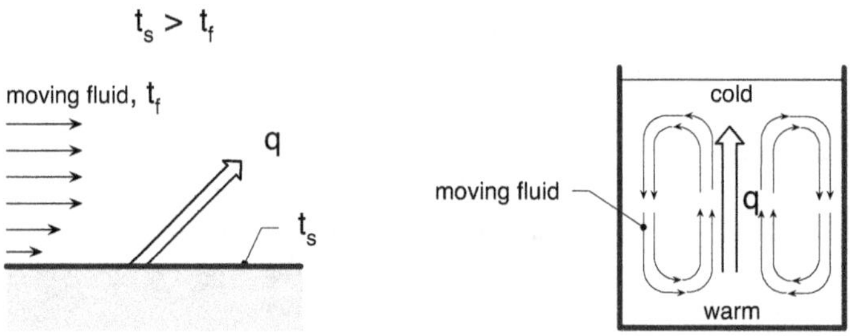

Figure 5.2: Convective heat transfer from a solid to a fluid, and within fluids

Radiation is the transfer of heat by electromagnetic waves through a gas or vacuum. Heat transfer by this mode therefore requires a clear line of sight connection between the surfaces involved. All objects above absolute zero radiate heat energy; it is the net radiative heat transfer between objects of different temperature that is the heat transfer of interest. Radiation is mostly of importance for heat transfer between solids across an air space or vacuum and within the voids of highly porous solids, but radiation between high-temperature gases is occasionally of practical importance.

State change, sometimes called phase change, occurs at a constant temperature but still entails the movement of energy (see Chapter 4). For example, evaporation of moisture absorbs energy at a surface and condensation on a surface releases energy. This energy is also called latent heat.

The mode of heat transfer often changes during the process of heat flow through and within building systems. Consider the following examples.

1. The sun transmits heat by radiation to the earth where it can be absorbed at the outer surface of a solid masonry wall. The heat absorbed is transported by conduction through the masonry and then transferred to the indoor air by convection and to the indoor surfaces by radiation.

2. Multi-layer assemblies have more complex chains: if the enclosure were a masonry veneer over a fiberglass-batt-insulated wood frame wall, the heat from the masonry would be transferred to the wood frame wall sheathing across the airspace by convection and radiation, heat flow through the sheathing would be by conduction, and flow through the batt insulation by some combination of conduction and radiation, followed by conduction through the gypsum wallboard interior and convection and radiation at the interior surface.

3. If the brick were wet from the previous summer day's rainstorm, then a significant portion of the absorbed solar energy would be used to evaporate (change the phase of) water, which would move outwards by convective mass transport and inwards by diffusion to the cooled inner wythes of the assembly, where it may condense and release energy. The same heat transfer modes as example 2 are also operating in parallel.

5.2 Terms, definitions, and units

Celsius temperature measures temperature based on a scale devised in 1742 by Anders Celsius, a Swedish astronomer. The Celsius scale divides the range from the freezing point of water (0) to the boiling point (100) into one hundred units — hence the (outdated) name "Centigrade" for this scale.

Cold is a relative term used to describe a condition of less heat, for instance a body with a lower temperature than others; in some respects, it is used to signify an absence of heat.

Conduction is the transfer of heat by direct molecular contact. This contact can be within a material or between two materials in contact.

Convection is the transfer of heat by the movement of a gas or liquid. When a mass moves, it takes its heat with it.

Energy (E) is, in a general sense, the capacity to do work.

SI:	Joules (J)
I-P:	British Thermal Unit (Btu)
Conversion:	1 Btu = 1055 J

Fahrenheit temperature refers to temperatures measured on a scale devised by Daniel Fahrenheit, the German physicist who invented alcohol and mercury thermometers in the early 18th century and who devised the temperature scale named after him. On the Fahrenheit scale, the freezing point of water is 32 and the boiling point is 212, likely based on 0 as the freezing point of a brine made of equal parts salt and water and 180 subdivisions between freezing and boiling of water.

Heat (E) is a form of energy associated with the vibratory movement of molecules. Units are as for energy.

Heat capacity (c) is a measure of the amount of heat that a unit mass of dry material[1] can store. It is defined as the heat energy required to raise the temperature of 1 g of material by 1 Kelvin. Some codes use the symbol HC.

SI:	kJ / (kg K)
I-P:	Btu / (hr· ft^2)
Conversion:	Btu / (hr · ft^2) = 3.1513 W/m^2

Heat flow rate (Q) is the time rate at which heat energy moves, and hence power. Heat flow can be based on a system (e.g., a building), or a component of the system (e.g., the building enclosure or window).

SI:	J/s, Watts (W), or kilowatts
I-P:	Btu / hr
Conversion:	Btu / hr = 0.29287 W
	W = 3.4145 Btu / hr

Heat flux (q) is the area density of the heat flow rate, i.e., the rate of heat flow per unit area. q = Heat flow/ Area = Q/A. The area referenced is usually perpendicular to the direction of heat flow.

SI:	W/m^2 = J /(s·m^2)
I-P:	Btu / (hr· ft^2)
Conversion:	Btu / (hr · ft^2) = 3.1513 W/m^2
	W /m^2 = 0.3173 Btu / (hr· ft^2)

Heat transfer is the movement of heat energy, always from a higher temperature to a lower temperature.

Kelvin temperature refers to temperatures on an absolute scale with the same gradations as the Celsius scale but with its origin at absolute zero, -273.15°C.

Overall heat transfer coefficient (U), U-factor, or **thermal transmittance** is a system measure of the amount of heat flow that will occur across a unit area of an enclosure system or other assembly for a unit temperature difference, i.e., a system conductance that typically includes both surface films. In Europe, k or "k-wert" is the symbol used.

SI:	W/m^2 = J /(s·m^2)
I-P:	Btu / (hr· ft^2)
Conversion:	Btu / (hr · ft^2) = 3.1513 W/m^2

[1] Moisture can significantly impact heat capacity of a material sample, and hence is accounted for seperately.

Radiation is the process of heat transfer from one body to another by electromagnetic radiation. No transfer medium (solid, liquid, or gas) is required.

Rankine is an absolute scale that uses the same gradations as the Fahrenheit scale but uses as its reference absolute zero, -459.69 °F.

Temperature (T or t) is a measure of the level of heat energy or the vigor of the molecular movement. Zero degrees Kelvin or Rankine represents absolute zero, i.e., no heat energy.

SI:	Celsius (°C) or Kelvin (K)
I-P:	Fahrenheit (°F) or Rankine (°R)
Conversion:	°C = (5/9) · (°F 32)
	K = °C + 273.15 or °F = (9/5) °C +32
	Rankine, °R = °F + 459.7

Thermal conductivity (k or λ) is a material property that indicates the quantity of heat flow across a unit area, through a unit thickness for a temperature gradient of one degree. The symbol λ is used in Europe.

SI:	W / (m · K)
I-P:	Btu · in /(hr · ft^2 · °F) or Btu/(hr · ft · °F)
Conversion:	0.144 Btu · in /(hr · ft^2 · °F) = 1 W/(m · K)
	6.9334 W / (m · K) = Btu · in/(hr · ft^2 · °F)
	1 Btu /(hr· ft ·°F) = 1.730735 W/(m·K)

Thermal conductance, C-factor is the time rate of heat flow through a unit area of a body induced by a unit temperature difference, or the conductivity of a material for a given thickness.

SI:	W / (m^2 · K)
I-P:	Btu / (hr · ft^2 · °F)
Conversion:	5.678 Btu / (hr · ft^2 · °F) = 1 W / (m^2 · K)

Thermal resistance (R-value) is the reciprocal of conductance. In many applications, this is an effective value that includes all heat transport modes. The units of R-value are usually I-P in North America, but the user should always confirm of SI is intended[2]:

Resistance = 1 /C = thickness / conductivity = l/k, (or l/λ).

SI:	RSI (m^2 · K) / W
I-P:	R (hr · ft^2 · °F) / Btu
Conversion:	RSI = R / 5.678
	R =5.678 RSI

Time (θ) is usually measured in units of seconds, but hours are often substituted in practice for lower intensity processes such as heat loss through the enclosure. The symbol θ is often used instead of the symbol "t" to avoid confusion with temperature.

[2] RSI is used in Canada to be explicit about the units, but this is not common elsewhere.

5.3 Conduction

The mostly commonly used fundamental heat flow relationship was first defined by Fourier as:

$$q = -k \cdot \text{grad } T \qquad \text{(Eq. 5.1)}$$

Along one dimension, as shown in Figure 5.3, this reduces to:

$$q_x = -k_x \cdot dT/dx \text{ , and}$$

$$q_x \cdot A = Q_x = - k_x \cdot A \cdot dT/dx. \qquad \text{(Eq. 5.2)}$$

The heat flux, q_x (W or J/s), in the x direction across the perpendicular y-z plane, is proportional to the temperature gradient along the x-axis. The coefficient, k_x (W / m·K), is defined as the *thermal conductivity* along this axis. To calculate flow, flux is of course multiplied by area. Although Fourier's law is precisely correct only for heat flow by conduction, its simplicity has encouraged its use for radiation, convection, and combined modes as well when possible.

The thermal conductivity in Equation 5.1 is assumed to be constant — this assumption is often a reasonable simplification for many building science applications. The variation of conductivity with temperature is often significant for many aerospace and industrial applications which operate over wider temperature ranges and may need to be accounted for in higher accuracy building applications. The negative sign is a consequence of the fact that heat flows from high temperatures to low temperatures. The same equation can be rewritten for the other two Cartesian directions or, if useful, in polar coordinates.

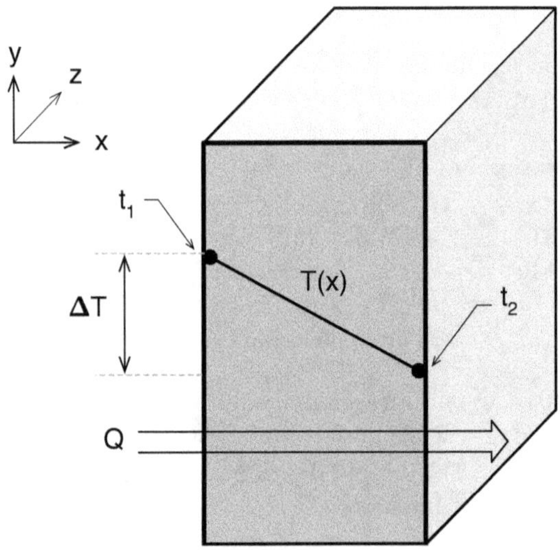

Figure 5.3: One-dimensional steady-state conductive heat transfer

In many practical building science applications, Fourier's law can be written as:

$$Q = A \cdot k / l \cdot (T_1 - T_2) \qquad \text{(Eq. 5.3)}$$

where Q is the rate of heat flow (W),

A is the area (m^2),

k is the thermal conductivity (W / m K),

l is the length of the heat flow path or thickness of the material (m), and

T_1 and T_2 are the temperatures on either side of the material of interest.

If one is examining a unit area of a building material, the equation can be further reworked to give the rate of heat flow per unit area, the *heat flux*:

$$q = Q/A$$
$$= k / l \cdot (T_1 - T_2) \qquad \text{(Eq. 5.4)}$$

The term k/l is the *thermal conductance*, and given the symbol C. Therefore, yet another form of the equation is:

$$q = C \cdot (T_1 - T_2) \qquad \text{(Eq. 5.5)}$$

We are often concerned with the ability of a material to resist flow rather than to conduct it. For example, insulation materials in the North American building industry are normally sold with a quoted thermal resistance or R-value. This value has the decided advantage that as the resistance to heat flow increases, the measure also increases. It is defined as the reciprocal of the conductance, so:

$$q = C \cdot (T_1 - T_2)$$
$$= 1/R \cdot (T_1 - T_2) \qquad \text{(Eq. 5.6)}$$

Note that, in numerical terms, $\Delta T = (T_1 - T_2) = (t_1 - t_2)$ and hence one can use either degrees Celsius, Fahrenheit, Kelvin when describing a temperature difference.

Table 5.1 lists many common building materials and their thermal conductivity at conditions representative of building enclosure applications. It is based on a number of sources, including the *ASHRAE Handbook of Fundamentals* [2021], ISO Standard 10456 [2007], and laboratory testing at the authors' lab.

The thermal performance of some "materials" is actually composed of several different materials and cannot be expressed in these terms. For example, hollow concrete masonry units are actually made of concrete with large voids of air. For these types of complex materials, a thermal conductance is listed in Table 5.1.

Table 5.1: Thermal conductivity of some common building materials

Material	Density (kg/m³)	Range of conductivity (W/m K)	Range of conductance (W/m² K)
Board / sheet products			
Plywood	400 - 600	0.08 - 0.11	
OSB	575 - 725	0.09 - 0.13	
Fibre cement compressed	1900	0.56	
Wood Fiberboard	230 - 565	0.05 0.09	
Glass-mat faced gypsum board	640 - 740	0.19 - 0.31	
High density fibreboard	875	0.135	
Wood particleboard	660	0.12	
Strawslab	260 - 350	0.085 - 0.11	
Corrugated metal deck			negligible
Finishes			
Ceramic tiles	2300	1.3	
Acoustic tiles - cellulose	262 329	0.048 0.053	
Acoustic tiles - glassfiber	51	0.034	
Gypsum wall board	600 - 800	0.15 - 0.21	
Sand plaster / lath		0.71	
Gypsum plaster / lath	1000	0.40	
Sand: cement plaster	1570	0.53	
Gypsum plaster w/perlite	720	0.22	
Gypsum plaster w/sand	1600	0.8	
Carpet fibrous underlay			2.73
Carpet rubber underlay			4.42
Terrazzo		1.8	
Hardwood flooring		0.16	
Siding / cladding			
Hardboard siding	640	0.094	
Wood siding - lap		0.1 - 0.12	

These values for conductivity, like most published data, can vary with manufacturing methods, temperature and moisture conditions, and even the test method employed. Some of the factors influencing thermal conductivity are discussed in Section 5.7.

Table 5.1: Thermal conductivity of some common building materials (continued)

Material	Density (kg/m³)	Range of conductivity (W/m K)	Range of conductance (W/m² K)
Plywood sheet siding	600	0.09	
Face brick – clay	1800 - 2200	1.0 - 1.4	
Face brick - concrete	2200	1.9	
Metamorphic stone	2600 - 3000	3.0 - 3.5	
Sedimentary stone	2200 - 2600	2.1 - 2.3	
Hollow profile vinyl / metal siding			8 to 9
Fibre cement sheet	1200 - 1400	0.14 - 0.30	
Metal - flush installed		40 - 80	negligible
Cement stucco	1750 - 2020	0.365 - 0.495	
Structural materials			
Softwood lumber	400 - 600	0.08 - 0.14	MC dependent
Hardwood lumber	720	0.15 - 0.18	
Cedar logs and lumber		0.098 - 0.12	
Concrete	2400	1.4 - 2.5	
Concrete (limestone)	1920	1.1 - 1.3	
Concrete (light)	1300	0.5 - 0.7	
Aerated concrete	400	0.05 - 0.15	
Aerated concrete	600	0.15 - 0.2	
Carbon steel	7680	42 - 65	50 recommended
Stainless Steel (300)	8000	13 - 16	
Stainless Steel (400)	8000	22 - 27	
Aluminum	2800	160 - 200	
Cement mortar	1800	0.7 - 1	
Adobe	1400 - 1800	0.4 - 0.8	
Concrete block 200 mm 50%	1840		5.2 normal
Concrete block, 200 mm 50%	1360		3.9 lightweight
Concrete block, 100 mm	1840		6.1
Hollow clay tile, 100 mm			5.5
Clay straw (function of density)	600 - 1400	0.15 - 0.5	
Durisol/Nexcem	400 - 500	0.072 - 0.085	

Table 5.1: Thermal conductivity of some common building materials (continued)

Material	Density (kg/m^3)	Range of conductivity (W/m K)	Range of conductance (W/m^2 K)
Insulations			
Aerogel	110 - 165	0.013 - 0.015	
EPS low-density	14 - 22	0.40 - 0.036	
EPS Graphite infused	18 - 29	0.038 - 0.0345	
EXPS Type 3 and 4		0.029	
Fibreglass batt insulation	12 28	0.036 - 0.048	
Stonewool batt	38 44	0.036 - 0.038	
Stonewool roof insulation	160	0.035	
Rigid polyurethane	28 34	0.024	
Rigid polyisocyanurate	26 36	0.022 - 0.029	
Phenolic foam (closed cell)		0.017 - 0.024	
Urea formaldehyde		0.031 - 0.032	
Low-density wood fiberboard	50 - 230	0.038 0.050	
Cellulose fiber	36 - 64	0.038 - 0.046	
Sawdust	145 - 160	0.05 - 0.08	
Strawbale	120 - 200	0.06 - 0.075	
Cork	100 - 150	0.041 - 0.045	
Sprayed asbestos		0.05	
Vermiculite, exfoliated	64 130	0.06 - 0.07	
Perlite, expanded	800	0.2 - 0.26	
Perlite, expanded	320	0.07 - 0.08	
Perlite, bonded/expanded	16	0.052	
Peat slab	240 - 480	0.058 - 0.101	
Jute resin bonded	420	0.065	
Eel grass batt	145 - 215	0.043 - 0.049	
Sheep's wool, fluffy	50	0.045	

Table 5.1: Thermal conductivity of some common building materials (continued)

Material	Density (kg/m³)	Range of conductivity (W/m K)	Range of conductance (W/m² K)
Roof materials			
Built-up bitumen roofing	1200	0.17	
Asphalt waterproofing	2100	0.70	
Asphalt shingles			12.9
Wood shingles			6
Crushed stone		1.60	
Thatch - straw	240	0.07	
Thatch - reed	270	0.09	
Other			
Fresh snow	190	0.19	
Compacted snow	400	0.43	
Ice at -1 °C and -20 °C	920	2.24 - 2.45	
Water at 20 °C	1000	0.60	
Earth, dry to damp	1400 - 2000	0.80 - 2.00	
Sand dry, loose	1400	0.30 - 0.80	
Sand and gravel, dense	1700 - 2200	1.8 - 2.20	
Air, still (conduction only)	1.2	0.027	
Argon, still (conduction only)	1.7	0.016 - 0.018	
Krypton (still, conduction only)	3.7	0.008 - 0.010	
Glass, soda lime	2500	0.80 - 1.00	
Copper	8960	380	
Lead	12250	35	
Bronze	8150	64	
Brass	8730	120	
Boron Arsenide	5220	1 300	
Polyetheylene (MDPE/HDPE)	950	0.20 0.40	

Notes:
1. Some C values suggest a thermal property that is practically independent of thickness.
2. For most building materials, the conductivity values have been calculated from the measured conductances of a single thickness. Do not extrapolate beyond reasonable limits.

5.4 Convection

One important mode of heat transfer in practice occurs when heat transfers between a moving fluid (gas or liquid) and another material when the two are at different temperatures. Sir Isaac Newton experimented with items heated in a forge and left to cool outdoors. He found that the temperature of the material approached the outdoor air temperature at an exponentially decaying rate. He demonstrated that the heat flux from a body to its surroundings due to convection is directly proportional to the temperature difference for small temperature differences. Newton's Law of Cooling is usually written:

$$Q = h_c \cdot A \cdot (T_s - T_\infty) \qquad \text{(Eq. 5.7)}$$

where Q is the heat flow rate (W),

h_c, is a proportionality constant, the *convective heat transfer coefficient* (W / $m^2 \cdot K$),

A is the surface area (m^2),

Ts is the surface temperature (K or °C)

T_∞ is the temperature of the fluid at a distance far enough from the surface that it is not to be affected by the heat flow.

The convective heat transfer coefficient is the property of an interface, not a material. While the coefficient is labeled convective, it is more properly a convection-conduction coefficient since conduction is always acting (between the solid to the fluid) at the same time as convection but is usually much smaller in magnitude and difficult to decouple from convection.

For building science purposes, the value of h_c typically can be assumed to be independent of the size of the temperature difference but, for special situations and especially during intense heating and cooling, and evaporation and condensation, it may be necessary to consider h_c as a function of temperature.

Just as for conduction, we can write Newton's Law of Cooling equation in flux terms as:

$$q = Q/A = h_c \cdot (T_s - T_\infty) \qquad \text{(Eq. 5.8)}$$

Note that the units of the convective heat transfer coefficient are the same as for conductance because there is no explicit distance over which the temperature drop occurs.

Natural convection is defined as convective heat transfer that occurs when the fluid motion transporting heat is driven by temperature-dependent density differences in the fluid (i.e., buoyancy). In many building science problems, the fluid is air. Hot air is less dense than cool air; thus, air rises alongside hot surfaces— such as radiators (more properly described as convectors) or solar heated exterior walls. If room air contacts a cold surface, such as a poorly insulated wall during cold weather, the room air will cool, contract, become denser, and then begin to fall. As the air falls downward along the cold wall surface, it will be cooled further and thus form a downward falling mass of cold air – often experienced as a draft. The convection coefficient acting at interior surfaces is sometimes designated by the symbol $h_{c,i}$.

Understanding issues such as interior surface condensation on walls and windows, cooling of exterior solar-heated surfaces, and human comfort are practical applications where convection (and hence the value of h_c) are important.

Laminar natural convection is a type of natural convective heat transfer with relatively small temperature differences, low velocity and hence smooth or laminar air flow. The following relations (shown in Figure 5.4) taken from the ASHRAE *Handbook of Fundamentals* [ASHRAE 2017] are useful for building science applications (all in $W/m^2 \cdot K$).

In the following relations, the temperature is in Kelvin or Celsius, and a representative dimension (H or L) is required: for horizontal surfaces L (m) is often taken to be 4 times the area ÷ perimeter of a surface, and for vertical surfaces H (m) is the height.

Convection to/from walls or other vertical surfaces (SI units):

$$h_c = 1.87 \cdot \Delta T^{0.32} \cdot H^{-0.05} \text{ or more simply} \qquad \text{(Eq. 5.9)}$$

$$= 1.42 \cdot (\Delta T/H)^{0.25} \qquad \text{(Eq. 5.10)}$$

A simple average value recommended by ISO 6946 [2017] for natural heat convection at interior walls is $h_c = 2.5\ W / m^2 \cdot K$.

Convective heat transfer upward from a warm horizontal surface or downward to a cold horizontal surface is more effective (h_c has a higher value) than heat flow in the opposite direction because the direction of buoyancy-induced flow is the same as the temperature gradient.

Heat flow with buoyancy (e.g., downward from a cold ceiling or upward from a warm floor) can be estimated using:

$$h_c = 1.32 \cdot (\Delta T/L)^{0.25} \qquad \text{(Eq. 5.11)}$$

An average value of $h_c = 5.0\ W / m^2 \cdot K$ is recommended by ISO 6946 [2017] when heat is flowing in the same direction as buoyancy.

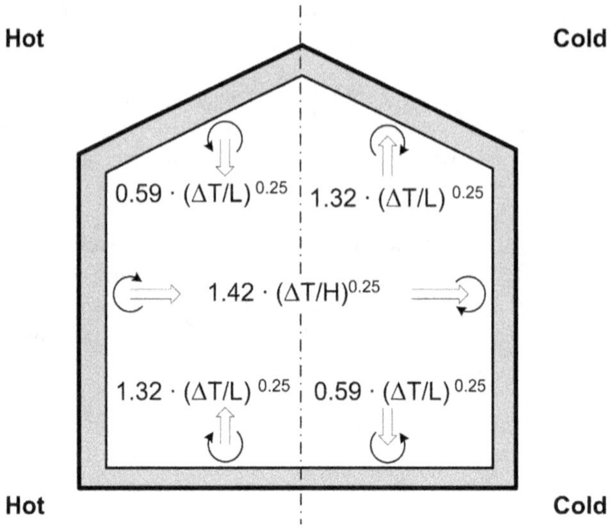

Figure 5.4: Heat transfer coefficients for laminar natural convection

Laminar natural convection flow can become **turbulent natural convection** if the temperature differences are large enough and the surfaces rough —for example, when solar heating or some other source of heat creates large temperature differences (more than approximately 10 °C) at the surface. Values for the heat transfer coefficient under turbulent convection of air for those situations include:

- horizontal surfaces with upward heat flow (e.g., sun-heated roofs)

  $$h_c = 1.52 \cdot \Delta T^{\,0.33} \qquad \text{(Eq. 5.12)}$$

- vertical surfaces with horizontal heat flow (e.g., sun-heated walls)

  $$h_c = 1.31 \cdot \Delta T^{\,0.33} \qquad \text{(Eq. 5.13)}$$

Natural convection heat flow against buoyancy (e.g., downward from a warm ceiling or upward from a cold floor) can be estimated using:

$$h_c = 0.59 \cdot (\Delta T/L)^{0.25} \qquad \text{(Eq. 5.14)}$$

An average value of h_c = 0.7 W / $m^2 \cdot K$ is recommended by ISO 6946 [2017] when heat is flowing against the direction of buoyancy.

Forced convection heat transfer involves fluid flow driven by some external means (e.g., wind or fans, not natural buoyancy). Heat transfer by forced convection occurs on the exterior of buildings, within ducts, and at interior surfaces impinged upon by air jets from HVAC. A useful and general convective-conductive heat transfer coefficient relationship for forced convection heat transfer with air flow over plates is:

$h_c = 5.6 + 3.9 \cdot V$, V= 1 to 5 m/s (Eq. 5.15)

$h_c = 7.2 \cdot V^{0.78}$, V= 5 to 30 m/s (Eq. 5.16)

where V is the air velocity.

The convective heat flow coefficients provided above are not necessarily very accurate. The nature of local flow paths and turbulence intensity can change the convective coefficient drastically, as can the presence of large local heat sources (such as radiators or air supplies near walls).

Example

The interior surface of an insulated exterior enclosure wall (3.9 m wide and 2.4 m high) is 18 °C and the interior air is 21 °C. Estimate the conduction-convection heat transfer coefficient at the interior. Compare the result to that for a 2.0 m tall window in the same room with a temperature 8 °C.

Answer

This is a situation of horizontal heat flow to a vertical surface under small temperature differences, hence:

$$h_c = 1.42 \cdot (\Delta T/H)^{0.25}$$

$$= 1.42 \cdot [(21\text{-}18)/2.4]^{0.25}$$

$$= 1.50 \text{ W/m}^2 \cdot \text{K}$$

The convective heat transfer coefficient is approximately 1.5 $W/m^2 \cdot K$.

By comparison, the window surface is much colder, and hence the coefficient for convective heat transport would be:

$$= 1.42 \cdot [(21\text{-}8)/2.0]^{0.25} = 2.27 \text{ W/m}^2 \cdot \text{K}.$$

The larger temperature difference results in a higher coefficient (i.e., more heat transfer per degree temperature difference), not just higher heat transfer. At some larger temperature difference, the flow would stop being laminar and would become turbulent, further increasing the coefficient.

These simple calculations demonstrate that not only is the convection coefficient sensitive to temperature differences but the variation may be significant. However, the importance of this variation is reduced by the fact that radiation, which also acts, is far larger in magnitude for this example.

Forced convection can be important for exterior building components exposed to the wind. The exterior convection coefficient may be designated by the symbol $h_{c,e}$. Figure 5.5 shows the range of total surface conductance (i.e., convection, conduction, and radiation) values that can be expected for typical cladding materials as a function of windspeed.

As can be seen from the intercept value in Figure 5.5 (i.e., no wind), radiation, conduction, and natural convection make up about 8 to 12 W/m^2/°C of the heat transfer coefficient, and forced convection makes up the rest. For an air velocity of 3 m/s (an average wind speed in most parts of the world), the convective component for brickwork is approximately 12 W/m^2.

Very smooth surfaces (such as glass, smooth metal, and painted wood) will have a lower coefficient because they induce less turbulence, and rough surfaces will have a higher value. Approximate values are:

- $h_{c,e} = 9 + 2.9 \cdot V$, V = 1 to 15 m/s for smooth surfaces, e.g. glass, metal
- $h_{c,e} = 12 + 5.8 \cdot V$, V = 1 to 12 m/s for rough surfaces, e.g. rough stucco

All of these correlations were obtained from wind tunnel data on small surfaces.

Older, more realistic, field studies by various German researchers suggested a coefficient of $h_{c,e} = 8.2 \cdot V^{0.51}$ for exposed windward and side walls and $h_{c,e} = 5.2 \cdot V^{0.36}$ for sheltered and lee walls [Schwarz 1978]. Hagentoft [2001] recommended $h_{c,e} = 5 + 4.5\ V - 0.14\ V^2$ for the windward side and (5 + 1.5 V) for the leeward. Finally, ISO 6946 recommends the relatively simple $h_{c,e} = 4 + 4\ V$, with V in meters per second [2017].

The turbulence of the approaching air flow is also important for exterior coefficients. Large variations in the nature and magnitude of flow can be expected due to cladding surface texture and exposure conditions alone.

In practice, surface films are a small proportion of the total thermal resistance of most enclosures. Therefore, an accurate exterior surface convection heat transfer coefficient is rarely, if ever, important for accurate heat loss calculations of modern enclosures. They may be important for research or specialist applications investigating the nano-climate near building surfaces however. Refer to any heat transfer text or the ASHRAE *Handbook of Fundamentals* for other situations or more accurate (but more complex) relations.

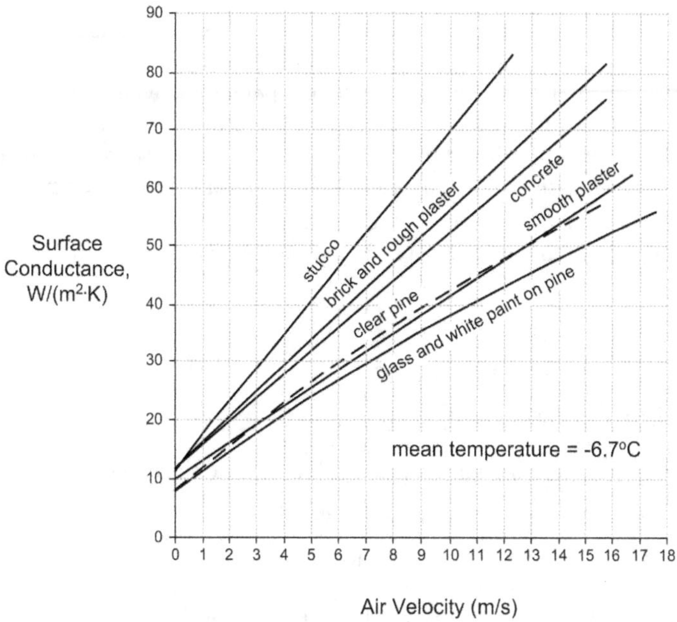

Figure 5.5: Equivalent surface heat transfer coefficients different surfaces [ASHRAE 2001]

5.4.1 Bulk convective heat transfer

When a fluid moves from one location to another, it carries heat energy with it. Hence, airflow through materials and assemblies also transports heat. Like all materials, air has the capacity to store heat, and air that flows through a material or out of a building thus carries heat energy with it.

The heat transfer of any fluid can be calculated by:

$$Q = c \cdot dm/d\theta \cdot \Delta T \qquad \text{(Eq. 5.17)}$$

where Q is the heat transfer rate (W)

c is heat capacity of the fluid (J / (kg·K)),

$dm/d\theta$ is the mass flow rate of the fluid (kg/s),

θ represents time, and

ΔT is the temperature difference (K or °C).

In most practical situations, air flow is measured using volumetric terms, i.e., cubic feet per minute, or liters per second. In volume terms, convection heat transfer can be predicted using

$$Q = c \cdot \rho \cdot dV/d\theta \cdot \Delta T$$

$$= c_v \cdot dV/d\theta \cdot \Delta T \qquad \text{(Eq. 5.18)}$$

However, the limitation of this approach is that the density of air changes appreciably with temperature and hence the volumetric heat capacity does as well. However, using the density of air at the temperature at which the volumetric flow rate is measured is accurate. Hence if the flow rate is measured at room temperature (e.g., air leaking out of a building), room temperature air properties should be used. For most calculations, assuming room temperature air (ρ=1.2 kg/m^3) results in a volumetric heat capacity of 1200 J/m^3 K. Ignoring temperature-induced density changes, the energy transfer by air can be estimated from:

$$Q = 1200 \cdot dV/d\theta \cdot \Delta T \qquad \text{(Eq. 5.19)}$$

where Q is the heat flow rate (W),

$dV/d\theta$ is the volumetric flow rate of indoor air (m^3/s),

ΔT is the temperature between the indoor air and the space which it is entering (°C).

The heat transfer due to heat flow through a porous body is somewhat more complex since it includes the bulk transport of fluid flow plus the heat transported from the surface of the solid constituents of the porous material. However, ignoring such interactions as a first estimation (valid for larger flows through cracks, holes, and highly porous materials), Equation 5.19 can be used to calculate the convective heat flow.

The air leakage rate through building enclosures has historically been measured in Air Changes per Hour (ACH). For example, a house may be described as having "an in-service air leakage rate of 0.30 ACH"[3]. Given an estimated or known air change rate, the volumetric flow of air per second can be calculated from:

$$dV/d\theta = (ACH) \cdot V \cdot (1 \text{ hour} / 3600 \text{ seconds})$$

Hence, heat flow as a function of ACH is:

$$Q = c \cdot \rho \cdot dV/d\theta \cdot \Delta T$$

$$= c \cdot \rho \cdot (ACH)\, V / 3600 \cdot \Delta T$$

For room temperature air with a density of 1.2 kg/m^3 and a heat capacity of 1000 J/kg, heat flow becomes:

$$Q = 1000 \cdot 1.2 \cdot (ACH) \cdot V / 3600 \cdot \Delta T$$

$$= 0.3 \cdot (ACH) \cdot V \cdot \Delta T \qquad \text{(Eq. 5.20)}$$

---

[3] See Chapter 7 for more on airtightness, airflow, and measurement. Note that many tests of building airtightness report values in terms of ACH at a test pressure of 50 or 75 Pa (i.e., denoted ACH50 or ACH75). For energy transfer calculations the reported rate of air leakage must be adjusted to match expected in-service pressure differences.

where Q is the rate of energy transfer due to air changes (W),

V is the building volume (m^3), and

ΔT is the temperature between indoors and outdoors (°C).

Example. Hot air, at 95 liters per second (200 cfm) and 30 °C (with a mass density, ρ, of 1.16 kg/m³) must be delivered to a 20 °C living room to balance the heat loss through its walls.

How much water will need to be delivered to the room via radiator-convector units if the temperature drop through the units is limited to 20 °C (e.g., entry water temperature of about 50 °C and exit of about 30 °C)?

Answer

The energy delivered by the air flow is:

$$Q = c \cdot dm/d\theta \cdot \Delta T$$
$$= 1000 \text{ J/kg K} \cdot (95/1000 \text{ m}^3\text{/s} \cdot 1.16 \text{ kg/m}^3) \cdot 10 \text{ K}$$
$$= 1102 \text{ W}$$

If we had simply used the approximation of room temperature air heat capacity of 1200 J/m³ K (i.e., a density of 1.2 kg/m³), the result would have been

$$Q = c_v \cdot dV/d\theta \cdot \Delta T$$
$$= 0.095 \cdot 1200 \cdot 10$$
$$= 1140 \text{ W}$$

This error is very small, and much less than the likely error in measuring or delivering the 200 cfm air flow in practice.

For water flow:

$$Q = c \cdot dm/d\theta \cdot \Delta T$$

Setting the heat flow rate equal to that delivered by air flow

$$1100 = 4200 \text{ J/kgK} \cdot dm/d\theta \cdot 20 \text{ K},$$

thus, $dm/d\theta = 0.013$ kg/s

Therefore, a flow of only 0.013 kilograms per second of water (0.013 liters/second or 0.2 gallons per minute) is required to deliver the same rate of energy as 95 liters per second (200 cfm) of air.

5.5 Radiation

The fundamentals of electromagnetic radiation are reviewed in Chapter 4. For practical, building-related heat flow calculations, radiation theory can often be considerably simplified. Such a simplified treatment of the transfer of heat energy by thermal radiation is considered here.

Since all materials, even cold ones, radiate energy, it is the net radiation exchange that defines the heat transfer. The shape or geometry can complicate this calculation. However, for the special and useful case of a relatively small surface exchanging radiation with a large area, the "world", (e.g., between a wall and the outdoors, or a roof and the sky, a human and her surroundings), the heat flow can be approximated using the Stefan-Boltzmann equation introduced in Chapter 4:

$$Q = \varepsilon \cdot \sigma \cdot A \cdot (T_s^4 - T_w^4) \qquad \text{(Eq. 5.21)}$$

where ε is the emmissivity of the emitting surface (see Chpt 4)

σ is the Stefan-Boltzmanm constant (5.67 x 10^{-8})

T_s is the surface temperature (K),

T_w is the temperature of the "world" (K) seen by the object, and

A is the small surface area (m^2).

Example. A human face, with an area of 0.04 m^2 and a surface temperature of 28 °C, is facing a large exterior wall at a temperature of 14 °C.

1. How much heat energy is emitted by the face?
2. What is the net rate of radiative heat exchange from the face to the wall?

Answers

1. Using the Stefan-Boltzman equation, the given temperatures, and assumed emissivity of 0.98 (provided in Table 4.3),

$$\begin{aligned} Q &= \varepsilon \cdot \sigma \cdot A \cdot T_s^4 \\ &= 0.98 \cdot 5.67\text{x}10^{-8} \cdot 0.04 \cdot (273+28)^4 \\ &= 18.25 \end{aligned}$$

Hence, the face would radiate 18.25 W.

2. Using Equation 5.21,

$$Q = \varepsilon \cdot \sigma \cdot A \cdot (T_s^4 - T_w^4),$$

$$\begin{aligned} Q &= 0.98 \cdot 5.67\text{x}10^{-8} \cdot 0.04 \cdot ((273+28)^4 - (273+14)^4) \\ &= 3.16 \text{ W} \end{aligned}$$

Thus, although the face gives off 18.25 W, the radiation from the world around it radiates to the face, and a *net* 3.16 Watts would be lost from the face to wall.

Now consider the common situation of two large parallel surfaces a small distance apart. It can be shown [Hutcheon and Handegord 1982] that the net radiation exchange between two non-blackbody plane surfaces of equal areas, (e.g., across a plane air space), can be found from the expression:

$$q_{1-2} = \frac{\sigma \cdot (T_1^{\,4} - T_2^{\,4})}{1/\varepsilon_1 + 1/\varepsilon_2 - 1} \qquad \text{(Eq. 5.22)}$$

where 1 and 2 designates the temperatures and emissivities of the two surfaces.

As the denominator in this equation is rather unwieldy, a term called the *emissivity factor*, F_E, is often introduced:

$$F_E = \frac{1}{1/\varepsilon_1 + 1/\varepsilon_2 - 1} \text{ such that} \qquad \text{(Eq. 5.23)}$$

$$q_{1-2} = F_E \cdot \sigma \cdot (T_1^{\,4} - T_2^{\,4}) \qquad \text{(Eq. 5.24)}$$

This latter form of the equation is widely used in practice.

The nature of the surface makes a significant difference to the radiative energy exchange. Consider, for example, the common case where both surfaces have a high value such as 0.9. F_E can be calculated as 0.8183. This is the most common scenario that describes many building materials. However, if both surfaces had relatively low emissivity, for example an ε value of 0.1, F_E would be 0.0526. This is *fifteen* times less than the high emissivity case. Even if only one surface has a low emissivity value of 0.1 and the other a value of 0.9, F_E, at 0.0989, is still eight times smaller than both surfaces having high emissivity.

It is clearly advantageous to use a low emissivity face on one or both faces of an airspace to reduce radiative energy exchange (to slow heat transmission) across it. There is only a small benefit to decreasing the emissivity of both surfaces. It is for these reasons that a low emissivity coating is applied to glazing systems, and that aluminized foil is used on one face of an exterior sheathing facing into an air space in a wall system. For many situations it is be convenient to use $F_E \cong \varepsilon_1 \cdot \varepsilon_2$ as an estimate (compare with the precise calculation above).

Another factor that must be considered is the angle of view, usually dealt with by the angle, or view angle, factor, F_A. Since radiation can travel in a straight line only, the area and angles of incidence between two exchanging surfaces play an important role in radiative heat exchange. For a small object exposed to a large object, the angle factor can be assumed to be one, as it was in the treatment above. For example, roofs exposed to the sky will have an angle factor of one. Vertical walls can be assumed to have an angle factor of just less than 0.5 for radiation exchange with either the sky or the horizontal ground around the wall. Figure 5.6 provides the view angle factor for a few common situations in building science: floors to walls, sloped ceilings to floors, and arbitrary triangular enclosures.

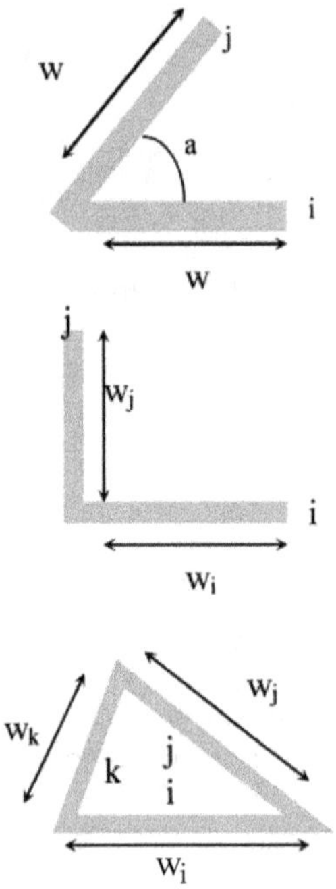

$$F_{ij} = 1 - \sin(\frac{a}{2})$$

$$F_{ij} = \frac{1 + (w_j / w_i) - [1 + w_j / w_i)^2]^{1/2}}{2}$$

$$F_{ij} = \frac{w_j + w_i - w_k}{2w_i}$$

Figure 5.6: View factors for common situations in building enclosures [Hagentoft 2000]

The potential complications of curved surfaces, small radiators and /or small receivers are enormous and are not dealt with in any depth here. Nevertheless, a complete description of radiative heat transfer may be important in some situations, and many good texts are available [e.g., Incropera & DeWitt 2017].

The practical equation that accounts for all of the above effects is:

$$q = F_E \cdot F_A \cdot \sigma \cdot (T_1^4 - T_2^4) \qquad \text{(Eq. 5.25)}$$

If one wishes to employ an equation of the same form as the conduction and convection equation, the above equation can be linearized and rewritten as:

$$Q / A = q = h_r \cdot F_A \cdot (T_1 - T_2)$$

where we define h_r as the *radiative heat transfer coefficient,* so that:

$$h_r = \frac{F_E \cdot \sigma \cdot (T_1^4 - T_2^4)}{T_1 - T_2}$$

$$= \sigma \cdot (T_1^2 + T_2^2)(T_1 + T_2) \qquad \text{(Eq. 5.26)}$$

which can be simplified to:

$$h_r = 4 \cdot F_E \cdot \sigma \cdot T_m^3 \qquad \text{(Eq. 5.27)}$$

where T_m is the mean temperature of the surface and its surroundings.

Example. The interior surface of an insulated exterior wall enclosure is 18 °C and the interior is 21 °C. Calculate the radiation heat transfer coefficient assuming that the surfaces have an emittance factor of about 0.90.

Answer

Using Equation 5.27,

$$h_r = 4 \cdot F_E \cdot \sigma \cdot T_m^3$$

$$T_m = (291+294)/2 = 292.5$$
$$h_r = 4 \cdot 0.9 \cdot 5.67 \times 10^{-8} \cdot 292.5^3 = 5.11 \text{ W / m}^2 \cdot \text{K}$$

Using Equation 5.26

$$h_r = F_E \cdot \sigma \cdot (T_s^2 + T_a^2)(T_s + T_a)$$

$$= 0.90 \cdot 5.67 \times 10^{-8} \cdot (291^2 + 294^2)(291+294)$$

$$= 5.11 \text{ W / m}^2 \cdot \text{K}$$

It can be seen that the radiation heat transfer coefficient is approximately 5.1 $W/m^2 \cdot K$ under normal room conditions. If the temperature of the air and the wall were to both increase by 3 °C, the coefficient would increase slightly (by 3%), to 5.27 $W/m^2 \cdot K$.

Note that it is important to use absolute Kelvin temperature when calculating radiation effects.

5.6 State change and Thermal Mass

As described in Section 4.3.1 and 4.3.2, materials release or absorb a material-specific amount of latent energy when the change state. As state changes occur at a constant or very narrow range of temperatures there is no temperature-dependent heat flux equation.

State changes are not strictly a heat transfer mechanism. However, the transport of water vapor, followed by condensation on a surface or interface will materially impact the heat transfer and temperature within enclosure assemblies because of the release of latent energy. The latent energy required for evaporation of water will likewise impact the heat transfer and temperature gradient across an assembly. Hence, the potential for state change should be considered when predicting and analyzing heat transfer across assemblies.

In heating and cooling systems the latent energy of steam and refrigerant gas state changes are fundamental to how these systems deliver or remove thermal energy.

Finally, the heat capacity of the materials that comprise an assembly is the reason that steady-state heat flow is an assumption in heat flow rather than a reality. If the interior or exterior temperature change and the heat capacity of assemblies is large, the temperature gradients and instantaneous heat flow will deviate from the steady-state methods described herein. Although there are some simple calculation methods available to assess thermal mass in enclosure heat flow, computer models are so readily available that investigations of thermal mass should in most cases avail themselves of these capabilities.

5.7 Effective Thermal Conductivity & Insulation Products

In a solid material such as steel, all heat transfer will occur in one mode: conduction. However, in almost all practical insulation products, heat transfer is occurring via several modes. Porous materials allow heat transfer by conduction through both the solid and gaseous components, radiation through the gas-filled voids, and even natural convection may play a role.

The different modes of heat transport are identified as a function of product density in Figure 5.7 for a specific fibrous glass product. As the density increases, radiation transport decreases because more fibers block radiation. However, more density increases the flow of heat by conduction through the solid glass fibres. The range of densities shown mean that porosity (air-filled pore fraction) is very high, and hence conduction of heat through the air changes imperceptibly with density. The minimum heat transfer rate occurs at a density of around 60 kg/m^3. This minimum will change with, for example, temperature (as this impacts the radiation curve) and fibre diameter (as this impacts radiation blocking).

It is technically difficult to separate out the contribution of each mode by measurement, and often of little practical value to do so. Hence, the term *effective thermal conductivity* is widely used to describe measurements that include all modes but assuming they behave like conduction.

Although it is immensely practical and convenient to lump all heat transfer modes together into one equivalent thermal conductivity, it can lead to confusion because radiation and convection heat transfer mechanisms are not always linear like pure conduction: radiation varies strongly with absolute temperature, natural convection depends on the magnitude of the temperature difference, and forced convection can overwhelm all other heat transfer modes if allowed to occur.

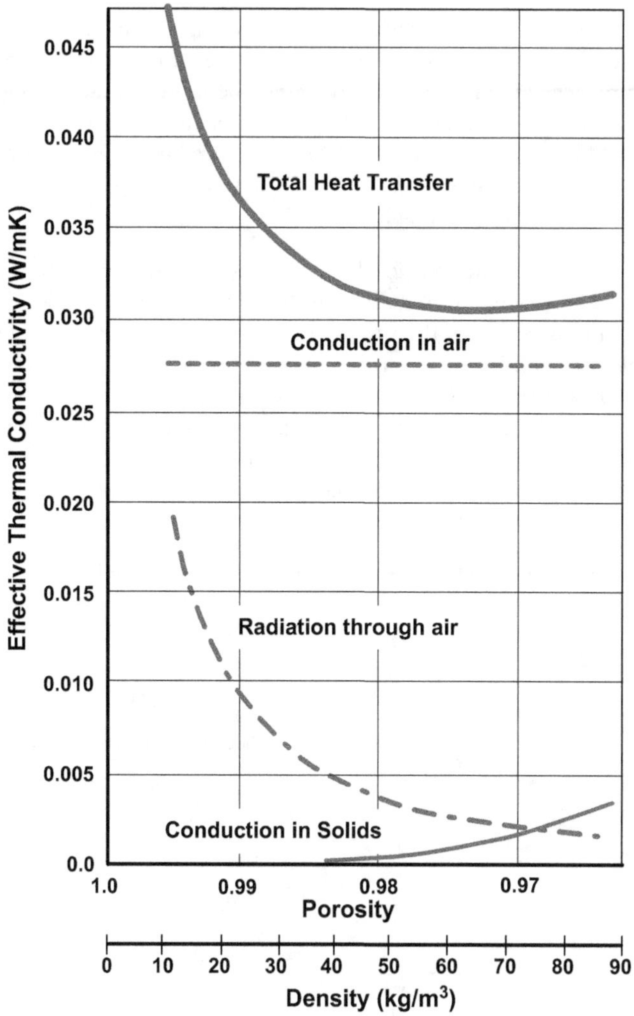

Figure 5.7: Different modes of heat transfer in porous insulation [Bankvall 1973]

Design and construction effort are required to ensure that the thermal conductivity values reported can be achieved in the field by understanding how convection and radiation are part of the effective thermal conductivity reported for insulation products.

5.7.1 Factors influencing effective thermal conductivity

Although thermal conductivity values can be found for a wide range of materials, thermal conductivity values for the same material can vary significantly because of many factors. The most significant factors are density, temperature, and moisture content.

Density impacts heat flow through a material as heat transfer by conduction occurs via the transfer of vibratory energy by direct molecular contact. At a very simple level, the more solid a material, the more conductive heat flow can occur. Insulating materials are all highly porous (gas filled voids) and low density. Highly conductive materials tend to have little porosity and high density. Hence, density is perhaps the single most important predictor for thermal conductivity of a material sample.

Figure 5.8 plots the thermal conductivity versus density of almost two hundred samples of over one hundred different building materials on a log-log plot. In this format it is clear that density plays a defining role over a broad scale. Materials with conductivities around 0.04 W/m· K (R-4/inch) have densities of around 100 kg/m^3 (6 pounds per cubic foot) or less. Polymers provide many of the data points in the medium-density range (500-2000 kg/m^3), but mineral-based lightweight concretes and mortar also extend into this range. High-density concrete and natural stone comprise the lower end of the over 2000 kg/m^3 range, with metals comprising the highest density samples.

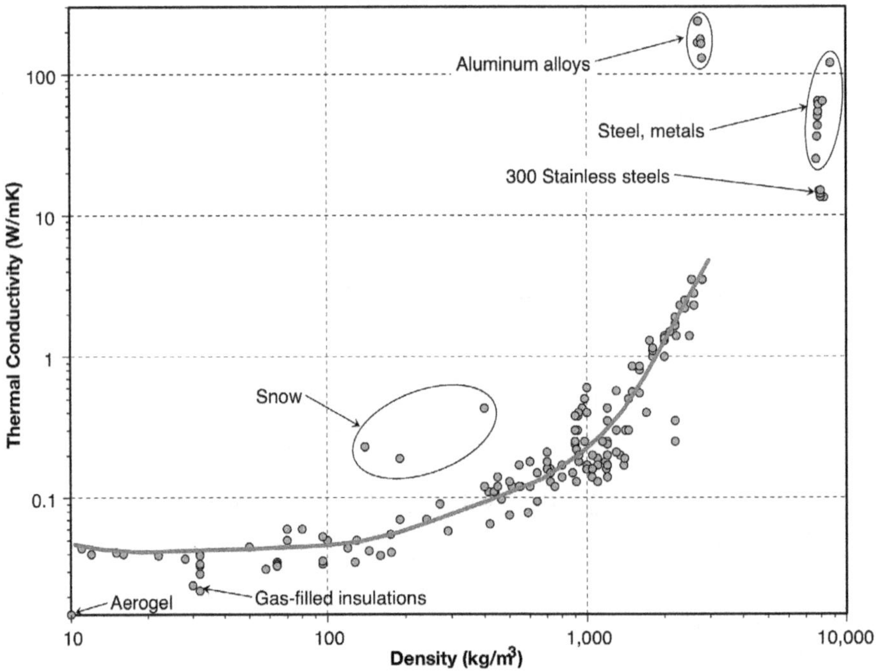

Figure 5.8: Thermal conductivity vs density for a wide range of building materials

Several further points stand out:

- very low conductivity can be achieved when foamed plastics have their pores filled with special gases rather than air,

- aerogel, a silica-based product, uses nano-sized pores and very low density to achieve exceptionally low conductivity
- aluminum is very conductive although light weight for a metal,
- 300-series stainless steels having surprisingly low thermal conductivity for such a dense and strong metal.

This type of plot reinforces the fact that high density and low conductivity do not coincide. Most strong materials are dense (the same density of molecular connections that provide strength, increase conductivity) and the search for an insulating structural material has been filled with failure and compromise. Another use of this information is to recognize that all things being equal, a lower density version of a material will almost invariably have lower thermal conductivity.

Temperature also affects apparent thermal conductivity, especially for low-density and low thermal conductivity materials. Figure 5.9 plots the mean temperature (the average of the temperature on either side of the sample) versus thermal conductivity for five different types of building insulation. It can be seen that over a realistic range of mean temperatures (for a cold climate building) of -5 °C to 42 °C, the thermal conductivity can vary by 15 to 20%.

This variation in apparent conductivity is primarily due to the variation of radiation within highly porous materials, not a change in conduction within the solid components. Therefore, the impact of temperature is far less for denser and more highly conductive materials (e.g., wood or brick) than for highly porous and lightweight products (e.g., fiberglass). To address this factor for commercially-available insulation products, the thermal conductivity of insulation products is usually rated at a standardized mean temperature test. For building insulations, a mean temperature of 24°C is chosen in North America and 10°C is used in Europe (this difference in rating also explains why European literature shows slightly higher thermal resistance for otherwise identical products in North America). Materials intended for use in cryogenic or high-temperature industrial applications should always be assessed close to the expected in-service temperature.

Moisture is a third important practical factor that may impact thermal conductivity. Water has a relatively high thermal conductivity (0.6 W/m·K), but also has high heat capacity and a very high heat of vaporization. Hence, the moisture content will drastically affect the conductivity of most materials; in general, the more porosity and the lower thermal conductivity a material exhibits when dry, the greater the negative impact of moisture. For many common natural insulation products (cellulose fiber, cork, wood) moisture content has real practical importance.

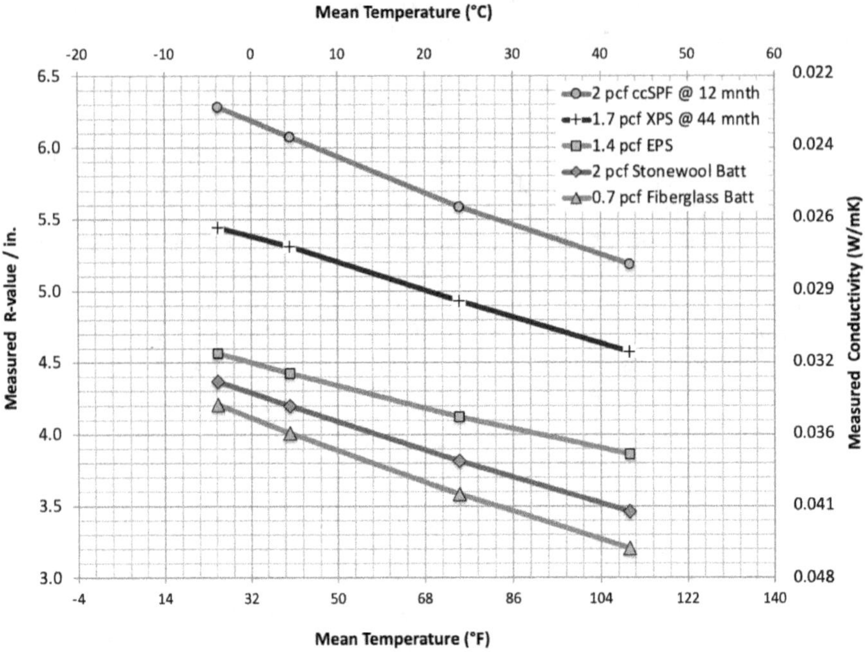

Figure 5.9: Temperature vs thermal conductivity for several insulation products

Testing of building materials usually occurs at "dry" conditions (often this means at a moisture content in equilibrium with room air), but in some cases, such as the design values in ISO Standard 10456 [2007], values for products at 80%RH are reported.

Heat transport can be very significantly enhanced if water in and next to damp insulation evaporates and then moves from the warm side of a material to the cold side transporting large amounts of energy [Hedlin 1988, Bomberg 1978]. It is for this reason that the often-published plots of thermal conductivity versus moisture content are only partially valid for materials with high resistance to vapor flow, and essentially useless for wet vapor permeable materials (Gertis & Holm 2014). The only effective means of assessing the performance of wet permeable insulation is to use a dynamic hygrothermal computer model, such as WUFI, that accounts for the latent energy of water evaporation and transport.

5.7.2 Insulation behavior

All materials and layers in a building assembly have some resistance to heat flow. However, some materials with a conductivity value lower than about 0.05 to 0.07 W / m· K (an R-value of better than about 2 per inch) are deliberately used in building assemblies for their ability to retard the flow of heat. These building products are called thermal insulations. Insulations are usually solid materials (so-called body insulation), but radiant barriers that control radiation heat transfer across airspaces are also available.

As noted above, most materials with high strength have relatively high density, and the strength of most building materials (e.g., concrete, wood, plastic) drops along with drops in their density. Hence, the need for low density (or more accurately, high porosity) reduces the structural capacity of most insulation. Accordingly, low-density, high-performance insulation layers—such as glass fiber batt—are used to control heat flow in most modern building enclosures, while high-density, high-strength, high-conductivity materials such as steel studs are used to support structural loads. In the past, building materials such as adobe, log, and brick were used in a manner that combined both insulating and load-bearing functions. Buildings constructed of these materials had thick walls, to provide a reasonable level of resistance to heat flow and to provide sufficient strength. Wood is perhaps one of the more interesting compromises between structural capacity and insulating ability.

Since conduction is a major mode of heat transfer and still air a low-cost insulator, insulation products tend to be low-density materials (i.e., porous materials with a large proportion of voids filled with air) and / or made of low-conductivity elements. For example, glass of 2500 kg/m^3 density and 1.0 $W/m \cdot K$ thermal conductivity is spun into fibers and formed into a batt of about 16 kg/m^3 (1 pcf) density and 0.040 $W/m \cdot K$ thermal conductivity. Fiberglass batt insulation is widely used as insulation despite the high conductivity of glass because of the very high percentage of pores filled with air. This type of thermal insulation is approximately 99.4% air.

Some foam plastic insulations have a lower percentage of air voids than glass fiber batt (often around 99% air) but are made of lower conductivity plastic material. Softwood of 500 kg/m^3 (30 pcf) density and 0.11 $W/m \cdot K$ thermal conductivity is used to produce cellulose insulation of 60 kg/m^3 (4 pcf) density and 0.038 $W/m \cdot K$ thermal conductivity.

As noted earlier, all three modes of heat transfer actually occur simultaneously within highly porous insulations. Referring back to Figure 5.8, it can be seen that at low densities, the effective conductivity is high since radiation can easily transport heat through the relatively open space. At higher densities, radiation is suppressed by the many fibers, but conduction becomes more important. Therefore, an optimum density can be chosen: one that varies with the type of material. For the glass fiber in question (a relatively coarse glass), the optimum technical density is about 60 kg/m^3. However, since the cost increases with density (because of increased material use and shipping costs), the density more commonly used for fiberglass is closer to 15 kg/m^3.

Smaller fiber diameters, although more challenging to produce, place more obstacles in the path of radiation flow per unit mass, and thus have a lower thermal conductivity at lower density. Higher density products are used (typically of 60 to 160 kg/m^3 density) when higher strength is required (as it is for low slope roofs and exterior semi-rigid products) and demands more fiber-fiber contact to transmit loads.

Stonewool fibers are generally larger diameter than glass, and hence conduction plays a larger role at lower densities. Stonewool products therefore tend to use higher densities to achieve the same thermal performance as spun glass. This extra density provides these products with greater strength and more resistance to convection and radiative effects. Even though more material is used for the same thermal resistance, stonewool products compete in applications that require these properties.

5.8 Insulation Products

Most commercially-available building insulation products are produced from foamed or fibrous forms of minerals and polymers (natural or plastic). Insulations are usually solid materials, but radiant barriers that control only radiation heat transfer across air spaces are also sometimes used, especially in glazing systems (e.g., the now commonplace low-e coating).

Insulation products are often chosen for attributes other than low thermal conductivity: mechanical properties, moisture tolerance, fire performance, and air and vapor permeance are often key. Table 5.2 lists the performance attributes associated with each of the four material categories.

Table 5.2: Four material categories, with a range of physical attributes, describe most insulation products used in buildings

Material Category	Examples	Moisture	Fire	Vapor Permeance	Air Permeance
Mineral fiber	Fiberglass, stone, slag	Tolerant	Non-combustible	high	high
Polymer Fiber	Cellulose, cotton, wool, straw, plastic	Sensitive	Combustible	high	high
Polymer foam	Polystyrene, polyurethane, polyisocyanurate	Tolerant	Combustible	Low-medium	low
Mineral foam	Foamglass, pumice, airkrete, aerogel	Tolerant	Non-combustible	Low-medium	low

Insulation products can be produced in at least five common physical forms: loose, batt, roll, board, and spray (Table 5.3). Each form has advantages and disadvantages, and some materials can be purchased in all five forms, and some only in one.

Table 5.3: Insulations can take on a wide range of forms, which may limit how they are used

Form	Installation	Limits to use
Loose	Poured, blown, packed	may settle, easily compressed
Batt	friction fit	held in place by friction, easily compressed
Roll	friction fit / mechanically attached	as for batts
Board	mechanically or adhesively attached	resistant to mechanical pressure
Spray	spray in place	sticks to adjoining surfaces, resilient

High-performance building enclosures generally require continuous thermal control layers on the exterior of the structure to ensure continuity. This layer can be products of foam board or spray foam, semi-rigid fiberglass or stonewool insulation: it must only be moisture tolerant. If organic-fiber products are used, they can only be used inside of the enclosure assembly's water control layer (Chapter 2) or they will be damaged in-service.

Many insulation products, including those with facers (e.g. the aluminum facer on polyisocyanurate foam intended for walls, faced EPS board), can act as part of the water and/or vapour, and/or air flow control so long as other requirements, such as structural support, durability, and continuity at joints, are met.

In North America, "R-value" is almost universally used to market insulation products. This is of course the inverse of the effective conductance of a specific layer thickness. Everett C. Schuman (1902-1995), a professor at the Pennsylvania State University, is generally credited with being the "father of the R-value" because of his efforts to popularize its use in building products like insulation. There is a historic plaque on the Penn State campus to commemorate the contributions of this early building scientist.

5.9 Calculating enclosure heat flow

Calculating the heat flow through building enclosures in practice makes use of the many simplifications presented earlier. The approach taken here is similar to that in multiple standards and guides, including ISO 6946 [2017], the ASHRAE *Handbook of Fundamentals* [2021], and Straube [2017]. There are also more specialized standards and guides for more detailed applications, such as two- and three-dimensional computer-based calculations (CSA Z5010). Several additional concepts are important for building science applications: heat flow through multiple layers, heat flow across surfaces and air spaces, the influence of radiation on exterior surfaces, thermal bridges (multi-dimensional heat flow) and heat flow through entire building systems. Each will be covered below.

Almost all approximate calculations are based on the assumption of steady-state one-dimensional heat flow. Underlying steady-state, one-dimensional heat flow are the assumptions that all material properties, heat flow, and temperature remain constant with time, and that the same quantity of heat flows through every parallel plane of a system. Just as in a series electrical circuit, all resistors have the same current (heat) flow and the voltage (temperature) drop will depend on the thermal (electrical) resistance.

5.9.1 Single layer

From Fourier's Law of conductive heat flow:

$$Q = k \cdot \Delta T \cdot A / l \qquad \text{(Eq. 5.28)}$$

but recall that k/l = C, and R = 1/C, so other useful forms can be derived:

$$Q = C \cdot A \cdot \Delta T$$

$$Q = A \cdot \Delta T / R \qquad \text{(Eq. 5.29)}$$

These relations are all dimensionally homogenous so can be used in SI or IP units. Recall that R-value = 5.678 RSI.

Example: A 2 m wide, 3 m high, and 50 mm thick piece of Type 4 extruded polystyrene (l= 0.029) has a surface temperature of 20 °C on one side and 40 °C on the other. Calculate

1. the R-value and RSI,
2. the conductance,
3. the heat flow, and
4. the heat flux.

Solution

1. The RSI and R values

$RSI = 1 / C = 1 / (k / l) = l / k$

$= 0.050 / 0.029$ W/m K

$= 1.72\ m^2 \cdot K / W$

As R= 5.678· RSI-value,

R-value =5.678 · RSI1.72

= R9.77

2. The conductance

$C = k / l$

$= (0.029\ W/m \cdot K) / (0.050\ m)$

$= 0.58\ W/m^2$

3. The heat flow rate

$$Q = k \cdot \Delta T \cdot A / l$$
$$= 0.029\ W/m\ K \cdot (40\ °C - 20\ °C) \cdot (3\ m \times 2\ m) / 0.050\ m$$
$$= 69.6\ W \text{ or } 69.6\ J/s$$

4. The heat flux

q = total heat flow per unit area = $Q/A = 11.6\ W/m^2$.

5.9.2 Multiple layers

Given the fundamental equations defined earlier, one can consider the heat flow through multiple layers or composite constructions. There are only two important facts to remember:

- The heat flux across the entire assembly must also pass through each layer (i.e., energy must be conserved, and heat capacity and phase changes are explicitly ignored in steady-state analysis).
- The sum of the resistances of each layer gives the total resistance; one cannot sum the conductances.

The Fourier conduction equation for a single layer can be extended to any layer *i* in a multi-layer assembly of flat materials (many types of building enclosures):

$$q = \Delta T_i / R_i = C_i \cdot \Delta T_i \qquad \text{(Eq. 5.30)}$$

For the practical situation with interior and exterior temperatures across an assembly of n layers:

$$q = Q/A = \frac{t_{int} - t_{ext}}{\sum_{i=0}^{n} R_i} \qquad \text{(Eq. 5.31)}$$

Note: $q \neq \sum_{i=0}^{n} C_i \cdot (t_{int} - t_{ext})$

The total thermal resistance of this multi-layer assembly is:

$$R_{total} = R_1 + R_2 + \ldots + R_n$$

where

R_{total} is the total one-dimensional thermal resistance of a multi-layer assembly, and

R_1 to R_n is the resistance of each of the assembly's layers, including air films, air gaps, and solid materials.

For a multi-layer assembly formed of different materials, and air spaces, and even complex framing, the thermal performance can be estimated provided that the thermal resistance of each of these layers can be found from tables or calculations.

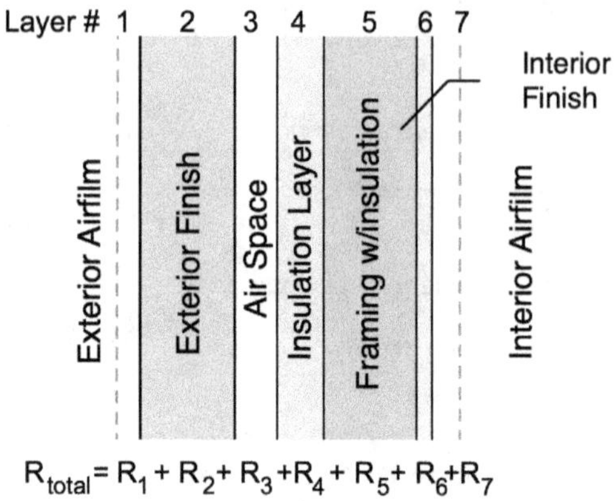

Figure 5.10: Example of calculating thermal resistance of multi-layer assemblies

The *overall coefficient of heat transfer* (U) or the *thermal transmittance* of a system of *n* layers from the environment on one side to the environment on the other (i.e., usually the interior air temperature to the exterior air temperature) is defined as:

$$U = \frac{1}{\sum_{i=0}^{n} R_i} \qquad \text{(Eq. 5.32)}$$

From Eq. 5.30, the temperature drop across layer i is:

$$\Delta T_i = q \cdot R_i \qquad \text{(Eq. 5.33)}$$

Combining Eq. 5.30 and 5.32 gives an equation for the temperature drop across any layer i in an assembly:

$$\Delta T_i = \frac{t_{int} - t_{ext}}{\sum_{i=0}^{n} R_i} \cdot R_i \qquad \text{(Eq. 5.34)}$$

or from Eq. 5.32 and 5.34,

$$\Delta T_i = U \cdot (t_{int} - t_{ext}) \cdot R_i \qquad \text{(Eq. 5.35)}$$

This last equation is one of the most useful. One normally calculates the assembly U-value to assess its ability to control heat flow. The use of the U-value is favored by mechanical engineers; they are primarily concerned with the design of the HVAC equipment and hence need to quantify the heat flow across the enclosure with little interest in the temperature drops across individual layers.

Building scientists are interested in temperature conditions within the enclosure to allow for the assessment of differential movement, durability, condensation and drying potential. It is a simple matter to calculate the temperature drop across every layer, or any layer of interest, using Equation 5.35. For example, to assess the influence of different outdoor or indoor conditions on the temperature drop across a particular layer of interest, only the temperature needs to be changed.

5.9.3 Heat transfer at surfaces ("Surface films")

The radiative and conductive-convective heat transfer modes operate in parallel at the exterior and interior surfaces of building components. Hence, the radiative and convective heat transfer coefficients are used in the form of an equivalent conductance or, often more practically, equivalent resistances. Since the flow is parallel, the two coefficients can simply be summed.

$$R_s = 1/\ (h_c + h_r) \qquad \text{(Eq. 5.36)}$$

Where R_s is the surface film resistance

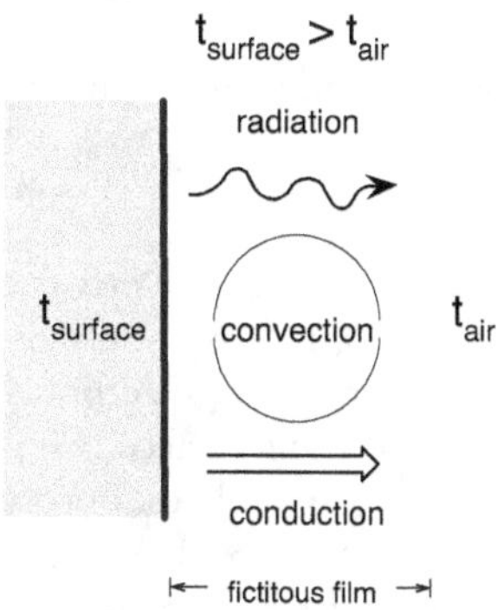

Figure 5.11: Surface film analogy of heat transfer at surfaces

These equivalent conductance coefficients are termed *surface film coefficients* and are often designated h_i for interior and h_o for exterior ($R_{s,i}$ and $R_{s,o}$ respectively in resistance terms). **It is important to recognize that a surface film is a fictious artifact, and does not exist with a physical thickness.** The term film is used simply so that a layer (of indefinite thickness) can be added to the conductive heat flow analysis described in the previous section.

Example. The interior surface of an insulated exterior wall enclosure element (2.4 m high) is 3 °C cooler than the interior air (t = 21 °C). Calculate the total effective conductance and surface resistance assuming that the paint has an emittance factor of about 0.90.

Answer

Since we have previously calculated each component, the solution is simple:

$h_r = 5.11$

$h_c = 1.50$

$h_i = h_r + h_c$

$= 5.11 + 1.50 = 6.61\ W / m^2 \cdot K$

$R_{s,i} = 1/h_i = 0.15$, or R-0.86 in IP units

The design values used for combined convection, radiation, and conduction are sometimes different (around 8 $W / m^2 \cdot K$) because they are based on different temperature differences between the interior and the wall surface.

The simplified relations for radiation and convection given earlier can be simplified further by assuming:

- a fixed temperature difference between the surface and the indoor air,
- a fixed mean radiant temperature of the indoors and outdoors, and
- an exterior wind speed.

The first two assumptions do not result in errors of surface films resistance of more than about 25%, but the third assumption is normally refined to consider average summer and winter wind speeds. Although an error of 25% may seem large, the resistance to heat flow at a surface is small compared to the heat flow resistance of most modern wall assemblies. Poorly insulated walls and windows have a lower overall thermal resistance, so surface effects are more important. A more precise calculation of surface films is therefore justified for such enclosures. When calculating surface temperatures, for assessing both thermal comfort and condensation risk, the exact value of the surface film coefficients becomes very important, and detailed investigations of convective and radiative effects may be warranted.

Total equivalent heat transfer coefficients for interior conditions can be as low as 2 $W/m^2 \cdot °C$ near three-dimensional corners (e.g., the intersection of two walls and a floor) and behind furniture, and 4 $W/m^2 \cdot °C$ near 2-D corners (e.g., ceiling-to-wall intersections, window glazing to sill, floor-wall) [Ge & Fazio, 2003]. Exterior values as low as 7 to 10 $W/ m^2 \cdot °C$ can occur in sheltered locations on calm days.

The thermal resistance of interior surface films and finishes is quite important for windows and curtain walls, since their low thermal resistance means that the interior film provides a significant contribution to the total thermal performance. Detailed analysis may be justified in special cases.

Based on a number of sources, the values for the surface heat transfer coefficient are listed in Table 5.4. For exterior and interior surfaces under average conditions, representative average total heat transfer coefficients of 17 and 8 W/m²·K respectively are used for most calculations. The combined effect of convection and conduction contributes about 15 and 3 W/m²· °C respectively [Sanders 1996] to these values. The overall equivalent surface conductances, $h_{i,o}$ and $h_{e,o}$, in Table 5.4 can be used in Fourier's heat flow equation without further modification.

It will be noted that the values are somewhat higher than predicted by the examples given above. The reason for the higher values is to provide a conservative measure of the thermal resistance of an assembly, because additional airflow is often generated inside buildings by mechanical equipment, and because historically the temperature of interior surfaces was larger than for well-insulated modern enclosures.

Table 5.4: Equivalent surface heat transfer coefficient (h_i, h_e) [ASHRAE 2021]

Surface position	Flow direction	h (W / m²·K)		R-value (h·ft²·°F/Btu)	
Still air (e.g., interior)		F_E= 0.90	F_E= 0.05	F_E= 0.90	F_E= 0.05
Horizontal (i.e., ceilings & floors)	Upward Downward	9.3 6.1	4.3 1.3	0.61 0.93	1.3 4.4
Vertical (i.e., walls)	Horizontal	8.3	3.4	0.68	1.7
Moving air (e.g., exterior)					
6.7 m/s (e.g. winter)	Any	34	34	0.17	0.17
3.4 m/s (e.g. summer)	Any	23	23	0.25	0.25
Average conditions	Any	17	16	0.33	0.33

Plane airspaces

Plane airspaces are commonly used in building assemblies. Heat is transferred across airspaces by a combination of conduction through still air, convection flows, and net radiation from the warm side to the cold. The modes of heat transfer vary in importance depending on the emissivities of the surfaces, the thickness of the airspace, and the absolute and relative temperatures of the two surfaces.

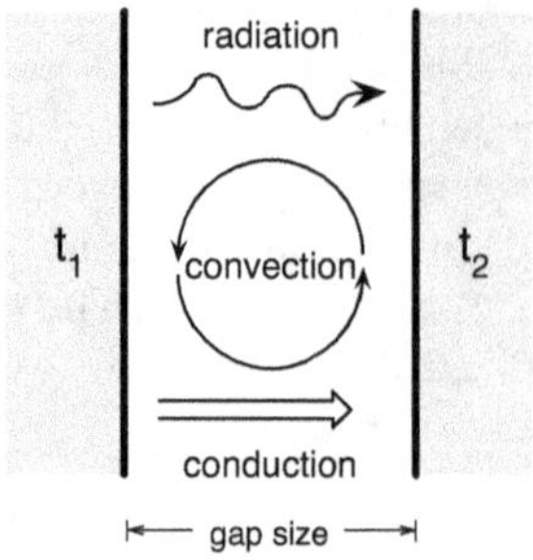

Figure 5.12: Heat transfer across plane airspaces

Simplified values for most situations encountered in practice are presented in Table 5.5. These values are all in terms of equivalent conductances so that they can be used directly in Fourier's heat flow equation. Note that these conductances are based on a mean temperature difference of 5.6 °C.

Table 5.5: Equivalent conductances for plane airspaces (W / m²·K)

Position of surface	Direction of flow	13 mm		20 mm		40 mm		90 mm	
		Emittance Factor F_E							
		0.05	0.82	0.05	0.82	0.05	0.82	0.05	0.82
Mean temperature 32 °C									
Horizontal	Up	2.78	7.69	2.56	7.69	2.38	7.14	2.13	7.14
	Down	2.44	7.14	1.72	6.67	1.06	5.88	1.45	5.56
Vertical	Horizontal	2.44	7.14	1.75	6.67	1.56	6.67	1.67	6.67
Mean temperature -18°C									
Horiz.	Up	2.78	5.55	2.63	5.56	2.44	5.26	2.17	5.00
	Down	2.00	5.00	1.41	4.35	1.18	3.70	1.82	3.45
Vert.	Horizontal	2.00	5.00	1.59	4.55	1.69	4.55	1.69	4.55

Source: ASHRAE *Handbook of Fundamentals* 2021, Chapter 26, Table 3.

Heat flow across an airspace can be found with a greater degree of accuracy by using detailed correlations of convection and radiation. ASHRAE [2021] recommends the following conduction-convection coefficient (h_c) and radiation coefficient (h_r) for horizontal heat flow across thin vertical plane airspaces in building enclosures:

$$h_c = 21.8 \cdot (1 + 0.00274\, t_{mean}) / w \text{ , for } w<13 \text{ mm} \qquad \text{(Eq. 5.37)}$$

$$h_r = 0.227 \times 10^{-7} \cdot F_E \cdot (t_{mean} + 273)^3 \qquad \text{(Eq. 5.38)}$$

where h_c, h_r are coefficients (W/m²·°C),

t_{mean} is the mean temperature in °C,

F_E is the emissivity factor, and

w is the width of the airspace (in mm).

However, a high degree of accuracy is rarely necessary or justified in light of the fact that many variables (e.g., variable cavity widths, blocked cavities, etc.) cannot be reliably predicted and the influence that the airspace has on the thermal resistance of modern enclosure assemblies is relatively small.

Opaque building enclosures in modern buildings tend to have U-values of less than about 0.5 W/ m²·K (R-values of over 10). The contribution of standard air gaps (high levels of emissivity) is thus modest. It can also be noted from Table 5.5, that there is very little change in conductance for gaps of 13 to 90 mm and modest changes between -18 and 32 °C. Hence, it is common for design values of air spaces to be standardized as a single value. For example, the National Building Code of Canada specifies the thermal resistance of a high emissivity air gap independent of temperature and specifies RSI0.18 (R-1.02) for 20 through 90 mm gaps.

5.9.4 Airflow through airspaces

In many practical situations, an airspace behind cladding is either intentionally or accidentally vented. Airflow through an airspace can change the heat flow characteristics, although significant flows are required to modify the equivalent conductances listed above.

The heat capacity of air is limited: it is about 1 kJ/kg K. Since a cubic meter of air weighs a little more than a kilogram and air flow rates in air spaces tend to be in the range of 0.1 to 10 m³/m²/hr (see Chapter 7), the additional heat transfer by ventilation flow is limited to the range of 0.1 to 10 kJ/hr · K or from 0.0028 to 2.8 W/m² · K. Only for large temperature differences and high flow rates will airflow through cavities have an effect on enclosure thermal performance. For most enclosures, the additional heat flow rate due to airflow through cavities will be small (well under 1 W/m²) and hence negligible in most situations

Consider the following example. If an airspace behind thin cladding (with intentional vent openings at the top and bottom) has a flow rate of 1 m³/m²/hr, it can be considered to be well ventilated. Under average weather conditions, with the cladding being only 3 °C warmer than ambient, such a flow rate will transfer heat at a flow rate of 0.8 W/m^2. This value is small, especially when compared to more common situations such as, for example:

1. a typical equivalent conductance for an airspace (heat flow = 7 $W/m^2 \cdot K$ x 3 K = 21 W/m^2), or
2. the average exterior surface film coefficient (heat flow = 17 W/m^2 K x 3 K = 54 W/m^2), or
3. the conductive heat flow across a 100 mm layer of concrete of (2.3 $W/m \cdot K$ / 0.1 m x 3 K) = 69 W/m^2.

Hence, typical ventilation flow rates behind conductive claddings will not transmit a significant amount of heat energy compared to the other mechanisms involved and can generally be ignored.

The situation that typically causes the largest temperature differences across an enclosure element occurs when a dark-colored, light-weight roof is exposed to the sun (see next section). If a substantial amount of ventilation below the roof membrane could be ensured (e.g., 10 $m^3/m^2/hr$) and the outdoor air were to be 50 °C cooler (an extreme by possible scenario), the heat transfer rate due to convection would be about 500 kJ/hr or 140 W/m^2.

While not insignificant, this heat transfer rate calculated for extreme conditions can be compared to a typical solar gain of over 800 W/m^2 under sunny conditions. A flow of 10 $m^3/m^2/hr$ would require a flow velocity of 0.66 m/s through a 25 mm slot (an area that would exceed normal code ventilation area requirements) at both the ridge and eave of a 6 m span roof. These flow velocities are possible but unlikely. By comparison, changing the color of the roof from black to white will reduce the heat flow by at least 50% (e.g., from 800 to 400 W/m^2).

Airflow around or through insulated layers is, of course, a different and significant problem. One reason for the different impact is that insulation by its nature reduces heat flow significantly, and heat transmission through an insulated layer may only be in the range of 0.3 W/m^2 °C. Another reason is that the temperature drop across the insulated layer is, again by its nature, much higher, and often almost the same as the total difference across the enclosure.

Consider the case of air movement from outdoors, to the inside of an RSI 3.5 (R-20) insulation layer, and then back out again. Using one-tenth of the airflow rate used in the above example, 1 $m^3/m^2/hr$, and an indoor-outdoor temperature difference of 20°C, the airflow would move 24 $kJ/m^2/hr$ or 6.6 W/m^2. This flow of heat is of the same magnitude as the 5.7 W/m^2 flowing through the insulated assembly (i.e., heat flow = $\Delta T/R$ = 20°/3.5 =5.71). Hence, airflow around or through a well-insulated assembly could be the most significant contributor to the transfer of heat. Of course, 1 $m^3/m^2/hr$ would be a significant amount of airflow around an insulating layer (Chapter 7), equal to a flow velocity of 27.8 mm/s in a 10 mm gap.

Radiation-induced exterior surface temperatures

Radiation, either solar heating by the sun or cooling by night-sky radiation, can have a very significant influence on the thermal exposure of the building envelope. Although rarely warranted, detailed procedures can be used to calculate the effect of solar heating and night-sky cooling on enclosure surface temperatures. The following fairly simple approach can be used. The calculated result will be relatively accurate with regard to energy transfer rates (e.g., W/m^2).

At the exterior surface of an enclosure, the net heat exchange involves the absorbed solar (short-wave) and terrestrial (long wave, emitted from the surrounding materials and atmospheric moisture and gases) radiation less the convective and radiative losses to the environment and the conductive losses to the interior. Assuming the sky acts as a radiative black body, this heat exchange results in a net heat flux at an exposed building surface of (Figure 5.13):

$$q = \alpha\, I_{sw} + F_A\, \varepsilon_{surface} \cdot \sigma \cdot (T_{sky}^4) - F_A\, \varepsilon_{surface} \cdot \sigma \cdot (T_{surface}^4) - h_c \cdot (T_{surface} - T_{air}) - U \cdot (T_{surface} - T_{in}) \quad \text{(Eq. 5.39)}$$

where I_{sw} is the total short-wave (solar) incident solar radiation,

$T_{surface}$, T_{air}, T_{in} and T_{sky} are the surface, air, interior and sky temperatures respectively,

F_A is the view angle to the sky, about 0.5 for walls and 1.0 for roofs, and

U is the overall heat loss coefficient for the enclosure element.

Evaporation of water or melting frost requires latent energy. If a phase change occurs, then an extra term for energy lost due to evaporation or melting can also be included in the above formulation.

Although there are many variable, the net heat exchange can be solved for a given set of conditions if a highly insulated surface (no conductive heat loss), little thermal mass (the surface cannot store a significant amount of thermal energy), and a convection coefficient are assumed.

Thermal radiation between walls and the surrounding terrestrial environment can often be based on exchange with surfaces assumed to be at or near air temperature. However, if a wall adjoins a large hot roof, the radiative temperature "seen" by the wall will be higher than if the adjoining surface were a grass surface, and the temperature of the adjoining surface should be substituted for air temperature in the calculations.

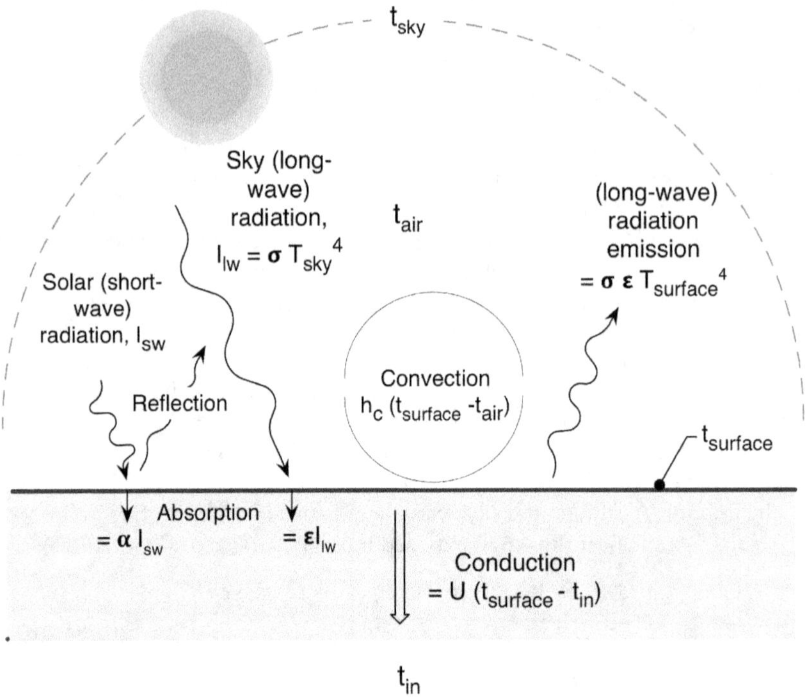

Figure 5.13: Radiation balance at a surface

Radiation exchange between an enclosure and the sky is complicated by the fact that clear skies can have an equivalent radiative temperature that is much lower than the air temperature near ground level. The air in the sky, of course, would not actually register a colder temperature if measured with a radiation-shielded thermometer: heat energy is being transferred by radiation from building surfaces to outer space and the atmosphere.

Night-sky radiation, as this longwave radiation loss is called, is the term $F_A\, \varepsilon_{surface} \cdot \sigma \cdot (T_{sky}^4 - T_{surface}^4)$ in Equation 5.39 and occurs all the time. However, this radiation loss is typically overwhelmed by solar gains in the daytime. The net radiation loss from a building surface to the night sky can reach 100 W/m², although under most conditions values of closer to 10 or 20 W/m² are representative.

The fewer clouds in the night sky and the lower the vapor and particulate content of the atmosphere, the more effective the radiation transfer and the colder the equivalent sky temperature "seen" by an enclosure. During most cold, clear nights the sky temperature will be more than 10 °C below the ambient air temperature. At high altitudes and in the desert, much larger temperature differences occur. In fact, sky temperatures as much as 20 °C lower than air temperatures are possible. Clouds and heavy fog practically eliminate the effect by blocking the passage of short-wave radiation.

The sky temperature at night is important since it affects the night cooling that surfaces experience. This cooling results in condensation (dew) and is important in many passive cooling applications and material durability problems. Dew forms on surfaces exposed to the sky that are cooled to below the condensation temperature of the ground level air. This wetting is particularly noticeable on non-absorptive lightweight roofs (e.g., metal) and wall cladding (e.g., vinyl siding) over well-insulated enclosures during clear nights following humid days.

Night-sky radiation reduces the temperature of exposed surfaces below air temperature, and this can readily be calculated. For example, if a wall surface temperature were 273K, and the sky temperature 263K, the net radiation to the sky (assuming F_A=0.5) would be:

$$q = F_A\, \varepsilon_{surface} \cdot \sigma \cdot (T_{sky}^4 - T_{surface}^4)$$

$$= 0.5 \cdot 0.9 \cdot 5.67 \times 10^{-8} \cdot (263^4 - 273^4)$$

$$= 19.65$$

The 19.65 W/m^2 of emitted radiation will cause the wall surface temperature to drop below the adjoining air temperature. Heat flow from the inside, if the wall were insulated to RSI3.5, will only be 5.7 W/m^2 at 273K (i.e. [293-273]/3.5) and the remaining deficit will be 10.15 W (19.65-5.7). If the temperature of the wall were to drop 3K below the air temperature, radiation to the sky will drop to 13.5 W/m^2, heat from the interior will provide 6.6 W/m^2 resulting in a deficit of 6.9 W/m^2. The deficit in heat flow will cause the temperature to drop, but convection and radiation from the surrounding environment will provide heat to the surface. It is the dynamic balance of heat loss to the sky by radiation and the heat gains from convection and radiation from terrestrial surfaces that restults in the actual surface temperature.

From the above it can be seen that knowledge of the sky temperature is needed for numerous calculations. As the sky temperature is not routinely measured, numerous researchers have attempted to provide simple correlations with environmental measures that are routinely recorded.

For a partly clouded night sky, Bliss [1961] proposed the relationship:

$$T_{sky} = T_{air}\,[\,0.8 + (T_{dewpoint} - 273)/250]^{1/4} \qquad \text{(Eq. 5.40)}$$

As this equation was derived for a sky with 50% cloud cover, the temperature of a completely clear sky will be several degrees cooler than this. If the dewpoint temperature of the air is not available, it can either be calculated from coincident relative humidity measurements or even assuming $T_{dewpoint} = T_{air} - 2$ is reasonable for many climates. Dry weather (or climates) and high-altitude locations with clear air typically have dewpoints of at least 5 and sometimes 10 °C below air temperature.

Hagentoft [2001] recommends a very simple equation (which implies a dewpoint temperature and is based on Celsius units) for calculations:

$$t_{sky} = 1.2\ t_{air} - 14 \qquad \text{(Eq. 5.41)}$$

The ISO standard 13791 [2012] recommends a slightly more complex, yet simple, equation:

$$T_{sky} = [9.36 \times 10^{-6}\ (T_{air})^6]^{0.25} \qquad \text{(Eq. 5.42)}$$

The sky temperature predicted above is for a clear sky and should be modified for cloud cover if it is present. Applying a simple factor (f=1-0.3c, where c is a cloud cover fraction) to the depression of sky temperature relative to air temperature, is one recommended estimate.

Since the angle factor between the sky and vertical walls is close to 0.5, the effect of night-sky radiation cooling is less pronounced for walls than for roofs, which would generally have an angle factor of 1.0 if horizontal. Any surface that is not directly exposed to the sky will not be affected by this radiation exchange. It follows that dew will not form on a surface that is "shaded" (below, say, a picnic table or a car) but will form on exposed surfaces such as the adjacent grass, wall cladding, or metal roof.

In more crowded area, the angle factor for walls will be less, as neighboring building fill more of the view: values of 0.41 for suburban areas and 0.33 for urban zones are suggested by ISO 13791.

Example. Estimate the surface temperature that might be reached by a bituminous roof (absorptance of 0.9) over a highly insulating and low-thermal mass substrate exposed to intense (1000 W/m²) sun on a calm, cloudless day with an ambient temperature of 20 °C.

Answer

As a low-slope roof is being considered, the angle factor, F_A, can be assumed to be equal to one. Because the roof is highly insulated, heat loss to the interior can be assumed to be zero.

Using Figure 5.11 the following equation can be developed:

$$\alpha I_{sw} + \varepsilon I_{lw} - \varepsilon\ \sigma\ T_{roof}^4 + h_c\ (T_{roof} - T_{air}) = 0$$

Setting long-wave absorptance equal to their emittances (Kirchoff 's Law) and gains equal to losses, we have

$$\alpha I_{sw} + \varepsilon I_{lw} = \varepsilon\ \sigma\ T_{roof}^4 - h_c\ (T_{roof} - T_{air})$$

Using Hagentoft's simple T_{sky} equation

$T_{sky} = 1.2\,T_{air} - 14$, T in Celsius

$T_{sky} = 1.2\,(20) - 14 = 10\ °C = 283\ K.$

$I_{lw} = \varepsilon_{sky} \cdot \sigma \cdot T_{sky}^4$, assuming $\varepsilon_{sky} \approx 1.0$

$= 1.0 \cdot 5.67 \times 10^{-8}\,(283)^4$

$= 364\ W/m^2$ incoming long-wave radiation

Given a solar intensity of 1000 W/m^2 and a solar absorption of 0.9, the total radiative gains can be calculated as:

$$\text{Gains} = \alpha\, I_{sw} + \varepsilon\, I_{lw}$$

$$= 0.9 \cdot 1000 + 0.9 \cdot 364$$

$$= 1227\ W/m^2$$

Losses are due to radiation to the sky and convective losses to the air, which can be evaluated as:

$$\text{Losses} = \varepsilon \cdot \sigma \cdot T_{roof}^4 + h_c\,(T_{roof} - T_{air})$$

Using $h_c = 1.52 \cdot (T_{roof} - T_{air})^{0.33}$ for convection from a horizontal surface (Eq. 5.13):

$$\text{Losses} = 0.9 \cdot 5.67 \times 10^{-8} \cdot T_{roof}^4 + 1.52 \cdot (T_{roof} - T_{air})^{1.33}$$

If an initial assumption is made that the roof is at the same temperature as the air (293K), then the losses are likely to be limited to the long-wave losses, that is, $0.9 \cdot 5.67 \times 10^{-8} \cdot 293^4 = 376\ W/m^2$. Clearly, the roof surface in this situation will heat up rapidly since the gains (1227 W/m^2) are much greater than the losses. By assuming that the roof has a temperature well above air temperature (as common sense would suggest), one can recalculate the losses. Simple iteration can be used to find the temperature, T=358 K (85 °C), for which the losses are equal to the gains. Hence, the roof in this example can be expected to heat to 85 °C, or some 65 °C above the exterior air temperature.

Various other conditions can be quickly analyzed using the same methods. If the solar radiation were reduced by half, to 500 W/m^2, the temperature rise of the roof would drop to 56 °C, or 36 °C above the ambient air temperature.

If the roof were colored bright white (absorptance of 0.3), the surface temperature would drop to 48 °C, representing a 28 °C rise. This is clearly a significant drop in temperature that has practical implications.

If a zinc roof were considered (an emittance of 0.2 and an absorption of 0.65), the temperature would rise to almost 99 °C, or 79 °C above ambient. Hence, emittance is as important as absorptance for calculating surface temperatures. In practice, the high temperatures reached by zinc roofs have required the development of special roof underlayment products tolerant of higher temperature than those traditionally and successfully used below dark-colored asphalt roofs.

The influence of conductive heat flow into the roof can also be considered by adding the conductive loss rate term:

$$\text{Conduction inward} = U \cdot (T_{roof} - T_{interior})$$

Since U is typically less than 0.3 W/m^2·°C for a well-insulated roof, heat loss to the interior will be only about 25 W/m^2 in the case described above. Including the conductive losses into the building will give a predicted surface temperature of only about 1 °C lower than that predicted for no conduction losses. For a poorly insulated roof (e.g., U = 1 W/m^2·°C), the difference in temperature will be 3 °C.

Thermal mass is often an important consideration. If the roof membrane were to be placed directly over a 200 mm thick concrete slab, then the heat capacity per square meter can be calculated as follows:

$$0.2 \text{ m} \times 1 \text{ m}^2 \times 2200 \text{ kg/m}^3 \times 900 \text{ J/kg K} = 396\,000 \text{ J / K}$$

Since the net heat flux is less than 900 W/m^2 (or 900 J/s/m^2) even when the roof is at the same temperature as the air, it will require:

$$q = (396\,000 \text{ J/K}) / (900 \text{ J/s}) = 440 \text{ s} = 7.3 \text{ minutes}$$

to warm the concrete 1 °C. Since peak solar radiation inputs last only a few hours at a time, it follows that a roof membrane installed on top of a concrete substrate will not reach nearly as high a temperature, or as quickly, as a membrane laid on top of thermal insulation.

Heat flow dynamics are somewhat more complicated than those assumed in the previous example (e.g., thermal mass, variable heat transfer coefficients, etc.), but the primary factors can be seen.

When predicting the radiative effect on surface temperatures, designers need to bear in mind that the thermal mass of the substrate is a much more important property than the insulation value, and that both the emittance and absorptance of the exposed overlay must be considered.

It is computationally convenient to develop equivalent temperatures that can be entered directly into a standard Fourier conductive heat flow analysis. One approach is to use an equivalent air temperature; this equivalent temperature is termed the *sol-air temperature* and can be calculated as:

$$T_{sol\text{-}air} = T_{air} + (\alpha\ I) / h_o$$

Where I is the incident solar radiation (W/m^2), and

h_o is the heat transfer coefficient at the outer surface (W/m^2 °C).

This procedure has the drawback that air temperature is no longer realistic but has the computational advantage that surface films play a role in the evaluation of the surface temperature. The value for the overall surface film coefficient is often assumed to be 17 $W/m^2{\cdot}°C$, although the value for this coefficient is quite variable, especially if the surface becomes hot. In fact, under solar heating, the surface film coefficient is likely to be as much as two times larger.

Another approach, somewhat more direct, is to use *equivalent surface temperatures.* Table 5.6 is a compilation of values for the most useful approximate surface temperatures based on calculation (using refined surface heat transfer coefficients) and supported by several years of field measurements [Straube and Burnett 1998]. Under this approach the effect of the surface film can be computationally neglected, and the temperature of the exterior surface can be entered directly into calculations.

Table 5.6: Approximate radiation-induced surface temperature design values (°C)

Situation	Thermally massive	Thermally lightweight
Roof: direct sun	$t_a + 42\ \alpha$	$t_a + 55\ \alpha$
Roof: sun + reflected / emitted radiation	$t_a + 55\ \alpha$	$t_a + 72\ \alpha$
Roof: exposed to night sky	$t_a - 5\ \varepsilon$	$t_a - 10\ \varepsilon$
Wall: winter sun	$t_a + 35\ \alpha$	$t_a + 48\ \alpha$
Wall: summer sun	$t_a + 28\ \alpha$	$t_a + 40\ \alpha$
Wall: exposed to night sky	$t_a - 2\ \varepsilon$	$t_a - 4\ \varepsilon$

Notes: t_a refers to the ambient air temperature, ε is the surface emittance, and α is the solar absorptance. Latitudes of 30 to 60°

While sol-air temperatures assume that all enclosures are lightweight and perfectly insulated, the values in Table 5.6 are based on two classes of reasonably well-insulated assemblies behind the exterior surface. The lightweight values assume that the surface layer or the immediate substrate has almost no heat storage capacity, while the thermally massive values assume a heat storage capacity equivalent to a 90 mm thick layer of masonry or concrete.

Furthermore, using some engineering judgment and the theory outlined above, the tabulated values can be modified. For example, a solid 300 mm thick, uninsulated stone wall will experience slightly lower peak temperatures (perhaps 5 °C), whereas a metal roof on a mountaintop in the Sahara Desert will likely experience night-time minimum temperatures perhaps 10°C lower than tabulated.

Thermally massive refers to walls with a significant amount of thermal storage capacity (e.g., brick veneer or equivalent) outside a low conductance material (e.g., insulation). Walls with significantly more mass (e.g., multi-wythe brick, rubble) or less thermal insulation ($U > 1$ W/m^2 °C) will be even less affected. Vinyl, metal panels, and EIFS are thermally lightweight. A thin roofing membrane (such as a metal or modified bitumen membrane) with thermal insulation as the substrate is an example of a thermally lightweight enclosure.

Although the values are for locations at a latitude of around 45°, they will apply with little impact on accuracy over a range of at least 30° to 60° from the equator. The values for walls are likely extremes for clear nights or sunny days and apply to east-west orientations in summer and south orientations in the winter.

5.10 Worked examples

5.10.1 Calculate the R value of an enclosure assembly

The process below can be used for many practical applications.

1. List each material in the wall or roof, its conductivity (k), and its thickness (l) in m.
2. Calculate the conductance of each layer using $C = k / l$.
3. Calculate the thermal resistance of each layer using $RSI = 1 / C$
4. Sum the individual thermal resistances to get the answer.

Usually the thermal resistance of the air films that exist on both the interior and exterior surfaces of an assembly are added to the wall as virtual layers. This makes the calculations more accurate. It is often more important to include surface films to better estimate the surface temperatures of the enclosure.

Example

Calculate the total thermal resistance, R, and overall heat transfer coefficient (U) of the wall shown below. Use conductivity values from tabulated values.

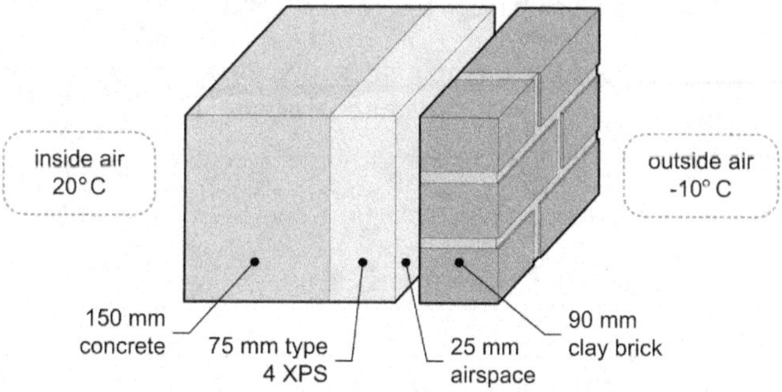

Answer

Layer material	Conductivity [W/mK]	Thickness [m]	Conductance [m²K/W]	Resistance [RSI]
Interior film note 1	N.A.	N.A.	8.3	0.120
Concrete	1.8	0.150	12	0.083
Type 4 XPS	0.029	0.075	0.39	2.56
Air space note 2	N.A.	0.025	N.A.	0.17
Brick	1.3	0.090	14.4	0.069
Exterior film note 1	N.A.	N.A.	34	0.029
		Total Resistance		RSI3.04
		Overall coefficient of heat transfer		0.33 W/m²/K

Notes

1. Since the interior and exterior films are fictitious, they do not have a thickness, and so no conductivity. Hence, tables typically contain only conductances or resistances for the layer. These films may have quite a range of values but, as can be seen, the effect of the value of the film resistance on the total resistance is small if the enclosure is a modern, insulated assembly.
2. The flow of heat through an air space is complicated by convection (air flows) and radiation, so tabulated values of conductance are used instead. Like surface films, these may have a range of values but they are not important to the overall accuracy for most modern walls. Interpolation and iteration should be used.

The total resistance is RSI 3.04 (or Imperial R 17.2), 84% of it provided by the insulation. The overall heat transfer coefficient is 0.33 W/m²/K.

5.10.2 Calculating steady-state heat flow through an enclosure system

Heat flow across an assembly is simply the temperature difference divided by the R-value times the overall heat transfer coefficient. The temperature difference is usually just ($t_{inside} - t_{outside}$).

1. Find the total thermal resistance of the enclosure as described earlier.

2. Find the overall heat transfer coefficient U, using $U = 1/R_{total}$.

3. Multiply the temperature difference across the assembly by U, i.e., $U \cdot (t_{inside} - t_{outside})$.

Of course, if the sun is shining on the wall, the outdoor temperature is not the correct one to use. The actual solar-heated surface temperature is more accurate, but it is sometimes difficult to find.

Note that the effect of heat storage, or thermal mass, is important for this particular wall since heavy, mineral-based materials can store a lot of heat.

Example

Calculate the amount of heat flow per unit area through the wall of the previous example when it is -10 °C outside with no sun and 20 °C inside.

$$\text{Heat flow} = U \cdot (t_{inside} - t_{outside})$$

$$= 0.33\ (20 - 10) = 0.33\ (30)$$

$$= 10\ \text{W/m}^2$$

Answer

Heat flux outward will be 10 W/m^2.

Example

Assume the wall in the previous example is exposed to bright sun. Calculate the heat flux.

Answer

The temperature of a very dark surface would be expected to be as much as 48 x 1 = 48 °C above the air temperature (see Table 5.6). For this example, we will assume dark red brick (thermally massive and solar absorptance = 0.80) and thus estimate a surface temperature of about 35 x 0.8 = 28 °C above the air temperature. This means that:

$$\text{Heat flux} = U \cdot (t_{inside} - t_{outside})$$

$$= 0.33\ [20 - (-10 + 28)] = 0.33\ (2)$$

$= 0.66 \text{ W/m}^2$

If the wall is dark red and is exposed to bright sunshine, the heat flux will be outward and only about 0.66 W/m^2 if we ignore dynamic effects.

5.10.3 Calculating steady-state heat flow through a building system

Energy flow through a building can be simply calculated so long as the assumptions of one-dimensional, steady-state flow are reasonably accurate. For approximate answers, this assumption is often acceptable. In cases where thermal mass and multi-dimensional heat flow are important, more elaborate techniques (usually computer-based) must be employed.

In essence, the simplest method calculates the transfer of heat through each component using the methods described above, and the total heat loss (or gain) is calculated as the sum of all the components. This can be accomplished in a step-by-step manner, as follows:

1. List the U-value for each element (wall, roof, window, etc.) along with the area of that element.
2. The product of the area and U-value of each element is the heat loss for that enclosure element per unit temperature difference (W/°C).
3. The sum of these products is the overall heat loss coefficient for the building.
4. To find the overall heat loss or gain, multiply the overall heat loss coefficient for the building by the temperature difference across the assembly at the conditions being considered.

If air leaks out of the building in winter, it is of course replaced with cold outdoor air. The replacement air must be heated to make it comfortable. The energy impact of air leakage is significant and must be considered.

It can be included by calculating an equivalent heat loss coefficient for air leakage. This can be found from Equation 5.19:

$$U_{air} = 0.3 \cdot n \cdot V$$

where U_{air} is the overall heat loss coefficient due to air leakage (W/°C), exactly analogous to U for enclosure assemblies.

n is the number of complete air changes per hour (ACH)

V is the total air volume of the building (in m^3)

Example

An industrial "big box" store is 8 m high, 32 m long, and 64 m wide.

The enclosure has a lightweight roof (U=0.36 W/m²· °C) and precast sandwich panel walls (U=0.4 W/m²· °C). The front of the store (which faces west) has a glass curtain wall 6 m high by 24 m long, with a U-value of 2.0 W/m² · °C. Six loading doors are at the back, each 5 m high and 3.5 m wide, with a U-value of 1.0. The building is estimated to have an air exchange rate under normal conditions of 0.5 ACH.

Ignoring the effect of the floor slab, sun, and any other doors, find the heating requirements when it is -10 °C.

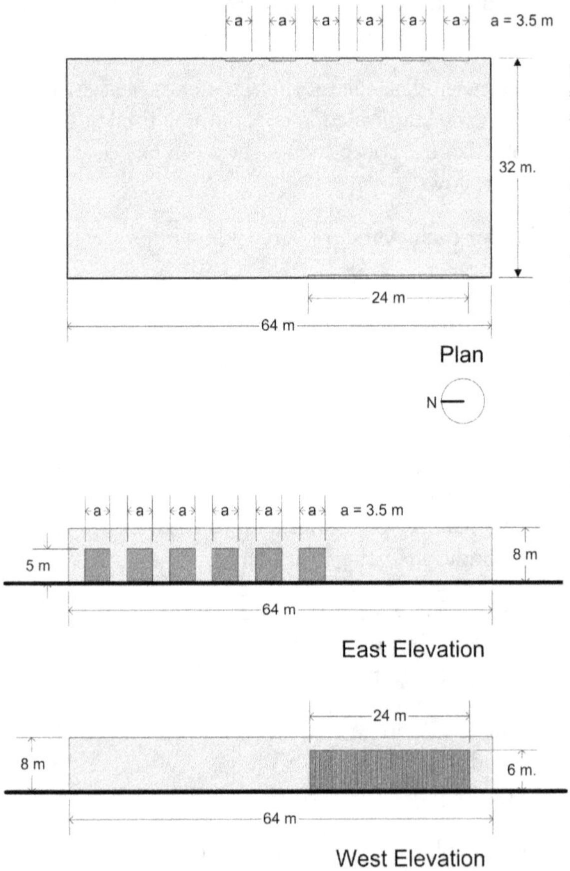

Building Volume
=32 x 64 x 8
=16384 m³

Roof Surface Area
=32 x 64
=2048 m²

Curtain Wall Surface Area
=24 x 6
=144 m²

Door Surface Area
=6(3.5x5)
=105 m²

Wall Surface Area
=(64x8)2 + (32x8)2 - (105+144)
=1287 m²

Building Surface Area
=2048 + 144 + 105 + 1287
=3584 m²

Answer

See the sketch of the building for details of the areas.

The heat loss coefficient for air leakage is:

$$0.3 \text{ n V} = 0.3\ (0.50)\ (16384) = 2458 \text{ W/°C}.$$

By tabulating the heat loss and area of each component, we can compute that the total heat loss is 4103 W/ °C. For an outdoor temperature of -10 °C and indoor temperature of 20 °C, the difference is 30 °C, and total heat loss is 123 kW.

Enclosure component	Area (m^2)	U Value ($W/m^2/°C$)	Q Heat loss (W/°C)	% Total heat loss
Walls	1287	0.40	515	13%
Roof	2048	0.36	737	18%
Doors	105	1.0	105	3%
Curtain wall	144	2.0	288	7%
Air leakage		*from above*	2458	60%
	Overall building heat loss coefficient		4103 W/°C	

The relative impact of each building component can also be seen from the tabular format. For this building, the U-value of the enclosure is unimportant relative to the impact of air leakage. It should also be noted that the lights and equipment in the store typically consume 30 to 50 W/m^2 of electricity, all of which is converted to heat. Hence, if we assume an average energy use of 40 W/m^2, the interior gains will be:

$$2048 \text{ m}^2 \times 40 \text{ W/m}^2 = 81920 \text{ W} = 82 \text{ kW}.$$

This energy will offset the losses of 123 kW, but a heating system of some type will be needed to make up the remaining (123-82) = 41 kW. HVAC design engineers would rarely include the heat produced by internal occupancy, as this may not always be available (e.g., a weekend or holiday night).

5.10.4 Calculating the temperature gradient through an enclosure

In many building science applications it is desirable to have some knowledge of the temperature gradient within the enclosure element being studied. This is especially important for studies of condensation and drying as well as thermal movement and stresses.

To find the temperature at any layer j within an enclosure, Equation 5.34 is often used.

1. Calculate the overall U-value of the assembly.
2. Calculate the total RSI (R-value) of all layers inside of the point of interest in the assembly, $R_{0\text{-}j}$,

3. Calculate the temperature drop from the inside to the point of interest using:
   $\Delta T_j = U \cdot (t_{int} - t_{ext}) \cdot R_{0\text{-}j}$
4. Calculate the temperature at the point of interest by subtracting the drop from the interior temperature.

To provide complete information about temperature within an assembly, the individual temperature drops across each layer should be calculated and then all individual temperature drops subtracted from the interior temperature. For example, using the wall system from the previous example, Table 5.7 can be developed. A spreadsheet-based approach is recommended to provide the analyst with an overview of the assembly.

Table 5.7: Calculation of temperature distribution through enclosure assembly

Layer Material	k (W / m·K)	t [m]	C_i (W / m²·K)	R_i (m²·K /W)	T (°C)	t (°C)
Interior temp						*20*
Interior film	N.A.	N.A.	8.0	0.120	1.58	
						18.4
Concrete	1.8	0.150	12	0.083	1.09	
						17.3
Type 4 XPS	0.029	0.075	0.39	2.56	33.68	
						-16.3
Air space	N.A.	25	N.A.	0.17	2.24	
						-18.59
Brick	1.3	0.090	14.4	0.069	0.91	
						-19.5
Exterior film	N.A.	N.A.	34	0.029	0.38	
Exterior Temp						*-20*

RSI total	3.04
U =	0.33 W/m²/K

5.11 Beyond 1-D Steady-state Heat Flow

One-dimensional, parallel and steady-state heat flow assumptions may not be suitable for the accurate prediction of space-conditioning energy consumption or local surface temperatures for three main reasons:

1. **Thermal bridging**: structural members (e.g., wood but especially metal members in light-frame construction and concrete and steel columns/walls in commercial construction), cladding attachment fixtures, and service entries are common examples that create a more effective path for heat flow than through the insulated enclosure. Only two- or three-dimensional analysis can predict and quantify this type of heat flow, termed a thermal bridge. The use of insulation to serve as a thermal break over the thermal bridges, and thermal breaks to interrupt themal bridges will effectively reduce this type of heat transfer.

2. **Thermal mass**: heat flow through building enclosures is not steady-state over longer periods of time than a few minutes or hours. Heat stored in the thermal mass of buildings and then released can reduce both heating and cooling requirements. Thermal mass also makes it possible to make more effective use of internal and solar gains. Simple steady-state methods of analysis cannot model these effects accurately. Hour-by-hour computer programs are widely available to aid in the assessment of dynamic effects.

3. **Air leakage**: Air leakage through the building enclosure can be a major source of energy loss. Airtight buildings use far less energy. Again, the simple, commonly used analysis techniques cannot assess the influence of air leakage on the thermal performance of a building system. The loss of energy by air leakage (convection) must be calculated in addition to the loss by transmission through the enclosure.

5.11.1 Thermal bridges

In the foregoing it has been assumed that heat flow is one-dimensional, i.e., that all building enclosure components are planar and extend to infinity in every direction. Surprisingly, this assumption is often reasonable for the thermal analysis of simple enclosure systems. However, heat flow deviates from the one-dimensional at corners, parapets, intersections between different assemblies, and within complex 3-D assemblies, such as light metal framing, aluminum tube frames, etc.

When heat moves through an enclosure element it flows through more than just the center of the panel. Additional heat will flow through areas of steel or concrete that penetrate the insulation layer for example. When heat flows at a much higher rate through one part of an assembly than through another part, the term *thermal bridge* is used to reflect the fact that the heat has "bridged over" or moved around the thermal insulation. Thermal bridges are inevitable in most enclosures and codes and standards increasingly require designers to account for them when judging compliance.

Thermal bridges become important when:

- they lead to cold spots within an assembly that might impact performance (e.g., surface condensation, durability) or result in comfort problems; or
- they are either large enough or intense enough (highly conductive) to meaningfully increase the total heat loss through the enclosure.

In practice, thermal bridging can routinely reduce the thermal resistance of some enclosure systems (e.g., steel stud framing with batt insulation) by 50% or more. On the other hand, some systems experience very little reduction (e.g., EIFS on concrete, ICF). To allow for a proper understanding of enclosure performance, realistic inputs for whole-building energy models, and a fair comparison of enclosure system choices, thermal bridging should be considered.

All enclosures should be designed to avoid or minimize the impact of thermal bridging. The most effective solution is the use of thermal or heat flow breaks, e.g., by providing a continuous insulation layer on the exterior of the enclosure support system (whether concrete, masonry, or framing) to reduce thermal bridging. Such an approach, suggested in Chapter 2, provides energy-saving benefits and improved resistance to exfiltration condensation

Many modern North American enclosure systems employ gypsum wallboard (GWB) and light-gauge steel framing ("steel studs") on the interior to provide a familiar finish and additional fire resistance, and/or to provide a space to easily run services such as power and communications. In many cases the space between the studs is also insulated and this is the most common significant thermal bridge in modern enclosure walls.

To calculate the thermal performance provided by a layer of 3-1/2" (90 mm), 4" (102 mm) or 6" (152 mm) steel stud, the significant thermal bridging caused by the heat flow through the studs and tracks must be considered. Studs that resist wind load tend to be thicker (18- or 20-gauge, in the range of 0.9 to 1.3 mm) whereas studs that support interior gypsum may only be 25-gauge. The thicker gauge steel does transmit more heat, but both types drastically reduce the nominal R-value of any insulation (fibrous or foam) installed between the framing elements. Hence, prescriptive tables in energy codes increasingly require a specific amount of insulation on the exterior of the studs to provide continuous insulation ("ci" in code short form).

For practical applications, steel stud framing and any insulation installed in the stud cavity can be simplified to a monolithic layer with an equivalent R-value (to which the R-value of the interior finish and any exterior layers can be added). Table 5.8 lists typical recommended values for design of common steel stud systems (extrapolated from Table A3.3.3.1 ASHRAE 90.1-2022).

Table 5.8: Clear-wall R-value for light gauge steel framing acting as a single layer

Cavity/Stud Depth		Rated Cavity R-value	Layer R_{cw} @ 16 inch centres	Layer RSI_{cw} @ 405 mm centres
inch	mm			
2.5	64	Empty	0.75	0.13
3.5	89	Empty	0.79	0.14
		R-13	6.0	1.06
		R-15	6.4	1.13
6.0	152	Empty	0.85	0.15
		R-19	7.1	1.25
		R-21	7.4	1.31
		R-24	7.6	1.34

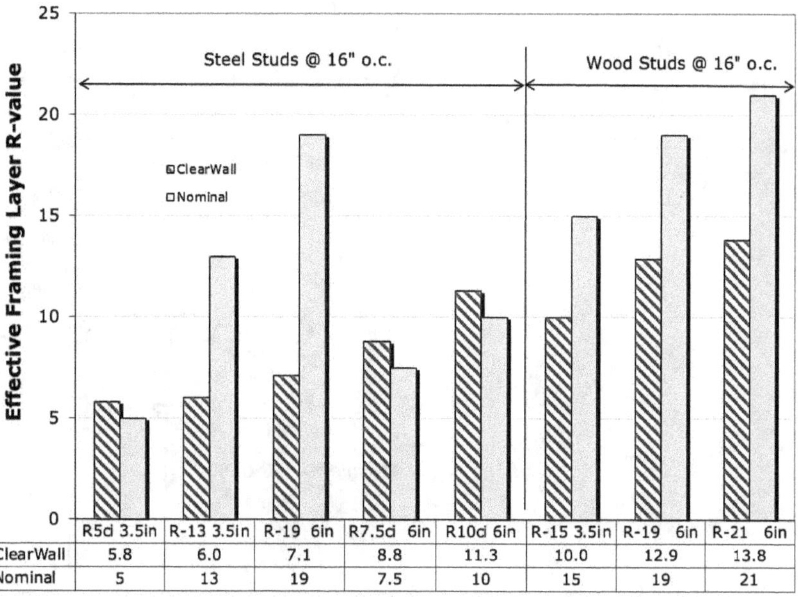

	R5ci 3.5in	R-13 3.5in	R-19 6in	R7.5ci 6in	R10ci 6in	R-15 3.5in	R-19 6in	R-21 6in
ClearWall	5.8	6.0	7.1	8.8	11.3	10.0	12.9	13.8
Nominal	5	13	19	7.5	10	15	19	21

Source: ASHRAE 90.1-2022, Table A3. ci denotes a layer of continous insulation outside framing with no penetrations

Figure 5.14: Equivalent R-values including thermal bridging for wood and steel framing layers

The magnitude of thermal bridging in the table should be noted: thermal bridging reduces the thermal performance of the installed insulation to between 32 and 42% of the rated value. Figure 5.14 graphically shows the significant influence of thermal bridging on heat flow through steel stud walls and the performance of walls with continuous exterior insulation.

It can also be seen that adding R-21 (RSI3.7) batt to a 6" (150 mm) steel stud, the thermal performance improves by only R-6.6 (7.4 – 0.85). It is for this reason that the higher cost of exterior rigid insulation on a per unit R-value basis is justified: close to 100% of the rated value of ci can be realized versus only about 1/3 of the batts.

Thermal bridging—especially that created by steel framing, at corners, at the intersection of walls with roofs or floors, or with projecting balconies and perimeter slabs—often leads to cold interior surface temperatures and the possibility of condensation. Figure 5.15 and Figure 5.16 illustrate how studs and corners can result in low surface temperatures. In the case of the steel framing shown, an exterior temperature of -20 °C can result in interior surface temperatures of 5 to 10 °C at studs, and below-freezing temperatures at floor-to-wall corners. One major function of continuous exterior insulation is to reduce thermal bridging in order to avoid low surface temperatures that can allow condensate and moisture-related problems to form (Figure 5.17).

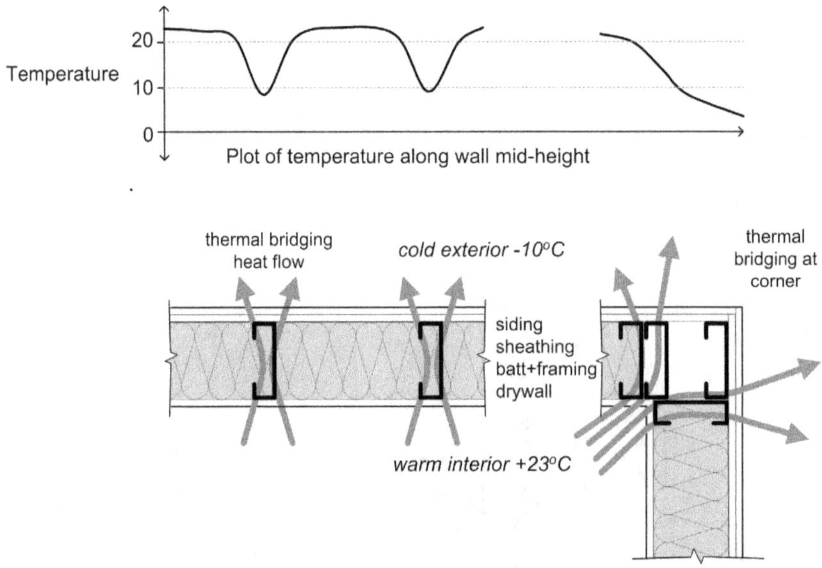

Figure 5.15: Surface temperatures at thermal bridges

Interior surface temperatures can be predicted using:

1. a good estimate of the thermal resistance of the thermal bridge,

2. the appropriate interior surface film coefficient (recognizing the impact of convection and radiation), and

3. knowledge of the likely interior relative humidity.

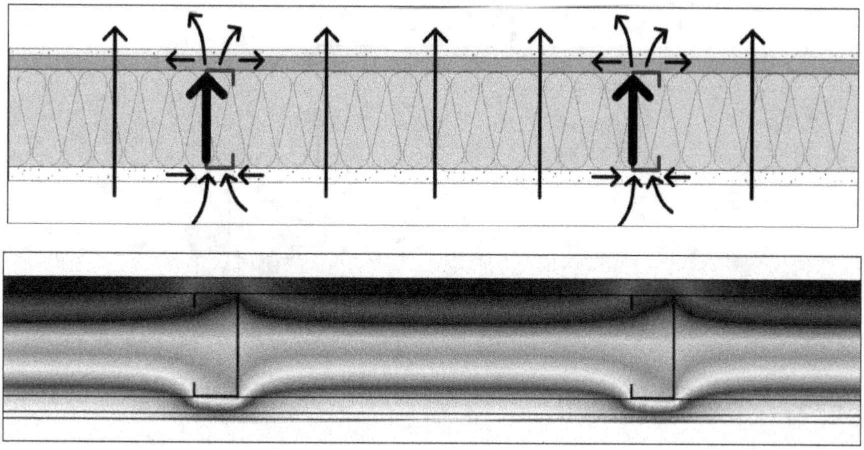

Figure 5.16: Two-dimensional heat flow in steel stud wall

The latter two components of the assessment often have significant uncertainty but are just as important as the heat flow through the thermal bridge.

Computer programs are available to aid in assessing heat flow through two- and three-dimensional assemblies. THERM, a widely used two-dimensional steady-state model from the Lawrence Berkeley Laboratory, Berkeley, California, can be downloaded from the internet at no cost.

The influence of thermal bridging is shown first, as a simple flow diagram, and second, as a contour map of temperature, in Figure 5.16. It is important to recognize that hand calculation methods aim to provide a prediction of heat flow: they cannot easily be used to identify local temperature extremes.

Hand and/or spreadsheet calculation methods for estimating the R-value/U-value of 2D enclosure systems fall into two categories: parallel path methods and the isothermal plane methods. Both are described in ASHRAE *Handbook of Fundamentals*. Various other combinations and weighted methods are also available but are not considered here. It should be borne in mind that accuracy and precision are not that useful for hand calculation methods: where accuracy is required, computer models should be used. However, the fast and general estimates provided by approximate calculations are often more than sufficient for making design decisions.

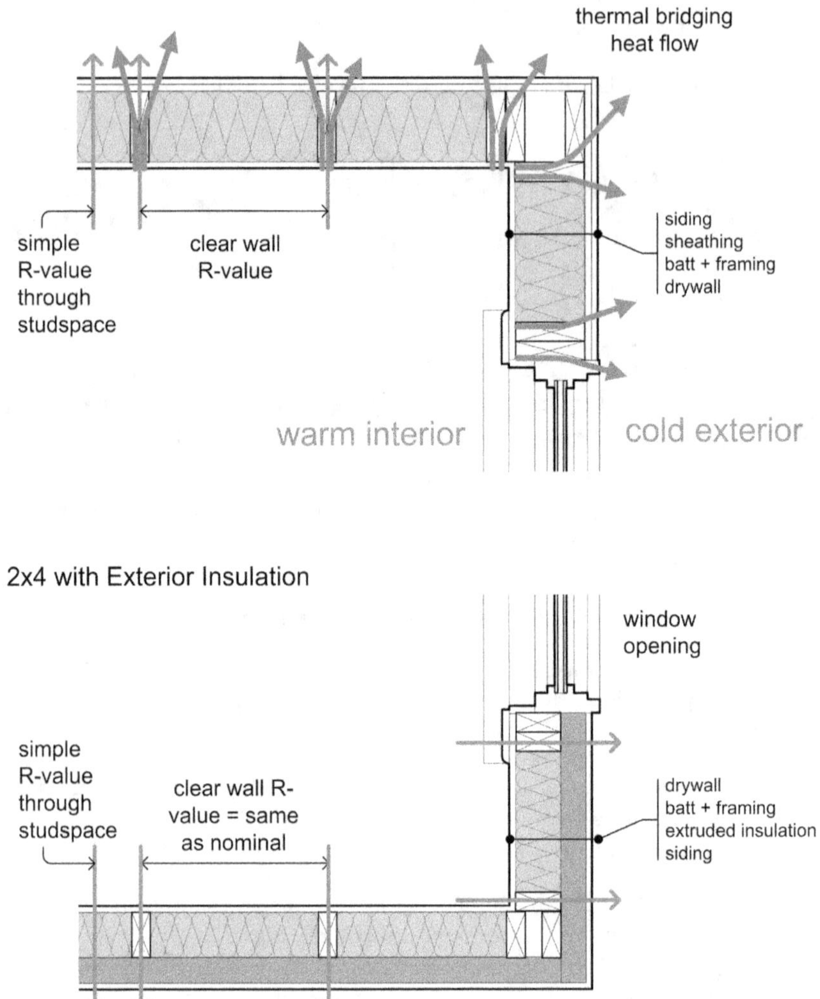

Figure 5.17: Thermal bridges and the role of exterior insulation as continuous insulation in light-weight framed structures

5.11.2 Parallel path method

The heat flow through envelopes with relatively non-conductive thermal bridges (e.g., wood studs, polymers, and even concrete or masonry) can be predicted with reasonable accuracy using the parallel path method. This method assumes that heat flows perpendicular to the surface of the enclosure, that is, that flow through different assemblies occurs along parallel paths. First, the conductance or overall thermal transmittance through each of the sections that constitute the total enclosure section (e.g., wood studs and batt insulation, concrete column and block infill) is calculated (Figure 5.18). Then the total heat flow is calculated, based on a weighted average of the areas.

For example, for an enclosure with a discontinuous layer involving two different materials (e.g., wood studs and batt insulation), the average conductance can be calculated from:

$$U_{avg} = U_1 \cdot \frac{A_1}{A_1 + A_2} + U_2 \cdot \frac{A_2}{A_1 + A_2} \qquad \text{(Eq. 5.42)}$$

where U_{avg} is the average overall transmittance,

U_1 and U_2 are the overall thermal transmittances through paths 1 and 2,

A_1 and A_2 are the areas perpendicular to the flow through paths 1 and 2.

Of course, the average R-value can then be calculated as $1/U_{avg}$ and any arbitrary additional parallel paths can be added.

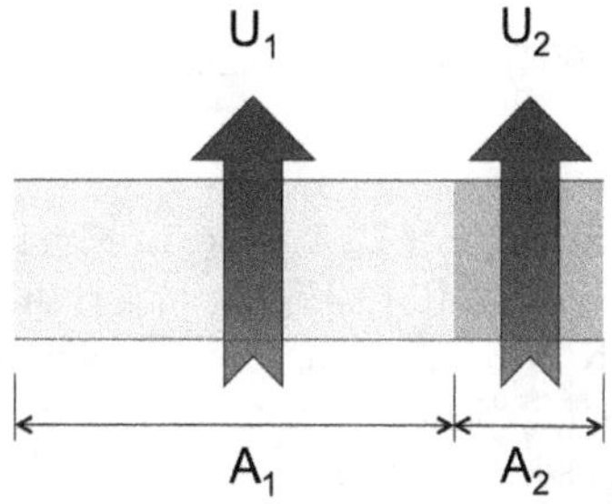

Figure 5.18: Parallel path method definitions

The parallel path method is easy to use and helpful for design, as it does not require as much effort or knowledge early in the design stage to assess the performance of details and enclosure assembly designs. For wood-framed wall systems it has been shown to be very accurate (+/- 2%) when compared to full-scale hot-box testing [Kosny et al 2006], and comparisons of computer-modeling of concrete slabs and masonry and precast walls [Straube 2017] have found acceptable accuracy (usually +/-10%) as well.

Example: A wood stud wall is comprised, from the inside, 13 mm gypsum wallboard, a 2x6 (140 mm) softwood frame filled with RSI3.35 batt, 11 mm OSB sheathing, and vinyl siding. Assuming wood has a thermal conductivity of 0.12 W/mK, and wood comprises 23% of the enclosure area (also called the framing factor), use the parallel path method to calculate the clear wall R-value.

Answer

The thermal resistance and U-value through each of the two paths, stud and cavity, is calculated, as shown in the table below.

Through Cavity		**Through Stud**	
Layer	RSI	Layer	RSI
interior film	0.12	interior film	0.12
13 GWB	0.10	13 GWB	0.10
RSI 3.35 Batt	3.35	140 wood	1.16
11 OSB	0.11	11 OSB	0.11
vinyl Siding	0.11	vinyl Siding	0.11
exterior film	0.03	exterior film	0.03
Total RSI	**3.81**		**1.66**
U-value	**0.262**		**0.604**

Using 23% for the proportion of the enclosure that is wood, the average U-value is calculated as:

$$U_{avg} = 0.23\ (0.604) + (1\text{-}0.23)\ (0.262)$$
$$= 0.341$$

This is equal to an RSI value of 2.93 (R-16.7), which can be compared to the installed R-value of the batt, RSI3.35 (R-19), and the center-of-cavity resistance, RSI3.81 (R-21.6). Although these values indicate a reduction in performance (of about 25% of the center-of-cavity), the impact of wood studs is far less than that of steel framing.

5.11.3 Isothermal planes

The isothermal planes method is more appropriate for envelopes containing highly conductive materials that penetrate insulated spaces (e.g., steel through insulation). Some sources (such as ASHRAE, ISO) recommend using both isothermal and parallel path methods. The Canadian National Building Code (2020) recommends a modified isothermal planes method combined with parallel path method. Detailed computer modeling has shown that such additional effort is not warranted for simple hand calculations as simple methods have results with sufficient and similar accuracy [Santos et al, 2020]. The use of tabulated values for steel stud framing is recommended.

For an enclosure with two sections (e.g., a concrete web connecting the inner and outer wythe of an insulated concrete sandwich panel), the average resistance of the insulation layer pierced by the concrete web could be calculated by:

$$R_{avg} = \frac{1}{\dfrac{A_{conc}}{A_{Total} R_{conc}} + \dfrac{A_{ins}}{A_{Total} R_{ins}}} \quad \text{(Eq. 5.43)}$$

The resulting average resistance is then used to calculate overall heat flow through the enclosure.

Although isothermal planes method may be used to include the thermal resistance of a complex layer, such as steel studs filled with batt or hollow concrete masonry filled with foam, the complexity of light steel framing geometry means other methods are still required to estimate the heat flow through the stud. The use of 1-D methods combined with tabulated equivalent layer resistance from testing or computer modeling. This is the approach shown earlier for use with steel framing (Table 5.8). Wood framing can take advantage of the simple parallel path method with good accuracy.

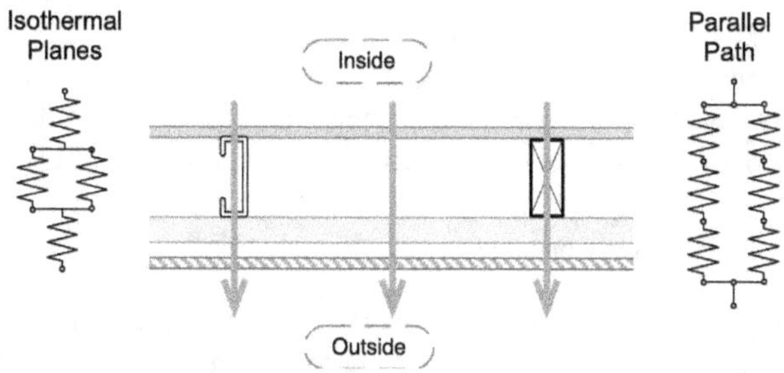

Figure 5.19: Parallel path and isothermal planes methods compared

5.11.4 ISO 14683 approach

A more recent methodology of calculating thermal bridging is to use linear and point-based thermal bridging factors ψ and χ, respectively according to the ISO 14683 [2017] and CSA Z5010 [2021] standards. These factors are mostly from two- and three-dimensional thermal computer models. They have been published for a number of assemblies [Higgins et. al. 2014, Pagan-Vazquez 2015 et al, Anon 2021], and are increasingly being provided by manufacturers.

Thermal bridging in building practice can usually be divided into two types: linear details that predominately exhibit two-dimensional heat flow and point details whose heat flow is primarily three-dimensional. Assigning the symbol psi (Ψ) to heat transmittance in two-dimensional details and the symbol chi (χ) to heat transmittance at a point thermal bridge results in a heat loss equation that accounts for thermal bridging for a given building enclosure component:

$$Q = [U_{cw} \cdot A + \Sigma(\Psi_i \cdot L_i) + \Sigma(\chi_j \cdot n_j)] \cdot \Delta T \quad \text{(Eq. 5.44)}$$

where

U_{cw} is the clear wall heat transmittance ($1 / R_{cw}$)

A is the area, including all details in the analysis area

Ψ_i is the linear heat transmittance value of detail "i"

L_i is the total length of the linear detail "i" in the analysis area

χ_j is the point heat transmittance value of detail "j", and

n is the number of point thermal bridges of type "j" in the analysis area.

This calculation method can be applied to all enclosure systems, but requires the development of specific thermal bridging factors, most of which have not been published yet. This methodology assumes that the psi- and chi-factors include all heat flow above and beyond the clear wall performance. This can be true for a specific detail, but changes such as the thickness of the insulation, interior air films, or support material results in published factors being more estimates than exact results. On a specific project it is quite feasible to calculate the psi or chi factors for each of the important enclosure details using computer simulations and then use Equation 5.44 to extend the results to the often dozens or hundreds of times each detail occurs on a building.

5.11.5 Limitations of approximate methods

While it is often accepted that 1-D calculations are appreciate and imprecise, it is very important to recognize that the approximate two- and three-dimensional methods described above not always provide an accurate representation of heat flow within the enclosure. The parallel plane method has the distinct advantage that it allows one to judge the likely relative temperatures within the enclosure. The calculation of temperature through the thermal bridge will provide a limit of the temperature profile through this section. The relative ease of lateral heat flow in the enclosure will determine the accuracy of the calculated temperature profile. Since steady-state, 2-D heat flow programs are available at such a low cost and are so fast and easy to use, non-computerized, multi-dimensional calculation methods are becoming less useful, and computer models should be employed for any applications requiring accuracy. Approximate methods remain useful for early-stage estimates and general predictions.

5.11.6 Dynamic heat flow

Heat flow is rarely a steady-state process. Although the air temperature usually does not change very quickly, the effect of the sun on heat flow can be rapid and thus potentially important.

Heat capacity is ignored in a steady-state analysis, as well as in almost all standardized (e.g., CSA, ASTM) tests, although it can often be very important to the actual thermal behavior of some types of buildings and enclosures.

It is a relatively simple matter to calculate heat flows and temperature fields using one of several available finite-volume software packages (including WUFI®) that incorporate both thermal conductivity and heat capacity. The calculation engines of some whole-building energy software packages include thermal mass with varying degrees of precision and even wider ranges of accuracy. Newer models such as EnergyPlus from the US Department of Energy even support temperature-dependent conductivity.

The heat flow within the soil around a building provides one excellent example of the potential importance of thermal mass. The same phenomenon occurs in thermally massive enclosures, although the time lags tend to be measured in hours, not months like below grade environments. Figure 5.20 compares the response of a building's interior temperature to the cyclical variation of outdoor temperature and solar gain. Thermal mass reduces the magnitude and delays the peak time.

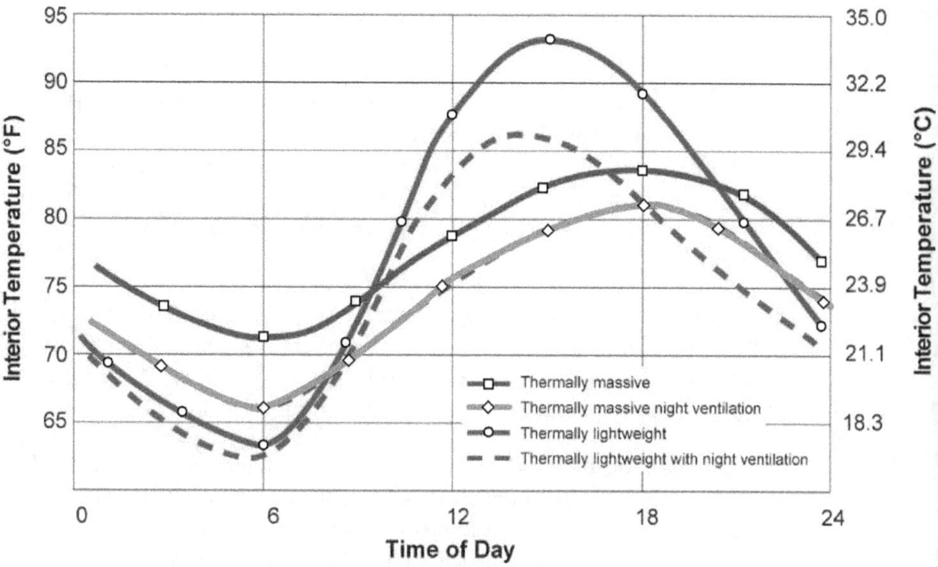

Figure 5.20: Temperature variation in a building with thermally-massive and lightweight enclosures

Although heat flow through an enclosure can be easily predicted, it is quite difficult to predict the energy impact of operating the building. A combination of complex interior surface films and interior finishes reduces the interaction of the thermal mass with the interior, mechanical equipment reduced the temperature swings over the day (often to meet modern comfort expectations), and the thermal mass of the interior furnishings and finishes also plays an important role.

The best simple measure of thermal mass is the heat capacity, or heat capacity per unit area of enclosure. However, thermal mass located to the exterior of much of the insulation will be far less effective. Hence, the ease or difficulty of heat flow into or out of the layer of thermal mass under consideration is very important. For example, thermal mass that is separated from the interior room air by only a surface film will be much more effective at moderating interior temperatures than thermal mass that is separated by an insulating layer such as carpet, or insulation. Similarly, thermal mass inside a layer of insulating material will respond much more slowly to exterior temperature variations than a layer with the insulation on the interior.

5.12 Definitions of R-value

There are many different definitions of R-value. The definition applied depends on the code, code official, and the different needs of energy modellers and designers. However, the most important distinctions between different definitions involve how thermal bridging is considered.

For opaque walls it is common to specify thermal resistance as an RSI (°C m^2/W) or R-value (°F hr ft^2/BTU) whereas the U-value (W/m^2 °C or BTU/ft^2 °F hr) is widely used for the thermal transmittance of vision glazing. Building codes of the past used an installed R-value/RSI requirement which only accounted for the insulation while window U-values included surface films but did include the thermal bridging effects of framing and edge-of-glass construction.

Codes and standards are changing to capture thermal performance better, but there is often little consistency in how the terms are described. A series of definitions were proposed thirty years ago [Christian & Kosny, 1995] to air clearer communication. There are at least six different R-value definitions:

- the **Installed R-value**, or nominal R-value which is simply the rated R-value of the insulation products in their installed condition (e.g. compressed batt insulation or not). The contribution of other materials is ignored but the R-value can be tested under tightly controlled conditions and be repeatedly delivered by the manufacturer,

- the **Assembly R-value** or **Center-of-Cavity** is calculated by assuming the assembly is one-dimensional and simply adding the thermal resistance of all layers as described above,

- the **Clear-wall R-value** (R_{cw}), or Clear Field accounts for the thermal resistance of all layers (Assembly R-value) but *also* includes the two-dimensional thermal bridging effect of standard repetitive framing (e.g.

wood studs or steel studs) using the parallel path, isothermal planes, or two-dimensional computer modeling,

- the **Whole-wall R-value,** (R_{WW}) which includes the Clear-wall R-value (R_{CW}) *plus* the thermal impact of conductive penetrations (e.g. floors, balconies) and any additional framing at openings (e.g. windows and doors), fasteners, and the effects of thermal bridges at changes in plane and other interfaces (e.g. foundation-to-above-grade wall, wall-to-roof, balconies, etc.) but *excludes* window area. For simplicity, the Clear-wall R-value is sometimes used when whole-wall R-value should be (i.e., thermal bridging is ignored), but this approach can significantly over-estimate the thermal performance of many commercial enclosure systems,

- the **Overall R-value** ($R_{overall}$) measures the performance of an entire enclosure type (such as wall, or roof) and includes the combined effect of Whole-Wall R-value (R_{WW}) plus the heat loss through windows, doors, and curtainwalls. It is important for understanding overall building performance and is implicit to the simple trade-off methods used to demonstrate code compliance (see Chapter 3), and

- the **Effective R-value** is used to describe an R-value that *may* include some or all thermal bridging, thermal mass and even air leakage. It has no universal, consistent definition. Hence, the meaning of effective R-value varies depending on both the user of the term, governing energy standards, and the context.

Any of these R-values might also be reported as a U-value ($U = 1/R$). To add to the complexity, U-values almost always include the resistance of surface films, whereas R-values may (the air-to-air R-value) or may not (the surface-to-surface R-value). It is for these reasons that those in the building industry must be quite careful when interpreting requirements and be specific and precise when communicating required thermal performance.

5.12.1 Windows and Overall R-value Tradeoffs

True thermal performance requires the enclosure designer to consider the influence of fenestration products (e.g., windows, curtainwalls, window walls, patio doors, etc.) on heat flow through the entire vertical enclosure. Fenestration R-values are much lower than that required of opaque walls in energy standards because in practice the available thermal performance of glazed systems is far less than that for opaque enclosure components. Because heat flows preferentially through low-thermal resistance components, much more heat flows through windows in most buildings, even those buildings with limited glazing area.

Designers of high performance buildings will generally consider the overall R-value as a measure of the enclosure thermal performance. Energy codes such as ASHRAE 90.1 and the Canadian National Energy Code for Buildings infer an overall R-value in their prescriptive compliance paths by assuming a maximum window-to-wall area and minimum component R-values.

The overall R-value of an enclosure wall assembly can be drastically changed by modifying the Window-to-Wall Ratio (WWR) [Love 2008, Ross & Straube 2014) and window performance. To compare the impact of WWR, glazing performance, and opaque wall performance, an equivalent overall R-value, which combines the influence of the opaque wall R-value with the fenestration systems (including doors) U-value, can be used as a single metric.

Overall equivalent transmittance, $U_{overall}$, (and $R_{overall} = 1/ U_{overall}$) can be calculated using the parallel path method as:

$$U_{overall} = (1\text{-}WWR) / R_w + WWR \cdot U_f \qquad \text{(Eq. 5.45)}$$

where

WWR is the Window-to-Wall Ratio,

R_w is the R-value of the opaque assembly (or 1/U), and

U_f is the U-value of fenstration systems.

Overall R-value is then simply

$$R_{overall} = 1 / U_{overall}.$$

An investigation of a range of different WWR and window and wall U-values is recommended to gain an understanding of how significant WWR is when window U-value is substantially lower than opaque enclosure U-value.

5.13 Solar control

The energy delivered to the exterior surface of a building by solar radiation is usually the most intense heat flux experienced by enclosures during service. Heat fluxes of up to 1000 W/m^2 can occur. Clear double glazing sometimes used in buildings allows a large proportion of the incident radiation to pass into the building. This energy is often a major source of interior heat gain, and it is usually the largest load when cooling buildings with large expanses of glazing. Accordingly, controlling solar radiation is an important part of controlling heat flow through the building enclosure.

5.13.1 Solar control for opaque enclosures

Solar radiation impinging on a surface can be absorbed, reflected, or transmitted. Opaque enclosures, by definition, have no transmission component. Hence, all of the solar energy is either absorbed or reflected by the surface. Even light-colored enclosure surfaces will absorb enough energy to increase their surface temperature significantly.

One may wish to reduce solar heat gain in order to minimize air-conditioning load, to increase material durability, or to reduce thermal movements. Shading (in the form of building components, or plantings) and lower solar absorption can be used to reduce the amount of solar heating.

5.13.2 Solar control for transparent enclosures

The control of heat flow through window and curtainwall systems is an important application of the theory in this chapter. The level of thermal resistance of almost all such products is very low relative to opaque enclosure elements. In buildings having enclosures with a large proportion of glazing, the control of radiative heat gains (solar gains), radiative losses (infra-red) and transmission losses (especially at night) will often dominate space conditioning requirements. Both knowledge and experience are required to ensure thermal comfort and a degree of energy conservation for the design of HVAC systems (especially cooling).

The analysis of heat flow through fenestration products is a specialized area of study. It can be seen from Figure 5.19 that the assessment of heat flow across a product can be complex because it must account for radiation absorption, reflection, emission, and transmission as well as conduction and convection.

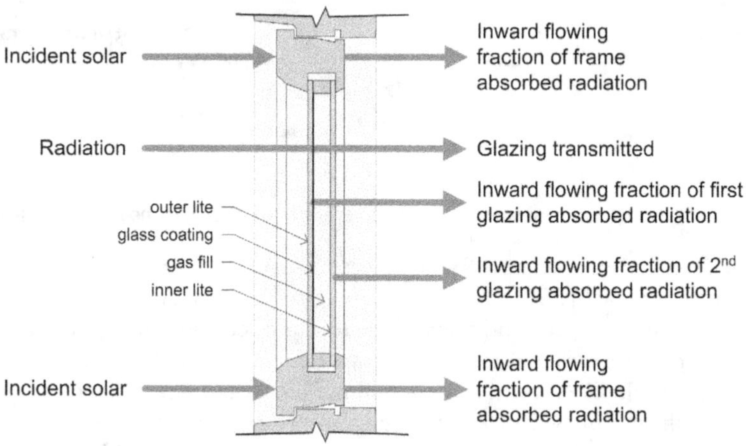

Figure 5.21: Components of solar radiant gain through a typical glazing and frame assembly

The level of control of solar gain through glazed components is measured using the Solar Heat Gain Coefficient (SHGC) which is defined simply as:

$$\text{SHGC} = \frac{\text{Solar energy entering building as heat}}{\text{Solar radiation falling on enclosure}} \qquad \text{(Eq. 5.46)}$$

For example, if the SHGC of a window is 0.50, and the sun is impinging on the window with an intensity of 500 W/m^2 (150 $Btu/ft^2/hr$) only 250 W/m^2 (75 $Btu/ft^2/hr$) will enter the building as heat. SHGC may be applied to only the glazing unit, to the fenestration product including frames, and, rarely, to other building components and shading devices. An example of a representative window product is shown in Figure 5.22.

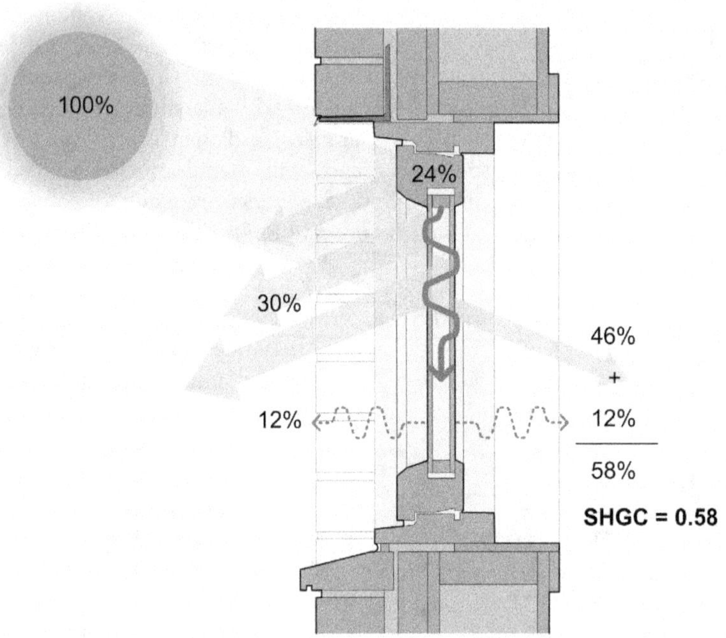

Figure 5.22: Components of energy flow for an example window solar heat gain coefficient

Of course, as the radiative and convective heat transfer modes are somewhat non-linear, the SHGC reported for a fenestration product is strictly accurate only for the specific conditions of the test or calculation that generated it: high levels of solar radiation and warm temperatures are generally specified. However, the non-linearities only add modest inaccuracy. A larger concern about the use of SHGC for predicting in-service performance is the angle between the sun's rays and the product. At steep angles, the SHGC decreases rapidly as light begins to reflect off the glass, and the frames and window surrounds begin to shade the glass from sun.

Given the definition of SHGC, it is straightforward to predict solar gain into a building:

$$Q_{solar} = SHGC \cdot A_{window} \cdot I \qquad \text{(Eq. 5.47)}$$

where Qsolar is the rate of solar energy entering the building,

A_{window} is the area of the fenestration product, and

I is the incident rate solar energy.

The location of shading devices (screens and Venetian blinds, etc.) on the interior is much less effective than if well insulated glazing units are used. For the highest performance shading should be applied to the exterior (Figure 5.23).

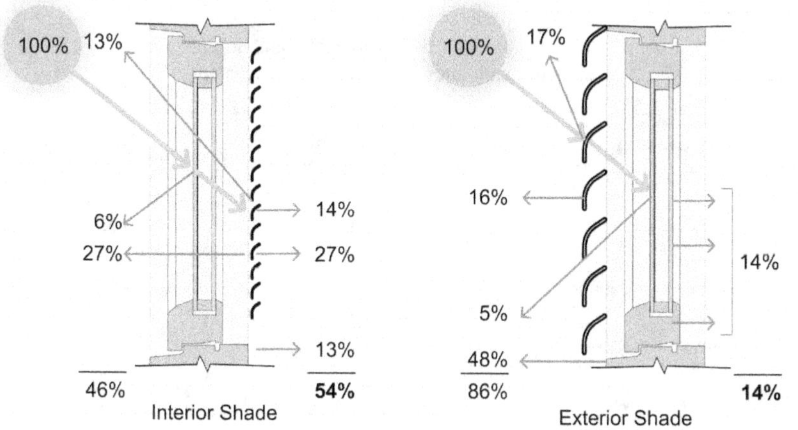

Figure 5.23: Influence of shading on heat flow through window

It may be interesting to note that simple heat transfer physics allows one to calculate the SHGC for opaque insulated components of a building such as the roof or wall. Using the methods described in Section 5.9, the surface temperature can be predicted and used to calculate the heat flow inward through the assembly with Fourier-law analysis. For example, consider a dark colored (α = 0.90) well-insulated roof (RSI5.0). Under bright sunshine the temperature is predicted to reach 55 $\cdot$ α or 49°C above the air temperature. If it is hot day (32°C) the membrane will reach 81°C. If the interior temperature is 24 °C, heat flow inward would be (81-24)/5 = 11.4 W/m², a very small fraction of the 800 W/m² that would represent sunshine. In SHGC terms this would be a SHGC of 11.4/800 = 0.014.

Modern glazings control radiation by the addition of various molecule-thick coatings and color tints that influence the absorption, reflection, transmission and emission as specific functions of frequency (Table 5.8). Although the values provided are representative, the industry is able to provide a vast array of combinations to meet specific project goals and performance targets. The SHGC considers all frequencies, long-wave and short-wave, as all convert to heat within the building. Another metric of interest is the visible light transmittance, VLT, which only considers the transmission of visible light through a product. This is defined in the same way as SHGC – as a fraction of incident radiation – but limits the range of frequencies to the visible range.

Gas fillers (argon, krypton, xenon) and special materials (polycarbonate, fiberglass) are used to control conduction and convection. Two-dimensional flow through window frames and spacers is also very important, and sometimes dominant, and hence should be assessed. Software such as the previously mentioned THERM, and WINDOW is readily available to evaluate window performance.

Table 5.8: Solar heat gain coefficient and visible transmittance of representative glazings

Glazing	Visible Light Transmittance (VLT)	Solar heat gain (SHGC)
Reflective blue-green	0.33	0.38
Film on clear glass	0.19	0.22
Green tinted, medium	0.75	0.69
Green low-e	0.71	0.49
Sun Control low-e green	0.36	0.23
Super low-e + clear	0.71	0.40
Super low-e + green	0.60	0.30

5.14 References

Anon. *Building Envelope Thermal Bridging Guide.* Version 1.6. Morrison Hershfield Limited / BC Hydro, Vancouver, 2021.

ASHRAE. *Handbook of Fundamentals.* American Society of Heating Refrigeration and Air Conditioning, Atlanta, GA, 2021.

Bliss, R.W., "Atmospheric Radiation Near the Surface of the Ground", *Solar Energy*, Vol. 5, Nr. 103, 1961.

Bomberg, M., and Shirtcliffe, C.J., "Influence of moisture and moisture gradients on heat transfer through porous building materials," *Thermal Transmission Measurements of Insulation*, ASTM STP 660, ed. R.P. Tyre, ASTM, Philadelphia, 1978, pp. 211-233.

Christian, J.E., and Kosny, J., "Toward a National Opaque Wall Rating Label". *Proc Thermal Performance of the Exterior Envelopes VI*, ASHRAE, Clearwater Beach, 1995.

Clear, R.D., Gartland, L. and Winkelmann, F.C., "An empirical correlation for the outside convective air-coefficient for horizontal roofs", *Energy and Buildings*, 35, pp. 797-811, 2003.

CSA Z5010-2021. *Thermal bridging calculation methodology*. Canadian Standards Association.

Ge, H., Fazio, P., "Experimentally establishing local convection film coefficients for a. standard metal curtain wall system". Proceedings of the 2nd International Conference on Building Physics, Leuven, Belgium, 2003.

Gertis, K., Holm, A. "Zur Messung der Wärmeleitfähigkeit feuchter Stoffe". *Bauphysik* 36, Vol 1, pp. 27–39, 2014.

Hagentoft, Carl-Eric, *Introduction to Building Physics*. Studentlitteratur, Lund, Sweden, 2001.

Hagishima, A. and Tanimoto, J. "Field measurements for estimating the convective heat transfer coefficient at building surfaces", *Building and Environment*, 38, pp. 873-881, 2003.

Higgins, J., Shane, C. and Finch, G. "Thermal bridging from cladding attachment strategies through exterior insulation". In *9th Annual North American Passive House Conference* September, 2014.

Hedlin, C.P., "Heat Flow Through a Roof Insulation Having Moisture Contents Between 0 and 1% by Volume, in Summer," *ASHRAE Transactions* 94, 1988, pp. 1579-1594.

Hutcheon, N. and Handegord, G., *Building Science for a Cold Climate*. 1982

Idso, B. S. and Jackson, D. R., "Thermal radiation from the atmosphere". *J. of Geophys. Res* Vol. 74, No. 23, pp. 5397 - 5403, 1969.

Incropera, F.P., and deWitt, D.P., *Fundamentals of Heat and Mass Transfer*. 8th Edition, John Wiley and Sons, Toronto, 2018.

ISO 6946:2017. Building components and building elements — Thermal resistance and thermal transmittance — Calculation method. International Organization for Standardization, Geneva, 2017.

ISO 10077:2012. *Thermal performance of windows, doors and shutters -- Calculation of thermal transmittance -- Part 2: Numerical method for frames.* International Organization for Standardization, Geneva, 2012.

ISO 10211:2007. *Thermal bridges in building construction Heat flows and surface temperatures Detailed calculations.* International Organization for Standardization, Geneva, 2007.

ISO 10456:2007. *Building materials and products — Hygrothermal properties — Tabulated design values and procedures for determining declared and design thermal values.* International Organization for Standardization, Geneva, 2007.

ISO 13791:2012. *Thermal performance of buildings - Calculation of internal temperatures of a room in summer without mechanical cooling - General criteria and validation procedures.* International Organization for Standardization, Geneva, 2012.

ISO 14683:2017. *Thermal bridges in building construction -- Linear thermal transmittance -- Simplified methods and default values.* International Organization for Standardization, Geneva, 2017.

Kosny, J, Yarboorough, D., Cholds, P. Mohiuddin, S., "Effects of framing on the thermal performance of wood and steel-framed walls". *Proceedings of the Fifteenth Symposium on Improving Building Systems in Hot and Humid Climates*, Orlando, FL, July 24-26, 2006.

Love J., Tian W., Tian Z, "Window-to-Wall Ratios and Commercial Building Environmental Control in Cold Climates", *Proc. 3rd SBRN and SESCI 33rd Joint Conference*, Fredericton, 8 pp., 2008.

NCMA. *Thermal Catalog of Concrete Masonry Assemblies*, Second Edition. National Concrete Masonry Association, NCMA Publication Number TR233A, Hendon, VA, December 2012.

Pagan-Vazquez, A., Chu, D., Straube, J., Ryan, R. and Lawton, M. *Development of thermal bridging factors for use in energy models.* Report for US Army Engineer Research and Development Center, Construction Engineering Research Laboratory, 2015.

Ross, B., Straube, J.F., "Enclosure Performance vs Building Form in Cold-Climate Commercial Buildings—What really Drives Energy Use?", *Proc. 14th Can Bldg Sci & Tech Conf.,* Toronto, Oct 2014, pp. 161-171.

Sanders, C., "Task 2: Environmental Conditions" *Annex 24*, Final Report, International Energy Agency, 1996.

Schwarz, B., *Die Wärme und Stoffübertragung an Außenwandoberflächen.* Doctoral Dissertation, Universität Stuttgart, Germany, 1978.

Santos, P., Lemes, G. and Mateus, D., 2020. "Analytical methods to estimate the thermal transmittance of LSF walls: Calculation procedures review and accuracy comparison". *Energies*, *13*(4), p.840.

Straube, J.F. and Burnett, E.F.P., "Field Measurements of Masonry Veneer Exposure Conditions", *Proc of the Eighth Canadian Masonry Symposium*, Jasper, Alta., Canada, May 31-June 3, 1998, pp. 208-221.

Straube, J.F. *Meeting and Exceeding Building Code Thermal Performance Requirements.* Canadian Precast/Prestressed Concrete Institute, Ottawa, 2017.

Chapter 6

Water Vapor Transport

6.1 Introduction

The moisture in air is one of the most important sources of wetting in buildings, and it therefore is often implicated in performance problems. The design professional needs to know how to design and construct buildings that are safe from damage caused by water vapor. To do so, it is essential to understand the physics of water vapor, its transport by diffusion and air movement, and its deposition and removal by condensation and evaporation.

Water vapor and its presence in air is discussed in Chapter 4. This chapter examines the transport of water vapor in air by diffusion and convection. Terms and units are defined first.

6.2 Terms, definitions, and units

S_d value is a European measure of the vapor diffusion resistance of a layer relative to a meter of air. Units are in meters, i.e., a layer with S_d=2.0 m has the vapor resistance equal to 2 meters of air. The vapor permeance of air varies with temperature and atmospheric pressure.

Vapor Permeability (μ) is a material property which indicates the quantity of vapor flow across a unit area that will flow through a unit thickness under a unit vapor pressure difference.

Units: Metric perm = ng / (s·m·Pa)

Imperial perm-inch = grain / (hr·in Hg·in)

US perm = 1.46 metric perm

Quoted values are typically averages from standard ASTM E96 wet-cup and dry-cup tests conducted at room temperature (see Chapter 8). Vapor permeability can vary, sometimes dramatically, with temperature, relative humidity, and moisture content, usually because other vapor transport mechanisms begin to act.

Vapor Permeance (M) is the permeability of a layer or product of a given thickness. Permeance (M) = permeability / thickness = μ /l

Units: Metric — PERM = ng /(s·m²· Pa)
Imperial — perm=grain/(hr·in Hg· ft²)
Conversion — US perm = 57.4 metric perm

Overall Vapor Permeance (V) is a little-used system measure of the amount of vapor flow that will occur across a unit area of an assembly of materials for a unit vapor pressure difference, i.e. a system permeance.

Units: as for vapor permeance

Vapor Flow (Q_v) is the rate at which water vapor moves. Vapor flow can be based on a system (e.g., building), a component of the system (e.g., the building enclosure), or per unit area.

Units: Metric — ng/s
Imperial — grains/hr

Vapor Flux (q_v) is the water vapor flow per unit area or q_v = Vapor Flow/ Area = Q_v /A. The area is perpendicular to the direction of vapor flow.

Units: Metric — ng /(s· m²)
Imperial — grains / (hr· ft²)

Vapor Resistance (R_V) is the reciprocal of vapor permeance. It can be calculated as thickness / permeance = 1 / M = l /μ.

Units: Metric — REP = s· m²·Pa / ng
Imperial — Rep = hr · in Hg · ft²/grain

Water Vapor Diffusion is the movement of water vapor, under a vapor pressure (concentration) gradient.

Water Vapor Diffusion Resistance Factor, a European measure for vapor diffusion permeability, is defined as the ratio between the vapor permeability of stagnant air and the vapor permeability of the material under identical conditions (the same temperature and pressure). The vapor permeability of air can be taken from Shirmer's equation in ng/Pa s m as:

$$\delta_{air} = (1.97 \times 10^5 \times T^{0.81}) / P$$

where T is in absolute temperature [K] and the barometric pressure, P, is given in Pascals.

Water Vapor Transmission (WVT) or **Water Vapor Transmission Rate** (WVTR) are different terms used to report permeance, but under specific temperature and humidity conditions. Although ASTM E96 suggests units for building materials, other materials (textiles and coatings) often use different units, and these sometimes appear in building literature. The use of WVT should be avoided because the conditions of the test are often not stated or unclear, and less confusing measures are available.

Units: Metric WVT =g/24 hrs/m^2
Imperial WVT =grains/24 hrs/ft^2

6.3 Water vapor transport

Water vapor moves through the building enclosure in two primary modes: diffusion and convection.

Diffusion is a mass and energy transport mechanism that is driven by differences in density. It can be viewed as any process that moves mass and energy from an area of higher concentration to an area of lower concentration. *Conduction* of heat is the diffusive transport of heat energy from more (high temperature) to less (low temperature) by the direct transfer of molecular energy. The random motion of molecules results in the diffusive movement of vapor along a concentration gradient.

Convection is the transfer of energy or mass by the bulk movement of a fluid (liquid or gas). Water vapor can be easily transported by the bulk transport of the air itself, and this convective transport mechanism, analogous to bulk convective heat transfer, is very important for building science applications.

The combined transport of mass by diffusion and convection is termed *advection.*

In practice, specific layers of building enclosure assemblies are often relied upon to control or limit vapor diffusion and convection. A layer that is intended to, or acts as, the primary control of vapor diffusion is the *vapor control layer*, or in common usage, the *vapor barrier* or *vapor retarder.* A layer that is intended to, or acts as, the primary control of air flow through an enclosure by convection is term the *air control layer* or *air barrier.* The difference between a vapor barrier and air barrier remains one of the most common questions, and enduring confusion in practice although clear explanations have long been available (Quirouette 1985).

6.3.1 Diffusion

The general fundamental diffusion mass flow relationship is based on a form of **Fick's Law**:

$$\frac{dm}{d\theta} = -D \cdot \nabla C \qquad \text{(Eq. 6.1)}$$

C is the concentration of some value (eg., vapor pressure),

∇ is the divergent operator, $\nabla = \left(\frac{\partial}{\partial x}, \frac{\partial}{\partial y}, \frac{\partial}{\partial z} \right)$,

$dm/d\theta$ is mass flow per unit time , and

D is a proportionality constant

For vapor diffusion the coefficient, D, is defined as the diffusivity. The negative sign indicates a loss as mass flows from high concentrations to lower concentrations.

For water vapor driven by vapor pressure gradients in one dimension, Fick's law can be written as:

$$\frac{dw_x}{d\theta} = -\mu \cdot A \cdot \frac{dP_w}{dx} \qquad \text{(Eq. 6.2)}$$

where w_x (ng) is the mass low rate of water vapor along x direction

$d\theta$ is time in seconds t,

A is the area (m^2) perpendicular to the flow

dPw (Pa) / dx (m) is the vapor pressure gradient and

μ (ng / m · Pa · s) a material-specific coefficient.

The material-specific coefficient μ is defined as the vapor permeability.

The negative sign is a consequence of the fact that vapor flows from high vapor pressures to low vapor pressures. The same equation can be rewritten for the other two Cartesian directions, in three-dimensional vector notation, or, if useful, in polar coordinates. Chapter 8 provides more detailed information about how vapor permeability is measured.

Although the vapor permeability varies with temperature and RH, an average vapor permeability, μ can be assumed for many practical building science situations, and Fick's law can be written as:

$$Q_v = A \cdot \frac{\mu}{l} \cdot (P_{w,1} - P_{w,2}) \qquad \text{(Eq. 6.3)}$$

where Q_v is the time rate of vapor flow,

μ is the vapor permeability,

l is the length of the flow path, (i.e., thickness of the material),

and P_1 and P_2 are the water vapor pressures on either side of the material of interest.

For a single layer of thickness l, the commonly-used property vapor permeance, M, is defined as vapor permeability divided by thickness:

$$M = \mu / l \qquad \text{(Eq. 6.4)}$$

The vapor permeability of a homogenous material is calculated from the measured vapor permeance of a sample with a known thickness.

6.3.2 Vapor Diffusion Resistance of Materials

ASTM E96 *Standard Test Methods for Water Vapor Transmission of Materials* is commonly used in the building industry to measure the vapor permeance of materials and calculate the vapor permeability of homogenous materials. In Europe, ISO 12572 is used. This test method imposes a known and stable difference in water vapor content across the test sample. The rate of water vapor diffusion is found by weighing the change in mass of a sample and its container over time. The results are normalized by area, difference in vapor content (in units water vapor pressure of vapor density, such as grains of water vapor per pound of air). Vapor permeability can then be calculated from the thickness of the sample tested.

The water vapor permeance of many organic materials, including wood and paper products, changes with the relative humidity to which it is exposed. This has been known for many years as a characteristic of many natural organic materials (see Chapter 8). Hence, it is critical to know the temperature and humidity of both sides of the sample when presenting and interpreting vapor permeance results, especially for humidity levels above about 60%. There is much less sensitivity to temperature.

The most commonly quoted ASTM E96 vapor permeance results use conditions of 0% RH on one side of the sample and 50%RH on the other side at room temperature (see also Figure 8.30). These test conditions are described variously as the dry-cup test, desiccant method, or Procedure A. If no conditions are included along with the results, this is the most likely test condition, but one cannot fully rely on the results given and assume this is the case.

The second most common test arrangement is the wet-cup test which exposes one side of the sample to 100%RH and the other to 50%RH at room temperature. The vapor permeance at these conditions can often be almost the same as the dry cup method, but for a lot of common organic building materials, significantly elevated.

In practice, precise vapor permeance data is rarely needed. Hence, several broad categories of vapor permeance have been proposed by Lstiburek [2004] (Table 6.1) have adopted by US codes and design guidance. A vapor barrier is often considered by Lstiburek to include Class I materials whereas Class II layers are labelled as a vapor retarder. The numerical values that divide these subdivisions are not precisely defined scientifically. For example, research has shown that layer permeances of 2 to as high as 5 US perms (120 to 300 metric perms) are sufficient vapor control for common building enclosure assemblies even in cold climates [Karagiozis 1993, Teesdale 1939, Wooley 1940].

Table 6.1: Vapor Control Layer Classification

Class	Permeance (US perms)	Permeance (Metric perms)	Description
I	Less than 0.1	Less than 6	Impermeable
II	0.1 to 1.0	6-60	Semi-impermeable
III	1 to 10	60-600	Semi-permeable
none	Over 10	Over 600	Permeable

Class I - less than 0.1 perms. Vapor impermeable, a vapor barrier

A Class I vapor control layer effectively divides the wall assembly into interior and exterior zones with regard to vapor diffusion. Vapor diffusion wetting and drying can occur on both sides of the Class I vapor control layer, but not across the vapor control layer.

e.g., polyethylene sheet, sheet metal, aluminum or polyethylene facing on any product.

Class II - 0.1 to 1.0 perms. Vapor semi-impermeable, a vapor retarder

e.g., the kraft facing on fibreglass batts, some vapor control paints, 2" of extruded polystyrene foam (XPS), over 2-3" of medium-density closed-cell spray polyurethane foam, and 3 to 5 inches of EPS

A Class II vapor control layer throttles or limits diffusion wetting but allows a small amount of diffusion to help dry small accidental wettings or construction moisture.

Class II vapor control layer provides sufficient vapor diffusion control to avoid condensation for most climates and occupancies.

Class III - 1.0 - 10 perms. Vapor semi-permeable

e.g., latex or enamel paints on drywall/gypsum board, more than about 6" to 8" (150-200 mm) of open-cell spray polyurethane foam

A Class III vapor control layer restricts vapor diffusion, but allows significantly greater vapor diffusion wetting and drying than a Class II vapor control layer.

Table 6.2 summarizes the vapor permeance and permeability of a range of building materials. To convert from metric permeance (in ng/Pa s m^2) to US perms simply divide by 57.3. Vapor permeance data is usually not as widely available as, for example, density, compressive strength, or thermal conductivity. Permeance, like all material properties, often exhibits a range of values. The same product from the same product can vary somewhat, and the products within a generic class (e.g., "building paper") may vary by an order of magnitude. Section 8.4.1 delves into more details about vapor diffusion through materials and the test methods used.

Table 6.2: Vapor permeance of some common building materials

Permeability ng/(s·Pa·m)

Material	**Density (kg/m^3)**	**Range of Wet Cup (50-100%)**	**Range of Dry Cup (0-50%)**	**Other test Type**	**Comments**
Board / sheet products					
Plywood	400 - 600	2 - 8	0.5 - 1.5	20 - 30	Other=saturated
OSB	575 - 725	1.5 - 3	0.5 - 1.5	4 - 5	Other=saturated
Vegetable fiberboard	270 - 300	32	31		RH dependent
Fibreboard - asphalt coated	300	22	21		
Gypsum wall board	700 - 900	25			
Exterior gypsum wallboard	900	25	12		
Gypsum plaster on gypsum lath		12 - 22			lath dependent
Finishes					
Glazed ceramic tiles		0.1			
Glazed brick				0.7	
Cement stucco	-	5 - 12	3 - 8		RH dependent
Cement-lime stucco				9.7 - 26	RH dependent
Face brick				2.5 - 20	
Earth plaster- see Straw clay/adobe					
Vapor barrier layers					
Polyethylene		4.74E-04	4.74E-04		
Alu foil, t>.025 mm		0	0		
PVC		0.004 - 0.008			
Polyester			0.001		
All metals: steel, copper, etc		0	0	0	
Insulations					
Air still				175	
Spray polyurethane	40		2.86 - 3.65		RH dependent
Expanded polyurethane			0.58 - 2.3		high density
Spray open cell polyurethane			70		low density

Table 6.2 continued: Vapor permeance of some common building materials

Material	Density (kg/m^3)	Range of Wet Cup (50-100%)	Range of Dry Cup (0-50%)	Other Test Type	Comments
	Permeability ng/(s·Pa·m)				
Insulations, continued					
Phenolic foam (no facing open cell)			38		
Polyiso-glass facer			2 - 3.2		
Foil-faced polyisocyanurate		0.01			
Fibreglass/rockwool		245			
Cellulose/sawdust/straw				110	RH dependent
EXPS Type 3	25		1 - 3		
EXPS Type 4	29		0.8 - 1.7		
EPS Type 1	16		4 - 7.5		
EPS Type 2	24 32		3 - 6		
Structural materials					
Wood – sugar pine				0.58 - 7.8	RH dependent
Concrete	2000 2400			2 - 6	Mix dependent
Concrete block	1600 1800			6 - 27.6	
Durisol material	400 850	26	20		
Adobe	1600 1800			17.5	
Straw clay	1200 2000			17 - 35	
Leichtlehm	400 1200			35 - 90	
Films, layers, sheets - All Values in PERMEANCE (ng/Pa s m^2)					
Latex paint, primer + sealer				200 - 350	
Primer + 1 coat oil paint on plaster				91 - 172	
2 coats flat oil paint				250	
3 coats exterior oil-based paint on wood				17 - 57	
Asphalt sheathing paper		400 - 1800	30 - 300	1000–4000	Other=saturated
BUR membrane		0.5	1		
Spun-bonded polyolefin housewrap				1000–3000	
Latex textured coating, 2-3 mm (EIFS lamina)				200 - 300	
Siloxane sealers, min				500–1500	

Notes:
Vapor permeability is difficult to measure accurately and highly sensitive to small changes in the material. Wettable, porous materials exhibit much higher permeability when wet or at high RH
Sources include ASHRAE, CMHC, NRC/IRC, IEA Annex 24, RDH Building Science Labs, etc

6.3.3 Analogy with Conductive Heat Flow

It can be observed that the form of Fick's Law for diffusive vapor flow is exactly the same as Fourier's Law for conductive heat flow. In fact, on a general level, conductive heat flow is a diffusive flow process, just like vapor, water, and air flow in porous media. Therefore, all of the same forms of equations can be used with different variable names as shown in Table 6.3.

Table 6.3: Comparison of conductive heat flow and diffusive vapor flow

	Heat		Vapor	
Driving Potential	Temperature	t,T	Vapor Pressure	P_W
Measures	Conductivity	k, λ	Permeability	μ
	Conductance	C	Permeance	M
	Overall Transmittance	U	Overall Vapor Transmittance	V
	Resistance	R	Resistance	R_V
Flow	Heat flow	Q	Vapor Flow	Q_v
	Heat flux	q	Vapor flux	q_v
Single Layer	$\frac{\lambda}{l} \cdot \Delta t$ $= C \cdot \Delta t$ $= \Delta t / R$	q	$\frac{\mu}{l} \cdot \Delta P$ $= M \cdot \Delta P$ $= \Delta P / R_V$	q_v
Multi-Layer	$U \cdot \Delta t$		$V \cdot \Delta P$	

For example, when examining a unit area of an enclosure, Fick's equation can be reworked to give the rate of vapor flow per unit area, the vapor flux, q_v:

$$q_v = Q_v/A = \mu/l \cdot (P_1 - P_2) \qquad \text{(Eq. 6.5)}$$

The term μ / l is termed the vapor permeance and given the symbol M. This prompts another form of the equation:

$$q_v = M \cdot (P_1 - P_2) \qquad \text{(Eq. 6.6)}$$

Just as for heat flow, the reciprocal of the permeance is defined as a vapor resistance, so:

$$q_v = \frac{1}{R_V} \cdot (P_1 - P_2) \quad \text{(Eq. 6.7)}$$

where the subscript v is used here to denote vapor flow.

6.3.4 Convective Vapor Flow

The convective flow of air can transport significant quantities of water vapor with it. The calculation of the rate of water vapor convection is simple given the flow rate of air. Recall that the humidity ratio is defined as:

$$W = \frac{m_V}{m_a} \quad \text{(Eq. 6.8)}$$

If the rate of air movement is known, the rate of water vapor movement can be found from

$$\frac{d\,w_V}{d\theta} = \frac{d\,w_a}{d\theta} \cdot W \quad \text{(Eq. 6.9)}$$

where w_V, w_a iis the mass of water vapor & air respectively,

$d\theta$ is the time interval, and

W is the humidity ratio (kg water vapor to kg air).

The mass flow rate of convective water vapor movement per unit area can therefore be found from:

$$m_W = \rho_{air} \cdot W \cdot Q_{air} \quad \text{(Eq. 6.10)}$$

where m_W is the rate of water vapor flow (kg/s),

ρ_{air} is the density of air (kg/ m³), and

Q_{air} is the volumetric flow rate of the air (m³/s).

6.4 Calculating steady-state 1-D vapor flow

Calculating vapor flow follows the same general principles as the calculation of heat flow. However, the results of the calculations tend to be much less accurate for several reasons:

- The values for vapor permeability are difficult to ascertain accurately and vary widely with moisture content and temperature. Variations of an order of magnitude are common for some materials. Extrapolating from tests on samples of different thicknesses further reduces accuracy.

- The moisture storage capacity of most building materials is so large that steady-state conditions almost never occur in practice. The processes of evaporation and desorption act as moisture sources and sinks within the enclosure.

Vapor diffusion is also often a relatively small contributor to vapor transport: in practical terms the bulk movement of air may transport up to one hundred times as much water vapor as diffusion.

Despite the inaccuracies and limitations of steady-state, diffusive vapor flow calculations, such calculations are still useful in that they provide a relative measure of vapor movement which, if carefully and thoughtfully applied, will usually be sufficiently correct to make design and analysis decisions.

6.4.1 Single layer

From Fick's Law:

$$Q_v = \mu \cdot \Delta P_w \cdot A / l \qquad \text{(Eq. 6.11)}$$

but recall that $\mu/l = M$, and $R_V = 1/M$ so,

$$Q_v = M \cdot A \cdot \Delta P_w = \frac{1}{R_V} \cdot A \cdot \Delta P_w \qquad \text{(Eq. 6.12)}$$

Example

A 2 m wide, 3 m high, and 50 mm thick sheet of extruded polystyrene ($\mu = 2.5$) has a vapor pressure of 1500 Pa on one side and 3500 Pa on the other. Calculate the vapor flow rate, vapor flux, permeance and R_V value.

Answer:

1. The vapor flow rate and the vapor flux

$$Q_v = \mu \cdot \Delta P_w \cdot A / l = M \cdot A \cdot \Delta P_w$$
$$= 2.5 \text{ ng/Pa s m} \cdot (3500 \text{ Pa} - 1500 \text{ Pa}) \cdot (3 \text{ m x } 2 \text{ m}) / 0.05 \text{ m}$$
$$= 600\,000 \text{ ng/s, and}$$
$$q_v = Q_v / A = 100\,000 \text{ ng/s m}^2 = 0.36 \text{ g/hr m}^2$$

2. The permeance of this foam layer is

$$M = \mu / l$$
$$= (2.5 \text{ ng/Pa s m}) / (0.050 \text{ m})$$
$$= 50 \text{ ng/Pa s m}^2$$

3. The R_v values

$R_v = 1 / M = 1/ 50$ ng/Pa s m²

$= 0.020$ Pa s m² /ng

6.4.2 Multiple layers

Most enclosure assemblies are, of course, comprised of multiple layers. Vapor flow is analogous to heat flow through multiple layers or composite constructions and similar calculation methods can be employed. The method described below is often called Glaser's method after a graphical method published many years ago [Glaser 1959]. The method has been used for much longer as a tabulation technique, however, and there are references that date back to the 1930s [Rowley 1939, Joy 1952].

Two important facts are fundamental to the method:

- the vapor flux across the entire assembly must also pass through each layer (i.e., conservation of mass)
- the sum of the *resistances* of each layer gives the total resistance; one *cannot* sum the permeances.

For any layer j

$$q_v = Q_v/A = \frac{\Delta P_j}{R_{v_j}} = M_j \cdot \Delta P_j \qquad \text{(Eq. 6.13)}$$

where q_v is the vapor flux through the whole assembly.

For an assembly of n layers:

$$q_v = \frac{P_{int} - P_{ext}}{\sum_{j=0}^{n} R_{v_j}} \qquad \text{(Eq. 6.14)}$$

Note: $q_v \neq \sum_{j=0}^{n} M_j \cdot (P_{int} - P_{ext})$

The overall vapor permeance (V) of a system of n layers is defined as:

$$V = \frac{1}{\sum_{j=0}^{n} R_{v_j}} \qquad \text{(Eq. 6.15)}$$

From Equation 6.12, the vapor pressure drop across layer j is:

$$\Delta P_j = q_v \cdot R_{v,j} \qquad \text{(Eq. 6.16)}$$

Combining equations gives:

$$\Delta P_j = \frac{P_{int} - P_{ext}}{\sum_{j=0}^{n} R_{v_j}} \cdot R_{v,j} \qquad \text{(Eq. 6.17)}$$

or from equations 6.14 and 6.16

$$\mathbf{\Delta P_j = V \cdot (P_{int} - P_{ext}) \cdot R_{v_j}} \qquad \text{(Eq. 6.18)}$$

This last equation is the most useful. One normally calculates q_v to assess the assembly's ability to control diffusive vapor flow. It is then a simple matter to calculate the vapor pressure drop across every layer, or any layer of interest, using Equation 6.17. To assess the influence of different environmental conditions, only the exterior and interior vapor pressure needs to be changed in this equation.

6.4.3 Surface transfer coefficients

The convective mass transfer at the exterior and interior surfaces of building components is so high, relative to the vapor resistance of the building materials, that it can usually be ignored. Similarly, the vapor resistance of still air is so low that accounting for convective effects within air spaces does not significantly increase the rate of vapor transfer across the enclosure. Therefore, unlike heat flow, no special allowances need be made at surfaces and across airspaces for most analyses. For vapor permeance laboratory measurements and research projects, the surface transfer may become important.

For most cases, the equivalent vapor permeances given in Table 6.4 can be used. For other cases, the convective-conductive heat transfer coefficient (not the radiative) can be converted using the following equation (see Chapter 8):

$$h_m = 6100 \cdot h_c$$

where h_c is in units of $W/m^2 \cdot K$, and

h_m is in $ng/(Pa \cdot s \cdot m^2)$.

Table 6.4: Equivalent water vapor permeances for surfaces and spaces

Condition	Equivalent permeance ng/Pa · s · m²	US perms
Interior surfaces	15 000	260
Still airspaces, per 10 mm (3/8")	18 500	320
Exterior surfaces	75 000	1300

6.4.4 Plane airspaces

Plane airspaces can be considered to have the vapor permeability of still air (varying from 183 to 196 ng/Pa · s · m over the range of 0 to 25°C) [ASTM 2016, Shirmer 1938]. The pure diffusion value for air is listed in Table 6.2, but is only valid for small airspaces with no convection. However, even small airspaces usually have some air movement that enhances mass transport. In any case, the vapor permeances of air spaces are very high (about 7500 metric perms / 130 US perms for a 25.4 mm / 1inch gap) and the precise values are usually unimportant for simple analysis.

6.4.5 Air movement and vapor transport

As mentioned elsewhere, diffusion is not commonly the source of severe moisture problems. As should be obvious from the calculations described above, the condensation quantities involved in most diffusion problems are small. Solar-driven inward vapor is the source of the largest practical diffusion-related condensation rates, but this type of condensation occurs only in specific situations for a limited amount of time. Cold weather diffusion condensation is normally only a problem in very cold weather for walls shaded from the sun, with little interior vapor resistance and no insulated sheathing, or very high interior humidity levels.

Air leakage condensation, by contrast, can result in significant accumulations of moisture and can occur under non-extreme temperature conditions. As for heat flow, convection is often far more important than diffusion.

The capacity of air to hold water vapor is high enough that bulk air movement can transport a significant amount of moisture. Convection is usually considered to be either forced or natural. Forced convection has an external force driving the airflow, such as wind pressure, or fans. Natural convection airflow is caused by density differences due to temperature or moisture content differences.

Simply and accurately calculating the quantity of air leakage condensation, however, is difficult. Given a flow rate, an upper limit for the amount of condensation (the maximum amount of wetting) can be calculated using the following procedure:

1. Calculate the temperature at every layer, j, through the enclosure.

2. Calculate the moisture content and dewpoint temperature of the interior air using the interior vapor pressure, $P_{ws,in}$ (or exterior air when considering warm-weather infiltration condensation).

3. Calculate the saturation vapor pressure (or moisture content) of the first surface in the enclosure that is below the dewpoint temperature of the stream of air, $P_{ws,j}$.

4. The maximum amount of convective condensation deposited on the condensation plane is the moisture content (W) of the indoor air less the saturation moisture content of air at this layer (W_{sat}).

5. Repeat steps three and four for each subsequent layer.

This method can be simply implemented using the ideal gas law described earlier. It is often easiest to base the assessment on the volume and temperature of interior air leaking outward. The maximum quantity of condensation transported by a flow of air from the interior to layer j is approximately:

$$Q_{condensation} = \frac{\Delta V \cdot (P_{w,in} - P_{ws,j})}{R_v \cdot T} \qquad \text{(Eq. 6.19)}$$

where $Q_{condensation}$ is in kg/s (or kg/hr)

ΔV is the volumetric flow rate of interior air outward (at interior temperatures) (m^3/s or m^3/hr)

$P_{w,in}$ is the interior water vapor pressure ($= RH_{in} \cdot P_{ws,in}$) in Pa,

$P_{ws,j}$ is the saturation vapor pressure at the temperature of layer j,

R_v is the gas constant for water vapor (461.5 J / kg·K)), and

T is the air temperature (K).

This method assumes steady-state diffuse (i.e., one-dimensional or spatially uniform) convection flow. The above method will almost always overestimate the actual amount of convective condensation, and hence the result should merely be considered a wetting potential. For example, large or concentrated flows of air will tend to elevate temperatures near the flow path and thus reduce the amount of condensation. Some moist air may pass through an assembly without contacting a condensation surface if the flow rate is high and the flow path is direct, e.g., large openings that pass straight through the enclosure (see Chapter 7). Modeling and laboratory testing have found "condensation efficiencies" of the order of 30 to 70% [Ojanen et al 1994].

Therefore, there are three primary reasons why condensation will be over-predicted by the simple method outlined above:

1. The stream of air may not actually contact a condensation surface,
2. The stream of air will cause heating of the surfaces and materials it contacts because of latent heat deposited by condensation and the sensible heat of the air, and
3. The stream of air tends to originate at cracks and openings but becomes diluted by dispersion, diffusion and mixing as it passes through the

enclosure (e.g., this is especially true for flow through mineral fiber insulations).

Air leakage can also cause drying, although the process for deposition and accumulation of condensate, and the process for the evaporation and removal of condensate, are not necessarily reversible. Evaporative drying due to air exfiltration is likely to be a much less efficient process than wetting because exfiltrating air is unlikely to reach 100% RH as it passes over a wetted layer in a wall assembly. Once a material is no longer saturated, exfiltration evaporation becomes an even less efficient means of drying (relative to the efficiency of the process of exfiltration condensation) because the drying front drops below the surface of the material and moisture diffusivity begins to retard drying. Nevertheless, air leakage may remove a significant amount of moisture in many situations.

Although the assumptions for air leakage rate, flow path, and condensation efficiency, etc., are likely to be only rough approximations, the simple approach described above does provide a rational basis for assessing the wetting and drying potentials of different wall assemblies for condensation due to exfiltration and diffusion. Dynamic hygrothermal computer models such as WUFI use similar simplifications to allow for the vapor and energy transport of air movement to be considered.

6.4.6 Vented spaces and equivalent vapor permeance

Many building assemblies employ airspaces which are vented to the exterior. The intent of venting the airspace is to facilitate the removal of water vapor by bulk air movement and diffusion, i.e., advection.

It can be shown that very small flows of air will transport significant quantities of water vapor. However, merely venting an airspace does not ensure that ventilation airflow occurs. Unless one can be sure that ventilation flow is occurring, it should be assumed that the cladding contributes to the vapor resistance of the assembly and the air space can simply be modeled as an unvented plane air space.

Building codes and regulations prescribe minimum values for the venting area for roofs and crawl spaces, but rules for walls are rarely codified. The 1990 *National Building Code of Canada* (5.2.1.2.(1)) required that where a layer outside the layer with the major thermal resistance significantly resists water vapor flow, then:

> *. . . . an air space ventilated to the outside or other method of equal effectiveness shall be provided for removing water vapor that may pass from the high vapor pressure side through the material with the major thermal resistance.*

This clause obviously applies to many wall cladding systems—masonry, vinyl and wood siding, and sheet metal all have low vapor permeance ratings—and a vented air space is provided in most wall assemblies designed with these claddings.

A former Canadian masonry construction standard (CAN3-A371-M84) required (§5.12.1) weepholes to allow water which penetrated the brickwork to drain out of the assembly. The same clause added, in a note, that:

> *Venting is often required in conjunction with weepholes to permit cavity walls and veneer walls to function properly.*

It is possible to calculate an equivalent vapor resistance of an enclosure layer which combines the effects of both diffusion and convection [Straube & Burnett 1995, Ten Wolde et al 1998]. The mass of water in air can be found from a form of the ideal gas law:

$$m_v = \frac{P_W \cdot V}{R_W \cdot T}$$

where m_v is the mass of water vapor (kg),

P_W is the water vapor pressure (Pa),

V is the volume of air (m^3),

R_W is the gas constant for water vapor (461.5 J/kg·K), and

T is the temperature (K).

Similar to the condensation calculation of Equation 6.19, the mass of water transported by an air volume exchange given a difference in vapor pressure, (assuming well-mixed air), is:

$$\Delta m_v = \frac{\Delta P_w \cdot \Delta V}{R_w \cdot T}$$

The transfer of water vapor through a wall layer by diffusion is the water vapor permeance. Using this equation, an air flow rate of 0.28 liters per second ($0.00028\ m^3/m^2\ s = 1\ m^3/m^2\ hr$), a vapor pressure difference of 1 Pa, and a mean temperature of 15 °C, the mass of water transferred, or the *equivalent vapor permeance due to ventilation only*, is:

$$\Delta m_v = \frac{1 \cdot 0.00028}{461.5 \cdot (273+15)} \cdot 10^{12}\ ng/kg$$

$$= 2100\ ng/\ (s \cdot m^2 \cdot Pa).$$

This value of permeance is over forty times that of the value for a 90 mm brick masonry veneer. This indicates that, at the very least, air flow can play a very important role in bypassing the vapor resistance of a brick veneer cladding.

To account for the resistance of the brickwork, a parallel path flow analogy can be used, and an equivalent vapor resistance and permeance calculated:

$$M_{v,equiv} = \frac{1}{R_{v,equiv}} = \frac{1}{R_{v,convection}} + \frac{1}{R_{v,diffusion}}$$

$$= M_{v,convection} + M_{v,diffusion} \qquad \text{(Eq. 6.20)}$$

In this example the vapor diffusion resistance of the brickwork is negligible. Even with a ventilation rate of only 0.1 m^3/m^2 hr, the transfer of vapor from behind the brickwork by ventilation flow is likely to be four to five times greater than by diffusion alone.

Section 7.5 of this text delves more deeply into the mechanisms and flow rates of ventilated roofs and claddings: the pressures and flows described below can be found there. A ventilation rate of 0.28 $l/m^2 \cdot s$ (1 $m^3/m^2 \cdot$ hr) through a cavity 50 mm deep results in a flow velocity of 0.0056 m/s: so small that it is exceptionally difficult to measure. Table 6.5 lists the equivalent vapor permeance of a perfectly vapor impermeable layer for various airflow rates that could realistically occur behind cladding or other layer. It can be seen from the table that even small rates of airflow result in an effective movement of water. Thus, cladding with low or zero permeance materials (such as metal, or glazed terra-cotta, or vinyl) behaves as if it is vapor permeable because of very small flows of air around that cladding.

Table 6.5: Equivalent vapor permeance for various air flow rates

Air Flow Rate $l / m^2 \cdot s$	Equivalent Vapor Permeance $ng / (s \cdot m^2 \cdot Pa)$
0.01	75
0.05	375
0.10	750
0.25	1 875
0.50	3 750
1.00	7 500
3.00	22 600

Because of the limited heat capacity of air, the effect of ventilation on temperature and energy flows is quite different and not nearly as significant. See the section on plane air spaces in the Chapter 5.

6.5 Multi-dimensional Vapor Diffusion

Water vapor diffusion is often considered to be one-dimensional. However, there are often practical situations in which the flow is two- or three-dimensional. Although large areas can bc analyzed using a parallel path method as for heat flow, the common practical case of a perforated vapor resistant layer may demand more complex analysis.

Measurements (for example by Sieffert [1960]) have shown that perforations through a thin vapor impermeable sheet will have a larger impact on diffusive transport than a 1-D analysis would support, i.e., for very small areas of perforations through thin highly impermeable layers, the increase in vapor permeance is likely to be much greater than the area of the perforations.

Salnina and Silarova [2009] reported the results of a number of measurements of thin films with perforations. These showed that the impact of perforations was proportionally more significant for low-permeance layers, that is, a highly impermeable layer will have its permeance increased very significantly with even 0.01% of its area perforated. However, the tests were conducted with air on either side of the membrane, and in practise, vapor barrier layers are commonly sandwiched between layers or at least placed in contact with a solid material.

Van der Spoel [2003] developed analytical solutions that predict the measurements of others reasonably well. His work emphasizes that it is not just the perforations of a vapor barrier film that impact vapor diffusion, but the vapor permeability and thickness of the layers on either side of such a thin sheet.

The available theory and results make it quite clear that very small perforation areas can have a major impact for highly impermeable layers (for example, those with a permeance of well under 6 ng/Pa s m^2, or 0.1 US Perms). Additional care in material selection and installation is warranted if such low permeance layers are an important part of a design. However, testing and analysis should also include the actual layers on either side of low permeance membrane to accurately account for the multi-dimensional flow that can arise.

For thin layers with moderately low vapor permeance (i.e., between 10 and 200 ng/Pa s m^2), small perforations (0.01% area) will only increases the permeance by a modest percentage.

6.6 Worked examples

The equations used to calculate vapor flow through an enclosure wall system are analogous to heat flow equations. Hence the tabular method described for calculating the distribution of temperatures within an enclosure in Chapter 5 can be modified and extended for calculating vapor diffusion.

Allowable condensation amounts depend on the nature of the material on which condensation is occurring. If the material is absorbent, the amount of moisture in the material should be kept within safe limits. See Chapter 8 for more information on performance thresholds. If the surface is not absorbent, the total amount that can be stored as a film on a surface ranges from 100 [Burch, 1995] to 500 g/m^2 [DIN 4108-3, Janssens 1998, Kuenzel et al 2023] depending on the angle and texture of the surface. Smooth vertical surfaces are at the lower end of this range and rough horizontal surfaces are at the upper.

The performance thresholds will vary with the material. For example, corrosion will proceed at a high rate at 100% RH regardless of the amount stored on the surface. Materials such as wood require a moisture content of over about 20% before mold will begin to grow (see Chapter 8): judgment is required to assess how deep moisture will be absorbed into the wood, and thus the mass of moisture required to reach this threshold.

The examples are presented sequentially, and each subsequent example adds more complexity and realism, although not necessarily greater accuracy.

6.6.1 Vapor pressure distribution

Example: Consider the wall assembly below with an interior temperature and relative humidity of 21 °C and 50%, respectively, and exterior temperature and relative humidity of 5 °C and 80%. What is the water vapor pressure at each interface due to vapor diffusion through the assembly? What is the relative humidity at each interface of the assembly based on the vapor pressure distribution calculated above? Use the permeability data given in this chapter.

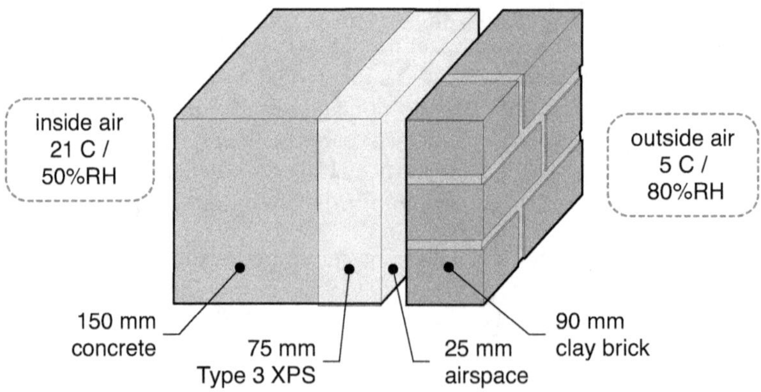

Solution

To determine the relative humidity at each interface in the assembly, we must first determine the vapor pressure and temperature. The vapor pressure is determined using the methods of this chapter and is summarized below:

- *Determine the vapor pressure at the interior and exterior of the wall assembly (the boundary conditions).*

Recall that the saturation vapor pressure at a given temperature is determined from:

$$P_{ws} = 1000 \cdot e\left(52.58 - \frac{6790.5}{T} - 5.028 \ln T\right)$$

where P_{ws} is the saturation vapor pressure and T is the temperature in Kelvin. The relative humidity, vapor pressure, and saturation vapor pressure are related according to:

$$\text{Ø (RH,\%)} = \frac{P_W}{P_{WS}}$$

from which the vapor pressure, P_w, at the interior and exterior can be calculated for the given temperature and RH boundary conditions.

Given the permeability and thickness of each material in the assembly (permeability values are found from tables earlier in this chapter), compute the vapor permeance.

The vapor permeance is calculated from:

$$M = \mu / l$$

where μ is the permeability of the layer material, l is the layer thickness, and M is the layer permeance

- *Calculate the vapor resistance, R_v, of each layer by inverting the vapor permeance calculated above.*

Thus:

$$R_v = 1 / M$$

- *Calculate the overall vapor resistance of the entire assembly as the summation of the vapor resistance of each layer of the assembly.*
- *Calculate the vapor pressure change across each layer of the assembly.*

The vapor pressure drop across a layer is a function of the total resistance of the assembly, the resistance of the individual layer, and the vapor pressure boundary conditions:

$$\Delta P_j = \frac{P_{int} - P_{ext}}{\sum_{j=0}^{n} R_{v_j}} \cdot R_{v,j}$$

where ΔP_j is the vapor pressure change across a given layer, $R_{v,j}$ is the vapor resistance of that layer, $\Sigma R_{v,j}$ is the overall vapor resistance of the entire assembly, and P_{int} and P_{ext} are the vapor pressures at the interior and exterior of the wall assembly.

- *Subtract the change in vapor pressure across the first layer, ΔPj, from the vapor pressure at the interior boundary condition.*

In this example, the interior has a higher vapor pressure than the exterior and thus vapor is driven to the exterior.

- *Subtract the change in vapor pressure for each layer from the vapor pressure at the boundary with the previous layer to determine the vapor pressure at every layer in the assembly.*

This calculation is performed most easily using a spreadsheet. Individual permeances for surface films and airspaces were discussed previously in the chapter, (note that there is no thickness associated with surface films).

Table 6.6 contains the results of the spreadsheet calculation used to determine the vapor pressure at each interface in this example.

Knowledge of the vapor distribution through the wall is of little use by itself. To make predictions, and to better understand the behavior of vapor within the wall assembly, we must determine the saturation vapor pressure and relative humidity at each interface. As the problem thus far considers only vapor diffusion and not air leakage, the temperature at each interface is the temperature to which the water vapor-air mixture is subjected. Thus, the temperature of the interface governs the capacity of the air to hold water, due to the interrelationship between temperature and saturation vapor pressure explored in detail in Chapter 4.

We use the methods of Chapter 5 to estimate the temperature change through each layer material and to calculate the temperature at each interface. The procedure is similar to that used to determine the vapor distribution through the wall assembly.

Briefly:

- *Determine the vapor resistance of each layer material. The resistance is the inverse of the layer permeance, which is the material permeability divided by the layer thickness.*

- *Calculate the overall vapor resistance of the entire assembly as the sum of the vapor resistance of each layer.*

Table 6.6: Vapor pressure distribution through the example wall assembly

Layer material	μ (ng/Pa·s·m)	t [m]	M_i (ng/Pa·s·m²)	$R_{v,i}$ (Pa·s·m²/ng)	ΔP_w (Pa)	P_w (Pa)
	Permeability (material property)	***Thickness (system property)***	$M = \mu / t$	M^{-1}	$\Delta P_v \cdot R_{vi}/\Sigma R_v$	***At interface***
Interior						*1249*
Interior film	-	-	15000	0.000067	0.3	
						1248
Concrete	2.6	0.150	17.3	0.058	302.4	
						946
Type 3 XPS	2.0	0.075	26.7	0.038	196.5	
						749
Air space	-	0.025	7200.	0.00014	0.7	
						749
Brick	10.0	0.090	111.1	0.0090	47.2	
						702
Exterior film	-	-	75000.	0.000013	0.1	
Exterior						*701*
			ΣR_v	**0.104**	$\Sigma \Delta P_w$	**547**

- *Calculate the pressure drop across each layer.*

  We employ the following relationship between the overall vapor resistance, the layer vapor resistance, and the vapor pressure boundary conditions:

$$\Delta P_j = \frac{P_{int} - P_{ext}}{\sum_{j=0}^{n} R_{v_j}} \cdot R_{v,j}$$

  where ΔP_j is the vapor pressure drop across layer j,

  $R_{v,j}$ is vapor diffusion resistance of that layer,

  ΣR_{vj} is the overall water vapor diffusion resistance,

  and $P_{w,int}$ and $P_{w,ext}$ are the interior and exterior vapor pressures.

- *Calculate the vapor pressure at each layer by subtracting the vapor pressure drop,* $\Delta P_{w,j}$ *from the interface vapor pressure at the previous layer.*

Recall from Chapter 4 that the governing equation that relates temperature to saturation vapor pressure is:

$$P_{ws} = 1000 \cdot e\left(52.58 - \frac{6790.5}{T} - 5.028 \ln T\right)$$

We can use this equation (or any other) to determine the saturation vapor pressure at each layer interface in the wall assembly (based on the interface temperature we have just calculated). Table 6.7 contains the results of the spreadsheet calculation used to determine the temperature and saturation vapor pressure at each interface in the wall assembly.

Finally, the relative humidity can be calculated at each system interface according to the RH equation, discussed previously:

$$\text{Ø (RH, \%)} = \frac{P_W}{P_{WS}}$$

Table 6.7: Temperature and saturation vapor pressure through the example wall assembly

Layer Material	k (W/ m·K)	t [m]	C_i (W/ m²·K)	R_i (m²·K /W)	ΔT (°C)	T (°C)	$P_{w,sat}$ **(Pa)**
Interior						*21.0*	*2497.4*
Interior film	-	-	8.0	0.13	0.7		
						20.3	2399.0
Concrete	1.8	0.15	12.0	0.083	0.4		
						19.9	2335.3
Type 3 XPS	0.029	0.075	0.39	2.59	13.5		
						6.4	996
Air space	-	0.025	5.9	0.17	0.9		
						5.5	909
Brick	1.3	0.090	14.4	0.069	0.4		
						5.2	886
Exterior film	-	-	34.0	0.029	0.2		
Exterior						*5.0*	*877*
			RSI_{total}	**3.06**			

Table 6.8 is a summary of the calculations and includes the relative humidity. The relative humidity is usually plotted because it can be used to evaluate the hygrothermal performance of an assembly. A relative humidity of 100% can be used to identify the occurrence of condensation in the wall, and high relative humidity initiates a number of deterioration mechanisms.

The temperature and vapor pressure have traditionally been plotted graphically on a cross section of the enclosure assembly being analyzed. A historic example graphic, from a 1946 National Bureau of Standards report [Dill & Cottony 1946] showing measured results is reproduced in Figure 6.1. The units are the same used in much of the United States today, and the graphical style is of its time, but the methodology is the same.

Table 6.8: Temperature, saturation vapor pressure, vapor pressure, and RH

Layer Material	T (°C)	$P_{w,sat}$ (Pa)	P_w (Pa)	RH (%)
Interior	*21.0*	*2497*	*1249*	*50*
Interior film				
	20.3	2399	1248	52
Concrete				
	19.9	2335	946	41
Type 3 XPS				
	6.4	996	749	78
Air space				
	5.5	909	749	82
Brick				
	5.2	886	702	79
Exterior film				
Exterior	*5.0*	*877*	*701*	*80*

6.6.2 Assessing interstitial condensation occurrence and quantity

Example

In the previous example, condensation (i.e., 100% RH or more) is not predicted to occur anywhere in the assembly. For the same wall assembly used in the previous example, what would happen if the exterior temperature were much lower, perhaps -14 °C and 80% RH?

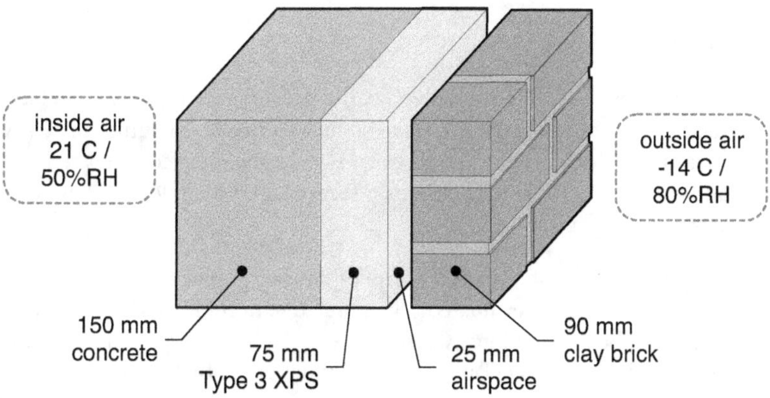

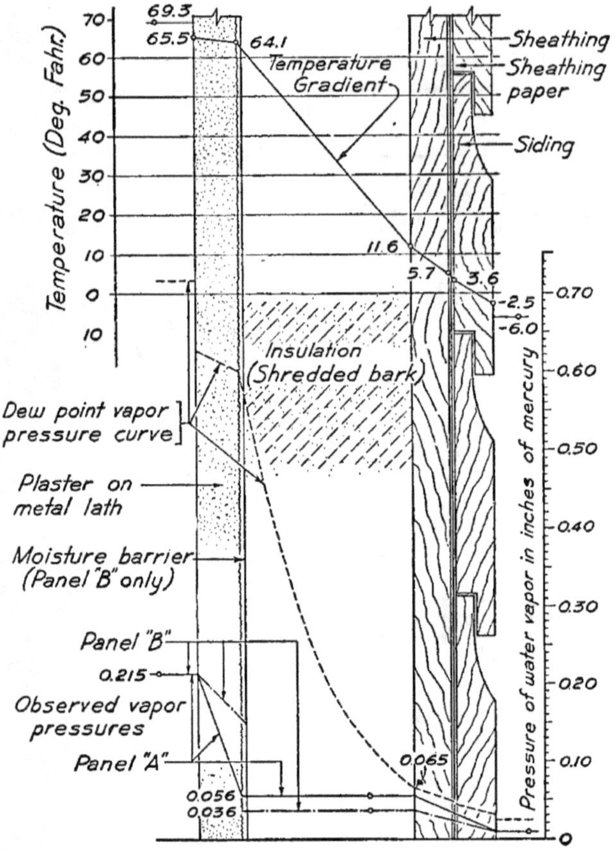

Figure 6.1: Plot of vapor pressure and temperature superimposed on enclosure cross-section [Dill & Cottony 1946]

Solution

If we follow the same procedure as the previous example to calculate the relative humidity at each interface, at an exterior temperature of -14 °C, we find that our analysis predicts an RH of 114% at the interface between the airspace and the brick. Table 6.9 presents a summary of the relevant data for this exterior temperature.

Humidity values of over 100% are, of course, practically impossible, and thus the wall must be re-evaluated. Since relative humidity cannot exceed 100%, the wall must be divided into two sections at the layer predicted to have over 100%. Then each section is analyzed independently in terms of vapor flow.

Relative humidity in excess of 100% indicates that condensation must occur at some condensation plane within the wall assembly. The previous examples assumed that the mass-flow of vapor into the wall assembly was equal to the mass-flow out of the assembly. However, for the case of condensation, this is no longer true as vapor is deposited within the wall assembly as condensate.

Table 6.9: Temperature, saturation vapor pressure, vapor pressure, and RH

Layer Material	T (°C)	$P_{w,sat}$ (Pa)	P_w (Pa)	RH (%)
Interior	*21.0*	*2497*	*1249*	*50*
Interior film				
	19.6	2287	1248	55
Concrete				
	18.6	2155	650	30
Type 4 XPS				
	-10.9	268	262	98
Air space				
	-12.9	229	261	114
Brick				
	-13.7	215	167	78
Exterior film				
Exterior	*-14.0*	*209*	*167*	*80*

As a simple but reliable rule for identifying the condensation plane can be stated: **the condensation plane will be the next upstream-facing solid surface on the cold side of the calculated dewpoint location**

The temperature distribution is assumed to be independent of vapor condensation. In reality this assumption is only an approximation, as the latent heat released when water condenses will cause an increase in the temperature. However, for low condensation rates, this is a reasonable approximation.

The solution procedure for a divided wall assembly is as follows:

- *Determine the condensation plane (see above).*
- *Divide the wall into two separate portions at the condensation plane.*
- *Set the vapor pressure at the condensation plane equal to the saturation vapor pressure for the interface temperature.*
- *Analyze each portion of the divided wall separately, using the methods described in the previous example, but using the temperature and vapor pressure at the condensation plane as a boundary condition (for example, the interior portion of the divided assembly uses the vapor pressure at the condensation plane as the exterior vapor pressure).*

For this example, vapor is flowing from the interior to the exterior, and the condensation plane is the back surface of the brick (facing the interior). As observed in Table 6.10 the wall is divided into two separate walls to recalculate the vapor pressure distribution through the wall assembly assuming that condensation is occurring on the back of the brick.

Table 6.10: Corrected vapor pressure distribution for a wall with condensation

Layer Material	M_i (ng/Pa·s·m²)	$R_{v,i}$ (Pa·s·m²/ ng)	T (°C)	$P_{w,sat}$ (Pa)	P_w (Pa)	RH (%)
Interior			*21.0*	*2497*	*1249*	*50*
Interior film	15000	0.000067				
			19.6	2287	1248	55
Concrete	17.3	0.058				
			18.6	2155	631	29
Type 3 XPS	26.7	0.038				
			-10.9	268	230	86
Air space	7200	0.00014				
			-12.9	229	229	100
	ΣR_v	**0.0954**		$\Sigma \Delta P_w$	**1020**	
		Flow to:	$\Delta P/\Sigma R_v$	10689.	ng/·s·m²	
			-12.9	229	229	100
Brick	111.1	0.0090				
			-13.7	215	167	78
Exterior film	*75000*	0.000013				
Exterior			*-14.0*	*209*	*167*	*80*
	ΣR_v	**0.0090**		$\Sigma \Delta P_w$	**62**	
		Flow away:	$\Delta P/\Sigma R_v$	6862	ng/·s·m²	
		Net Accumulation:		3827	ng / s·m²	
				0.0138	g / hr·m²	

Example

Determine the condensation rate on the back surface of the brick for the wall assembly analyzed in the previous example at an exterior temperature of -14 °C and 80% RH.

Solution

We can calculate the amount of vapor that passes through a given wall assembly according to the following equation:

$$I_{inflow} \text{ or }_{outflow} = \frac{P_{int} - P_{ext}}{\sum_{j=0}^{n} R_{vj}}$$

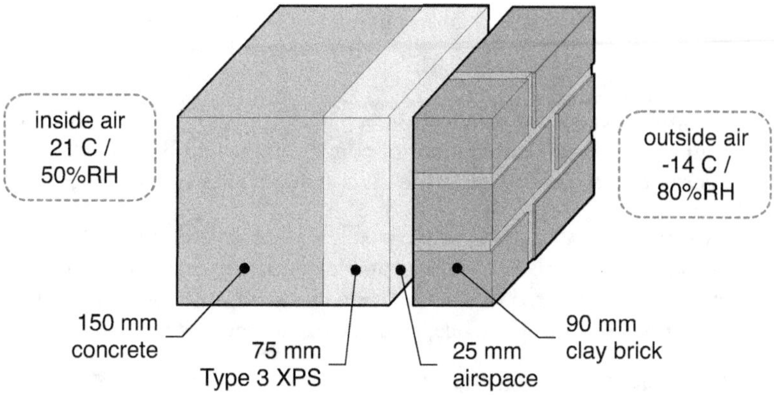

where I_{inflow} or $I_{outflow}$ is the mass-flow rate of water vapor,

ΣR_{vj} is the overall vapor resistance of the assembly, and

P_{int} and P_{ext} are the vapor pressures at the interior and exterior surface of the assembly.

For an example that does not involve condensation, the mass-flow rate of vapor into the wall from the interior will equal the mass-flow rate of vapor from the wall to the exterior (or vice versa for vapor driven inward). However, for a wall that experiences condensation, the mass-flow rate into the assembly is no longer equal to the mass flow rate out of the assembly.

We can estimate the condensation rate as the difference between the mass-flow rate to the condensation plane, and the mass-flow rate from the condensation plane. Table 6.10 contains data on the mass flow rates through the two sections of the wall assembly, labeled *flow to* and *flow away*. *Flow to* refers to vapor driven to the condensation plane and *flow away* refers to vapor driven from the condensation plane. The following is an example calculation to predict the mass flow rates to and from the condensation plane as well as the condensation rate.

Again, as the interior vapor pressure is higher than the exterior vapor pressure, we know that vapor is driven to the exterior.

$$I_{flow\ to} = 1/(0.0954\ Pa \cdot s \cdot m^2/ng) \cdot (1249\ Pa - 229\ Pa)$$

$$I_{flow\ to} = 10{,}689\ ng/s \cdot m^2$$

A similar process is used to calculate the vapor outflow from the brick-air interface.

$$I_{flow\ away} = 6862\ ng/s \cdot m^2$$

$$I_{flow\ to} - I_{flow\ away} = \text{Condensation Rate}$$

$$= 10{,}689 \text{ ng/s}\cdot \text{m}^2 - 6862 \text{ ng/s}\cdot \text{m}^2$$

$$= 3827 \text{ ng/s}\cdot \text{m}^2$$

$$= 0.0137 \text{ g/hr}\cdot \text{m}^2$$

The predicted rate of condensation, 0.0137 g/hr· m^2, is vert small. Over the course of one day, this amount of condensate would be a layer about 1 micrometer thick. Note that the units are given per square meter of wall area.

How much condensation is acceptable? As described in Chapter 8 and the introduction to this section, acceptable condensation depends on the type of materials, the desired level of performance, and the probability of the boundary conditions occurring. Hence, a degree of judgment is required.

Example: Ventilated Claddings
The previous example assumed that the brick veneer was unvented. If the veneer were to be well vented, a permeance of 1000 ng/pa·s·m^2 would be a reasonable estimate of the equivalent vapor permeance for the exterior brick. In this case, the use of a well-vented veneer would eliminate the possibility of condensation under these conditions (-14 °C, 80% RH).

Answer

Table 6.11 contains the results of the spreadsheet calculation used to determine the temperature and saturation vapor pressure at each interface in the wall assembly for the case in which the brick is well vented. Note that the vapor pressure drops considerably from the previous example in which the brick was not vented. The potential for condensation is thus eliminated, under these conditions, as the relative humidity at every interface (including the airspace-to-brick interface that was at 114% RH for the unvented case) has dropped significantly.

Table 6.11: Well-ventilated brick veneer

Layer Material	M_i (ng/Pa·s·m²)	$R_{v,i}$ (Pa·s·m²/ ng)	T (°C)	$P_{w,sat}$ (Pa)	P_w (Pa)	RH (%)
Interior			*21.0*	*2497*	*1249*	*50*
Interior film	15000.	0.000067				
			19.6	2287	1248	55
Concrete	17.3	0.058				
			18.6	2155	601	28
Type 4 XPS	26.7	0.038				
			-10.9	268	180	67
Air space	7200.	0.00014				
			-12.9	229	179	78
Brick	1000.	0.0010				
			-13.7	215	167	78
Exterior film	*75000.*	0.000013				
Exterior			*-14.0*	*209*	*167*	*80*
	ΣR_v	**0.096**		**$\Sigma \Delta P_v$**	**1082**	

6.6.3 Air leakage condensation

The amount of condensate predicted for cold weather condensation is often insignificant (as was observed in the previous example). Air leakage condensation is usually much more important.

Example
Calculate the volume of condensation that would form if exfiltration at a rate of 0.5 liter/s m² wall of occurred under conditions of 21 °C, 50% RH indoors and -5°C , 80% RH outdoors for the wall assembly used in the previous examples. How accurate is this estimate likely to be?

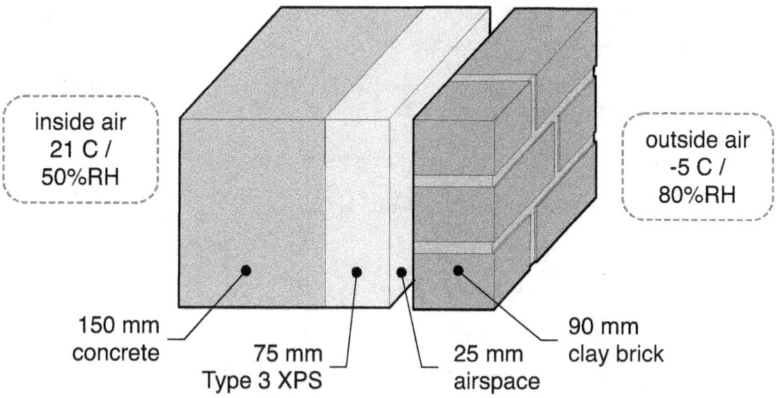

Solution

It is difficult to accurately predict the total amount of air leakage condensation; however, an upper limit (maximum possible) can be determined as follows:

- *Calculate the temperature at every layer through the assembly, as has been done in the previous examples.*
- *Calculate the moisture content and dewpoint temperature of the interior air using the interior vapor pressure, $P_{w,in}$ (or exterior air when considering warm-weather infiltration condensation).*

To find the air moisture content from the vapor pressure:

$$W_w = \frac{0.622 \cdot P_w}{P_t - P_w}$$

$$W_{in} = \frac{0.622 \cdot 1249Pa}{101325Pa - 1249Pa}$$

$W_{in} = 0.00776$ kg/kg

To identify the condensation plane, the dew-point of the interior air may be calculated:

$$t_d = \frac{4030}{18.689 - \ln(\frac{Pw}{133})} - 235 \quad [°C, Pa]$$

$$t_d = \frac{4030}{18.689 - \ln(\frac{1249Pa}{133})} - 235$$

$t_d = 10.0$ °C

The first upstream-facing solid surface below the dewpoint is the interface between the airspace and the brick, and hence this is the plane on which condensation is expected to occur.

- *Calculate the saturation vapor pressure (or moisture content) of the first surface in the enclosure that is below the dewpoint temperature of the stream of air, Pws,j.*

The vapor pressure at the predicted condensation plane (-4.2 °C and 100% RH) is

$P_{ws} = 452$ Pa

The moisture content of this air is:

$$W_{out} = \frac{0.622 \cdot 452Pa}{101325Pa - 452Pa}$$

$W_{out} = 0.00279$ kg/kg

- *The maximum amount of condensate transported by convection and deposited on the condensation plane is the moisture content (W) of the indoor air less the saturation moisture content of air at this layer (W_{sat}).*

The mass flow rate of vapor due to 0.5 lps/m² of exfiltration assuming 1.2 kg/m³ density for indoor air is:

$$\rho_{mixture} = 1.2 \text{ kg/m}^3$$

$$\frac{dm}{d\theta} = 1.2 \text{ kg/m}^3 \cdot 0.5 \text{ l/s} \cdot \text{m}^2 \cdot 1 \text{ m}^3/1000 \text{ l}$$

$$\frac{dm}{d\theta} = 0.0006 \text{ kg/s} \cdot \text{m}^2$$

Under ideal conditions, moisture will condense on the back of the brick at:

$$\text{Condensation rate} = (W_{in} - W_{sat}) \cdot \frac{dm}{d\theta}$$

$$= (7.76 \text{ g/kg} - 2.79 \text{ g/kg}) \cdot 0.0006 \text{ kg/s} \cdot \text{m}^2$$

$$= 4.97 \cdot 10^{-3} \text{ g/s} \cdot \text{m}^2 \text{ or } 17.9 \text{ g/hour} \cdot \text{m}^2$$

- *Repeat steps three and four for any subsequent layers.*

There are no other layers on which condensation could form in this case.

Table 6.12 contains the results of the spreadsheet calculation used to determine the temperature at each interface and the saturation water content.

Discussion

Although condensation is not predicted to occur due to diffusion at -5 °C (Table 6.13 below), the previous calculation showed that air leakage through the assembly is predicted to cause condensation with a maximum accumulation rate of 17.9 g/hour· m². This is significantly greater (more than 1000 times) than the amount of condensation predicted due to diffusion for the case when the exterior temperature falls to -14 °C. Hence, not only is air leakage condensation likely to occur at higher temperatures (which occur for many more hours per year), but the intensity of condensation (rate of accumulation) is also likely to be much greater. Given sufficient air leakage, a significant accumulation of water can condense on interstitial surfaces even if only a small percentage of the moisture available in the leaking air stream actually condenses.

This estimate is the absolute maximum amount of condensation that can occur at the interface; in reality, the actual amount of condensation will be somewhat less for reasons described earlier in the chapter.

Table 6.12: Temperature, dew point, and saturation water content

Layer material	T (°C)	$P_{w,sat}$ (Pa)	P_w (Pa)	Dew point interior air (°C)	Moisture content (g/kg)
Interior	*21.0*	*2497*	*1249*	*10.0*	*7.76*
Interior film [note 1]					
	19.9	2339	1248		
Concrete					
	19.2	2239	746		
Type 4 XPS					
	-2.7	503	419		
Air space					
	-4.2	**452**	**418**	W_{sat}	**2.79**
Brick					
	-4.8	432	339		
Exterior film					
Exterior	*-5.0*	*424*	*339*		

Table 6.13: Temperature, saturation vapor pressure, vapor pressure, and RH at –5 °C

Layer material	T (°C)	$P_{w,sat}$ (Pa)	P_w (Pa)	RH (%)
Interior	*21.0*	*2497*	*1249*	*50*
Interior film [note 1]				
	19.9	2399	1248	53
Concrete				
	19.2	2239	746	33
Type 4 XPS				
	-2.7	503	419	83
Air space				
	-4.2	452	418	92
Brick				
	-4.8	432	339	79
Exterior film				
Exterior	*-5.0*	*424*	*339*	*80*

6.6.4 Interstitial condensation under summer conditions

Example

Determine the vapor and temperature distribution through the wall assembly given above for a hot, humid, summer day with exterior conditions of 30 °C and 80% RH.

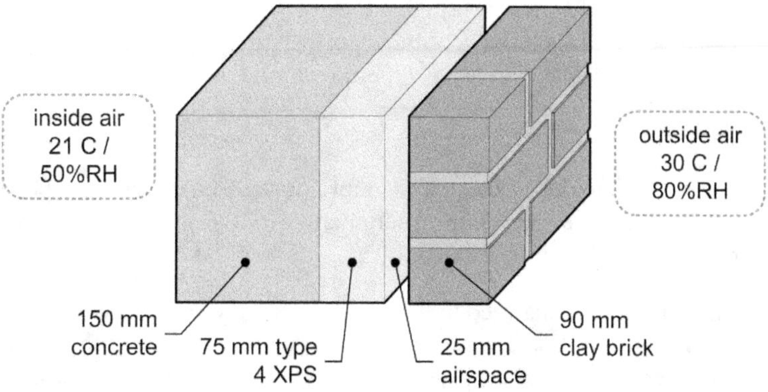

Solution

Under these conditions, contrary to the previous examples, water vapor is driven inward, as the exterior is now the high vapor pressure side. Table 6.13 contains the results of the spreadsheet calculation used to determine the temperature at each interface and the saturation water content. Note that condensation is unlikely in this scenario.

Table 6.14: Temperature, saturation vapor pressure, vapor pressure, and RH for a summer day

Layer material	T (°C)	$P_{w,sat}$ (Pa)	P_w (Pa)	RH (%)
Interior	*21.0*	*2497*	*1249*	*50*
Interior film [note 1]				
	21.4	2554	1250	49
Concrete				
	21.6	2593	2443	94
Type 4 XPS				
	29.2	4070	3218	79
Air space				
	29.7	4189	3221	77
Brick				
	29.9	4238	3407	80
Exterior film				
Exterior	*30.0*	*4259*	*3407*	*80*

Example
In most cases, walls are heated above the ambient air temperature by solar radiation. Calculate the vapor and temperature conditions for a wall heated by solar exposure using the wall system from the previous examples.

Solution

If we calculate the surface temperature of a red-brick veneer exposed to the sun on the same warm humid summer day (30 °C and 80% RH), a temperature of about 45 °C can be expected.

This situation is analyzed in Table 6.15 using the spreadsheet method of the previous examples. The solution process is identical, with the exception that the exterior temperature (boundary condition) is now the surface temperature of the brickwork, heated by the sun (the surface film is omitted). It can be seen that the increase in temperature results in a general decrease in relative humidity throughout the wall. This scenario represents many wall systems and is especially relevant for roof assemblies with non-absorbent roofing membranes.

Table 6.15: Temperature, saturation vapor pressure, vapor pressure, and RH for a wall heated by solar exposure

Layer material	T (°C)	$P_{w,sat}$ (Pa)	P_w (Pa)	RH (%)
Interior	*21.0*	*2497*	*1249*	*50*
Interior film [note 1]				
	22.0	2653	1250	47
Concrete				
	22.6	2761	2443	88
Type 4 XPS				
	43.1	8711	3218	37
Air space				
	44.5	9339	3221	34
Brick				
	45.0	9605	3407	35
Exterior	*30.0*	*4259*	*3407*	*80*

Example
The results of the previous example will be quite different if one assumes the cladding or roofing is wet from previous rain events, condensation, or built-in moisture. Given the right conditions, this inward-driven water vapor can also condense in the interior of the wall assembly.

Calculate the vapor and temperature conditions for a wall with wet cladding heated by solar exposure. Assume the same wall assembly and summer conditions as in the previous example.

Solution

The solution process for this situation is identical to that used for the previous examples except that for this case, we assume the back of the brickwork to be at a relative humidity of 100% and use that value as the exterior boundary condition (the ambient exterior humidity is not used in this calculation). We are also assuming that the brickwork is heated by solar exposure, and thus we impose the temperature onto the brickwork itself.

Table 6.16 is a summary of the temperature and vapor pressure calculations through the wall assembly. The analysis indicates relative humidity exceeding 100% at two interior interfaces, which indicates that the wall must be divided for accurate analysis (though not necessarily twice; one division may be sufficient).

Table 6.16: Temperature, saturation vapor pressure, vapor pressure, and RH

Layer material	T (°C)	$P_{w,sat}$ (Pa)	P_w (Pa)	RH (%)
Interior	*21.0*	*2497*	*1249*	*50*
Interior film				
	22.0	2653	1254	47
Concrete				
	22.6	2761	6147	223
Type 4 XPS				
	43.1	8711	9327	107
Air space				
	44.5	9339	9339	100
Brick				
	45.0			
Exterior film				
Exterior	*30.0*	*4259*	*3407*	*80*

As in the previous examples, mass-flow into a wall cannot be equated to mass-flow out of a wall if condensation is predicted. In this case, the wall can be divided either at the airspace-to-XPS interface or at the XPS-to-concrete interface. However, as the vapor pressure due to the inward vapor drive is significant, one division is not sufficient to reduce the relative humidity at the other interface to less than 100%, and thus the wall must be divided again.

The solution procedure for a wall divided twice is identical to that for a wall with a single division; therefore the solution procedure is not discussed further.

Table 6.17 and Table 6.18 present the results of the analysis spreadsheet. The first table reflects the wall after the first division and is included to demonstrate the need for the additional wall division.

Table 6.17: Corrected vapor-pressure distribution, after first division

Layer material	M_i (ng/Pa·s·m²)	$R_{v,i}$ (Pa·s·m²/ ng)	T (°C)	$P_{w,sat}$ (Pa)	P_w (Pa)	RH (%)
Interior			*21.0*	*2497*	*1249*	*50*
Interior film	15000.	0.000067				
			22.0	2653	1254	47
Concrete	17.3	0.058				
			22.6	2761	5773	209
Type 4 XPS	26.7	0.038				
			43.1	8711	8711	100
	ΣR_v	**0.0953**		**$\Sigma \Delta P_v$**	**7457**	
		Flow away:	**$\Delta P_w/\Sigma R_v$**	78,339	ng/s·m²	
			43.1	8711	8711	100
Air space	7200.	0.00014				
			44.5	9339	9339	100
Brick						
			45.0	-	-	-
Exterior						
	ΣR_v	**0.00014**		**$\Sigma \Delta P_v$**	**628**	
		Flow to:	**$\Delta P_w/\Sigma R_v$**	4,519,694	ng/s·m²	
		Net Accumulation:		4,441,356	ng/s·m²	

Thus, we observe that condensation occurs on two planes: the interface between the airspace and the XPS, and the interface between the concrete and the XPS. A large amount of moisture is driven inward due to the solar heating and the vast majority of moisture that condenses within the assembly (total of 4,493,500 ng/s·m² or 16.2 g/hr·m² distributed over the two condensation interfaces). The largest accumulation occurs at the interface between the airspace and the XPS (4,361,036 ng/s·m² or 15.7g/hr·m²) where it can be harmlessly drained away. Only 132,466 ng/s·m² or 0.5g/hr·m² condenses on the concrete. The concrete can easily and safely store significant quantities of water and will also not be damaged by high humidity levels (but freeze thaw damage could occur if several kilograms of water were stored in the brick veneer during freezing conditions).

Table 6.18: Final corrected vapor pressure distribution, after second division

Layer material	M_i (ng/Pa·s·m²)	$R_{v,i}$ (Pa·s·m²/ ng)	T (°C)	$P_{w,sat}$ (Pa)	P_w (Pa)	RH (%)
Interior			*21.0*	*2497*	*1249*	*50*
Interior film	15000	0.000067				
			22.0	2653	1250	47
Concrete	17.3	0.058				
			22.6	2761	2761	100
	ΣR_v	**0.0578**		$\Sigma \Delta P_v$	**1513**	
		Flow away:	$\Delta P/\Sigma R_v$	26,192	ng/s·m²	
			22.6	2761	2761	100
Type 4 XPS	26.7	0.038				
			43.1	8711	8711	100
	ΣR_v	**0.0375**		$\Sigma \Delta P_v$	**5950**	
		Flow away:	$\Delta P/\Sigma R_v$	158,658.	ng/s·m²	
			43.1	8711	8711	100
Air space	7200	0.00014				
			44.5	9339	9339	100
Brick						
			45.0	-	-	-
Exterior						
	ΣR_v	**0.00014**		$\Sigma \Delta P_v$	**628**	
		Flow to:	$\Delta P/\Sigma R_v$	4,519,694.	ng/s·m²	
		Net Accumulation:		**4,493,503**	ng/s·m²	
				16.2	g/hr·m²	

Discussion

It can be seen that if the sun strikes the wall very large vapor pressures are generated (over 9000 Pa) and significant amounts of condensation can occur. Since the concrete is quite moisture-tolerant and can harmlessly store a large amount of water vapor, the assembly is unlikely to be harmed. However, if the back up wall were made of exterior gypsum sheathing over batt insulation with a low-permeance vapor interior retarder (e.g., polyethylene or vinyl wallpaper), the condensation would occur near the inside of the assembly and could cause serious damage and/or mold growth. Note that the XPS in the worked example acts to dramatically reduce inward flows. Such inward vapor drives can be important for all walls with permeable cedar-shake-clad roofs and new concrete walls with vapor-permeable insulation placed on the inside.

Diffusion drying

Diffusion not only causes wetting, it can also dry wall assemblies. Modern enclosures attempt to completely control air flow and its related wetting, (see later chapters). This means that air flow is practically eliminated as a means of both wetting and drying. Drainage removes bulk liquid water, especially from parts of saturated or non-absorptive materials, and can be a powerful drying mechanism. However, most damage mechanisms (e.g., corrosion, rot and mould, freeze-thaw) can continue even if all moisture is removed by drainage [Straube 1997]. Hence, if moisture damage is to be avoided, additional drying is not only necessary, it is often critical to good moisture performance. Since air leakage drying is eliminated, and drainage is insufficient, diffusion is the only other mechanism available for drying. For this reason, understanding diffusion is important to the understanding of the performance of modern building enclosures.

Example

Consider the wall assembly below. If the concrete has a density of 2200 kg/m^3 and contained about 3% moisture by weight when it is closed in, it will contain 9.9 kg of water per m^2 (2200 x 0.15 x 0.03). Inspection of the sorption isotherm (see Chapter 8) shows that the RH of the concrete remains at 100% until the moisture content drops to about 1%. This sorption isotherm performance is a common and important feature of many porous building materials.

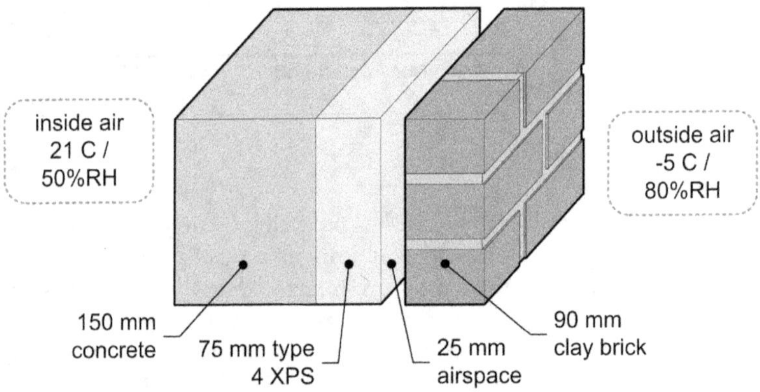

Assuming a painted drywall finish over a layer of 6 mil poly is added to the interior of the wall and that the RH remains at 100% through the body of the material, how long will it take to dry 6.6 kg/m^2 of moisture (i.e., drop the internal RH to below 100%) if interior conditions are 21 °C /50% RH and exterior are -5 °C / 80% RH.

Solution

The solution process for this example is much the same as for the previous examples. The intent is to determine the drying rate of wet concrete, through the wall assembly. The concrete is set to a relative humidity of 100% (on both sides), and the wall is divided into two separate walls with the wet concrete as a boundary condition for both. Using the methods previously discussed, we can determine the mass-flow rate through the divided wall. The sum of these flow rates is the drying rate of concrete. Table 6.19 contains the results of a spreadsheet analysis used to calculate the drying rate of the concrete wall. It should be noted that the high vapor pressure generated by the drying concrete forced a second wall division, as condensation is predicted to occur on the back of the brick work.

Note that the vapor permeability of the "6 mil" Poly is given as 0.000474 ng/Pa·s·m. Since 6 mille is equivalent to 6/1000 of an inch or 0.15 mm, the permeance can be calculated as 3.1 ng/Pa·s·m^2.

The drying rate of the concrete is about 3014 ng/ s·m2 (0.011 g/hr·m^2) to the interior and 44,742 ng/s·m^2 (0.161 g/hr·m^2) towards the exterior. Note that the drying in the exterior direction is mostly redistribution of the water onto the back of the brickwork, since 32,363 ng/s·m^2 (0.116 g/hr·m^2) condenses. If the brickwork is not ventilated, and the condensate is absorbed rather than drained, the drying of the wall would only be 12,378 ng/s·m^2 (0.045 g/hr·m^2) to the exterior. More realistically, the condensed water will wick (move by capillary suction) to the face of the brick and be evaporated. In all cases, the concrete does benefit from this drying, so the total drying rate will be:

$$44{,}742\ \text{ng/s·m}^2 + 3014\ \text{ng/s·m}^2 = 47{,}755\ \text{ng/s·m}^2$$
$$=0.17\ \text{g/hr·m}^2$$

Given the rate of water vapor flow from the concrete to the interior of the wall (neglecting condensation on the back of the brickwork), we can estimate the time it would take for the concrete to dry:

To drop from 3% MC to 1% MC, 6.6 kg/m^2 must be lost,

$$\text{Drying duration} = 6600\ \text{g/m}^2\ /0.17\ \text{g/hr·m}^2\ /24\ \text{hrs/day}$$
$$= 1600\ \text{days}$$

Thus, it would take 1600 days at the rates calculated above to dry the wall. Of course, this is far longer than the temperature is likely to remain at – 5 °C. Also, most concrete walls are not enclosed immediately. Hence, for more accurate analysis, each of the monthly average outdoor and indoor conditions could be entered and the rate of drying plotted.

Table 6.19: Vapor pressure distribution for drying by diffusion

Layer material	M_i (ng/Pa·s·m²)	$R_{v,i}$ (Pa·s·m²/ ng)	T (°C)	$P_{w,sat}$ (Pa)	P_w (Pa)	RH (%)
Interior			*21.0*	*2497*	*1249*	*50*
Interior film	15,000	0.000067				
			20.0	2344	1249	53
Interior Paint	275	0.0036				
			20.0	2344	1260	54
Dry Wall	1572	0.00064				
			19.2	2228	1262	57
6 mil. Poly	3.12	0.32				
			19.2	2228	2228	100
	ΣR_v	**0.325**		$\Sigma \Delta P_v$	**979**	
		Concrete drying to the interior:	$\Delta P/\Sigma R_v$	3014	ng/s·m²	
			19.2	2228	2228	100
Concrete	17.3	0.058				
			18.5	2135	2135	100
Type 4 XPS	26.7	0.038				
			-2.8	500	457	91
Air space	7200.	0.00014				
			-4.2	451	451	100
	ΣR_v	**0.0376**		$\Sigma \Delta P_v$	**1684**	
		Concrete drying to the exterior:	$\Delta P/\Sigma R_v$	44742.	ng/s·m²	
			-4.2	451	451	100
Brick	111.1	0.0090				
			-4.8	432	339	79
Exterior	*75000.*	*0.000013*				
			-5.0	424	339	80
	ΣR_v	**0.0090**		$\Sigma \Delta P_v$	**112**	
		Wall drying to the exterior:	$\Delta P/\Sigma R_v$	12,378	ng/s·m²	
			Drying rate of the concrete wall:	47,755	ng/s·m²	
			Condensation rate on the back of the brick:	32,363	ng/s·m²	

The vapor barrier in this case reduces the drying rate considerably and as shown in previous examples, is not really necessary to avoid damaging condensation.

A more accurate method of calculating the drying time of the concrete wall is to compute the yearly drying rate as the sum of the drying rate of each month of the year using the average monthly temperature and relative humidity. The calculations were performed using the same spreadsheet as was used for the previous example. Table 6.20 below, summarizes the monthly drying rate for a given year based on the indicated average monthly conditions. The yearly drying rate is the sum of the monthly drying rates. Additionally, the drying rate for the average month was calculated from the averages of the temperature and RH and converted to a yearly drying rate (multiplying by twelve months). This average yearly drying rate is slightly more than 1% greater than the drying rate computed as the sum of the individual months and thus is a good approximation.

Table 6.20: Monthly average drying rate

Month	Exterior temp. (°C)	Exterior RH (%)	Interior temp. (°C)	Interior RH (%)	Drying rate (g/mth·m²)
January	-6.0	80	20.0	40	118.5
February	-5.0	78	20.0	45	116.3
March	0.0	75	21.0	50	115.4
April	6.0	68	22.0	55	107.3
May	12.0	66	23.0	55	104.7
June	17.0	66	24.0	60	97.8
July	21.0	67	25.0	60	90.1
August	20.0	72	24.0	60	78.5
September	15.0	75	23.0	55	87.5
October	8.0	76	22.0	50	101.1
November	3.0	80	21.0	45	109.2
December	-2.0	81	21.0	45	120.2
		Annual Total Drying Rate (g/year·m²):			1246.6
		Average Monthly Drying Rate (g/mth.·m²):			103.9
Average:	7.4	74	22.2	52	104.9
		Average Annual Drying Rate (g/year·m²):			1259.3

As before, for the concrete to drop from 3% MC to 1% MC, 6.6 kg/m² of moisture must be lost,

$$\text{Drying duration} = 6600 \text{ g/m}^2 \ / \ 1246.6 \text{ g/year·m}^2$$

$$= 5.24 \text{ years}$$

After 5 years, 6600 g/m² – 5 ·1246.6 g/year·m² = 367 g/m² of moisture remains. It will take between three and four months to completely dry this moisture, depending on which month is considered (as some months have much greater drying rates than others). Thus, the predicted drying time is somewhere around 1900 days, which is higher than the first calculation (1600 days); the latter was based on the drying rate of a winter month. As expected, there is some variation due to monthly conditions, and the highest drying rates are experienced during the winter months.

Although this calculation can be carried on in even greater detail, the assumptions required generally do not justify the additional effort. One-dimensional computer models that account for capillary flow, diffusion, and full storage functions provide a much more accurate prediction of drying time without much more effort than a daily-average, diffusion-only calculation.

6.7 References

ASTM E96-16, *Standard Test Methods for Water Vapor Transmission of Materials*, ASTM International, West Conshohocken, PA, 2016.

Burch, D., "An Analysis of Moisture Accumulation in the Roof Cavities of Manufactured Housing," *Airflow Performance of Building Envelopes, Components, and Systems, ASTM STP 1255*, Mark P. Modera and Andrew K Persily, Eds., American Society for Testing and Materials, Philadelphia, pp. 156-177.

David R. Lide, ed., "Physical Constants of Organic Compounds", in *85th ed CRC Handbook of Chemistry and Physics*, CRC Press, Boca Raton, FL, 2005.

Glaser, H., "Graphisches Verfuhrung zur Untersuchung von Diffusionvorgangen," *Kaltetecknik*, Vol. 11, 1959, pp. 345-355.

Joy, F. "Basic Concepts of Water Vapor Migration and Their Application to Frame Walls." STP 119 *Symposium on Thermal Insulating Materials*. ASTM, 1952.

Lstiburek J. W. "Understanding vapor barriers". *ASHRAE Journal*. Aug 2004 Vol 46, pp. 40-51.

Janssens, A. *Reliable Control of Interstitial Condensation in Lightweight Roof Systems: Calculation and Assessment Methods*. Doctoral dissertation, Katholieke Universiteit Leuven, Belgium, December 1998.

Karagiozis AN, Kumaran MK. "*Computer model calculations on the performance of vapor barriers in Canadian residential buildings*". Transactions of ASHRAE. 1993, Vol. 99 pp. 991-1003.

Künzel, H., Tanaka, E. and Zirkelbach, D., 2023. "Tipping point for condensation water drainage on surfaces and interfaces of insulated wall assemblies", *Proc of 56th International Conference of the Architectural Science Association,* Tasmania, Australia.

Ojanen, T., Kohonen, R., Kumaran, M.K., "Modeling Heat, Air, and Moisture Transport through Building Materials and Components," *Manual on Moisture Control in Buildings*, Heinz Trechsel, editor. American Society for Testing and Materials, Philadelphia, 1994, pp. 18-34.

Quirouette R.L. *The difference between a vapor barrier and an air barrier*, Building Practice Note 54, IRC/NRCC, Ottawa, 1985.

Rowley, F.B., "A theory covering the transfer of vapor through materials", *ASHVE Transactions*, Vol. 45, p. 54, 1939.

Seiffert K. "Messungen vom diffusionswiderstandsfaktoren (Measurements of diffusion resistance factors)", *Kaltetechnik* 1960; 12(7):187–190.

Shirmer, R., "Diffusionszahl von Wasserdampf-Luft-Gemischen und die Verdampfungeschwindigkeit," *Verfahrenstechnik*, Vol. 7, 1938, pp. 170-177.

Slanina, Petr, and Šárka Šilarová. "Moisture transport through perforated vapor retarders." *Building and Environment* 44, no. 8, 1617-1626, 2009.

Straube, J. and Burnett, E., *Vents, Ventilation, and Pressure Moderation.* Building Engineering Group Report for CMHC, Ottawa 1995.

TenWolde, A., Carll, C.G., Malinauska, V., "Air Pressures in Wood Frame Walls", *Proc.of Thermal Performance of the Exterior Envelopes of Buildings VII*, Clearwater Beach, Fl., pp. 665-675, Dec., 1998

Teesdale, L.V. *Condensation problems in modern buildings.* Forest Products Lab, 1939.

van der Spoel, W. "Water vapor transport through perforated foils". *Research in building physics: proceedings of the 2nd international conference on building physics.* Leuven: Lisse, Exton, PA: AA Balkema (pp. 207-15), 2003.

Wooley, H.W., *Moisture condensation in building walls*, National Bureau of Standards, Building Materials and Structures Report BMS63, 1940.

6.8 Notes

Chapter 7

Air Transport

7.1 Background

Understanding air movement is a crucial and intrinsic part of heat and moisture control across and within building enclosures in all climates. For instance, a large proportion of the annual space conditioning energy consumption for most modern, well-insulated buildings is due to uncontrolled air leakage. Wintertime condensation of water vapor in exfiltrating air, or summertime condensation of infiltrating air, is one of the two major sources of moisture in the above-grade enclosure (driving rain being the other). Airflow through the enclosure can also carry spores generated by mould or pollutants released by materials within the enclosure into the air and thereby cause health problems. Uncontrolled air leakage through the enclosure is therefore often a major cause of performance problems affecting comfort, health, energy consumption, durability, etc.

Water vapor diffusion, while amenable to simple analysis, is rarely as significant a moisture transport mechanism as airflow is for modern building enclosures. Wintertime exfiltration and the consequent condensation, however, is acknowledged to be a common cause of building performance problems in cold weather. Warm weather infiltration condensation is often the cause of problems during warm and humid weather (which can occur during summer even in many cold climates).

There are five main reasons why controlling airflow is important to building performance:

1. To control moisture: water vapor in the air can be deposited within the enclosure and cause serious problems.
2. To save energy: air leaking out of a building must be replaced with outdoor air which requires energy to condition it. Various sources estimate that 30% or more space conditioning energy consumption in a well-insulated house is due to air leakage through the building enclosure. Both convective circulation and wind washing reduce the effectiveness of thermal insulation and thus increase energy transfer.

3. To improve comfort: cold drafts and the excessively dry air from large, uncontrolled ventilation flows affect human comfort.
4. To safeguard health: wind-cooled interior finishes promote condensation, which supports biological growth which, in turn, can affect the quality of the indoor air. In addition, the effects of any off-gassing from materials within the enclosure assembly which may be harmful to health, can be avoided if these gases are excluded from the interior by controlling the direction and amount of airflow. The flow of undesirable or dangerous odors, gases, and fumes between the exterior and interior (e.g., between a busy street and a classroom) and between zones within buildings (e.g., the parking garage and the apartment) must also be controlled.
5. To control fire and smoke: the movement of air can spread fire and transfer smoke, often with fatal consequences.

There are other circumstances that require the control of airflow, for example airborne sound transmission and resistance to bio-terrorism assault, but the previous four categories encompass most situations.

For the reasons given above, building designers need to understand and control these airflows.

For airflow to occur there must be both (Figure 7.1):

- a flow path or opening connecting two points, and
- an air pressure difference between the two points.
- Although these prerequisites are obvious and simple to state, in practice it is often both difficult to establish the existence and nature of the flow path and to assess the pressure difference.

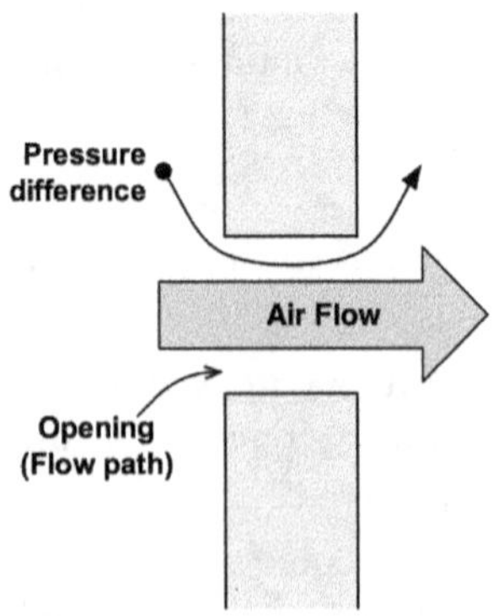

Figure 7.1: A pressure difference and flow path are required for airflow

Three primary mechanisms generate the pressure differences required for air to flow within and through buildings and thus building enclosures:

1. wind,
2. stack effect or natural buoyancy, and
3. mechanical air handling equipment and appliances.

Figure 7.2 illustrates these three effects and shows some of the possible flow paths.

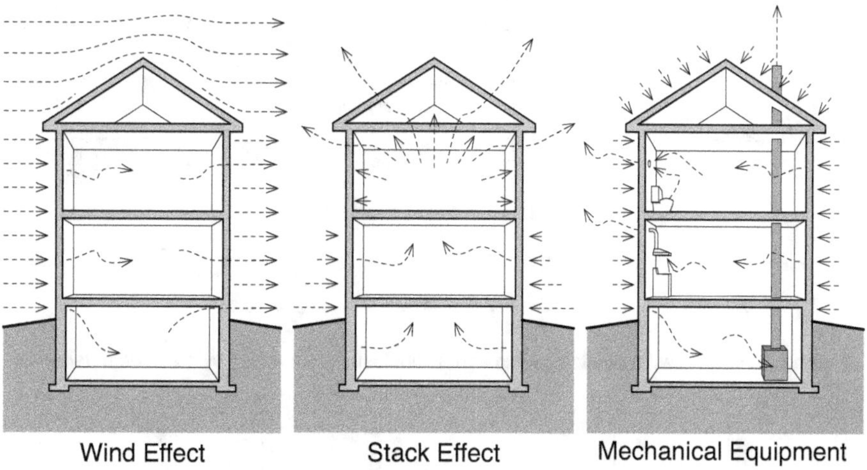

Figure 7.2: Forces driving airflow through building enclosures

Real airflows are highly complex: airflows occur between building zones, within ducts, both through and within interior partitions (Figure 7.3 and Chapter 11), and through the enclosure driving by the combination of all forces acting.

Unintended airflow across building enclosures is termed air leakage and the direction of flow defines if the leakage is labelled exfiltration (for flow outward) or infiltration (inward flow). The flow path may be simple and direct or quite long and tortuous (Figure 7.4). For building or zone energy consumption the flow path has little impact (and some HVAC engineers ignore exfiltration because this is "just" exhaust air flowing through an unplanned path), whereas the difference with respect to hygrothermal enclosure performance and indoor comfort can be profound.

The basic physics required to understand, describe, and approximately predict airflow through and within the building enclosure are covered in the following sections. The physical forces that generate the air pressures needed to drive airflow are also reviewed. Flow through enclosures and within intentionally ventilated cavities is then covered in some depth. Many applied aspects of air barriers and airtightness are addressed in Chapter 11.

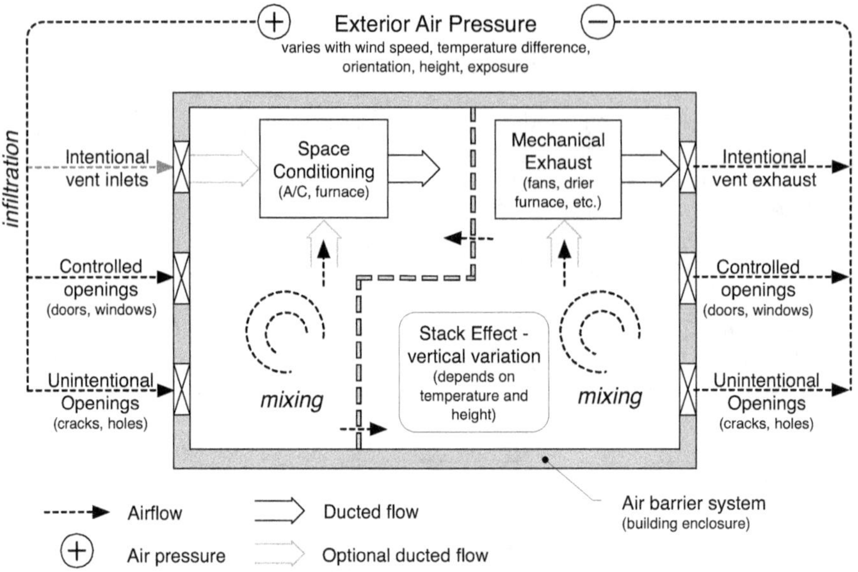

Figure 7.3: Airflows within buildings and across enclosures are driven by pressure differences

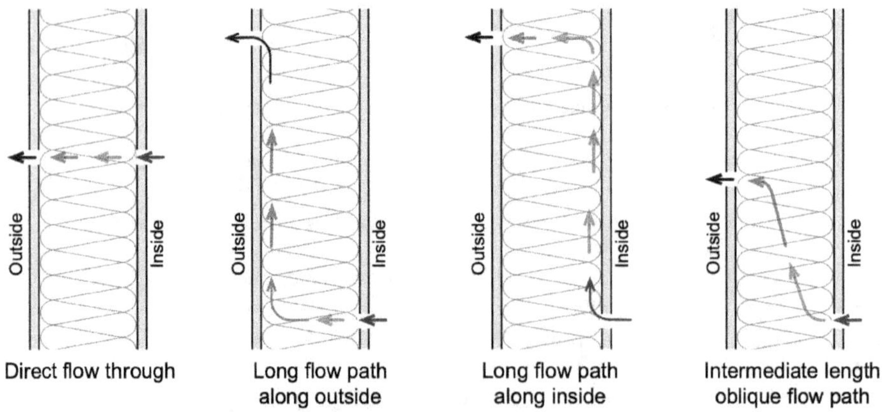

Figure 7.4: Example exfiltration air leakage paths across building enclosures

7.2 Basic physics

Idealized airflow follows the simplified laws of fluid mechanics, i.e. the air can be considered an incompressible ideal gas. The mass of airflow into an enclosure element or building zone must also flow out, but both the water vapor and heat energy content can be gained or lost as the airflows through.

For incompressible fluid flow, Bernoulli's equation can be used to relate velocity, pressure and location:

$$P_1 + \frac{1}{2}\rho_1 v_1^2 + \rho_1 \cdot g \cdot z_1 = P_2 + \frac{1}{2}\rho_2 v_2^2 + + \rho_2 \cdot g \cdot z_2 \quad \text{(Eq. 7.1)}$$

where, in consistent units,

P is pressure,
ρ is density,
v is velocity,
g is gravitational acceleration, and
z is elevation.

This equation essentially states that energy must be conserved for streamline flow. The first term is static pressure, the second term is velocity pressure (essentially kinetic energy), and the third term is potential energy in the form of a pressure head.

The second term is quite useful for translating air velocity to a pressure. The *stagnation pressure* is defined as the pressure exerted by a flow decelerated to zero speed, and is given by[1]:

$$P_{stag} = \frac{1}{2}\rho v^2 \quad \text{(Eq. 7.2)}$$

where P_{stag} is in Pascals

ρ is density in kg/m³, and

v is the air speed in m/s.

As an example, using room temperature air density of 1.2 kg/m³, a 20 m/s wind would have a stagnation pressure of 240 Pa.

To evaluate the wind force imposed on an object, for example cars, airplanes, buildings, support cables, flag poles, etc., a drag coefficient (C_D) is applied to the stagnation pressure to calculate the force imposed, that is:

$$F_{Drag} = C_D \cdot A \cdot \frac{1}{2}\rho v^2 \quad \text{(Eq. 7.3)}$$

A well-designed car will have a C_D of less than 0.3 and a billboard might have a C_D closer to about 1.2. Note that to assess the pressure on the face of an element due to wind loads (rather than the net force on the entire object), the drag coefficient C_D is replaced with a pressure coefficient Cp (see Chapter 3).

[1] In inch-pound units, the pressure imposed by wind is usually measured in pounds per square foot, with the air speed in miles per hour. These are not consistent units, and so a unit-conversion factor is used: thus wind pressure is given in the US ASCE7 standard as $P = 0.00256 \cdot V^2$.

7.2.1 Flow regimes

Depending on the area of study, fluid flow behavior is variously classified as streamline, incompressible, steady, etc. For building science applications, it is sufficient to consider two broad regimes of flow: turbulent and laminar. When fluid flow behavior is dominated by the inertia of the particles, it is termed *turbulent* flow. When fluid flow is dominated by the viscosity of the fluid, it is termed *laminar.*

Airflow through small cracks and pores, or driven by small pressure differences, tends to be laminar flow. Flow that passes through larger openings, shallow and wide cracks, or driven by high pressure differences, tends to be turbulent flow. Airflow behaves differently in the two flow regimes, and the transition between one regime and another is often characterized by variable flow rates and erratic, hysteretic behavior.

7.2.2 Laminar flow

Laminar fluid flow through a porous medium can be described by Darcy's equation. The many solutions developed for heating ducts or groundwater flow through soil can also be applied to laminar airflow building science problems. In a Darcy formulation, the airflow rate, Q, is linearly related to the driving air pressure difference, ΔP, by a constant, e.g.:

$$Q = K \cdot A \cdot \Delta P \qquad \text{(Eq. 7.4)}$$

The proportionality constant, K, is a material property termed *air permeance.* This equation can be formulated in volume units (e.g. m^3/s), but should, strictly speaking, be based on mass flow (e.g. kg/s) since this would account for fluid density differences. Note that this expression for laminar airflow is very similar to the equations for conductive heat and diffusive water vapor flow.

7.2.3 Turbulent flow through orifices

The volumetric fluid flow rate of a fluid through a sharp-edged orifice as a function of a pressure difference can be described by the fundamental relationship:

$$Q = C_d \cdot A \cdot \sqrt{\frac{2\Delta P}{\rho}}, \qquad \text{(Eq. 7.5)}$$

where Q is the volumetric flow rate,

A is the area of the orifice,

ρ is the density of the fluid,

ΔP is the pressure difference, and

C_d is the *discharge coefficient* (not to be confused with the drag coefficient!).

This orifice flow equation presumes that the flow is turbulent. Units commonly used for flow, pressure and density are m^3/s, Pascals, and kg/m^3 respectively. This relationship can be derived from Bernoulli's basic flow equation.

Measured results will deviate from this calculation because it ignores losses due to friction and turbulence. A discharge coefficient C_d must be applied to match actual measurements although it can be derived analytically only for some unique situations. The discharge coefficient is the product of two other coefficients: the contraction coefficient, C_c, and the velocity coefficient, C_v. The former coefficient accounts for the fact that the flow narrows as it flows through the orifice. The second coefficient accounts for losses due to friction and turbulence. For turbulent (high-speed) flow through a circular sharp-edged orifice, Kirchoff calculated a discharge coefficient of $C_d = \pi/(2+\pi) \approx 0.61$. This value is still widely used if experimental results are unavailable.

7.3 Forces driving airflow

The forces driving airflow across the building enclosure include wind, stack effect, and mechanical pressures. Each is discussed below.

7.3.1 Wind

From both a structural and a hygrothermal design point of view, the wind is a critically important force to consider. Hence, every enclosure designer should understand wind and its effect on buildings. Chapter 3 describes the wind and methods of predicting its effects. Refer to this chapter for review, as necessary.

7.3.2 Stack effect

Stack effect is the term used to describe temperature-difference induced buoyancy pressure differences between columns of air within a building and the outside. The importance of stack effect was recognized as high-rise buildings became more common, but scientific study was limited until the work of Tamura, Wilson, and Shaw in the 1960's [Tamura & Wilson 1966-1, 1966-2, 1967, Shaw 1977].

Stack effect pressures are generated by differences in air density with temperature, i.e., hot air rises and cold air sinks. The air within a building during the wintertime acts as if it is a bubble of hot air in a sea of cold air in much the same way as a hot air ballon. In the summertime the situation is reversed, although air temperature differences are usually smaller.

The density of dry air, ρ_a, varies with temperature (see Chapter 4) approximately as:

$$\rho_a = \frac{352.0}{T_a} + \frac{344.8}{T_a^2} \ [kg/m^3] \qquad (Eq. 7.6)$$

where T_a is the air temperature in degrees Kelvin.

The greater the height of the column of air, the greater the potential difference in pressure. The pressure difference generated by a column of air h meters high with a temperature difference (in Kelvins) between indoor and outdoor air is:

$$\Delta P = \left[\left(\frac{352.0}{T_i} + \frac{344.8}{T_i^2}\right) - \left(\frac{352.0}{T_o} + \frac{344.8}{T_o^2}\right)\right] \cdot h \cdot 9.81 \qquad (Eq. 7.7)$$

where ΔP is measured in Pascals

h is measured in meters,

T_o and T_i are the outdoor and indoor temperatures respectively, (in Kelvin).

or more approximately, at standard temperature and pressure,

$$\Delta P = 3465 \cdot \Delta h \cdot \left(\frac{1}{T_o} - \frac{1}{T_i}\right) \ [Pa] \qquad (Eq. 7.8)$$

For example, if the air in a one-meter high cylinder, closed at the top, open at the bottom and containing room temperature air (20 to 22°C), is taken outdoors at an outdoor temperature of -10 °C (14°F), an outward pressure of 1.34 Pa would act at the top (Figure 7.5). The horizontal plane at which the pressure equals the outdoor pressure (i.e., the plane at which the difference is zero) is called the Neutral Pressure Plane (NPP). In this example, the pressure difference at the bottom must be zero since it is connected to the outdoors.

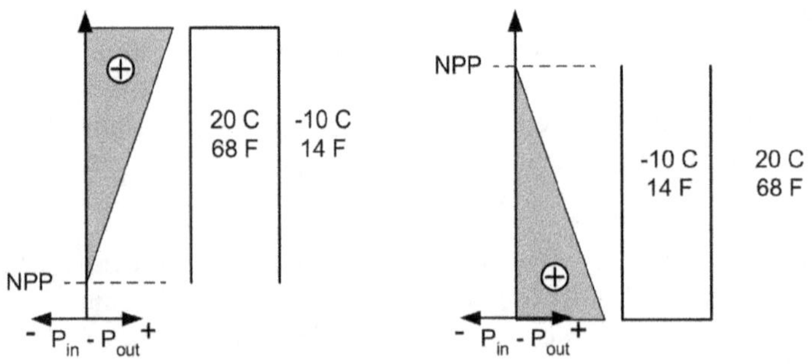

Figure 7.5: Stack effect pressure distribution in simple cylinders

If the cylinder remains outdoors, the air it contains slowly cools down to the exterior temperature and no pressure difference will exist. If the cylinder is then inverted and brought back indoors to room temperature, the pressure at the closed bottom end of the cylinder will again be 1.34 Pa but acting outward on the closed bottom as the cold air falls downward relative to the indoor air.

Note that the pressure difference across the cylinder walls varies linearly with height, starting at the Neutral Pressure Plane (NPP).

In the above examples, no flow occurs because no flow path is provided. When applied to buildings, some airflow will occur because all buildings have some degree of leakage across the enclosure. The leakage will not change the temperature indoors (because the mechanical conditioning equipment operates) and thus the leakage of air will not change the stack effect pressures so long as space conditioning equipment operates and maintains temperature.

If an open-ended cylinder containing room temperature air is used, any temperature difference between the cylinder and the surrounding air will cause flow, and the warm air will immediately be removed and replaced with cool outdoor air. However, if a heating coil is added to the cylinder to maintain the air temperature at 20 °C (68 °F), air flowing into the bottom will be heated. Friction will slow airflow and result in a constant pressure drop along the height of the cylinder (Figure 7.6). Note that the NPP is now located at mid-height and that airflow is involved. Obviously, the smaller the volume of air flowing through this cylinder, the smaller the amount of heat energy that will be required to maintain the interior of the cylinder at 20 °C. The same effect will result in reversed airflow if the temperature difference is reversed.

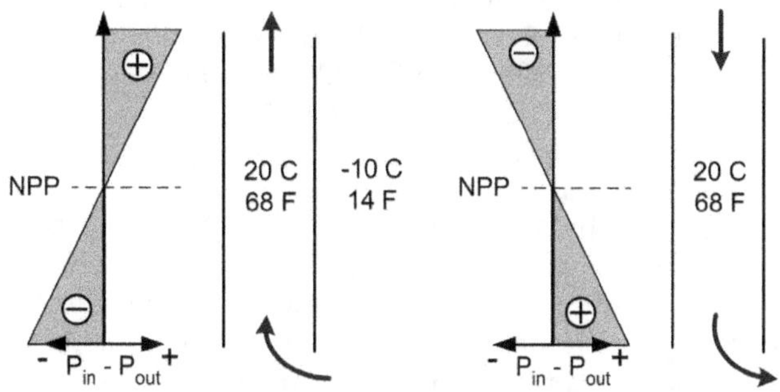

Figure 7.6: Stack-effect driven airflow through heated and cooled cylinders

Real buildings behave as if they are large and complicated vertical tubes. At the simplest level, they act like perforated tubes with closed ends. The ends of the cylinder, i.e., the roof and the slab-on-grade, can sometimes be reasonably assumed to be reasonably airtight. The majority of the air will flow through imperfections in the air barrier of the walls, which are assumed to be uniformly distributed over the height of the building. Given these assumptions the NPP will lie at the mid-height of the building (Figure 7.7).

If there are more openings at the top or bottom of the building (e.g., a mechanical penthouse on the roof, open doors or parking garage in the bottom), the NPP will be moved closer toward the larger opening. In tall buildings, the measured location of the NPP has been found to vary from about 30% to 70% of the height of the building.

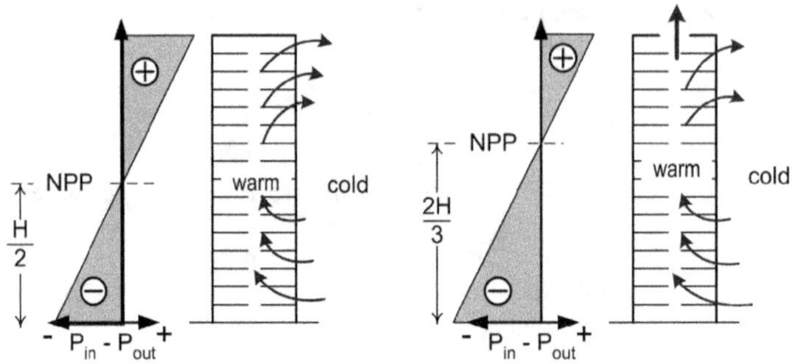

Figure 7.7: Stack-effect driven airflow through typical multi-story building

Note that these are gross simplifications of the distribution of air leakage and are only considering stack effect. Wind pressure variations over space and over time, and equally variable mechanically-induced pressures are superimposed on the stack effect pressure. Flow resistance within the building and non-uniform air leakage characteristics of the enclosure are further complications.

In houses, a significant amount of air may be exhausted through ventilation fans (in bathrooms and kitchens), vented furnace and fireplace chimneys, etc. Since the amount of air flowing into a building must be equal to the amount flowing out, the result of exhausting air by fans or chimneys is to place the house under a more negative pressure.

The role of chimneys and the location of the neutral pressure plane has long been identified [Wilson 1960] but the significance of stack-effect driven air leakage to wide spread air leakage condensation problems took longer [Orr 1974, Marshall Macklin Monaghan 1983, Burnett 1981]. Another factor to consider is that wood-framed ceilings below ventilated attics are usually not as airtight as compact low-slope roofs with watertight membranes. Together, these factors were found to result in a NPP of 80 to 100% of the building height (Figure 7.8) for single-family wood frame homes.

The consequence of a NPP near the ceiling of a house is that the walls experience mostly infiltration during cold weather. Since infiltrating air results in few moisture problems, this is a fortuitous situation that helps explain why many older low-rise houses with poor air barrier systems and chimneys suffered no interstitial condensation problems from uncontrolled air leakage in cold weather. Of course, energy is still wasted and thermal comfort and indoor air quality are often compromised. In extreme cases (especially when aided by wind pressures), sufficient quantities of inward flowing exterior cold air can also cool the interior wall surfaces to such an extent that condensation and frost form on the interior. This scenario also results in a larger than normal inward acting pressure on the basement floor, which can drive soil gases and radon through an air leaky sab on grade.

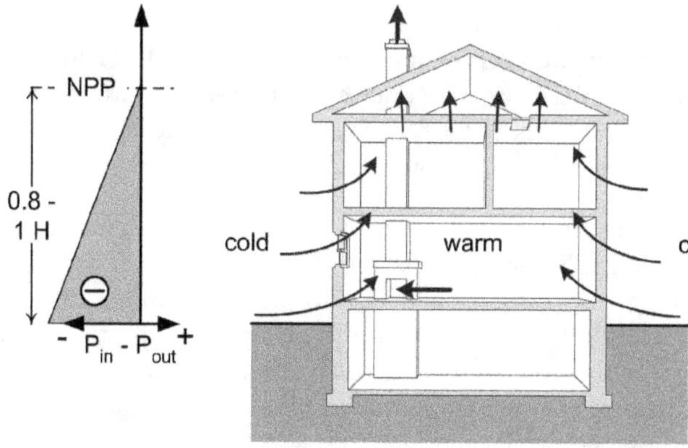

Figure 7.8: Stack effect in a house with exhaust (chimney) and leaky ceiling

Modern houses rarely use chimneys and natural draft heating appliances and so are more likely to have a NPP near mid-height. However, because modern houses are so much more airtight that the intermittently-operated exhaust fans in bathrooms and especially kitchens can, when operated, result in so much negative pressure that the NPP rises above the roof of the home.

During January, the average outdoor air temperature is -5°C (23°F) or less in many cities with cold climates. In such conditions an indoor temperature of 20 °C (68 °F) will result in a 1.1 Pa pressure difference for every meter of height. Over a 30 m tall, 12-story apartment building, this results in a pressure difference of more than 30 Pa. If the floors are relatively leaky and the walls are reasonably tight (a typical new building with good airflow control), the NPP is likely to be located close to building's mid-height. Therefore, an outward acting pressure of 15 Pa can be expected at the top along with a 15 Pa inward acting pressure at the bottom.

The significance of this result can be appreciated by comparing it to the result of the example wind pressure calculations in Chapter 3. Under average conditions the maximum inward-acting wind pressure across the wall of a similar building will be about 14 Pa. Exfiltration pressures are usually of greater interest during very cold weather (e.g., below -10°C), since exfiltration condensation is a major source of interstitial wetting and the temperature differences that drive stack effect are the largest. The average suction (that is, outward acting) pressure due to wind is likely to be only 7 Pa in the example building, half that of the calculated stack pressures. In hot weather (e.g., over 35°C) the temperature difference between inside and out may only be about 10°C, and hence stack effect pressures will be smaller.

Therefore, stack effect pressures are at least as important as wind-induced pressures for mid-height buildings. While occasional wind storms (with a velocity three times average, perhaps 10 m/s) may cause relatively large pressures (85 Pa positive and 43 Pa suction in the example), these events do not last a long time. Gust pressures can easily exceed 100 Pa in such events, but rarely exceed 300 Pa more than once every few years in most locations. These wind-induced pressures occur for a very small, practically insignificant percentage of the time when considering their effect on airflow.

Stack effect pressures act as long as there is a temperature difference. Stack pressures increase linearly with height, whereas wind pressures increase at a decreasing rate above 10m. Therefore, for buildings above 10 to 15 stories in cities with average January temperatures below freezing, stack effect pressures are likely to be the most significant force driving airflow across the enclosure. The proliferation of super-tall (usually defined as over about 300 m) buildings has created practical challenges due to stack effect pressures of over 100 Pa that act for periods of months.

7.3.3 Controlling stack effect

It is not practical to control the temperature difference that causes stack effect pressures, but large stack-effect pressure differences across the enclosure of tall buildings can be almost eliminated [CMHC 1996]. This can be achieved by providing an air barrier system between floors. If the floors are essentially airtight, stack effect pressure across the enclosure can be reduced to a few Pascals for a 3 m (10 ft) floor-to-floor height and a 25 °C (40 °F) temperature difference. Despite code requirements for fire stopping between floors, in practice this is not done well enough, and additional airtightening measures need to be taken.

Since vertical building shafts such as stair wells and elevators (and the doors connecting these shafts to corridors) cannot easily be sealed, the pressure variation with height within these vertical shafts is different from the pressure variation over the height of the building. The pressure differences that arise between the elevator/stairwell core and each particular floor can be dealt with in office buildings by using doors from a foyer to the floor. This has the advantage of superior sound and smoke control, as well as additional security opportunities.

In multi-unit residential occupancies, the demising walls between the apartments and the corridor should be airtight to control the flow of sound, odors, and smoke (during fire). Hence these demising walls will, and should, resist some of the air pressure imposed by stacked effect. Air will still flow from the lobby and garage levels up the building (with obvious air quality effects) but airtight vestibules and isolating dwelling units can limit this flow and its impact. Ricketts [2014, Ricketts & Straube 2014, Ricketts et al 2015] conducted a detailed field study of a multi-storey building and the airflow caused within it by wind, mechanical, and stack pressures. This unique study showed that complex pressures and even more complex air flow paths can conspire to have significant negative consequences for energy consumption and indoor air quality.

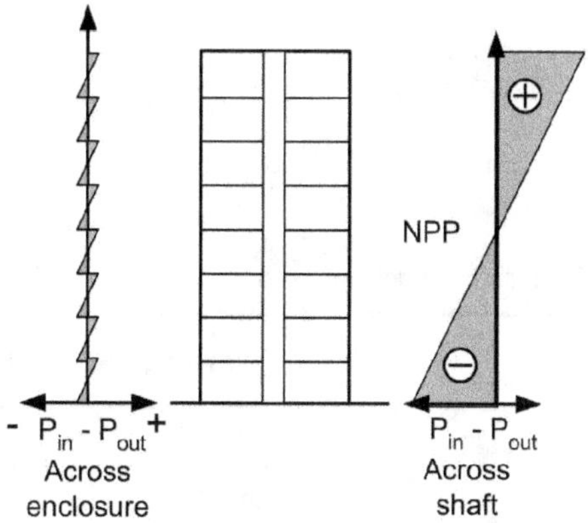

Figure 7.9: Controlling stack effect with horizontal air barriers

Vertical air supply shafts can also be eliminated (ventilation and exhaust airflow openings to the exterior can be provided at each floor), or the flow controlled at each floor outlet (via modulating dampers that automatically adjust for the stack effect pressure). In an ideal situation, air would be provided independently to each unit or floor.

7.3.4 Mechanically-induced Air Pressures

The purpose of fans and blowers is to cause the movement of air within and through buildings. The mere movement of air does not generate large pressures. However, if more air is exhausted from a building than is supplied, a net negative pressure is created and if more air is supplied than exhausted, a net positive pressure is generated.

Bathroom exhaust fans, clothes dryers, built-in vacuum cleaners, dust collection systems, and range hoods all exhaust air from a building and hence create negative pressures inside the building. If the enclosure is airtight or the exhaust flow rate high, large negative pressures can be generated. These negative pressures could cause several problems:

1. By driving inward air leakage through the enclosure, outdoor air may transport moisture into the enclosure or the building, resulting in potential indoor air quality problems and interstitial condensation during hot, humid outdoor weather conditions.

2. The negative pressures can cause back drafting of combustion appliances (see below).

3. The efficiency of most air handling devices will decrease with increasing back pressure.

4. Local exhaust can cause air pressures across partitions and finishes within a building between rooms and spaces.

To provide an understanding of the flow rates involved, a partial listing is provided in Table 7.1.

Table 7.1: Typical flow rates imposed by some mechanical equipment

Device	Flow rate (lps)	Flow rate (cfm)
Bathroom fan	20 - 50	40-100
Range hood	25 – 125	50 - 250
Gas/oil furnace (operating)	15 – 30	30 - 60
Downdraft range system	< 700	< 1500
Clothes dryer	40 – 75	80-150
Built-in vacuum cleaner	25 - 100	50 - 200

Most air handling equipment has a label showing the airflow rate. Most equipment, especially residential fan systems, are not rated with realistic back pressures and thus over-estimate the flow rate they provide. Dust collection systems, commercial range hoods and other industrial air handlers can move large volumes of air rapidly and also tend to resist back pressures effectively. Hence, they can seriously disturb the pressure-flow relationship within a building and should be handled on an individual basis. In modern airtight buildings it is increasingly common to select (or be required by codes) dedicated mechanical units or operable openings to the outside to allow outdoor air to be supplied to offset large volumes exhausted.

Positive pressures are often intentionally imposed by mechanical designers of commercial and institutional buildings [Bohac et al 2016]. The intent of this old practice of building "pressurization" is generally to:

1. avoid comfort complaints due to cold air leaking in through enclosures with flawed air control layers, and

2. control ventilation air quality (RH, particulates, odour, etc.) by ensuring all ventilation air in the building has passed through mechanical HVAC systems with no dilution by outdoor air that may enter via leaks, operable windows, doors, etc.

The common approach mechanical system designers use to achieve a "pressurized" building is by supplying some arbitrary excess quantity of outside air greater than the designed exhaust air flow. The challenge is ensuring that sufficient "excess" airflow is provided to generate an outward pressure (the goal of the practice) but not so much as to force damaging quantities of exfiltration (a waste of energy, a source of condensation,) or unduly high pressure differences (the cause of difficulties closing doors, a generator of whistling noises), etc. It is unlikely a mechanical designer can select an appropriate excess air supply rate without knowing how air leaky a building enclosure is (by test or experience). Hence, more modern building designs often attempt to measure pressure differences and control airflow supply and exhaust rates to limit air pressure differences to a desirable range (typically 1 to 4 Pa would be ideal, while pressures over 10 Pa often increase problems).

Designing an HVAC system for a specified airtightness target and then testing the airtightness after construction is current best practice. The HVAC system still needs to be commissioned of course. This is an area of current change in the building industry because building enclosures are becoming tighter, air tightness testing is increasing, and building mechanical systems are more sophisticated and capable than ever.

Other HVAC-related problems can occur when air is forced through ducts that leave the building enclosure or pass outside the primary air barrier system (e.g., the bad practice of placing ductwork in vented attics or vented crawlspaces). In this common scenario, any leaks in the ductwork (and most ductwork is surprisingly leaky) will result in a net exhaust of air and hence a net negative inward pressure on the building enclosure (Figure 7.10). The consequence during hot-humid weather is often significant condensation near the inner air-conditioned layers of enclosures. The reverse can happen if leaky ducts outside the air barrier are under a net suction pressure (return ducts).

Mechanical systems routinely unintentionally induce air pressures within buildings and across specific parts of the enclosure. One example is the unducted ceiling return air plenum widely used in commercial buildings. In this common design approach, the conditioned space may indeed be pressurized but the above-ceiling plenum may still be operating under a negative pressure. By drawing return air from the plenum, the air handler induces negative pressure in the plenum and positive pressures in the conditioned space, both of which adjoin the enclosure (Figure 7.11). Consequently, pressure differences across the enclosure are imposed, but can vary in direction over the enclosure. Ducted returns and/or highly air porous ceilings can be used to avoid this condition.

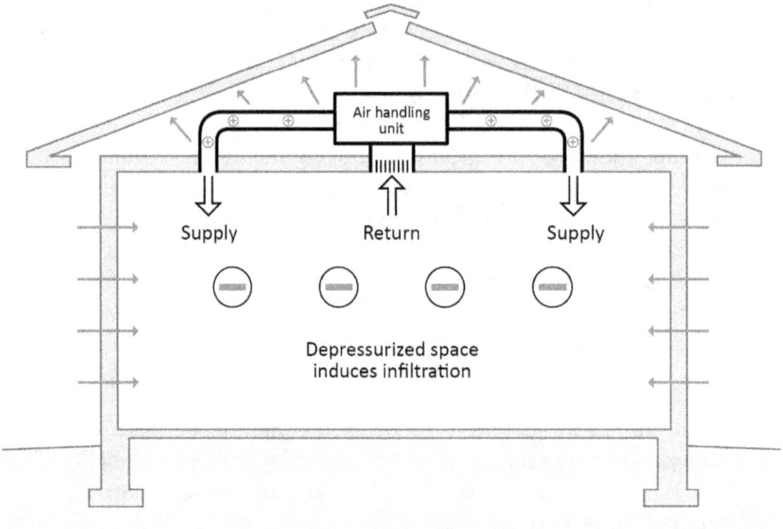

Figure 7.10: Air flow through leaky ducts *outside* the building enclosure can induce significant pressures *across* the enclosure

7.3.5 Combustion appliances

The use of combustion appliances in buildings, especially natural-draft water heaters, furnaces, and open fireplaces can have an important effect on airflows and pressures. All fuel-burning appliances require combustion air. Strictly speaking, combustion air has three components:

- *stoichiometric air*, which is the theoretical volume (actually mass) of air required for the combustion process.

- *excess air*, which is the "little bit of extra air" that appliance manufacturers require to ensure that the amount of air available for combustion is adequate.

- *dilution air*, which is the air required to dilute the flue gases enough to allow their passage through the vent.

The term "makeup air" is generally applied to the sum of these three components of combustion air, along with cooling air, which is the air required to cool the room in which the appliance is located. The amount of makeup air required may be significant. For example, a 30 kW (100 000 Btu/hr) furnace requires about 10 lps (20 cfm) of makeup air.

A combustion appliance (i.e., a boiler, furnace, or water heater) can be vented using natural thermal buoyancy of the exhaust (called natural draft) or be assisted with a fan (induced draft)[2].

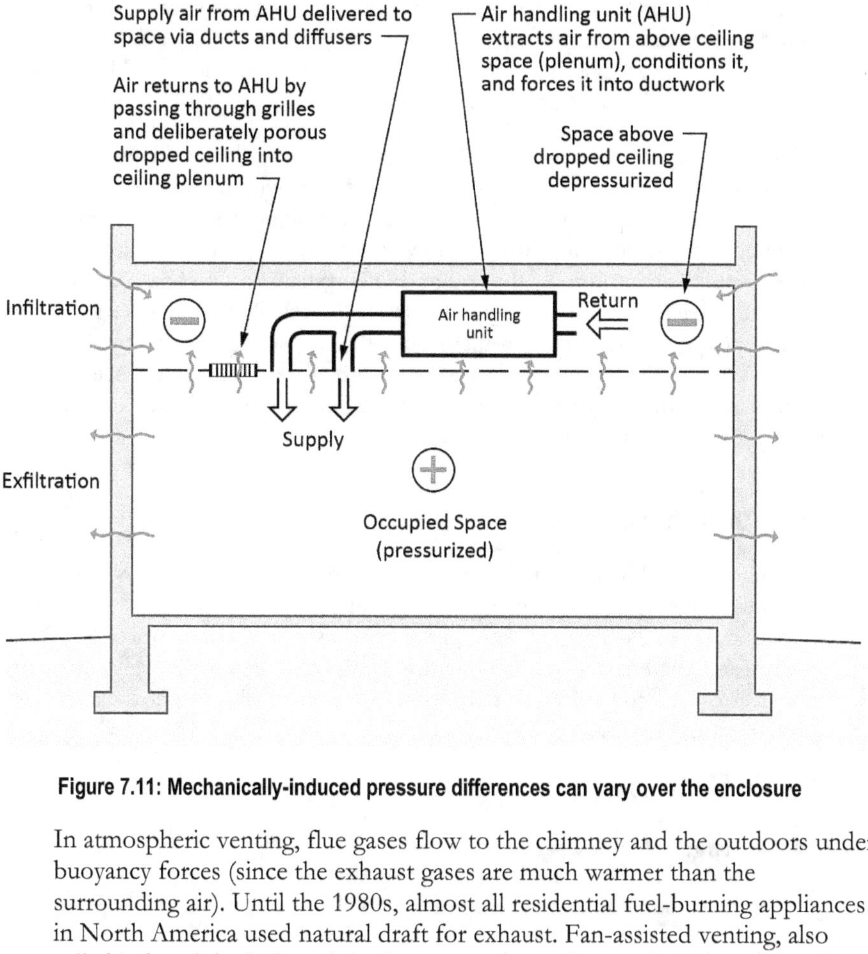

Figure 7.11: Mechanically-induced pressure differences can vary over the enclosure

In atmospheric venting, flue gases flow to the chimney and the outdoors under buoyancy forces (since the exhaust gases are much warmer than the surrounding air). Until the 1980s, almost all residential fuel-burning appliances in North America used natural draft for exhaust. Fan-assisted venting, also called induced draft, forced draft or power-vented, can use either a sidewall vent or a chimney. A boiler or furnace that has been designed with narrow flueways (to increase efficiency by improving heat transfer) often requires fan-assisted venting to help extract flue gas out of the appliance. In both cases make up air is supplied from inside the building.

---

2 Some appliances, such as room gas heaters and fireplaces, are not exhausted at all – these "unvented" appliances should be avoided in all spaces because of the serious impact and risks on indoor air quality.

Sealed-combustion appliances draw combustion air directly from the exterior to the burner. Because the fuel is burned in an environment that is entirely separated from the interior air of the living space, sealed-combustion appliances are unaffected by pressure differences inside the building. They are therefore the best choice for airtight buildings and homes that require high air quality. Sealed-combustion appliances can exhaust waste to a side wall via a short pipe (the most common scenario), or up through the roof. Most sealed-combustion appliances use fan-assisted venting.

Backdrafting, the reverse flow of exhaust gases through the exhaust flue and into the conditioned space, can occur if pressures within the space containing the combustion appliance and the flue connection are less than the pressures driving exhaust flow outward [Steel 1982, Dumont & Snodgrass 1990]. Natural draft appliances, especially those that do not exhaust a significant quantity of heat, can backdraft if pressures as low as 10 Pa act. Power-vented appliances are almost immune to this problem since the pressure generated by the fan is likely over 50 Pa. Sealed combustion units are also not affected since the pressure conditions within the flue and the burner are connected to the outdoors, not to the interior pressures.

7.4 Flow through enclosures

In practice, airflow through building enclosures covers the range of laminar to turbulent, including the transition zone, and almost always occurs through a number of parallel or series flow paths consisting of cracks, orifices and permeable materials. For most practical problems, airflow through cracks and orifices can be considered to fall within one of the following flow categories (see Figure 7.12):

- orifice flow,
- large crack flow, and
- tortuous and small crack flow.

Engineering judgment must be used to guide the choice of the appropriate flow category and corresponding flow equation. The dimensional ratios shown in Figure 7.12 are more important than the actual dimensions in these definitions.

Airflow through materials with relatively large pores (over about 0.1 mm) under normal pressure ranges is often assumed to be diffusive i.e., the flow velocity is related linearly to the pressure drop. Diffusive flow rarely occurs in most practical building enclosure situations, but it is often assumed to occur for the purposes of calculation because of the difficulty of defining the number, size, and distributions of cracks and orifices.

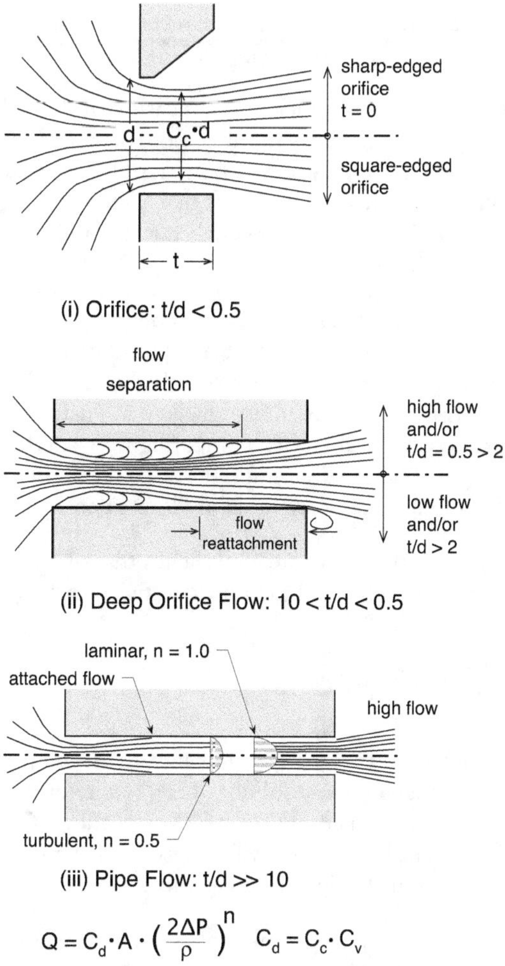

Figure 7.12: Typical airflow regimes through cracks and orifices

Buildings enclosures as a whole or as components (e.g., windows, walls) are often tested for their airflow characteristics by imposing a series of pressure differences, monitoring the flow rate at each pressure, and fitting the data to a standard equation with test-specific coefficients.

A general power law has been found to fit the data from most such leakage tests [Baker *et al* 1987]. This equation has the form:

$$Q = C \cdot (\Delta P)^n \qquad \text{(Eq. 7.6)}$$

where the square root has been replaced by the flow exponent, n,

C is a *flow coefficient*

The units of this relation can be whatever is desired, as the flow coefficient will incorporate conversion factors.

In this equation, the flow coefficient is a measure of the leakage of the tested enclosure assembly and includes the area, flow path, flow regime, friction, and temperature-density effects. A flow exponent of 0.5 indicates that the flow is completely turbulent, while an exponent of 1.0 indicates that the flow is completely laminar (i.e., it follows Darcy's Law for fully saturated fluid flow).

This power law form is widely used to relate the pressure drop and flow rate through building enclosure assemblies and components. Although the power law is not theoretically satisfying (since it is not dimensionally homogeneous and is not based on theoretical derivations) [Etheridge 1977], it has found wide acceptance in the industry. Care must be taken to ensure that density differences are accounted for when comparing flow coefficients (C) from different tests (e.g., cold weather infiltration tests are based on high density cold air, and exfiltration tests are based on lower density interior air).

As the flow through the enclosure approaches that of a perfect orifice, the flow exponent will approach 0.5 and the flow coefficient will be approximately equal to:

$$C = \frac{C_d \cdot A}{\left(\rho/_2\right)^n}\ , \text{for } n = 0.5$$

Bear in mind, however, that the flow through an assembly will almost never be through a single opening or crack but rather along a tortuous path through several parallel paths, each comprised of a series of openings, cracks, and porous materials. Therefore, some value between 0.5 and 1.0 is usually measured: a value of n=0.65 is often assumed.

Although the value of the flow exponent should not be pressure-difference sensitive, this is not always the case in measurements of buildings or enclosures since air pressure differences may change the enclosure, enlarging or closing openings. If a layer bulges and opens as more pressure is applied, it is possible to measure a flow exponent of more than 1.0, although in fact, it is the area that is changing with imposed pressure, not the flow exponent. This can occur, for example, when an enclosure contains an unsupported membrane such as a housewrap.

Airflow within or through permeable materials (as opposed to assemblies) can be calculated using Darcy's Law. In such cases flow is slow enough to be considered laminar and the flow exponent in the power law equation will be approximately 1. It can be seen from the material property data in Table 7.2 that most solid materials have a flow exponent of close to unity. In these cases, the air permeance value, K, can be approximated by:

$$K = C/A \text{ or } C = K \cdot A$$

Table 7.2 Air permeance of common building materials ($Q = K\,\Delta P^n$) [Bumbaru *et al* 1988]

Material	K (x 10^6)	n
200 mm concrete block	150	0.990
12.7 mm drywall, unfinished	0.266	0.995
12.7 mm interior moisture-resistant drywall	0.120	1.000
12.7 mm sheathing grade gypsum & taped joints	0.200	1.000
11 mm asphalt-impregnated fiberboard	11.27	0.995
11 mm fiberboard	11.47	0.990
8 mm plywood sheathing	0.11	0.944
16 mm waferboard	0.101	0.979
12.7 mm particle board	0.210	0.996
15 x 127 m tongue & groove boards (8 joints / m)	1670	0.564
30 lb roofing felt	2.535	0.996
15 lb unperforated asphalt building paper	3.607	1.000
15 lb. perforated asphalt sheathing paper	6.629	0.947
Tyvek on 25 mm fiberglass (Glasclad™)	6.877	0.987
Tyvek™ on 11 mm fiberboard	6.5	1.000
38 Sprayed in-place cellulose	1320	0.970
75 mm loose-fill vermiculite	1030	0.979
152 mm fiberglass batt insulation	610	0.949
25.4 mm Type I expanded polystyrene	250	0.900
25.4 mm Type II expanded polystyrene	1.63	0.993

Notes:
Some materials with no measurable leakage, i.e. $K < 8 \times 10^{-8}$ m /(Pa·s): 38 mm Type IV extruded polystyrene, 9.5 mm plywood sheathing, foil-faced foam insulations, foil-backed drywall, taped and finished gypsum drywall w/ 2 coats of latex paint and, of course, roofing membranes, steel sheeting, etc.

Materials with an air permeance of more than about 3×10^{-6} m/(Pa·s) are generally inadequate for use as part of an air barrier system. See Chapter 11.

Modern products sold in North America as housewraps and air barrier membranes have an air permeance of under 0.02 l/(s m²) at a pressure difference of 75 Pa (see Chapter 11 for more on air barriers): assuming a flow exponent of one, this is an air permeability of $K=2.65 \times 10^{-6}$ m/(Pa·s).

A very common situation that must be considered is the effect of a flaw in the primary plane of air resistance (or air barrier, see Chapter 11). One reasonable approach is to apply the orifice flow equation for flow through the flaw and

Darcy's equation for flow through the remainder of the materials in the assembly. Knowing that the quantity of airflow across each layer of the enclosure is the same, a system of equations can be developed and solved as for an electrical series circuit or conductive heat flow [Kronvall 1980]. This can occasionally be useful for estimating the fraction of total pressure acting across each layer of an assembly of leaky layers.

Approximating the type and size of "flaws" is a very inexact art and complex flow analysis is rarely useful or justified – the designer is typically better served by ensuring a well-detailed, buildable, workmanship-tolerant, inspectable and testable air barrier system.

7.5 Airflow in cavities: Ventilation

A practical use of airflow calculations is the prediction of airflow through wall cavities and airspaces. Ventilation of air through attics, crawlspaces, and wall cavities is commonly employed for a variety of reasons, but primarily for removing moisture.

The flow mechanics approach described in this section was introduced many years ago [Liersch 1981] but the theory was developed and confirmed by laboratory and field measurements of air flow resistance and driving pressure [Straube & Burnett 1995] and a multi-institutional ASHRAE-sponsored study of drying [Burnett et al 2004]. Subsequent research in New Zealand [Bassett & McNeil 2009] and recent investigations in Europe [Falk 2014, Falk & Sandin 2012, Langman et al 2016] have also taken and confirmed this approach.

A ventilation flow system balances the driving forces and the resisting forces. Figure 7.13 presents a simplification of ventilation flow mechanics through a cavity (either vertically or horizontally). The resistance to flow from points A to B and C to D is due to the vents. From points B to C, the flow resistance is due to friction with the cavity walls. The flow is driven by pressure differences between A and D: these are spatially different wind pressures and thermal and moisture buoyancy.

Ventilation flow can be seen to be analogous to flow through an orifice into a wide rectangular duct and out again through an orifice. Predicting the resistance to ventilation airflow in this simplified model is developed by first examining the flow through the cavity and then the flow through the vents.

Resistance to airflow in the cavity (between points B and C in Figure 7.13) is theoretically dependent primarily on three characteristics:

1. flow velocity
2. roughness of the sides, and
3. the size (depth) and shape of the cavity.

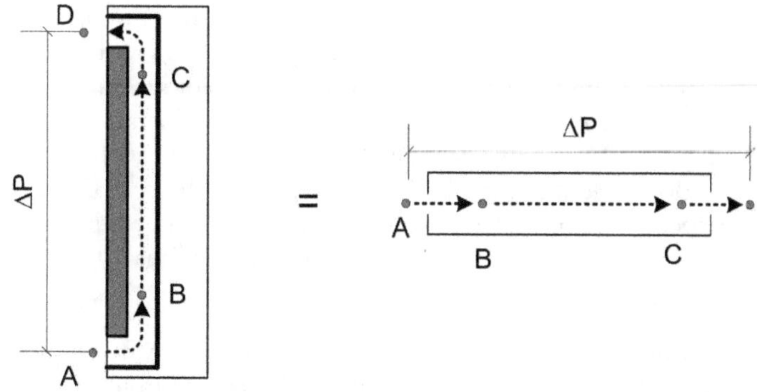

Figure 7.13: Airflow model of an airspace in a wall

In practice, a fourth characteristic, the number and size of obstructions and degree of baffling, can be very important. Because this fourth characteristic depends mostly on workmanship, it is difficult to quantify. Nevertheless, its impact on flow can be approximated. Friction varies significantly with velocity – the friction depends on whether the flow is laminar or turbulent.

In laminar flow the shear between particles is relatively powerful and causes the air to flow smoothly. In turbulent flow, the inertia of individual air particles exceeds the shear between particles. For internal flows (flow where the air is confined on all sides) the transition between laminar and turbulent flow occurs when the Reynolds number lies between 2000 to 3000. The Reynolds number is a dimensionless measure of the ratio of viscous to inertial forces. For standard air conditions the Reynolds number can be found from the following equation [ASHRAE 2017]:

$$Re = 66\,400 \cdot D_h \cdot v \qquad \text{(Eq. 7.7)}$$

where, Re is the dimensionless Reynolds number,

D_h is the hydraulic diameter (or equivalent diameter)[m], and

v is the velocity of the flow [m/s].

The hydraulic diameter of an airspace can be defined as:

$$D_h = 4 \cdot A / p \qquad \text{(Eq. 7.8)}$$

where $A = b \cdot d$ is the cross-sectional area of an airspace b wide and d deep [m], and

$p = 2 \cdot (b + d)$ is the perimeter [m].

Therefore, the hydraulic diameter of a wide rectangular cavity is:

$$D_h = 4\,b \cdot d \div 2\,b = 2\,d \qquad \text{(Eq. 7.9)}$$

Figure 7.14 [Straube & Burnett 1995] plots the flow rate through cavities versus the Reynolds number and indicates those regions of flow velocity or flow rate that are laminar and those that are turbulent. This plot shows that flow can normally be assumed to be laminar over the typical velocities expected in cavity ventilation (i.e., 0.1 m/s for deep cavities and up to 1.0 m/s for smaller cavities). Both field [Schwarz 1973, Popp et al 1980, Jung 1985] and lab measurements [Pinon et al 2004] confirm that for most practical systems the flow velocity will be well below 1 m/s (200 fpm) and thus can be assumed to be laminar.

In laminar flow, the pressure drop across the cavity will vary linearly with velocity. If the flow increases sufficiently, it becomes turbulent and friction will then increase faster than flow rate. This fact can be used in design to naturally limit ventilation flows to velocities that are less likely to entrain water into the cavity through the vents.

The Darcy-Weisbach equation is commonly used to give the pressure drop due to friction in fluid flow through pipes and airflow through pipes and ducts:

$$\Delta P_{pipe} = f \cdot (L / D_h) \cdot P_{stag} \qquad \text{(Eq. 7.10)}$$

where f is the friction factor (this accounts for flow velocity and pipe roughness),

L is the length of the pipe,

D_h is the hydraulic diameter, and

P_{stag} is the stagnation pressure of the airflow, which can be calculated for room temperature air as:

$P_{stag} = ½ \cdot \rho \cdot V^2 = 0.6 \cdot V^2$ for $\rho_{air} = 1.2$ kg/m³ @ 20°C.

For laminar flow the friction factor varies simply with the flow rate but also depends on the shape of the duct. For the two extremes of shape [Kronvall 1980, White 1991]:

$f = 64/(Re \cdot \gamma)$, circular ducts, and

$f = 96/(Re \cdot \gamma)$, channel flow (as in a cavity). (Eq. 7.11)

A blockage factor, γ, has been included above to account for very rough and/or partially obstructed cavities. This factor is approximately equal to the average reduction in cross-sectional area caused by protrusions from the side of the cavity. For clear cavities, such as those formed behind panel cladding systems, $\gamma=1$. Since many brick veneers have partially blocked cavities, a γ value of less than even 0.5 may be appropriate. Measurements of stucco applied over building paper held off the wall with 20 mm furring strips showed values of $\gamma=0.5$.

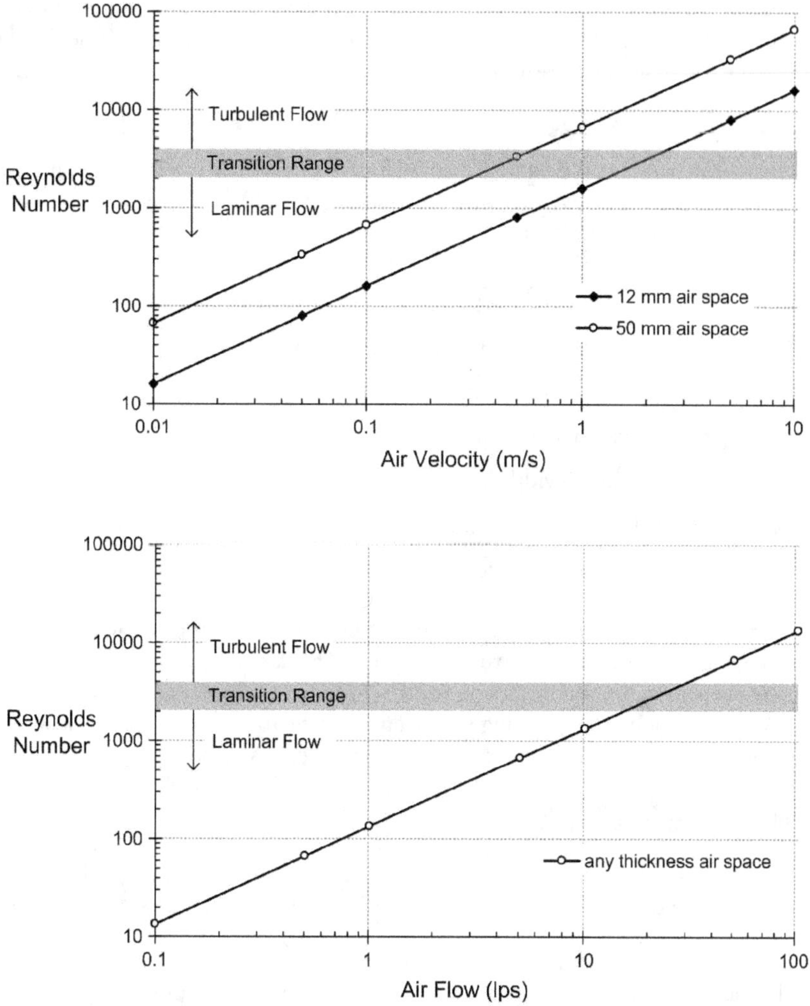

Figure 7.14: Laminar and turbulent flow ranges in ventilated cavities

For turbulent flow, the friction factor varies considerably with flow and roughness. However, a simplified approximation developed by Altshul-Tsal [ASHRAE 2021] is:

$$f = 0.11 \cdot \left(\frac{\varepsilon}{D_h} + \frac{68}{Re}\right)^{0.25}, \text{ for } f > 0.018 \qquad \text{(Eq. 7.12)}$$

where ε is the absolute roughness. This is defined as the average height of the projections divided by cavity width.

For ventilation flow through an enclosed cavity, it can be assumed that the flow will be laminar (see above). Flow velocity is merely flow volume divided by the cross-sectional flow area:

$$v = Q/A \qquad \text{(Eq. 7.13)}$$

Assuming laminar flow and rectangular cavity cross section the pressure loss air flowing through a cavity (from B to C in Figure 7.13) can be estimated as [Straube & Burnett 1995]:

$$\Delta P_{cavity} = \frac{V\,h}{692\,\gamma\,D_h^2}\;\frac{Q\,h}{4610\;b\,d^3} \qquad \text{(Eq. 7.14)}$$

where ΔP_{cavity} is the pressure drop along the cavity [Pa]

d is the cavity depth [m],

h is the cavity height [m],

b is the cavity width [m],

γ is a blockage factor, and

Q is the volumetric flowrate [m^3/s].

A blockage factor, γ, has been included above by the author to account for very rough and/or partially obstructed cavities. This factor is approximately equal to the average reduction in cross-sectional area caused by obstructions, such as cladding attachments, furring strips, etc. This factor can be chosen based on judgement, previous experience, and lab test results.

7.5.1 Flow through vents

Two practical building-related cases should be considered for ventilated enclosures: discrete vents that act as orifices, and continuous slots.

The change in direction from horizontal to vertical flow and spreading out laterally is expected to impact discrete vents. For continuous or highly elongated slots airflow mostly must change from horizontal to vertical. Deep discrete vents (such as head joints in masonry) or vents in thin materials preceding or followed by complex flow paths with directional changes (as exist in some curtainwall systems and metal panels) likely add additional flow resistance. For the situation where a continuous (or semi-continuous) slot is provided as a vent, the cavity tends to provide a significant proportion of the flow resistance, and the change in direction at the vent becomes important.

Following HVAC analysis principles, recommended as early as Liersch [1981], has since been used to characterize the complex flow of air through vents, screens, and around bends. It is well known that the pressure drop due to entry, exit, vents, grilles, and direction changes directly with the stagnation pressure (that is, with the square of the flow rate). Thus, the pressure drop across a vent and around a change in direction can be calculated as a fraction of the velocity pressure (stagnation pressure):

$$\Delta P_{vent} = C \cdot (\tfrac{1}{2} \cdot \rho_{air} \cdot V^2) \quad \text{(Eq. 7.15)}$$

where ΔP_{vent} is the pressure drop across the vent [Pa]

C is the *fitting loss coefficient*,

ρ_{air} is the density of air [kg/m^3] and

V is the average velocity through the vent, [m/s].

Recall that for a rectangular slot the average velocity becomes:

$$V = Q / (b \cdot d)$$

Q is the flow rate [m^3/s],

b and d the dimensions of the vent [m].

For standard conditions air has a density of about 1.2 kg/m^3, so another useful form of the equation is:

$$\Delta P_{vent} = C \cdot 0.6 \cdot V^2 = 0.6 \cdot C \cdot (Q / (b \cdot d))^2$$

European designers have long used the fitting loss coefficients shown in Figure 7.15 in practice. Values from HVAC design can also be considered but these have usually been developed for higher turbulent flow and may not be especially accurate.

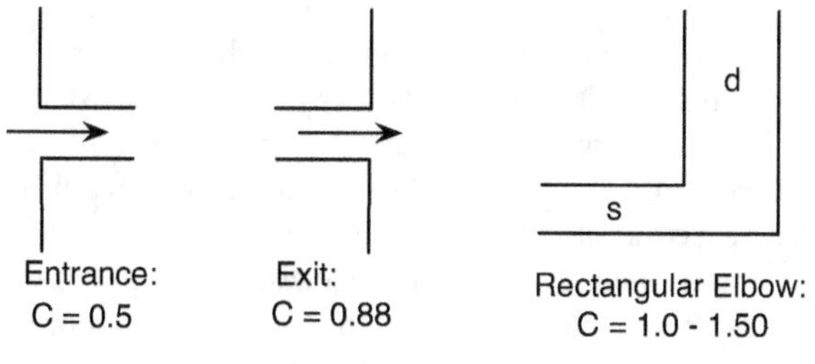

Figure 7.15: Friction coefficients for duct flow

The flow performance of some discrete openings in the form of simple circular, oval, and rectangular perforations have been measured in the lab [Straube & Burnett 1995]. The work of Pinon et al [2004], reports on numerous laboratory measurements of rainscreen-type vents and cavities. Their work considered, for example, continuous slot vents of 9.5 mm (3/8") and 19 mm (¾") width, combined with cavities of 19, 50, and 100 mm (¾", 2", and 4"). Falk & Sandin [2013] provide measured results for a specific perforated metal screen.

A useful compendium of laboratory test results is summarized in Pinon et al [2004]. Some of these results are presented in Table 7.3. Note that it is not always easy to separate the pressure losses due to the vent entry and the change or direction and expansion of flow from lab results. However, measuring the pressure drop from A to B and C to D in Figure 7.13 is relatively easy to measure and these are the basis of the results reported below.

	[mm]	[mm]	C	C	C
Vent type	**Vent size**	**Cavity Depth**	**Inlet**	**Outlet**	**Inlet+Outlet**
Circular	27	19	2.13	2.01	4.14
Rectangular	9.5 x 57	19	2.25	2.02	4.27
		50	2.19	1.38	3.57
		100	2.29	1.19	3.48
Slot	10	19	1.81	1.93	3.74
		50	2.24	1.71	3.95
Slot	19	19	1.84	3.02	4.86
		50	1.67	2.06	3.73
		100	2.21	2.61	4.82

Table 7.3: Fitting Loss Coefficients (C) measured by Pinon et al [2004]

The fitting loss coefficient values shown vary around a value of 2 for the inlet and 2 for the outlet. They also are within the range predicted from the literature. There are two important notes to interpreting this data. First, the fact that a single small circular vent has a coefficient of the same magnitude as that of a continuous 19 mm slot seems surprising. However, the pressure drop caused by the two vents type will in fact be vastly different, as the velocity used to predict the pressure drop is calculated by dividing the total volumetric flowrate by the area of the vent: in the case of the two walls described, the ratio of venting area is almost 40:1, and thus the predicted pressure drop is 1600 times more! A second observation is that loss coefficient variations of ± 0.5 about the average total of 4 do not have a large impact on the predictions: the uncertainty of construction tolerances and vent area are more significant to the predicted pressure drop.

7.5.2 Wall system flow

At equilibrium, the pressure drop across two vent holes and cavity will equal the pressure drop due to external driving forces P_{drive}:

$$\Delta P_{drive} = \Delta P_{vent,entrance} + \Delta P_{cavity} + \Delta P_{vent,exit} \qquad \text{(Eq. 7.16)}$$

For many wall systems the entrance and exit vents will be of the same type and can be lumped together (for other systems the flow may enter and leave via different vent types) and the above equation simplified for laminar flow to:

$$\Delta P_{drive} = 2 \cdot \Delta P_{vent} + \Delta P_{cavity} 0.6 \cdot C \cdot (Q / (b\ d))^2$$

$$= 2 \cdot 0.6 \cdot C \left(\frac{Q}{A_{vent}}\right)^2 + \frac{Q\,h}{4610\,\gamma \cdot b \cdot d^3} \quad \text{(Eq. 7.17)}$$

In the case of discrete vents, the value of b is the horizontal spacing between vents. Note that it is possible, with experimental data, to modify both the fitting loss coefficient and the blockage coefficients.

Therefore, with knowledge of the driving forces and the characteristics of the wall system, the ventilation flow can readily be calculated.

7.5.3 Ventilation Driving Pressures

Predicting the forces driving ventilation is more difficult, or at least less accurate, than predicting or measuring flow resistance. Nevertheless, there is guidance to estimate the spatial variations that are imposed by wind (see Chapter 3) and the buoyancy / stack effect within wall cavities. Numerous field measurements of cavity velocity, temperature, and/or pressures [Schwartz 1973, Straube & Burnett 1995, Finch & Straube 2007, Falk & Sandin 2013, Langmans et al 2016] have shown that natural driving pressures are usually in the range of 0.25 to 5 Pa.

For example, a brick veneer wall with a 25 mm (1") airspace and open head joints top and bottom at 600 mm (24") on center, might be expected to have flows of between 0.1 and 1 litres per second per m^2, resulting in from 5 to 50 air changes per hour (ACH) under natural driving pressures [Sandin 1993, Vanpachtenbeke et al 2017].

The series of equations presented have been verified by field measurements [Bassett & McNeil 2006 2009]. Although not as widely field validated, the same approach can also be used to assess flow through ventilation spaces in cathedral ceilings, below roofing tiles, or through raised flooring among other applications. The driving force of thermal buoyancy will be revisited in Section

7.6 Performance Impacts of Airflow

The impact of airflow on the hygrothermal performance of building enclosures can often be very important, even at low rates of airflow. Increased energy loss, condensation, comfort complaints, reduced insulation performance, and indoor air quality problems can all be caused by airflow. Some of the thermal effects are discussed in and in Chapter 4. Bomberg and Onysko [2002] provided an excellent review of the inter-related issues of airflow and moisture in wood frame buildings.

Air flowing through an enclosure is a common problem resulting in condensation, and hence has been studied extensively [Wilson & Garden 1965, Kohonen 1985, Ojanen & Kohonen 1989, Ojanen & Kumaran 1996, Ojanen

& Simonson 1995, Derome 2000, Janssen & Hens 2003, Ge et al 2019, Viljanen et a 2022]. All of the preceding citations should be studied by the serious student.

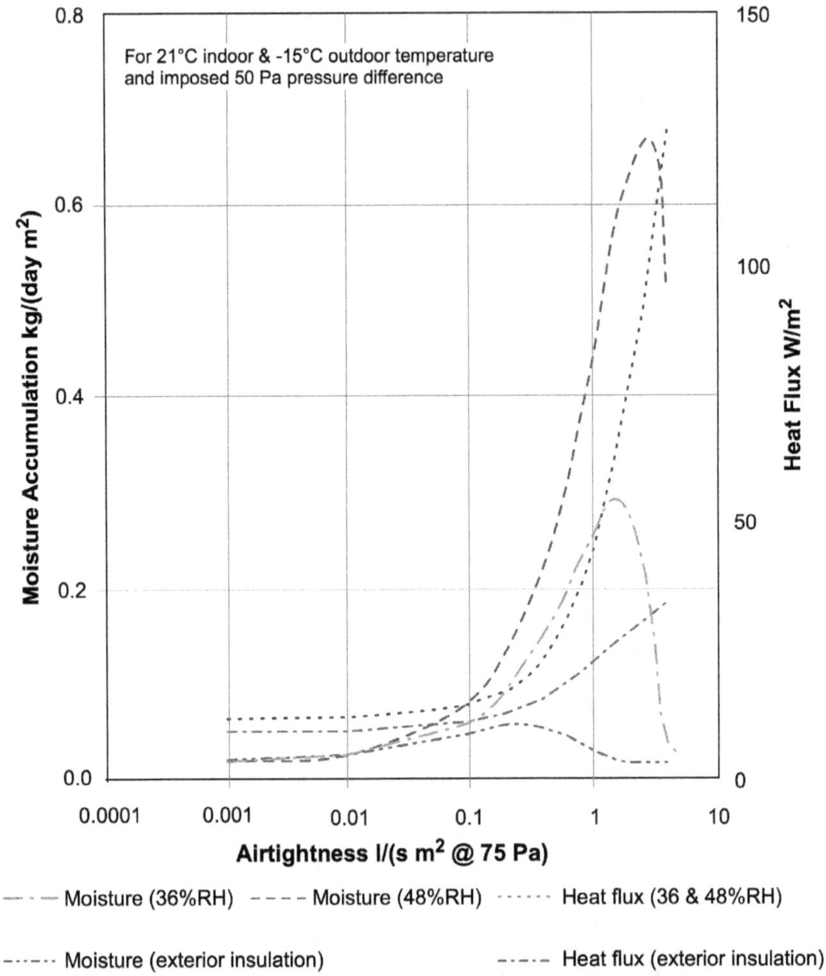

Figure 7.16: Air leakage condensation, heat and flow as a function of air leakage rate, indoor humidity, and exterior insulation (adapted from [Ojanen & Kumaran 1996])

Approximate methods for predicting air leakage condensation were presented in Section 6.4.5. However, the actual performance of many assemblies with air leakage is complex and difficult to predict accurately. The condensation rate depends in a non-linear way on factors such as airflow rate, flow path, nature of dynamically varying pressures, and other factors. Modelled and measured behaviour of enclosures with some of these factors can be found for specific examples in the literature. For example, the results of early modeling work [Ojanen & Kumaran 1996] showed the influence of several factors (Figure

7.16), including the dramatic reduction in condensation possible by using exterior insulation.

7.7 Natural convection

Thermal buoyancy (i.e., hot air rises) and differential wind pressures can cause numerous natural and forced convective airflow paths *within* building enclosures (Figure 7.17). This is a concern when it occurs through or around insulation layers as these internal airflows can short-circuit thermal insulation. The attendant increase in heat transfer [Bankkvall 1978, 1987a,b, Harrje et al 1985, Gullbrekken *et al* 2017] and risk of moisture deposition [Kalamees & Kurnitski 2010] increases with thick layers of insulation and larger temperature drops.

Natural convection within and around insulation have long been recognized as an important consideration [Wolf et al 1966, Bankvall 1978a]. Airflows driven by natural convection (stack effect) are primarily affected by two factors: the magnitude of the temperature difference and the difference in height. As derived earlier for stack effect pressures in building, the pressure driving convective flows can likewise be calculated from:

$$\Delta P = 3645 \cdot \Delta h \cdot \left(\frac{1}{T_1} - \frac{1}{T_2}\right) \qquad \text{(Eq. 7.19)}$$

To ensure that no flow paths connect airspaces on the warm side of the insulation to the cold side, insulation should always be placed in tight contact with at least one surface. Semi-rigid cavity insulation must be firmly attached to one side of the airspace and batt insulation is manufactured oversized so that when it is compressed by the drywall it does not allow for any small gaps on either side. Research has shown that significant heat losses (10 to 30%) can result from connected air gaps of only a 1-2 mm width (1/16") [Lecompte 1990, Bankvall 1972, Brown et al 1993]. Full bed or serpentine adhesive patterns are preferred to spot adhesive ("daubs") when adhering foam board insulation to vertical surfaces (e.g., for EIFS or interior insulation retrofits) for the same reasons. It should be noted that this is counter to considerations of water drainage in gaps.

Convective flow within the hollow voids of concrete block, and light wood framing, cold-formed steel, and hollow aluminum framing can be a very effective means of transporting heat and moisture. This can be a potential problem at any insulated or empty parapets (Figure 7.18) that project past the insulated perimeter of the enclosure. Solutions involve compartment separators in the form of mortar-filled courses (for masonry), air impermeable foam plugs, or perhaps best of all, insulating the exterior on all sides of the parapet. Ideally, the air control layer should be installed in a location that controls unwanted flow.

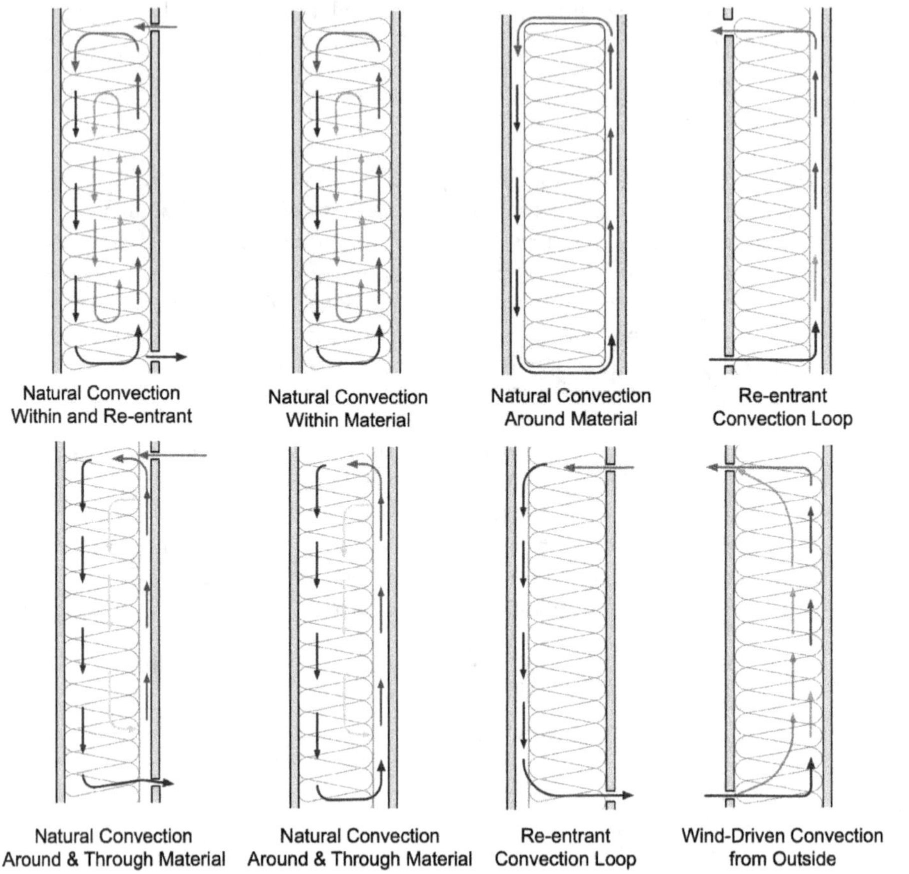

Figure 7.17: Types of natural and wind-driven convection airflow through walls

Natural convection within low-density air permeable insulation installed within framed walls is possible [Bankvall 1972, Huttunen & Vinha 2013], but the insulation products available are generally designed to avoid this. The modified Rayleigh number (Ra_m) is used to quantify the risk of natural convection within an air permeable insulation. For typical air properties, ISO 10456 defines

$$Ra_m = 3\times10^6 \cdot d \cdot \kappa \cdot \Delta T / k \qquad \text{(Eq. 7.20)}$$

where d is the depth of the insulation (m) and

κ is the air permeability of the insulation in m^2.

For vertical installation, it is recommended that Ra_m be less than 2.5. Thicker monolithic layers of very air permeable (low-density fibrous) insulation exposed to large temperature differences are susceptible. Using higher density fibrous insulations and dividing insulation into two or more layers (such as using insulating sheathing in addition to studspace insulation) avoids the potential for increased heat flow due to natural convection.

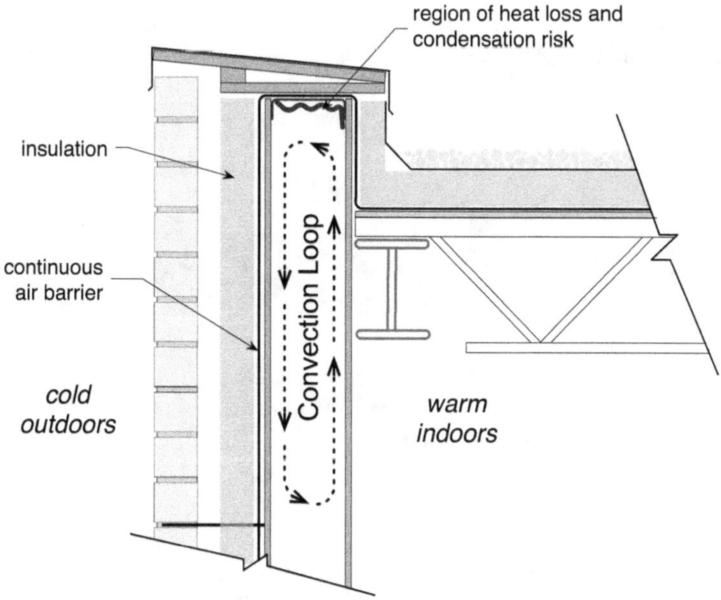

Figure 7.18: Natural convection in hollow stud framed parapets

Natural convection within horizontally installed insulation layers, such as in attics, can also occur (especially where one face is open to the air and the other against the ceiling finish as is common in vented attics). Research has shown that the thermal resistance of low-density attic insulation can be significantly degraded by natural convection under large temperature gradients (cold winter nights and hot summer days) [Silberstein et al 1990, Wilkes 1991, Whalgren 2007]. For horizontal attic insulation exposed to an air gap above, Ra_m may be as much as 15.. As for walls, thicker insulation layers, higher air permeability, and larger temperature differences increase the risk. The use of multiple layers of batt insulation with staggered joints, or higher-density (presumed to be lower air permeance) blown cellulose insulation over low-density batt is recommended to minimize these effects.

7.7.1 Wind Washing

High velocity air flowing over the surface of insulation layers behind the cladding can also increase the amount of heat loss by greatly reducing the surface film coefficient (a small impact) and, more significantly, by penetrating the structure of low-density fibrous insulations (e.g., batt insulation is very vulnerable) (Figure 7.19) [Bankvall 1978b, Cummings 2012, Silberstein et al

1991]. This phenomenon is often called wind washing [Timusk et al 1991], especially in North America (Figure 7.19). Building corners and parapets are especially susceptible because the wind induces very steep pressure gradients over the building face in these areas. Pressure gradients of 100 Pa/m can form, and even small airflow paths can allow excessive airflows [Uvslokk 1996].

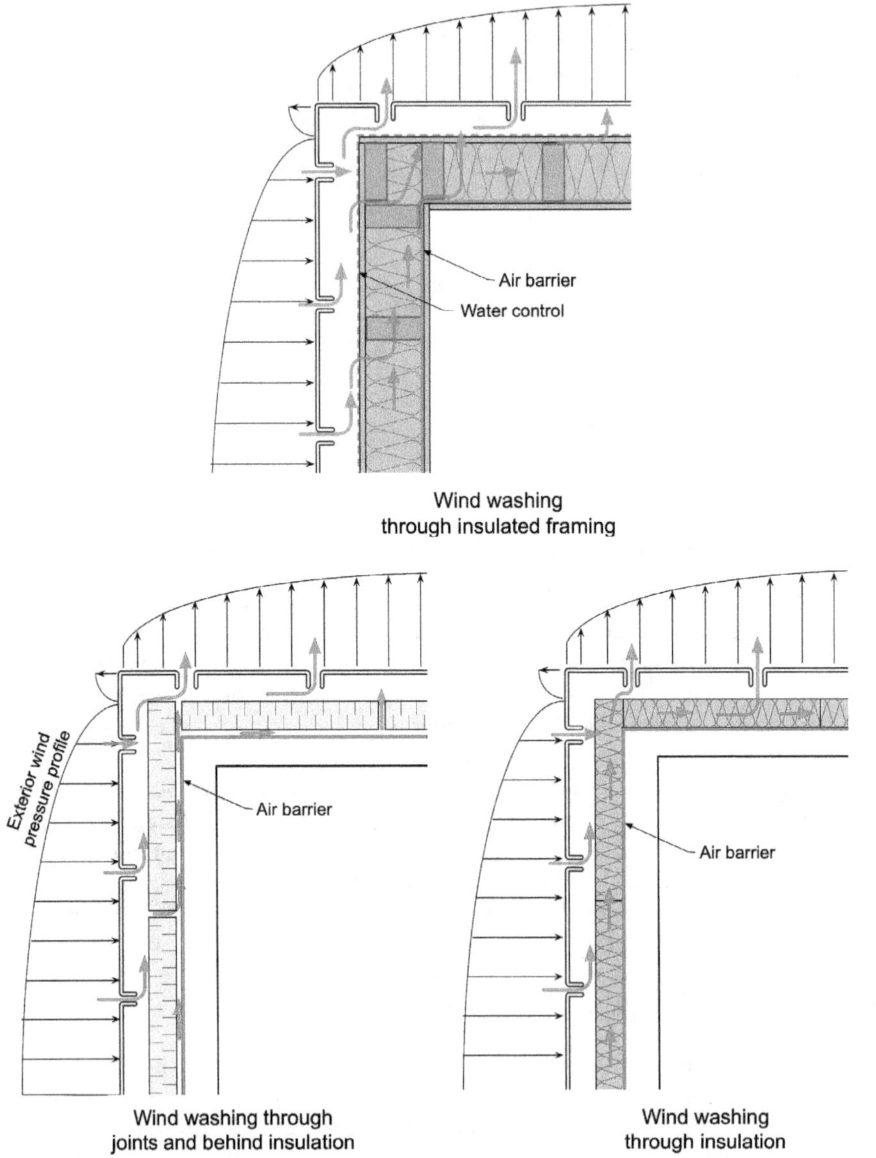

Figure 7.19: Wind-washing is wind-induced forced convection within or around insulation layers

Air impermeable layers placed outside low-density fibrous insulation can control this form of heat loss. In Scandinavia, secondary, outer layers of airflow resistance are called wind barriers or convection barriers and widely used to protect low-density batt insulation placed between wood framing, because no sheathing or water-resistant membrane was historically used. North American designs commonly use housewraps and sheathing, rendering the need for a wind barrier moot. Norwegian research has recommended limiting the maximum air permeance of wind barriers barrier to between 25 and 100 x 10^{-6} m^3/Pa [Uvslokk 1996]. More recent analysis using highly conservative assumptions in Finland have results in air permeance of 10 x 10^{-6} m^3/Pa [Ojanen 2020]. High-density mineral fiber insulations [Sedlbauer & Kuenzel 1999], rigid foam insulations, and housewraps can provide this level of control [Van Straaten et al 2016] if properly installed. The effect of moisture and temperature movement should be investigated when assessing the ability of a material to control wind washing.

In-plane airflow resistors provide *compartmentalization*, which helps to confine air leakage to limited areas of the envelope, reduces wind washing effects, and can also improve pressure moderation performance (discussed later). Compartmentalization should be provided in all assemblies, either provided by tight separators at discrete intervals (e.g., sheet metal) or by the distributed resistance of low-permeance materials (e.g., dense-pack cellulose and foam). Corner separators are often the most useful because of the high-pressure gradients acting around corners (Figure 7.20).

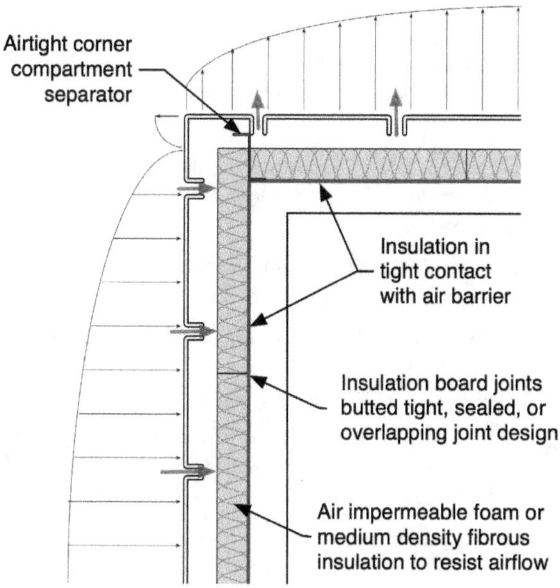

Figure 7.20: Recommendations to avoid windwashing impacts

Framing members can also provide resistance to in-plane flow. Wood blocking, or draft stops, have long been used in wood framed construction to prevent the spread of fire and smoke. Wood framing elements are generally not sufficiently airtight at corners because drying shrinkage causes small gaps between the framing and the siding (or drywall).

7.8 References

ASHRAE Handbook of Fundamentals. Atlanta, 2021.

Baker, P.H., Sharples, S., Ward, I.C., 1987. "Airflow Through Cracks," *Building & Environment*, Vol. 22, No. 4, pp. 293-304.

Bankvall, C. 1972. *Natural convective heat transfer in insulated structures*. (Report 38). Div. of Building Technology, Lund Institute of Technology.

Bankvall, C.G., 1978. "Natural convective heat transfer in permeable insulation". *Thermal Transmission Measurements of Insulation*, ASTM STP 660, R. P. Tye, ed., American Society for Testing and Materials. pp. 73-81.

Bankvall C.G., 1978. "Forced Convection: Practical Thermal Conductivity in an Insulated Structure under the Influence of Workmanship and Wind," *Thermal Transmission Measurements of Insulation*, ASTM STP 660, R. P. Tye, ed., American Society for Testing and Materials, pp. 409-425.

Bankvall, C. G., 1987. "Air Movements and the Thermal Performance of the Building Envelope", *Thermal Insulation: Materials and Systems, ASTM STP 922,* F. J. Powell and S. L. Matthews, Eds., American Society for Testing and Materials, Philadelphia, 1987, pp. 124-131.

Bankvall, C.G., 1987. "Thermal Performance of the Building Envelope as Influenced by Workmanship," *Thermal Insulation: Materials and Systems. ASTM STP 922,* F. J. Powell and S. L. Matthews, Eds., American Society for Testing and Materials, Philadelphia, pp. 679-684.

Bassett, M. R., & McNeil, S. 2006. "Measured ventilation rates in water managed wall cavities". In *Proceedings from the 3rd International Building Physics Conference*, Montreal, Quebec, Canada pp. 403-10.

Bassett, M.R., McNeil, S., 2009. "Ventilation Measured in the Wall Cavities of High Moisture Risk Buildings", *Jrnl of Building Physics*, Vol. 32, No. 4, pp. 291-303.

Bohac, D. Quinell, J. Novacheck, J., Hewett, M. Lutz, A., Fitzgerald, J. 2016. "Measured Commercial and Institutional Building Pressures", *Proc. of ASHRAE Buildings XII Internation Conference*, Clearwater, FL.

Brown W.C., Bomberg M.T., Ullett JM., 1993. "Measured thermal resistance of frame walls with defects in the installation of mineral fibre insulation". *J. of Thermal Insulation & Building Envelopes* Vol 16, pp. 318–339.

Bumbaru, D., Jutras, R., Patenaude, A.,1988. *Air Permeance of Building Materials*. Report by AIR-INS Inc. for Canada Mortgage and Housing Corp., Ottawa.

Burnett, E.F.P., Straube, J.F., and Karagiozis, A.N. ASHRAE Project RP-1091, 2004. *Development of Design Strategies for Rainscreen and Sheathing Membrane Performance in Wood Frame Walls*. ASHRAE Atlanta, November.

Burnett, Eric F.P., 1981. *Moisture Related Problems in Housing in Newfoundland.* Canada Mortgage and Housing Corporation, Ottawa.

CMHC, 1996. *Controlling Stack Pressure in High-Rise Buildings by Compartmenting the Building.* Report by Morrison-Hershfield for Canada Mortgage and Housing Corporation, Ottawa.

Cummings, J., and C.R. Withers, Jr. 2012. "Envelope thermal failures due to wind washing in two- story homes." *Proc. 3rd Building Enclosure Science and Technology Conference*, Atlanta, GA.

Derome, D., 2005. "Moisture accumulation in cellulose insulation caused by air leakage in flat wood frame roofs", *J. of Thermal Envelope and Building Science* Vol 28 (3), 269– 287.

Dumont, R.S, Snodgrass, L.J., 1990. "Investigation of Chimney Backflow Conditions: A case study in a Well-Sealed House," *ASHRAE Trans* Vol 96(1).

Etheridge, D.W., 1977. "Crack Flow Equations and Scale Effects," *Building & Environment,* Vol. 12, pp. 181-189.

Falk, J., & Sandin, K., 2013. "Ventilated rainscreen cladding: Measurements of cavity air velocities, estimation of air change rates and evaluation of driving forces". *Building and Environment*, 59, 164-176.

Falk, J., 2014. *Rendered rainscreen walls - cavity ventilation, ventilation drying and moisture-induced cladding deformation.* Ph.D. Thesis, Lund University, Lund.

Finch, G., Straube, J.F., 2007. "Ventilated Wall Claddings: Review, Field Performance, and Hygrothermal Modeling", *Proc. of ASHRAE Buildings X*, Clearwater, FL.

Harrje, D. T., Dutt, G. S., & Gadsby, K. J., 1985. "Convective loop heat losses in buildings". *ASHRAE Transactions*, 91(2), pp. 751-760.

ISO, 2007. ISO 10456:2007 *Building materials and products — Hygrothermal properties — Tabulated design values and procedures for determining declared and design thermal values.* International Standards Organization, Geneva.

Janssens, A. & Hens, H. 2003. 'Interstitial condensation due to air leakage: a sensitivity analysis', *J. of Thermal Envelope and Building Science* 27(1), pp. 15–29.

Kalamees, T., Kurnitski, J., 2010. "Moisture Convection Performance of External Walls and Roofs". *J. Build. Phys.* Vol. 33, pp. 225–247.

Kronvall, J., 1980. *Airflows in Building Components.* Doctoral Dissertation, Lund University, Lund, Sweden.

Langmans, J., Desta T.Z., Alderweireldt, L., Roels S., 2016. "Field study on the air change rate behind residential rainscreen cladding systems: A parameter analysis". *Bldg & Environ.*, Vol 95.

Lecompte, I. G. N., 1990. "Influence of Natural Convection in an Insulated Cavity on the Thermal Performance of a Wall," *Insulation Materials, Testing, and Applications, ASTM STP 1030*, D. L. McElroy and J. F. Kimpflen, Eds., American Society for Testing and Materials, Philadelphia, pp. 397-420.

Liersch, K.W.,1981. *Belüftete Dach- und Wandkonstruktionen 1: Vorhangfassaden : Bauphysikalische Grundlagen des Wärme- und Feuchteschutzes.* Bauverlag, Wiesbaden.

Marshall Macklin Monaghan Limited, 1983. *Moisture Induced Problems in NHA Housing.* Canada Mortgage and Housing Corporation; consultants, J.H. VanPoorten, J. Timusk, J. Melchers. Ottawa.

Ojanen, T. and Simonson, C., 1995. "Convective Moisture Accumulation in Structures with Additional Inside Insulation," *Proc. of Thermal Performance of the Exterior Envelopes of Buildings VI*, Clearwater Beach, Fla., Dec 4-7, pp. 745-752.

Ojanen, T., and Kohonen, R., 1989. "Hygrothermal Influence of Air Convection in Wall Structures," *Proc. of Thermal Performance of the Exterior Envelopes of Buildings IV*, Clearwater Beach, Fla., Dec 4-7, pp. 234-242.

Ojanen, T., and Kumaran, K. 1996. "Effect of Exfiltration on the Hygrothermal Behaviour of a Residential Wall Assembly", *Journal of Building Physics* January, Vol. 19, pp. 215-227.

Ojanen, 2020. *Ventilated façade concept for Paroc - Principle design guidelines.* VTT Research Report VTT-R-01215-20.

Orr, H.W., 1974. "Condensation in electrically heated houses" *2nd International CIB/RILEM Symposium on Moisture Problems in Buildings*, Rotterdam, Sept.

Pinon, J.P., Burnett, E.F.P., Davidovic, D. and Srebric, J. 2004. "The Airflow Characteristics of Ventilated Cavities in Screen Type Enclosure Wall Systems (RP-1091)", *Proc of Performance of Exterior Envelopes of Whole Buildings IX*, ASHRAE, Atlanta.

Popp, W., Mayer, E., Künzel, H.1980. *Untersuchungen über die Belüftung des Luftraumes hinter vorgesetzten Fassadenbekleidung aus kleinformatigen Elementen*, Fraunhofer Institut für Bauphysik, Forschungsbericht B Ho 22/80.

Ricketts, L., *Airflow in High-Rise Multi-Unit Residential Buildings.* M.A.Sc. Thesis, Civil Eng Department, University of Waterloo, 2014.

Ricketts, L., Finch, G., Straube J.F. 2015. "Pressure differences and airtightness in tall buildings–theory & reality." In *Proceedings of the Fourth NIBS Building Enclosure Science Technology Conference, Kansas City, MO, USA*, pp. 13-15.

Ricketts, L., Straube, J.F., 2014. "A field study of airflow in mid to high-rise multi-unit residential buildings". *Proc of 14th Canadian Conf on Bldg Science and Technology*, Toronto.

Sandin, K., 1993. "Moisture conditions in cavity walls with wooden framework", *Building Research Information*, Vol. 21, No.4, pp. 235-238.

Schwarz B. 1973. "Witterungsbeanspruchung von Hochhausfassaden" (in German). *HLH (Heizung, Lüftung/Klimatechnik, Haustechnik)*, Bd. 24, Nr. 12, 376-384.

Sedlbauer, K. Künzel, H.H. 1999. "Luftkonvektionseinflüsse auf den Wärmedurchgang von belüfteten Fassaden mit Mineralwolledämmung". wksb 44. Jahrgang, Heft 43, Seite 53-59.

Shaw, C.Y., Tamura, G.T. 1977. "The Calculation of Air Infiltration Rates Caused by Wind and Stack Action for Tall Buildings." *ASHRAE Transactions*, vol. 83, pt. 2, pp. 145- 158.

Silberstein, A., E. Arquis, and D.J. McCaa. 1991. "Forced convection effects in fibrous insulation". *In Insulation materials: Testing and applications*, 2nd vol. ASTM STP1116. ASTM West Conshohocken, PA, pp. 292–309.

Silberstein, A., Langlais, C., & Arquis, E., 1990. "Natural Convection in Light Fibrous Insulating Materials with Permeable Interfaces: Onset Criteria and Its Effect on the Thermal Performances of the Product". Journal of Thermal Insulation, Vol 14(1), pp. 22–42.

Steel, F., 1982. *Airtight houses and carbon monoxide poisoning.* Canadian Building Digest 222. NRCC, Ottawa.

Straube, J., Burnett, E., 1995. *Vents, Ventilation, and Pressure Moderation.* Building Engineering Group Report for CMHC, Ottawa.

Straube, J., van Straaten, R., & Burnett, E. 2004. "Field studies of ventilation drying". *Thermal Performance of the Exterior Envelopes of Whole Buildings X International Conference.* Clearwater Beach, Florida. December.

Tamura, G.T., Wilson, A.G., 1967. "Building pressures caused by chimney action, and pressure differences caused by chimney effect in three high buildings". *ASHRAE Transactions* 73, No. 2.

Tamura, G.T., Wilson, A.G., 1967. "Building Pressures Caused by Chimney Action and Mechanical Ventilation". *ASHRAE Transactions* 73, No. 2.

Tamura, G.T., Wilson. A.G., 1966. "Pressure differences for a nine-story building as a result of chimney effect and ventilation system operation", *ASHRAE Trans.*, Vol 72, No. 1. pp. 180-189.

Timusk, J., Seskus, A.L., Ary, N., 1991. "The Control of Wind Cooling of Wood Frame Building Enclosures", *J. of Thermal Insul. and Bldg Env*, Vol 15, pp. 8-19, July.

Uvslokk, S., 1996. "The Importance of Wind Barriers for Insulated Timber Frame Constructions", *J. of Thermal Insul. and Bldg Env*, Vol 20, July, pp. 40-62.

Van Straaten, R., J.F. Straube, T. Trainor, A. Habellion, 2016. "Wind Washing Effects on Mineral Wool Insulated Sheathings", *Proc. Of Buildings XIII*, Clearwater Beach, FL., Dec.

Vanpachtenbeke, M., Langmans, J., Van den Bulcke, J., Van Acker, J. Roels, S., 2017. "On the drying potential of cavity ventilation behind brick veneer cladding: A detailed field study". *Bldg and Environ, 123*, pp.133-145.

Viljanen, K., Puttonen, J., Lü, X., 2022. "Hygrothermal performance of highly insulated external walls subjected to indoor air exfiltration". *Journal of Building Physics*, 46(2), pp.121-175.

Wahlgren P., 2007. "Overview and literature survey of natural and forced convection in attic insulation". *Journal of Building Physics.* Vol 30, pp. 351–370.

White, F., *Viscous Fluid Flow.* 2nd ed. McGraw-Hill, New York, 1991.

Wilkes, K. E., Wendt, R. L. Delmas, A., and Childs, P. W., 1991. "Thermal Performance of One Loose-Fill Fiberglass Attic Insulation," *Insulation Materials: Testing and Applications. 2nd Volume. ASTM STP 1116.* R. S. Graves and D, C, Wysocki, Eds., American Society for Testing and Materials, Philadelphia.

Wilson, A.G. Garden, G.K.,1965. "Moisture Accumulation in Walls Due to Air Leakage". *RILEM/CIB Symposium on Moisture Problems in Buildings*, Helsinki.

Wilson, A.G., 1960. "Influence of the house on chimney draft", *ASHRAE Journal*, Vol. 2, No. 12, p. 63.

Wolf, S., Solvason, K.R., Wilson, A.G., 1966. "Convective Air Flow Effects with Mineral Wool Insulation in Wood-Frame Walls". *ASHRAE Transactions*, Vol 72, Pt. 2.

7.9 Notes

Chapter 8

Moisture Storage and Transport in Porous Media

Moisture is one of the most important factors affecting building envelope performance, including durability, since it is involved in almost every material deterioration mechanism. The ability of an enclosure assembly to dry and store moisture, as well as its resistance to wetting are of great importance in assessing its vulnerability to moisture-related damage and deterioration. Understanding and predicting moisture movement within and through the enclosure is therefore crucial to the control and understanding of moisture-related problems.

Chapter 4 reviewed the nature of the water molecule itself, and its interaction with air in vapor form. This chapter reviews the interaction of water with porous materials, i.e., the physics of moisture storage, wetting, and drying. The result of microscopic physics on macroscopic building material and element behavior is discussed.

8.1 Moisture and porous materials

8.1.1 The nature of porous materials

Although most materials appear solid to the eye, many natural building materials, e.g., wood, brick, gypsum, stone and concrete, are actually quite porous, (porosity is the volume of air in a macroscopic sample of the material). Apparently solid and heavy materials such as concrete and brick can be 50% air by volume. By comparison, materials such as metals and some plastics have little to no porosity.

One simple view of porous materials is to consider them made of a large number of small solid perfectly spherical and equal sized particles packed cubically as shown in Figure 8.1. The spaces between the particles create the pores in this idealized porous material.

Figure 8.1: Cubical packing of spherical particles

Figure 8.2 plots particle size versus surface area under this simple assumption. It can be seen from the figure that some materials have very large internal surface areas. For example, brick typically has an internal surface area of from 0.1 to 5 m^2/g [Arnott and Litvan, 1988] and cured cement gel has internal surfaces totaling at least 10 to 100 m^2/g [Sereda 1970].

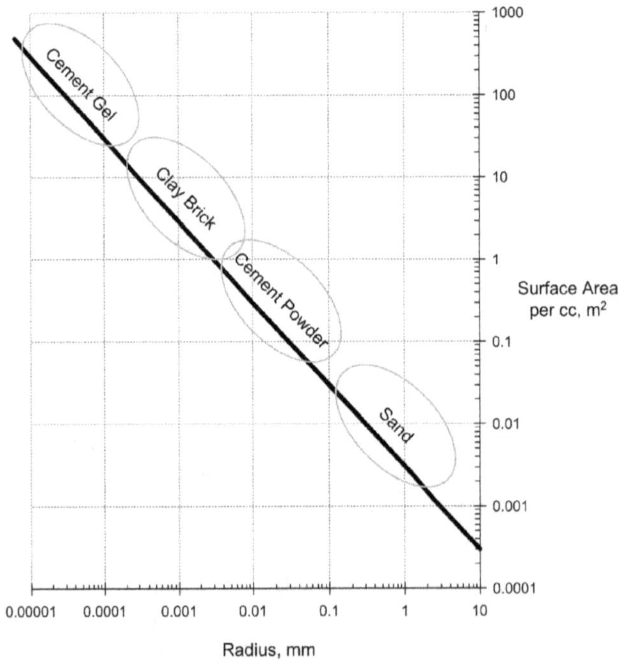

Figure 8.2: Internal surface area versus particle size

Real materials are composed of a wide variety of particle shapes and sizes of course. Hence, the pores would be expected to be very irregular in shape. Scanning electron photographs confirm that pores are in fact highly irregular and that the solid components are agglomerations of crystals and fibres. These facts mean that real materials will have an even greater surface area and a wide range of pore sizes. Nevertheless, on an average basis, experimental results show that the cylindrical capillary pore model is a useful and reasonable model for most moisture physics. All of the theory that follows presumes that cylindrical pores are involved.

The larger the percentage of volume occupied by small pores, the larger the internal surface area. Because of their large internal volume and surface area, some porous building materials can store their own weight in water (or ice) in their internal pore volume with one-quarter of this amount as adsorbed water vapor on the pore walls at high relative humidities (Table 8.1).

Table 8.1: Porosity properties of some common building materials

Material	Density (dry) kg/m3	Open porosity (%)	Moisture @ 95% RH (M%)	Wcap (M%)
Concrete	2100-2200	15-18	4-5	6-8
Brick	1600-2100	11-40	3-8	6-20
Cement mortar	1800-1900	20-30	5-7	14-20
Fiber cement sheet	1350 - 1450	40-50	15-25	30-35
Gypsum (exterior)	650	70	5-10	50-100
Softwood	400-600	50-80	20-30	100-200
Fiberboard	240-380	60-80	20-25	100-200
Wood chipboard	700	50-70	15-20	100-150
Expanded polystyrene	32	95	5	> 300

Note: these values are approximate estimates and from a variety of sources [e.g., Whitley et *al* 1977, Kumaran 1996, Lohmeyer 1996, Pel 1996, ASHRAE 2017, Kuenzel 1994, RDH Labs]

The total porosity, ψ_T, of a completely dry porous material comprised of solid materials and air is normally defined as the ratio of the total air volume (V_a) to the total volume of the material sample (V_T) or that fraction of the porous volume that is not solid. In terms of fractions, X_i is the volume fraction of the ith component of the total non-solid volume. The total non-solid volume in a porous material can usefully be divided into the volume fraction of pores that are (Figure 8.3):

- open, interconnected, and with access to the exterior, X_O,
- dead-end pores connected to exterior X_D, (a subset of X_O), and
- closed or sealed pores with no access to the exterior, X_C.

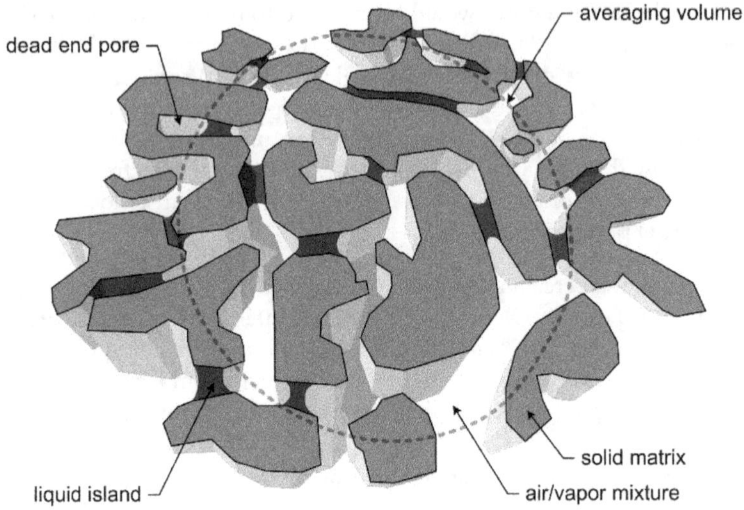

Figure 8.3 : Porous media definitions

In the assessment of moisture storage, it is the open porosity, $\psi_O = X_O \cdot \psi_T$, that is of interest. Table 8.1 summarizes the range of open porosity that can be expected for a number of building materials. It can be seen that there is a wide range, even within a product class.

An effective porosity $\psi_{eff} = (X_O - X_D) \cdot \psi_T$, is important to describe liquid permeability to reflect the fact that dead end pores do not participate in the flow of liquid water through the material but can impact moisture storage.

There are no experimental methods that can accurately and definitively define each of the porosity fractions. However, the open porosity can be approximately measured by various means, including mercury intrusion porismetry, helium and nitrogen gas adsorption, and vacuum saturation [Haynes 1973, Hall and Huff 2021]. Most methods cannot identify closed pores and sometimes fail to measure very small pores (e.g., less than a few nanometers). Another problem arises when measuring materials whose pore structures change with water content because of shrinkage, swelling, or chemical bonding (e.g., wood, gypsum and some plastics).

The pore size distribution for a representative clay brick is shown in Figure 8.4 as calculated from mercury intrusion testing [Arnott 1990, Litvan 1975]. The cumulative percentage of total volume intruded for each of the pore radius ranges is plotted along with the percentage of water absorbed during a vacuum saturation test (25.5% total volume or 88.3% of accessible pore space), and the 24 hour cold water immersion test (18.9% of total volume or 65.4% of accessible pore space). The total open porosity (ψ_o) was 28.9%. Hence, as shown by others [Peake and Ford 1982], vacuum saturation and immersion will often give different results than mercury porismetry.

The liquid transport of salts to a layer, and the consequent deposition when the liquid water evaporates, can effectively change the pore space topology of a material. This can greatly decrease the vapor permeance of the material (by at least an order of magnitude) and result in reduced drying capacity.

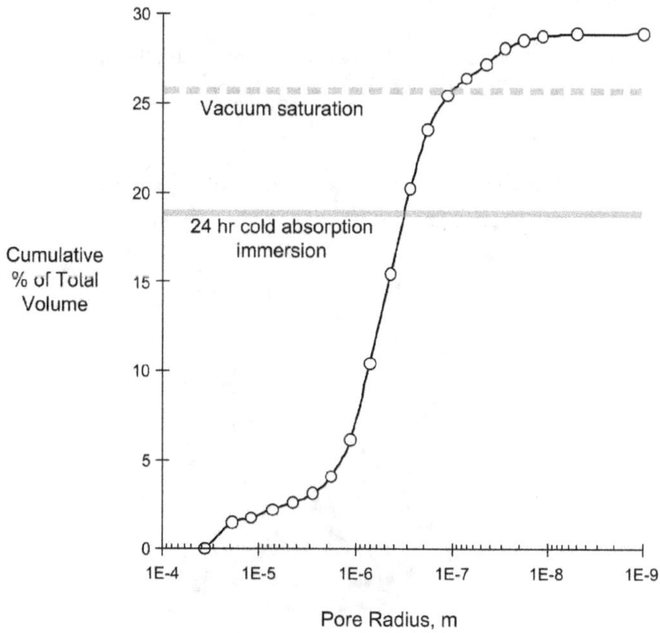

Figure 8.4: Pore size distribution of a representative clay brick [Arnott 1990]

Note: A water molecule is approximately 3 x 10^{-10} m in diameter

8.1.2 Capillary moisture physics

In a porous body which does not have all of its pores filled with water or air (i.e., it is neither fully saturated nor completely dry), it is common to consider three material phases: the solid phase, and the two fluid phases: gas (air, including water vapor), and liquid.

The pressure difference between fluid phases that is maintained within a pore space at equilibrium is, by definition, the capillary pressure. The **Laplace-Young equation** [Hiemenz and Rajagopalan 1997, Hall and Huff 2021] is:

$$P_{cap} = P_{non} - P_{wet} = \sigma \cdot J \qquad \text{(Eq. 8.1)}$$

where P_{cap} is the capillary pressure,

P_{non} is the pressure of the non-wetting fluid (e.g., air), P_{wet} is the pressure of the wetting fluid (e.g., water),

σ is the interfacial or surface tension between the fluids,

and J is equal to twice the average curvature:

$J = 1/r' + 1/r''$

where r' and r" are the two radii of the meniscus with the material of the pore walls (Figure 8.5).

It is commonly assumed that the contact angle with all pore walls is the same (e.g., r' = r"). This allows for a simplification:

$$J = 2 \cdot \cos\theta / r \quad \text{(Eq. 8.2)}$$

where θ is the contact angle of water with the material, and

r is the capillary radius.

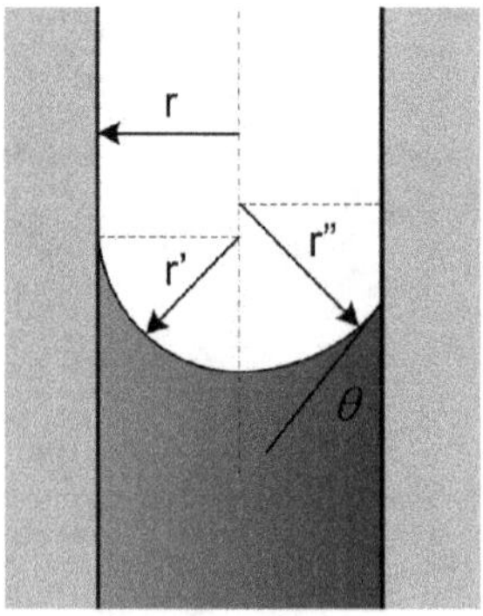

Figure 8.5: Radii and contact angle definitions for the calculation of capillary pressures

For liquid water and gaseous air as the wetting and non-wetting fluids respectively, Equation 8.1 can therefore be written as:

$$P_{cap} = P_{air} - P_{water} = \sigma \cdot J \quad \text{(Eq. 8.3)}$$

where P_{cap} is the capillary pressure.

The most commonly experienced macro-scale example of liquid attraction to solid surfaces is seen when water adheres to a glass (silicon dioxide) surface (Figure 8.6). The resulting meniscus can be seen with the unaided eye. Different materials have different interactions with water depending on their pore sizes and pore size distribution, and surface tension.

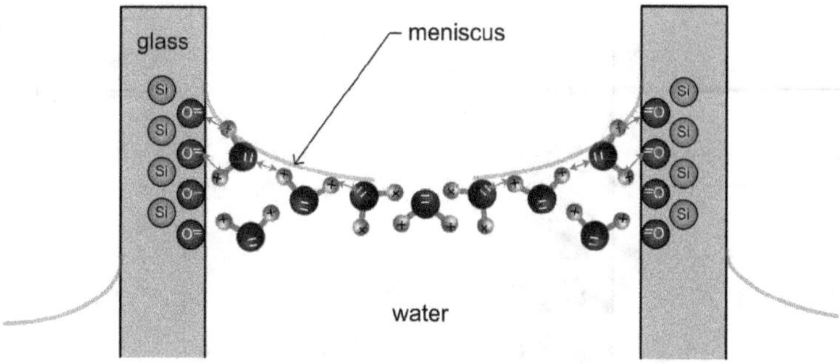

Figure 8.6: Capillary attraction to glass-walled container

A capillary suction pressure formula (the Washburn [1921] equation) can be derived by relating all pressures to atmospheric (Pa= 0) and ignoring the weight of the liquid within the curved portion of the liquid column (a reasonable approximation for small pores and high surface tensions):

$$P_{cap} = \frac{-2\,\sigma\cos\theta}{r} \qquad \text{(Eq. 8.4)}$$

where P_{cap} is the capillary pressure in Pascals

σ is the surface tension (σ = 0.072N/m or J/m^2 for distilled water at 20 °C), and

r is the pore radius in meters.

The negative sign indicates that the capillary force is a suction for contact angles of less than 90° (e.g., for wettable materials). While the contact angle of water with most building materials is assumed to be nearly 0°, its value is affected by a number of variables. Practical experiments with construction materials show a range of values in the 0 to 30° range [Bao & Wang 2017].

Equation 8.4 indicates that water will be drawn into a cylindrical pore of radius r, with a capillary suction pressure of Pcap (see also Figure 8.7). The very small capillary pores (e.g. pores in the range of 10 to 1000 nm) that comprise such a significant part of the volume of concrete and wood will generate suctions of 100 kPa to 10 MPa. It can now be appreciated that the pressures involved in wicking water in the smaller pores of a capillary material are far greater than either gravity forces or wind pressures (Chapter 12).

For cracks or non-cylindrical pores, an equivalent hydraulic radius should be used in the Washburn equation:

$$r_{eff} = \frac{2\,A_{pore}}{perimeter} \qquad \text{(Eq. 8.5)}$$

which is eqaul to w for a long crack of width w.

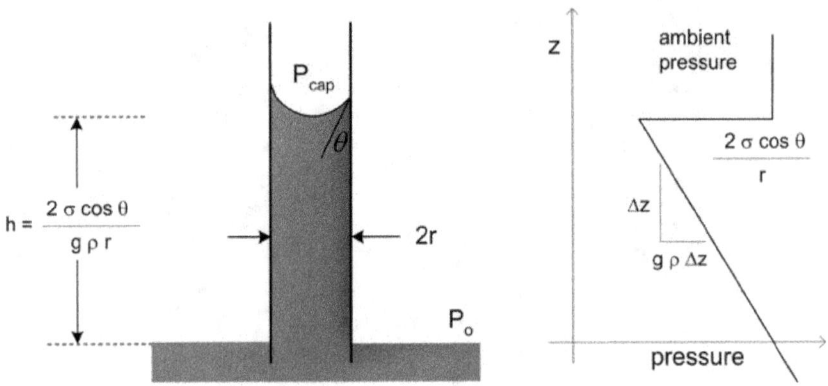

Figure 8.7: Capillary suction of water in a cylindrical tube

Simple substitution into Equation 8.5 also shows that a 0.3 mm wide crack (an allowable crack width in reinforced concrete design codes) will result in a suction pressure of as much as 480 Pa (Table 8.2). Smaller cracks, say 0.05 mm, will create larger suctions (2880 Pa).

Table 8.2: Capillary Suction Pressures

Gap width [mm]	Pressure [Pa]	head [mm]	Pressure [psf]	head [inch]	Windspeed [mph]
3.0	48	4.90	1.00	0.19	8.8
2.0	72	7.35	1.50	0.29	11
1.0	144	14.7	3.01	0.58	15
0.5	288	29.4	6.02	1.16	22
0.3	480	49.0	10.0	1.93	28
0.2	720	73.5	15.0	2.89	34
0.1	1440	147	30.1	5.78	48
0.05	2880	294	60.2	11.6	69
0.01	14400	1469	300.8	57.8	153
0.001	144000	14694	3007.5	578	485
0.0001	1440000	146939	30075.2	5785	1533

Using 140° as the contact angle, appropriate for a highly water repellent material one can calculate that a pressure of 375 Pa would be required to force water into a crack 0.3 mm wide. This magnitude of repulsive pressure can only be overcome by a strong wind gust or 35 mm of standing water. Pressures of over 1000 Pa would be required to force water into a 0.1 mm wide hydrophobically treated crack -- such wind pressures are much less likely to occur. Even the largest capillary pores or air entrainment pores in concrete will resist very high hydrostatic pressures once treated with a water repellent.

When considering very small pores sizes, the radius in the Kelvin and Washburn equation may need to be adjusted to reflect the size of the adsorbed water layer (see Figure 8.8). Since the thickness of the adsorbed layer is likely to be no more than about 1 to 3 nm, its influence will be negligible for pore radiuses larger than about 10 to 25 nm.

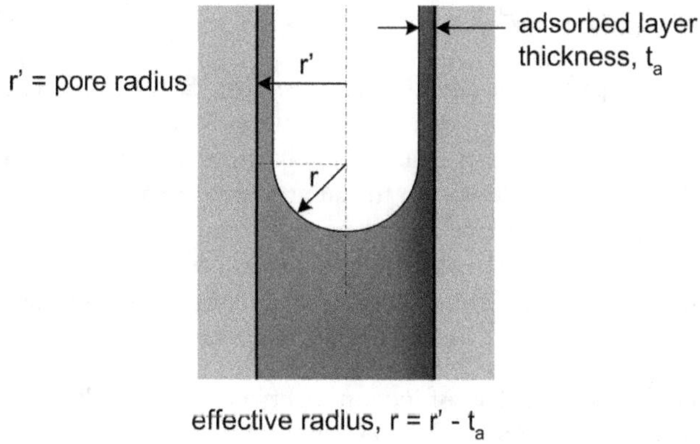

Figure 8.8: Effective radius for small capillaries

8.1.3 Vapor pressure in porous materials

The pressure of water is reduced in pores (or over any other sufficiently curved water surface) by an amount described above by the Laplace-Young equation. The curved surface indicates that forces other than surface tension are acting on the water molecules, and these forces increase the probability that a water vapor molecule will be captured into the liquid phase.

Therefore, within the structure of microporous materials (loosely defined as materials with a significant proportion of the pore volume comprised of pores with diameters less than 1 micrometer) there is no simple relationship between maximum vapor pressure (absolute air moisture content) and temperature such as there is over flat water or in large pores. The reduction in vapor pressure over a curved meniscus (see also) can be found from the Kelvin equation, which was developed from thermodynamic principles by Lord Thompson-Kelvin [Hiemenz and Rajagopalan 1997]:

$$\ln\left(\frac{P_{cap}}{P_{wv}}\right) = \frac{-2\,\sigma\cos\theta}{r\cdot\rho\cdot R_{wv}\cdot T} \qquad \text{(Eq. 8.6)}$$

where, ρ is the density of water,

P_{cap} is vapor pressure over the curved water surface in a pore,

P_{wv} is the water vapor pressure over a flat surface at the same temperature,

R_{wv} is the water vapor gas constant, and other terms are as before.

By noting that the fraction P_{cap} / P_{wv} can be considered to be relative humidity when $P_{wv}=P_{wv,sat}$, Equation 8.6 can be rearranged to give the RH of the ambient air that will cause condensation or saturation of the air within a pore of a given radius (this point is called capillary condensation):

$$P_{cap} / P_{wv} = RH = e^{\left(\frac{-2\,\sigma\cos\theta}{r\cdot\rho\cdot R_{wv}\cdot T}\right)} \qquad \text{(Eq. 8.7)}$$

Liquid water open to the air will evaporate until the air reaches its maximum capacity at that temperature: this is the saturation water vapor content of the air. The saturation vapor pressure at any given temperature is merely the density of water vapor molecules in the air at which the amount captured by collisions with the water surface are balanced by the amount escaping the surface. Because of the increase in kinetic energy of the water molecules with temperature, more molecules can break free as the temperature increases. Hence the saturation vapor pressure increases with temperature in an exponential manner.

This equation is plotted as a function of effective pore radius at room temperature conditions in Figure 8.9. Significant reductions in saturation vapor pressure only occur within very small pores. Pores with radii in the range of 1 to 10 nanometers are predicted to experience condensation at ambient air relative humidities ranging from 35 to 90%.

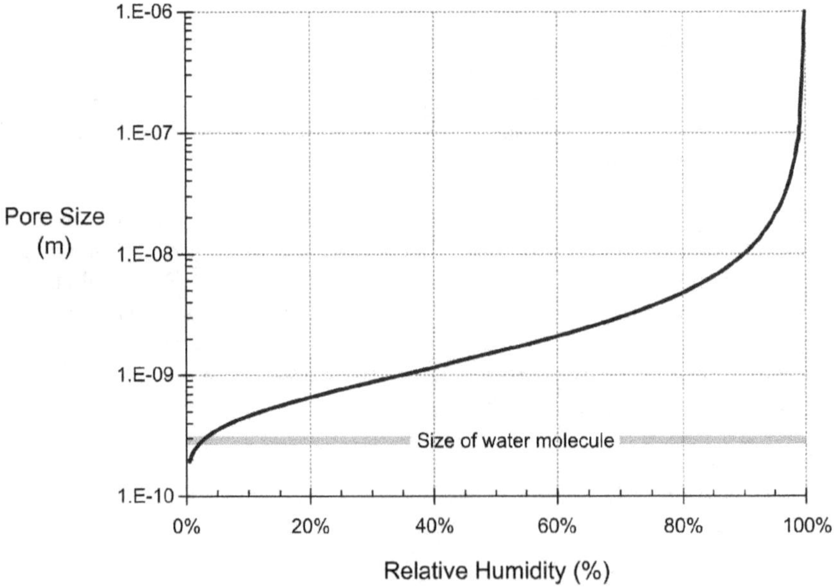

Figure 8.9: Ambient relative humidity at which capillary condensation is predicted to occur by the Kelvin equation

In this range of pore sizes the thickness of adsorbed water layers plays a significant role and cannot be ignored (Figure 8.8). Note that most materials do not contain a significant volume of pores with an equivalent radius of less than about 10 nanometers, although this small percentage of pores may be disproportionately important to behavior.

The fact that capillary condensation does occur in very small pore spaces as predicted by Kelvin's equation has been proven by physical observation in the laboratory [Fisher *et al* 1981]. That the curvature is accurately predicted and that the condensate has some properties similar to bulk liquid water has also been confirmed by physical measurement [Kohonen and Christenson 2000].

8.1.4 Wettable materials and hydrophobicity

All surfaces can be defined as either wettable or non-wettable. Non-wettable surfaces or materials are also called hydrophobic. This property can be defined on a macroscopic level by observing the contact angle between water and the surface. Wettable surfaces have a surface structure that attracts polar water molecules more strongly than other nearby water molecules in the drop. Hydrophobic surfaces actually repel water molecules or have no affinity for them; therefore the inter-water forces are larger than the water-solid forces and a nearly circular drop forms (non-circular because of gravity forces).

A hydrophobic or non-wettable surface is defined as one for which the interfacial contact angle is greater than 90° (Figure 8.10). A hydrophobic treatment (e.g. wax, oil, silicone) will change the contact angle to as much as about 140° but maintaining that high contact angle is difficult because of surface soiling and weathering affects. Surface texture also plays a significant role.

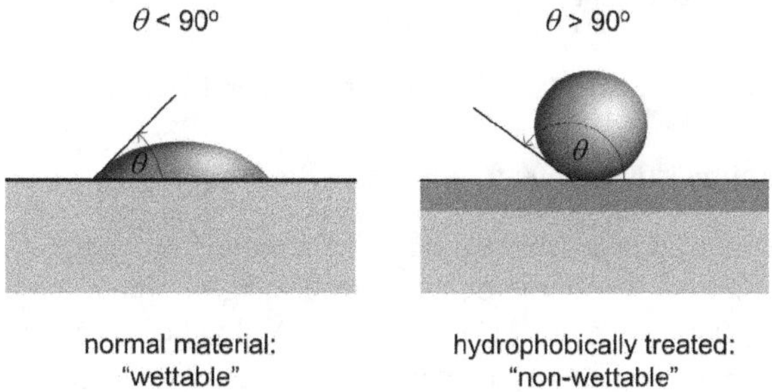

Figure 8.10: Macroscopic contact angles and wettable materials

It is normally assumed that the contact angle of water with most porous building materials is zero. However, this angle is difficult to measure on the microscopic level (i.e., within a small pore), and the macroscopic properties viewed with the naked eye are strongly influenced by surface roughness. It may have a value of as much as 50 degrees and, as discussed later, may change as a function of the wetting history.

Water repellent treatments can change the behavior of the surface of a material by adding a thin, usually monomolecular hydrophobic layer to the internal surface area of a porous body (as penetrating sealers such as silane and siloxane do to concrete and brick). Coatings provide a thicker (several thousand to millions of molecules thick) hydrophobic coating on the exterior (e.g. waxes, acrylics, and silicones).

Unlike coatings, some penetrating repellents render a porous material hydrophobic even below the surface. Therefore, if cracks form in service after treatment, significant water pressures can be resisted by the hydrophobic material surface exposed by the crack (Figure 8.11). Not only do capillary suctions no longer act to draw water into a hydrophobic material, but significant hydrostatic pressures are also required to push water into cracks and pores. This is a significant fact because most porous building materials can be rendered hydrophobic with some form of treatment, as simple as wax or as durable as silicone or silane.

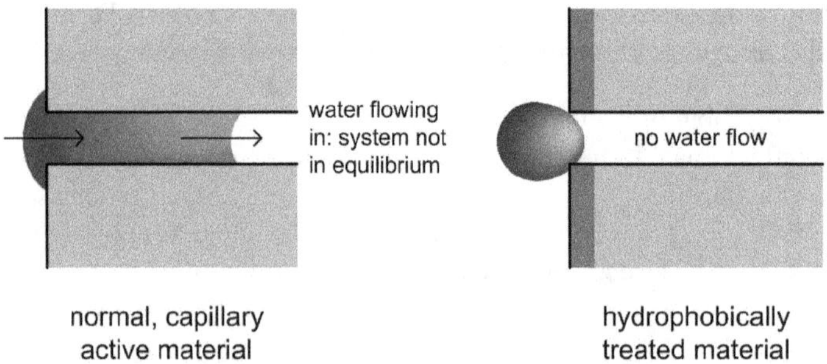

Figure 8.11: Behavior of pores and cracks

8.1.5 Measures of moisture content

It is important to note the method by which moisture content is reported. Most sources use a gravimetric measure, e.g., kg of water per kg of material, denoted m (kg/kg dry) or M% in this text. Volumetric measures also are used, cubic meters of water per cubic meter of material, denoted here as ψ (m^3/m^3) or as V%. Mixed measures are also commonly used, e.g., kg of water per cubic meter of material, or liters of water per kg of material. This commonly used measure will be denoted here as w (kg/m^3).

It will be shown in the following sections that equilibrium RH (%) and the moisture content of a porous material are related. Since RH is reasonably easy to measure and performance thresholds are sensitive to it, it is often used as an approximate measure of moisture content.

The degree of saturation, S, is the percentage of open pores that are filled with water. It may be expressed as a decimal fraction or percentage, just as relative humidity is. The degree of saturation is very useful when measuring freeze-thaw conditions and other situations where the materials are highly saturated. As discussed earlier, one must be sure of the method used to define "fully saturated", since many common laboratory methods do not fill all of the open voids.

8.2 Moisture performance thresholds

As stated in the introduction, moisture is important to building science because it is involved in many performance problems and material deterioration mechanisms. The relationship of moisture content to various performance thresholds is very briefly summarized in this section.

Porous materials are susceptible to moisture damage because the large surface area accessible to water molecules allows processes to occur over a large surface area. The ability of most porous materials to store moisture can also increase the time required to dry the material, thereby increasing contact time.

Non-porous materials, such as steel and most plastics, store small amounts of moisture on their surfaces as adsorbed water, liquid water in drops, and frozen water as ice and frost. However, the amount of water stored on the surface of non-porous materials is very small, and none can be stored within the material. This moisture can, however, still support surface corrosion and fungal growth. In fact, the practically non-existent ability of steel and plastics to store water often means that deterioration can begin with even a very small amount of surface water deposition.

For many building products, safe storage capacity is exceeded once the relative humidity exceeds about 80% [Sereda and Feldman 1970]. This can be explained by the availability of moisture in thick but loosely adhered adsorbed layers described later.

8.2.1 Corrosion

Corrosion is an electrochemical process. Metals oxidize by giving up electrons, thereby allowing electronegative oxygen to bond to them. All metals oxidize, but the oxides of ferrous metals (iron-based) is larger in volume and has inferior mechanical properties than the base material. The corrosion product that most are familiar with is the red rust that can be scrapped off corroded steel products.

The primary factors affecting the rate of ferrous metal corrosion are the availability of moisture, the presence of chlorides and acids, time, and temperature. The ISO 9224 standard provides means of assessing the likely corrosion rates for common metals as functions of the above variables.

The Time of Wetness (TOW) is the primary and most important variable. Although surface corrosion can begin at relative humidities of as low as 60%, the rates are negligible under normal conditions. Hence, the TOW is defined as the number of hours per year above 80%RH and freezing at the surface.

Chlorides, from the sea, deicing salts, soil, and building materials accelerate the rate of corrosion and enhance its initiation. Elevated salt concentrations can be found several km from shorelines and several hundred meters from major highways. Acids reduce the relative humidity at which corrosion can begin and accelerate the rate of corrosion, often significantly (by several orders of magnitude). If the pH is less than about 10, high corrosion rates can be expected [Sereda 1976, ISO 1992]. Zinc is especially sensitive to acidic (less than pH=7) environments. Acids can be provided from acid rain, industrial processes within or near buildings, and from building materials themselves (e.g., historic phenolic foam). Gypsum can generate very acidic byproducts (pH=1.0) if it exposed to moisture with little oxygen.

The initially high pH of cement- or lime-based concrete, mortar, and grout (i.e., pH > 12) results in a passive oxidation layer on embedded steel that effectively resists corrosion. This passive layer can be defeated by chlorine ions if the pH drops below about 11. Zinc also corrodes quickly when exposed to very high (over pH 13) environments.

Carbonation is the process by which atmospheric carbon dioxide reacts with alkali compounds in cement-based products and eventually leads to a reduction in pH and depassivation. Carbonation of concrete is therefore a significant durability issue for reinforced concrete exposed to moisture. Carbonation occurs most rapidly at an environmental RH of between 50 and 80%RH without liquid water immersion [CEB 1986]. Saturation or submersion also limits the access of carbon dioxide to concrete, and hence carbonation is slowed or stopped. Carbon dioxide diffusion barrier coatings are sometimes applied to control carbonation, but high quality, impermeable concrete and sufficient cover to the steel is the most common approach.

Once initiated, steel corrosion within concrete, masonry mortar, and grout tends to reach a maximum rate in the range of environmental humidity of 85-90% RH, but corrosion may still occur down to 60% RH [CEB 1986]. Recent research suggests that significant rates of active corrosion are actually unlikely below an internal RH of about 80% [Enevoldsen *et al* 1994]. Other complementary research also shows that for all practical purposes, corrosion of metal in concrete stops at saturation levels of less than 50% [Lopez and Gonzalez 1993].

Water is not available as an electrolyte below freezing. As the temperature of the water increases the corrosion rate will also increase. The Arhenius relation predicts a doubling in corrosion rate for every 10 °C (18 °F) increase in temperature.

Zinc-based coatings are often used because they form a strong oxide film (like aluminum, see below) when they corrode which provides a physical barrier. Zinc has the additional benefit that when connected with metals below it in the electrochemical series, it will sacrificially protect that metal.

Organic coating films such as epoxy and paint are sometimes used to provide corrosion protection by restricting the access of moisture and oxygen to the base metal. These barriers will work well if the coating is installed and maintained in perfect condition. If they fail, they can accelerate the rate of corrosion (by concentrating the electrochemical potential at a few small spots) and increase the consequences of corrosion. A thin film of corrosion over a wide area often causes less damage that a small amount of corrosion deep into a piece.

Aluminum is widely used in the building industry as a corrosion-resistant material. The oxide film that forms when aluminum oxidizes acts as a protective barrier to further corrosion. The film is only 3 to 10 nm thick but has good mechanical properties and is stable over the range of pH 4 to 8.5. If an aluminum surface is exposed to mechanical abrasion, the oxide film may be constantly removed and corrosion will occur relatively quickly. Highly alkaline environments (e.g., cement and lime-based mortars and concretes) will attack aluminum, so some form of protection or separation is often required until the alkalinity naturally reduces over time due to carbonation.

8.2.2 Mold growth and wood decay

The primary requirement for both mold growth and wood decay is moisture.

Fungal growth requires spores, nutrients appealing to the mould, temperature and moisture conditions conducive to growth, and air [FPL 1999, Adan 1996, Ashton 1970, Baker 1969].

Spores within a building must be assumed. Mould spores exist in the outdoor in significant concentrations, especially in summer and in areas with vegetation, as well as in materials delivered to construction sites. If conditions for fungal growth are favorable, there are likely mould spores available even indoors.

Nutrients for fungi are also usually available in buildings. Many building products are organic – wood and paper, many glues and paints, etc. – and so acts as a ready food source. Cellulose has the advantage that it can be broken down into sugar. However, mould can even grow on ceramic tile since small amounts of captured airborne dust, dirt and even soap provides food for the nutrients. The larger the internal surface area of the material, and the more predigested it is, the more quickly and easily mould will bloom. It is for this reason that the paper facing on gypsum and the cellulose in ceiling tiles provides an ideal source of food, while the original source for these products, solid wood, is not nearly as attractive.

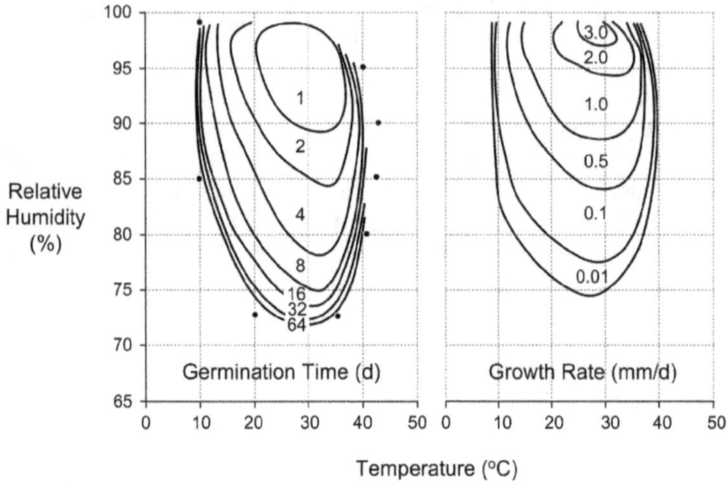

Figure 8.12: Temperature-time-moisture relationship for *Aspergillus restrictus* on a high nutrient substrate [Krus *et al* 2001]

Temperature is important for the growth of most organisms, including mold. Ideal conditions for mold growth are species dependent, but tend to be in the range of 20 to 35 C. Most molds will grow much more slowly outside this range, and growth essentially stops below 5 °C or above 50 °C. Mold spores can survive outside of this range—from well below freezing to over 60 °C — and so mold growth will resume when the temperature becomes favorable again.

Moisture is necessary for all living organisms. Fungi typically require a surface relative humidity of 75 to 80%. Some molds can grow with less moisture, but these species are less important to building problems, and the growth is very slow, even on nutrient rich surfaces. As the amount of moisture increases beyond the optimum conditions (often in the range of 90 to 95% RH), the susceptibility of some materials to deterioration often decreases. For example, when fully saturated, fungal growth slows or stops because of a lack of oxygen. This explains how buildings built on submerged wood piles in cities such as Venice and Amsterdam, have survived for hundreds of years.

The most common variables that play a role in mold growth and their effect are summarized in Figure 8.13.

Alkalinity and radiation also play a role – most fungi require the pH of the growth substrate to remain within the bounds of about 5 to 8. This sensitivity to alkalinity is the reason that lime washes, with a fatal pH of 10 to 13, were used in the past to control fungal growth. Exposure to ultraviolet radiation may slow or kill fungi, but this depends on the intensity and duration. For exterior cladding, the UV intensity is often high enough to stop growth. Exposure to direct sunlight typically reduces fungal growth since the light warms (and hence dries) the surface and the UV intensity can be high.

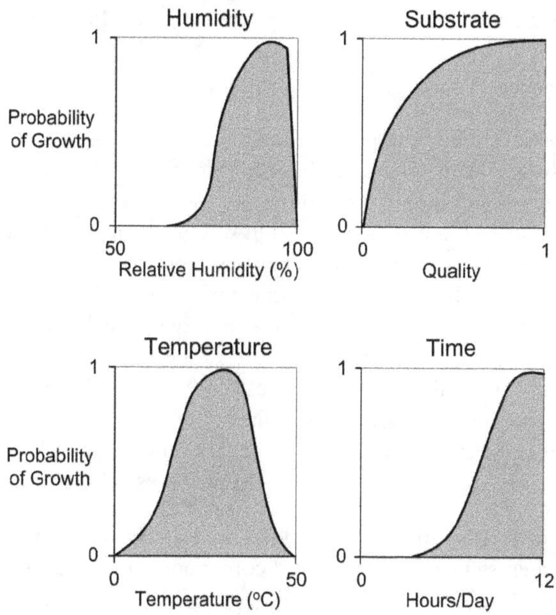

Figure 8.13: Relative importance of different factors affecting mold growth [Krus *et al* 2001]

8.2.3 Wood decay

Wood decay is a special type of fungal infection that attacks cellulose. Brown rots (inaccurately named dry rot) consume the cellulose and hemicellulose in wood. White rot attack focuses on the lignin.

The moisture and temperature conditions required for decay fungi to thrive on wood are more stringent than fungal growth. More moisture must be available, about 28% moisture by weight or around equilibrium with 95% RH. A significant amount of literature can be found to define the nuances and the details of influences in much more detail.

8.2.4 Freeze-thaw damage

Freeze-thaw damage can only occur in materials which are more than about 90% saturated, (i.e., well over 98% RH, and often 100% RH), while temperatures are well below freezing. Although it is often thought that freeze-thaw damage is caused by the expansion of water upon freezing, this is only the case for large pores and voids. The rate of cooling and the temperature appear to have a significant affect, as does the presence of salts and air entrainment.

Within small pores a complex interplay of forces is acting, and expansion of water is not sufficient to explain freeze thaw damage. For an introduction to this complex subject the reader is directed to Litvan [1980] & Mensinga [2009].

8.2.5 Subfluorescence

Many mineral building materials contain salts of various types, especially those based on sodium and potassium. In addition, the deicing salts often used in cold climates, soils adjoining buildings, and sea water are all sources. This salt may enter porous materials dissolved in water. In the case of a porous material containing salty water, the liquid water will tend to flow toward a drying surface (often the exterior but the drying surface may be on the inside or within the enclosure at the boundary with a non-capillary active layer). When the water evaporates the salts must remain behind. If the evaporation occurs at the surface, the salts are visible and the process and the product are termed efflorescence. If the evaporation occurs within the porous material (because of the relative rates of capillary transport and evaporation), the salts recrystallize within the pore space and the process is termed subfluorescence.

When salt crystals form out of solution they expand. Very large stresses can be imposed when salt crystals form and constrained within porous materials (in the order of 1 to 100 MPa, depending on the salt, the pore topology, and the stiffness of the solid material) [Winkler & Wilhelm 1970]. The damage appears to be similar to freeze-thaw damage, but the visible presence of salts and the usual lack of cold weather are obvious indicators that differentiate this damage mode.

Even if the salts do not cause damaging expansive stresses, they may block a significant number of pores. This results in a reduction in vapor permeance. This can increase the moisture content of material further inside the enclosure and eventually result in failure from other moisture related causes or differential hygric stresses.

Subfluorescence has been shown to be the cause of several commonly experienced masonry, stone, and stucco-on-masonry failure mechanisms [Kuenzel 1990] and is especially common in historic buildings.

The effect of dissolved salts on moisture transport also has many practical applications but has not been widely studied.

8.3 Moisture storage in porous materials

As noted earlier, the ability of a material or an assembly to store moisture may be a critical consideration in moisture control for several reasons.

Storage provides the buffer in time between the deposition and removal of moisture. The amount and duration of moisture storage can be an important measure of an enclosure's durability. If the volume of stored water exceeds the safe level for a material, and is present for long enough, deterioration can occur. In general, the most important features are how much moisture can be stored and for how long without crossing a performance threshold.

8.3.1 Liquid water and ice storage mechanisms

Assuming that water molecules can penetrate all of the open pore volume (a reasonable assumption given the small size, less than 0.3 nanometers, of a lone water molecule), the maximum amount of moisture that can be stored by a material will be governed by both its porosity and the topology of the space. Air can also be trapped in dead end pores, and thereby limit the maximum volume of water that can be absorbed. The largest amount of moisture can be stored if a material is first placed in a vacuum chamber, then placed in water, and then atmospheric pressure is reapplied; this procedure eliminates trapped air and ensures that all pores interconnected to the exterior by any path are filled with water. The amount of water stored in such a vacuum saturation test is called the maximum moisture content (w_{max}). The reported vacuum saturation moisture content for materials with very small pores can vary depending on the technique used, especially the degree of vacuum and length of immersion time [Peake and Ford 1982].

The equilibrium moisture content of a material in contact with water at atmospheric pressure is variously termed the *capillary saturation* moisture content, *the free saturation*, or just the capillary moisture content (w_{cap}). The value of w_{cap} will vary slightly for each sample depending on the test method chosen. In general, test methods ensure that water is available over the entire surface of the sample without generating hydrostatic forces that might drive water into the material.

In some test protocols a material sample is simply immersed in a bath of water at room temperature. The amount of water absorbed will asymptotically approach some value, sometimes also called the cold water saturation moisture content. Water is actually being forced into the material in these tests by hydrostatic pressure. However, for materials with small pores the size of this driving force is small relative to the capillary suction forces and hence this may not affect accuracy by a meaningful amount.

It may require a few minutes to a few weeks for the capillary saturation moisture content to be reached, although test standards may specify a fixed duration. Nevertheless, by leaving a sample immersed in water for a long time (as long as years), the vacuum saturation moisture content can be approached.

The most accurate tests for capillary saturation are those that place a thin (to limit gravity forces) sample in a vapor-sealed container (so that evaporation does not play a role) in constant capillary contact with water (so that no hydrostatic pressures act).

Once a material has reached capillary saturation, all capillary forces will be in equilibrium. Therefore, the material will not be able to store any more moisture unless some external force (e.g., hydrostatic head, large air pressures, etc.) forces water into the pores.

A sample can also be placed in boiling water (sometimes called the boiling saturation) as a means to accelerate the room temperature or "cold" water saturation test or as a simple means of reaching the maximum saturation value of a vacuum test. Repeated or long duration boiling will increase the amount of water that is absorbed, and even long-term immersion in cold water may allow the moisture content to approach the maximum moisture content, w_{sat}. For example Arnott [1990] showed that a 5-hour immersion in boiling water results in approximately the same amount of water absorption as a vacuum saturation test for a wide range of clay bricks. Both test methods, however only fill about 90% of the pore space of clay bricks accessible to mercury intrusion methods.

It can be concluded that capillary moisture content as measured by most tests lies somewhere in the spectrum between the true capillary moisture content and the vacuum saturation moisture content (in which high levels of vacuum are used in conjunction with long soak times).

The moisture content above which macroscopic-scale liquid moisture transport begins to occur is termed the critical moisture content (w_{crit}). The critical moisture content is difficult to measure but generally occurs at a moisture content in equilibrium with an environment of between about 93 and 98% RH. The critical moisture content is important because it marks the moisture content below which the liquid water is discontinuous between pores at a microscopic level. Therefore, below w_{crit}, moisture moves in a porous material primarily by the mechanisms of vapor diffusion and adsorbed layer diffusion, while above this level capillary conduction is the dominant mode of moisture transport. Liquid transport can still occur over short distances (along connected groupings of very small pores) below the critical moisture content.

The critical moisture content might also be expected to be important for air transport through porous materials because it is assumed that above w_{crit} the liquid water in the majority of pores is interconnected, thereby blocking the passage of airflow. The air permeability is therefore expected to drop dramatically as the moisture content approaches w_{crit}. However, recent experimental evidence strongly suggests that air permeability drops dramatically at w_{cap}, not w_{crit} [Descamps 1997]. This is likely the case because capillary transport of liquid water will occur through the smaller pores at low moisture contents, while air will preferentially flow through the largest pores. Since the largest pores become filled only at capillary saturation, it is to be expected that as w_{cap}, is approached the air permeability will drop to almost zero.

8.3.2 Capillary suction curves

The moisture storage in a porous material between a dry state and capillary saturation can be defined by a capillary suction curve, or a water retention curve. This curve is based on the suction pressures that hold water within pores. As can be seen from the Laplace-Young (Equation 8.1) as the atmospheric pressure changes, the curved water surface in the pore will seek a new equilibrium. If the exterior air pressure exceeds the capillary pressure predicted by the Washburn equation (Equation 8.4) liquid water will be expelled from pores with this radius and larger. By successively increasing the air pressure acting on an initially saturated porous material, the total amount of water remaining in a porous material can be compared to the applied pressure. This is the intent of a pressure plate test [Penner 1963]. Such pressure plates have found increasing use in building material characterization [Krus 1996, Janz and Johnnesson 2001].

Figure 8.14 is a representative plot of the results of a porous material in a pressure plate test. It can be seen from the plot that the major wetting and drying curves follow different paths. For the same capillary pressure, there are clearly two different moisture contents. Several reasons for this hysteresis have been postulated [Pel 1996]. The so-called "ink bottle effect" describes the response of moisture to the irregular shape of real pores and capillaries. Figure 8.16 illustrates a hypothetical situation whereby hysteresis would be expected since two equally stable capillary pressures are possible, each with a different moisture content.

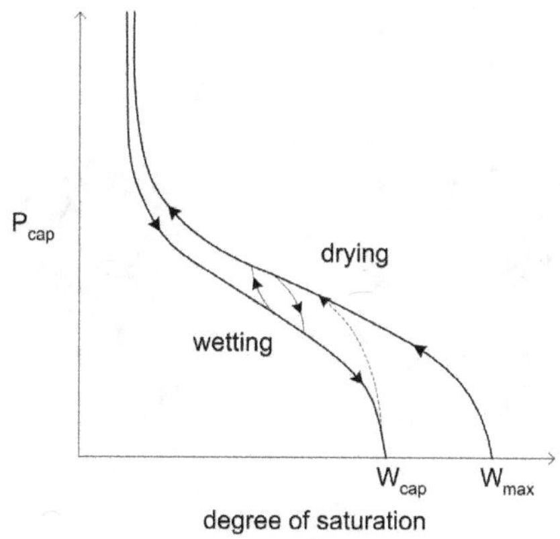

Figure 8.14: Typical capillary pressure curves

Another proposed explanation for hysteresis is a contact angle that changes depending on whether the capillary meniscus is advancing or retreating (Figure

8.15). The advancing angle is generally larger than the retreating angle in macroscopic investigations and the same behavior may occur within pores. Different advancing or retreating contact angles would also help explain hysteresis in the capillary condensation region of sorption isotherm.

Finally, the behavior of air in the material will depend on whether wetting or drying is occurring. Trapped air in dead-end pores can retard wetting but have little effect on (and may even accelerate) drying.

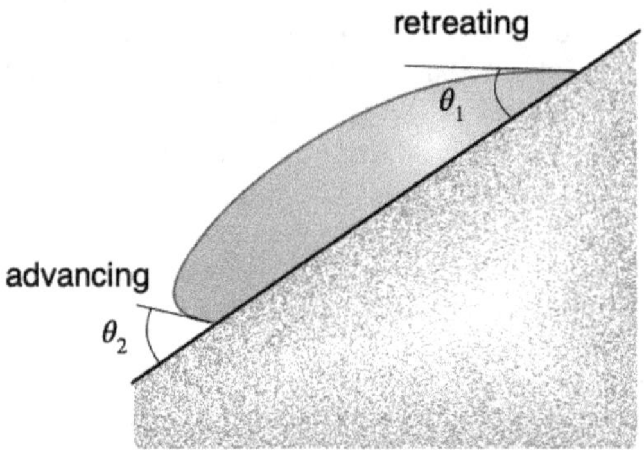

Figure 8.15: Advancing and retreating contact angles [Hiemenz and Rajagopalan 1997]

It is often assumed that one can trace an average curve between wetting and drying curves. However, the difference can sometimes be quite large and a significant error might accumulate in detailed step-by-step computer simulations. This may be important in wet/dry cycling of materials from almost saturated to almost dry, such as that which occurs in thin porous cladding elements.

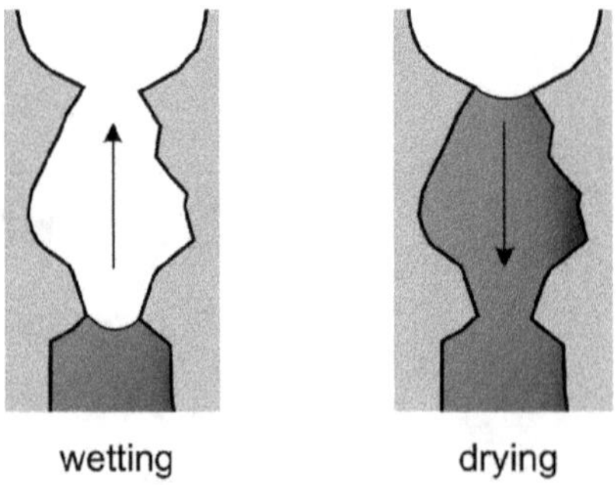

Figure 8.16: Moisture storage hysteresis due to the ink bottle effect

8.3.3 Vapor storage mechanisms

As stated earlier, a significant amount of water can be stored in porous bodies in the adsorbed state. This section explores how porous materials store moisture in the vapor or adsorbed phase.

Because the molecular attraction of water to a surface only acts strongly over a short distance, only a few layers of water molecules can be held. Hence, moisture stored by adsorption is usually only significant when materials have large surface areas, i.e., the internal surface area connected to the ambient air is very large per unit volume or per unit mass. This is the case for most common building materials. Even a single or thin layer of adsorbed water molecules will result in significant amounts moisture storage within materials with a very large internal surface area.

All hygroscopic materials attract and adsorb water vapor until they reach equilibrium with the surrounding air. Depending on the dynamic equilibrium established, the equilibrium moisture content varies widely with material, temperature, and RH. The amount of moisture stored by adsorption as a function of the vapor concentration (i.e., vapor pressure) in the surrounding air and temperature is shown in Figure 8.17 for a typical hygroscopic material. The vapor pressure is merely a convenient means of measuring the absolute amount of moisture in the air; the humidity ratio can also be used since the two are directly related.

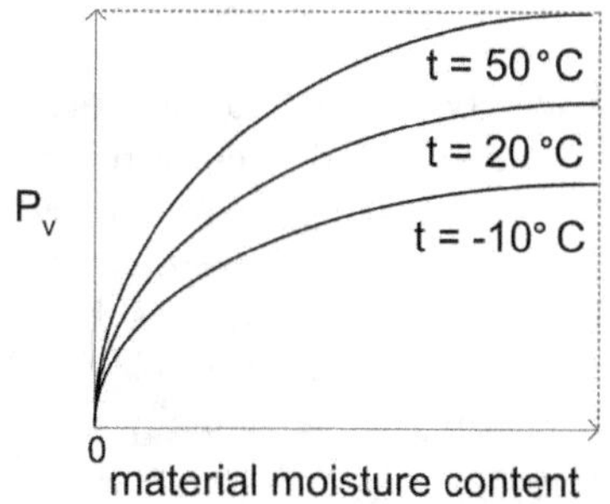

Figure 8.17: Surface vapor pressure of a typical hygroscopic material as a function of temperature and moisture content

Expressing the response of a hygroscopic material in terms of RH instead of vapor pressure will generally collapse the family of lines shown in Figure 8.17 to the single line shown in Figure 8.18: such a diagram is called the sorption isotherm. Wetting (adsorption) and drying (desorption) follow slightly different paths in many porous materials with high surface areas. A common reason given to explain hysteresis is simply that not enough time been given for equilibrium to be reached. However, careful measurements still show some small amount of hysteresis, and the same, as yet unproven, explanations used to explain hysteresis of the capillary retention curve are applied to sorption. This hysteresis usually is not large enough or significant enough to be a concern, and hence average curves are used.

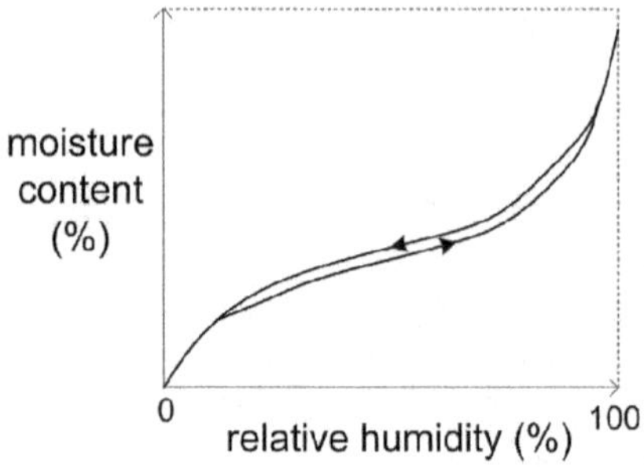

Figure 8.18: Typical sorption isotherm of a hygroscopic material

Although the sorption isotherm is a very useful and fundamental descriptor of material response, it has not been defined for many building materials and the accuracy and repeatability of available results are poor. The selection and quality of data is increasing, driven by the need created by hygrothermal modeling. Sorption isotherms for a range of useful building materials can be found in Nevander and Elmarson (1994), Kumaran et al (2002, 2006) and Schumacher et al (2020). Subtle differences in manufacturing or raw material sources can result in significant differences in the sorption isotherm if the differences affect the pore structure.

When a material has adsorbed all the moisture it can by adsorption, further moisture will be stored in pores and cracks within the material by capillary suction (e.g., absorption). For example, wood will adsorb up to approximately 30% moisture content at 98% RH by adsorption, but by placing it in contact with liquid water it can hold two to four times this amount (see Table 8.1). Liquid water (e.g., 100% RH) must be available for absorption to occur. In practical situations, very high relative humidity levels cannot occur without liquid condensation forming. Even in laboratory settings, a consistent RH of 95% is difficult to achieve within a chamber without small localized temperature and vapor pressure gradients resulting in surface condensation. Therefore, in practice, adsorption and absorption are often acting together in the RH range of more than about 95%, albeit at different places and times over a sample. This is even more significant in real enclosures because the gradients of moisture and temperature are far greater than in a controlled laboratory setting.

Knowledge of the size and distribution of pores can help predict the vapor pressure within a material. The most widely accepted adsorption theory, termed BET after its developers Brunauer, Emmet, and Teller [1968], describes and predicts the amount of water vapor that can be expected to be adsorbed onto the internal surface area of various materials up to about

50%RH. Rounsley [1961] developed the following equation by using statistical thermodynamics to improve and extend the BET model:

$$\frac{X}{X_1} = \frac{C \cdot \emptyset}{1+(C-1)\cdot \emptyset}\left(\frac{1-\emptyset^n}{1-\emptyset}\right) \qquad \text{(Eq. 8.8)}$$

where X is the equilibrium moisture content (M%),

X_1 is the moisture content in a mono-layer of adsorbate (M%),

ϕ is the relative humidity,

C is a constant, comprising the heat of adsorption, heat of condensation and the universal gas constant (typically 20 to 50),

n is the maximum number of layers that can be adsorbed (typically 4-6).

Equation 8.8 represents the most probable distribution of adsorbed molecules. The constant C and the number of layers can be found from simple, low-RH adsorption tests. El-Sabaawi and Pei [1977] and later Tamon, Okazaki, and Toci [1981, 1983] combined this background adsorption theory with Kelvin's Law to develop a predictive equation for sorption isotherms given experimental information on adsorption and pore size distribution.

Their models accurately predicted the complete sorption isotherm from saturation to complete drying for several materials. These results support the physical validity of their assumptions: adsorption occurs in all pores following the BET/Rounsley model until, as the Kelvin equation predicts, condensation begins to occur within the smallest capillary pores. Experimental results also confirmed that the reduction in pore size because of the thickness of the adsorbed film must be considered when using Kelvin's equation to assess condensation within the pores. By accounting for both mechanisms (one dominant at low RH and the other at high RH up to saturation) the sorption isotherm can be predicted for some materials (Figure 8.19).

Physical measurement of the thickness of the adsorbed film have been attempted to confirm the BET sorption theory, but the literature does not provide consistent values. Some measured data [Beaglehole and Christenson 1992] shows that one full layer of water molecules (0.3 nm or 3 Angstroms thick) forms on mica and silicon at about 60%RH. Another study [Freund *et al* 1999] produced physical measurements of 20 nm and added that the water formed in droplets at low RH until merging into a single layer at an RH above 55%. BET theory would predict about than 3.5 layers of water (about 1.1 nm thick) at 60%RH. Hence, the small mass changes measured during adsorption of water vapor to small, non-porous and flat surfaces which tend to agree with BET theory, (see Panella *et al* 1999) are considered more reliable.

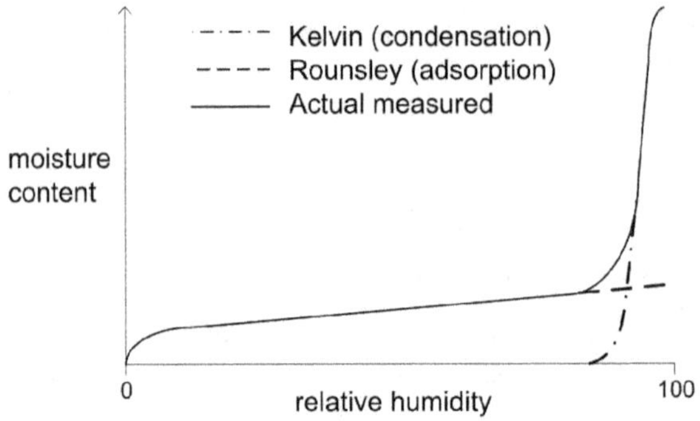

Figure 8.19: Typical predicted sorption isotherm according to Kelvin Equation and modified BET theory

Temperature does play a small role in most materials in defining the sorption curve, so the concept of a true sorption isotherm is somewhat flawed. However, in the temperature range of many building science problems, the variation with temperature is not that significant. The isotherm for wood (taken from the US FPL Wood Handbook [1999]) at different temperatures is plotted in Figure 8.20.

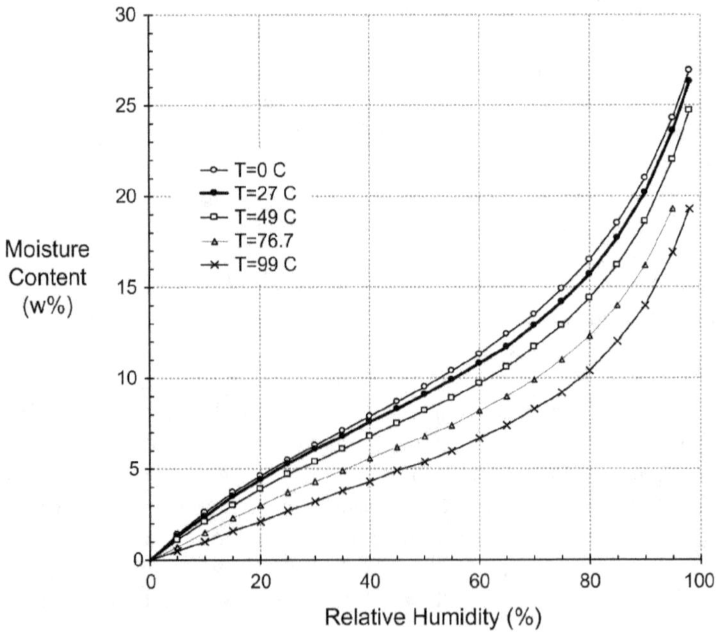

Figure 8.20: Average sorption isotherm for wood as a function of temperature

It is clear that the temperature dependency at temperatures less than 50 °C is small. If one is examining the moisture content of wood sheathing below a black roofing membrane or zinc coated steel, however, temperatures can easily exceed 75 °C during the day and drop to well below 20 °C at night. In cases such as these temperature-dependent sorption curves might need to be accounted for.

8.3.4 The moisture storage function

From the discussion above, the moisture storage within porous materials is clearly the result of several mechanisms, with different behavior due to each. There is a consensus in the modern literature for the qualitative description of the moisture storage function [e.g., Kuenzel 1994, Nevander and Elmarson 1994, Geving & Thue 2002], sometimes called the moisture retention curve.

Table 8.2: Moisture storage mechanisms

Moisture from	Storage location
free water vapor	in pore volume (porosity)
Adsorbed water vapor	on pore walls (specific surface area)
Capillary condensed water	held in very small pores
Capillary bound liquid water	held by capillary suction in pores
Unbound liquid water	held by containment

Figure 8.21 shows a typical moisture storage function. Three clear regimes can be identified:

1. the hygroscopic regime, divided into three sub-regions,
2. the capillary regime, and
3. the over-saturated regime.

In the first sub-regime, labeled A in Figure 8.21, a single layer of adsorbed water molecules forms over the internal surfaces of the material. The strength of the polarity of the water molecule ensures that this is a strong bond.

In sub-regime B, the water molecules in the air are attracted to the first layer of adsorbed molecules and form subsequent layers. Depending on the material (e.g., its contact angle and pore sizes), region B may include from one to many additional layers of adsorbed molecules. Since each layer after the first is held less strongly, the slope of the RH versus moisture content curve decreases in this region.

In sub-regime C, the layers grow to such a size (e.g., at least 4 or 5 layers) that they begin to interact and interconnect and significant capillary condensation forms in the mid-sized pores. By applying Kelvin's equation to the pore-size distribution of many materials, a rapidly increasing slope in the sorption isotherm is predicted in this region. For example, from Kelvin's equation, it can be seen that capillary condensation will occur in the smallest pores at mid-range relative humidities (e.g., 35% RH for 1 nanometer pores and 90% RH for 10 nanometer pores). As a result, it can be noted that the greater the proportion of pores smaller than 10 nm, the higher the moisture content in the mid-RH range in the sorption isotherm. Therefore, calcium-silicate brick, which has 10% of its pores less than 10 nm in size, adsorbs significantly more moisture at lower RH's than clay brick, which has almost no pores with a radius less than 10 nm. (Compare Figure 8.9 and Figure 8.22 and note that the total pore volume of concrete is less than half that of clay brick and calcium silicate).

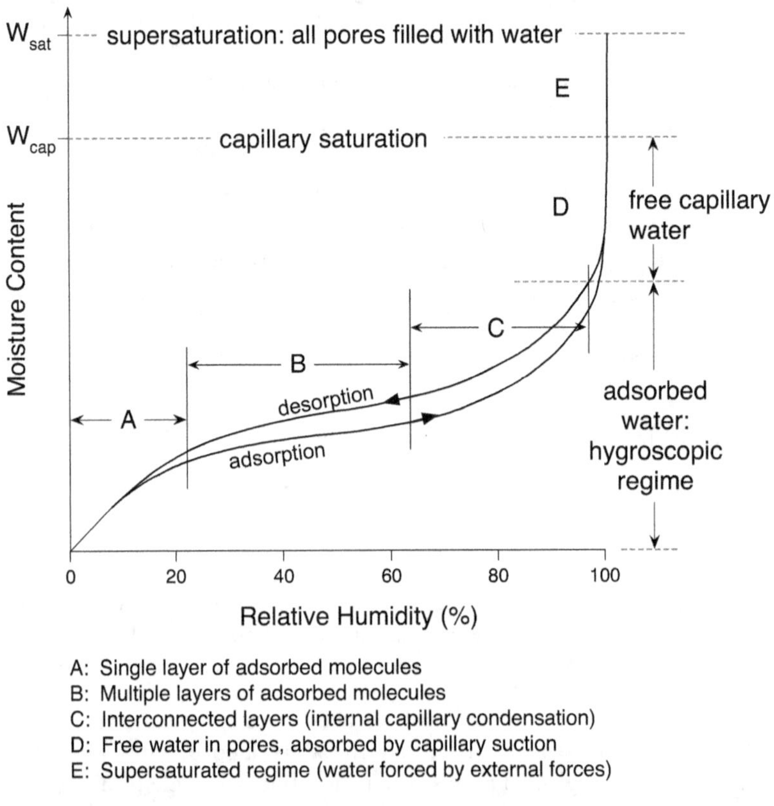

Figure 8.21: Regimes of moisture storage in a hygroscopic porous material

At the highest relative humidities, all but the largest pores are practically filled with water and the capillary regime begins. Kelvin's equation indicates that condensation will occur in pores of 100 nanometer radius at relative humidities less than 98.9% RH. This region is also difficult to measure.

The capillary regime, labeled Region D, is somewhat arbitrarily designated as that part of the moisture storage function above the critical moisture content. Physically, it is presumed that a continuous liquid phase forms within the pore structure. It is characterized by a very steep change in moisture content as the relative humidity changes.

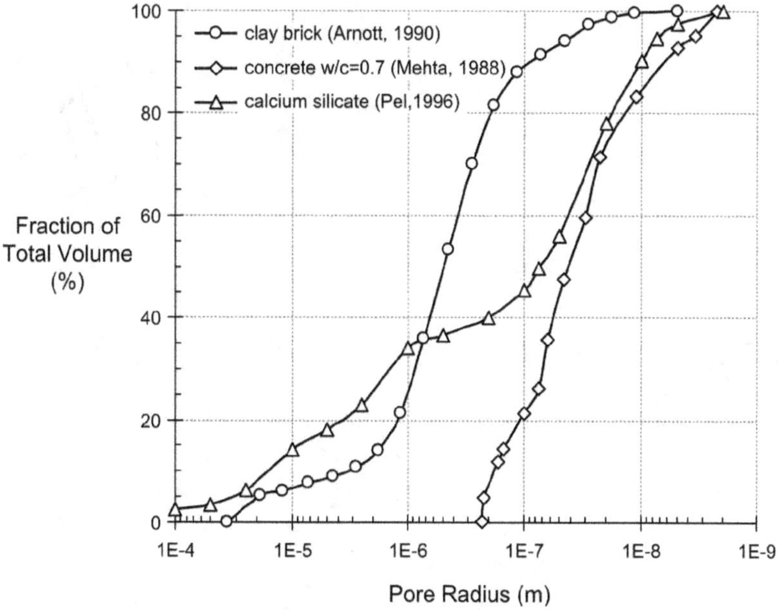

Figure 8.22: Pore size distribution comparison of clay brick, concrete, and calcium silicate

Region E, the supersaturated regime, is characterized by an infinite gradient at 100% RH [Krus and Kuenzel 1996]. It is defined as that region between the free capillary saturation moisture content and the fully saturated moisture content. In this region there are no internal forces driving water into or out of a material (since the moisture content is above the capillary saturation content), and the moisture content is independent of RH. External forces, not capillary suction, must force water into a material above its capillary saturation moisture content.

The sorption isotherms for a number of common building materials are plotted in Figure 8.23 for reference.

Significance of the regions

Some performance thresholds for various building materials and deterioration mechanisms have already been discussed. The description of moisture storage in materials presented above explains the underlying reasons for many of these performance thresholds.

The upper portion of the hygroscopic regime is important because the adsorbed moisture can be both loosely bound and of a sufficient quantity to support moisture-related deterioration mechanisms, e.g., mould and fungal growth and corrosion, etc. This upper region is ideal for those deterioration mechanisms that require oxygen as well as moisture.

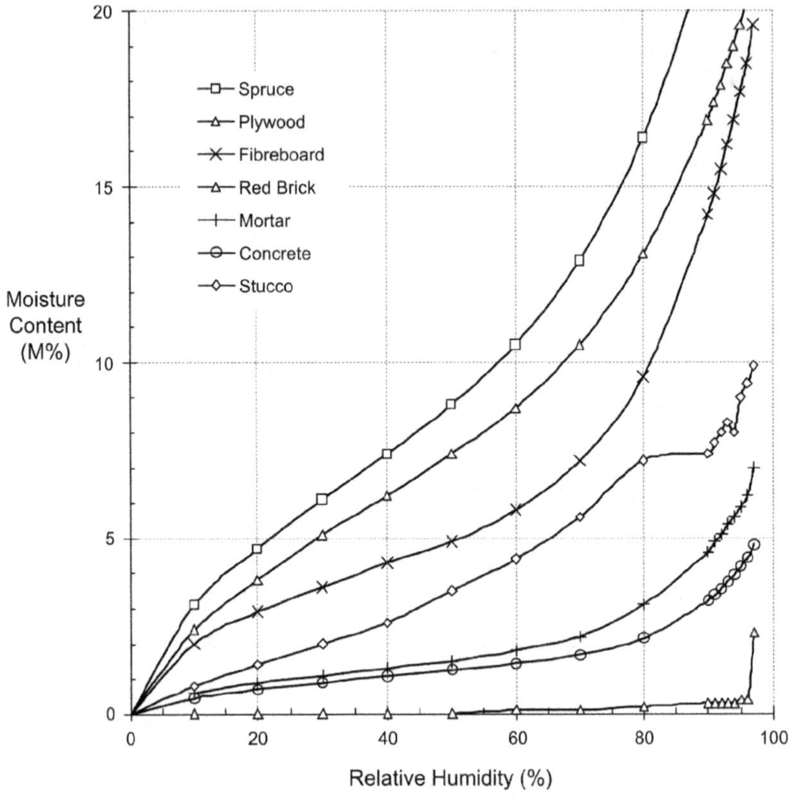

Figure 8.23: Sorption isotherm for several building materials [Kumaran 2002]

Although building materials generally operate in the hygroscopic regime, if wetting occurs the capillary regime (Region D) also becomes important. In fact, with regard to moisture-related damage, the capillary regime is often critical. Free water can freeze and also provides fertile breeding grounds for biological growth and electrolytes for corrosion, although oxygen is often limited if pores are completely filled with water (i.e. at the upper end of Region D).

Cyclical wetting and drying, temperature gradients, and freeze-thaw action all result in complex forces which may fill pores that might not be filled in a simple laboratory test of capillary saturation. This over-saturated regime (Region E) can be explored in a laboratory setting by employing various degrees of vacuum saturation. This region is important for some material deterioration mechanisms. For example, bricks and concrete will often experience freeze-thaw failure after only a few cycles if they become over-saturated. Corrosion and biological growth are usually restricted in this range because the water filled pores restrict the diffusion of oxygen.

Moisture storage in materials also causes small movements that often interfere with the performance of a material's function in an enclosure. If restrained, the movements can generate stresses sufficient to cause material failure. For example, wetting and drying due to adsorption explains why wooden doors tend to be tight in the humid summer and loose-fitting in the dry winter. Drying shrinkage stresses in concrete and the initial expansion of clay bricks can also be explained by the manner in which moisture is stored.

Materials such a wood, mortar, and concrete begin life with all of their pores filled with water. As the material dries, the water vapor pressure in the pores drops and adsorbed water will be tempted to leave the surface of the pores. However, the strength of the bonding between water molecules which comprise the adsorbed layers results in tension forces as drying progresses. It is these internal tension forces that cause drying shrinkage stresses and the consequent strain or cracking in wood, concrete, etc.

Conversely, brick begins life at 1000 °C, completely dry. When water vapor enters the pores, compressive forces are developed as water molecules forces themselves onto the brick material (silica-based) and other adsorbed water molecules. This internal compressive force causes the initial expansion of brick, and the wetting expansion of other porous materials.

8.3.5 Measuring the moisture storage function

The complete moisture storage function can be described in the form of a sorption isotherm, with the fast-changing and difficult to measure region above about 95% RH defined in greater detail by the suction curve. The moisture content vs. suction curve at free capillary saturation (w_{cap}, RH=100%) exhibits an infinite slope to the fully saturated moisture content, w_{sat}.

The most common methods of measuring the moisture storage function in building physics is to divide the moisture storage function into the hygroscopic region, measured by sorption tests, and the capillary suction or retention curve, measured with a pressure plate apparatus. These tests are described below.

The sorption isotherm is measured by conducting a series of gravimetric measurements of a sample after it has reached equilibrium with a specific relative humidity environment. When the series of tests is conducted starting from a completely dry sample to high humidity, the result is an adsorption isotherm. If the series begins with high humidity and proceeds to a desiccant, the result is a desorption isotherm. This test has been standardized for North American purposes in ASTM C1498. It should be noted that oven drying may cause chemical and physical changes to some materials (particularly gypsum)—these materials should be dried at lower temperatures or with a desiccant.

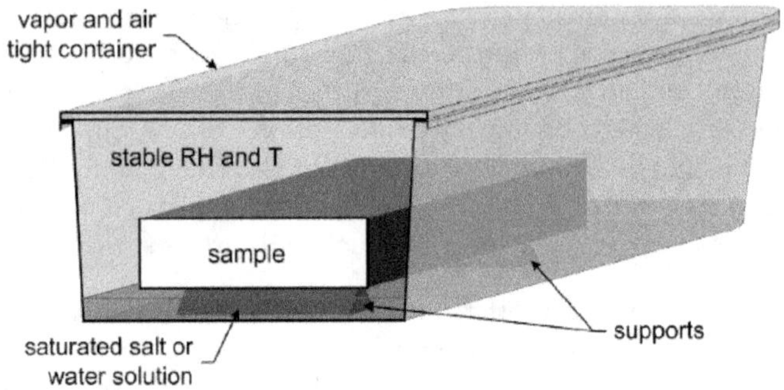

Figure 8.24: Apparatus for conditioning sorption samples

Saturated salt solutions are often used to provide a known RH level in the test chamber. Each saturated salt solution has a different (and lower) equilibrium vapor pressure than over water. ASTM Standard E104 [ASTM 1985] has standardized this method of generating RH. Some example salts and their equilibrium levels are given in Table 8.3.

The steep gradient in the storage function above about 95% RH requires precise control of RH to measure usefully small increases in moisture content. An additional difficulty is that at high relative humidity levels small and unavoidable temperature gradients can cause condensation on sample surfaces (resulting in liquid wicking and an over estimation in adsorption). Hence it is practically impossible to measure this part of the storage function in the same way as for the hygroscopic region, and other experimental means are used such as the pressure plate test.

Table 8.3: Equilibrium RH values for various saturated salt solutions at 25 °C
(temperature insensitive salts over a range of 0 to 30 °C shown in italics)

Salt	RH (%)	Salt	RH (%)
Lithium Chloride	*11.3*	*Sodium Chloride*	*75.3*
Potassium Acetate	22.5	Potassium Chloride	84.3
Magnesium Chloride	32.8	Barium Chloride	90
Potassium Carbonate	43.2	Potassium Nitrate	93.6
Magnesium Nitrate	52.9	Potassium Sulfate	97.3

In the pressure plate test, saturated porous material samples are placed in contact with a special saturated porous ceramic disk within an airtight chamber

[Krus 1996, Janz and Johnnesson 2001]. The pressure in the chamber is increased in steps, driving out some of the water in the sample, which is collected in the bottom of the chamber and measured. It can require more than a week for water within a small, fine-pored sample to reach equilibrium with the imposed air pressure. For pore sizes of one micrometer, 100 nanometers, and 10 nanometers, pressures of about 144 kPa (20.9 psi) 1.44MPa (210 psi) and 14.4 MPa (2100 psi) respectively are required to drive water out of the pores. These pressures equate to relative humidities of 99.9, 98.9, and 89.9% respectively. Using a pressure of 6.9 MPa (1000 psi) allows one to complete the moisture storage function begun with a sorption isotherm up to 95%. The relative accuracy of each of the methods is still a matter of some debate in the research community. Capillary suction curve measurement has been standardized for soil samples (ASTM D2325 and its replacement ASTM D6836) and for building materials (ASTM C1699).

Although the pressure plate apparatus has found wide use, measuring the properties fine-pored materials requires a considerable amount of time (often months). Descamps [1997] demonstrated that mercury porosimetery results (a relatively fast test) and Kelvin's equation can be used to directly calculate the storage function in the capillary region. Krus's [1995] measurements of sandstones showed some significant differences between the pore size distributions predicted by the two methods. The presence of high levels of salts was also shown to change the results of the pressure plate measurements.

8.4 Moisture transport in porous materials

Moisture transport through porous building materials, especially under combined moisture and temperature gradients, is a complex phenomenon. In fact, no widely accepted theory of combined heat and moisture flow exists, although consensus on the majority of issues is developing. The greatest difficulty exists in extrapolating a model based on the microscopic pore structure of a material and the individual transport phenomenon to the total macroscopic moisture transport behavior. The majority of the more recent building physics applications appear to have abandoned this approach in favor of the measurement of aggregate macroscopic material properties that can be used in practical numerical finite volume models.

Moisture, as a liquid, an adsorbate, or a vapor can move through all porous materials. The coupling of heat and moisture transport and the driving potentials are often different for each phase and each transport mechanism. It is generally agreed that liquid water flows into porous materials driven either by capillary suction and/or by external pressures (e.g., a head of water or, water plus an imposed air pressure). Water vapor transport is driven by differences in vapor pressure. Adsorbed water is driven by mass concentration gradients in the adsorbed layer.

The most important moisture transport mechanisms in porous media for each moisture phase (vapor, adsorbate, and liquid) are summarized in Table 8.4 and discussed in detail below.

Table 8.4: Summary of moisture transport mechanisms in porous media

Mechanism	Driving Potential
Diffusion	Vapor concentration
Effusion	Vapor concentration
Surface Diffusion	Adsorbed film thickness (RH)
Capillary transport	Suction pressures
Osmosis	Solute Concentration

8.4.1 Water vapor diffusion

As described in earlier chapters, water vapor moves through the building envelope in two primary modes; diffusion and convection. Diffusion is analogous to heat conduction in that the random motion of molecules results in the movement of vapor along a species (i.e. water vapor or carbon dioxide in building science) mass concentration gradient. Convection is the transport of water vapor by the bulk movement of the air itself.

The general, fundamental diffusion mass flow relationship (which can be applied to much more than just water vapor diffusion through air) is based on a form of **Fick's Law**, first proposed in 1855:

$$\frac{dm}{d\theta} = -D \cdot \nabla C \qquad \text{(Eq. 8.9)}$$

C is the concentration (e.g., mass density) of some value (e.g., vapor pressure),

∇ is the divergent operator, $\nabla = \left(\frac{\partial}{\partial x}, \frac{\partial}{\partial y}, \frac{\partial}{\partial z}\right)$

$dm/d\theta$ is mass flow per unit time, and

D is a proportionality constant defined as the *diffusivity*.

The negative sign is a consequence of the fact that mass flows from high concentrations to low concentrations.

Fick's Law governs the diffusion of vapor through materials (including air) driven by differences in mass fractions between species. If the pores are of the same order or larger than the mean path length of vapor molecules between collisions (with themselves or with a solid material), diffusive vapor flow through a porous material can be accurately described by Fick's Law.

However, if many collisions with the sides of the pore occur, the process is governed by Knudsen diffusion (also called effusion), and will be slower than vapor transport through air (Figure 8.26). Although Fickian diffusion governs most situations in the building industry, Knudsen diffusion is also covered below.

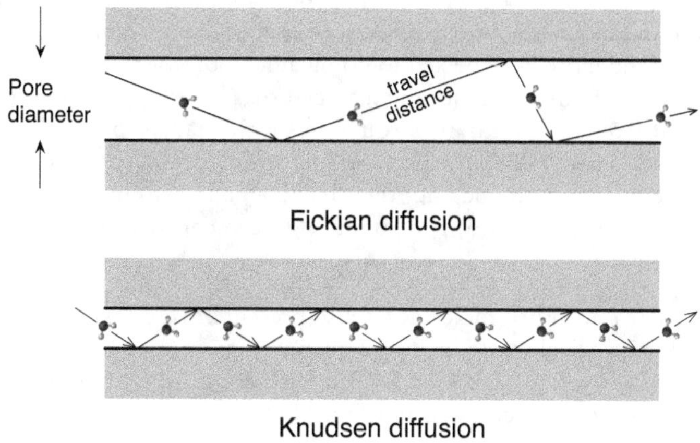

Figure 8.25: Fickian and Knudsen diffusion processes

Knudsen diffusion

The Knudsen number:

$$N_K = \frac{l_m}{2 \cdot r} \qquad \text{(Eq. 8.10)}$$

is the ratio of the mean free path length of a free water vapor molecule between collisions (l_m) to the pore diameter (r). When the pore sizes are large and the vapor pressure high, $N_K << 1$ and hence Fickian diffusion governs. For the case where pores are very small and vapor pressure is low, $N_K >> 1$ and interactions between molecules almost cease and flow is governed by collisions with the pore wall. In these cases Knudsen diffusion, is important. [Welty *et al* 2001, Quenard *et al* 1990].

Knudsen diffusion can be described by:

$$q_{v,K} = D_{v,k} \nabla C_A = 9.7 \cdot r \cdot (T/M_a)^{1/2} \cdot \nabla C_A \qquad \text{(Eq. 8.11)}$$

where, M_A is the molecular mass (g/mole) of the gas

$D_{v,k}$ is the Knudsen diffusion coefficient

∇ is the divergent operator, $\nabla = \left(\frac{\partial}{\partial x}, \frac{\partial}{\partial y}, \frac{\partial}{\partial z}\right)$

C_A is the concentration of the gas,

T is absolute temperature (K), and

r is the pore radius of interest (m).

The mean path length between collisions of a vapor molecule in air can be calculated given the density of vapor molecules (vapor pressure) and the temperature. This path length is less than 100 nanometers for most building science applications (i.e., water vapor in air at room temperatures). Fickian diffusion can therefore be expected to dominate in pores with a radius greater than about 100 nanometers. Since the diameter of a water vapor molecule is of the order of 0.3 nm, vapor transport in pores with a radius of 1 nm (e.g., only 6 molecules across will fill the pore) will be strongly influenced by interference with the pore walls. Some important building materials have a significant volume of pores in the order 1 to 10 nm, e.g., cement paste, calcium-silicate bricks, etc.

If a layer of adsorbed molecules is also present in very small pores, Fickian diffusion is unlikely and Knudsen transport may govern. Knudsen transport will however not be significant at high RH's even if the vapor pressure is low (e.g., low vapor molecule density) since the small pores in which it might be expected to be important will be filled by capillary condensation (which will occur at relatively high RH in pores with radiuses of less than approximately 10 nm.

Although it is often ignored, Knudsen transport would be expected to dominate the vapor transport in pores of less than 10 nm in size at low relative humidities and low vapor pressures. This combination of factors, and the low flow rate, means that Knudsen transport has little practical import in building science.

The relative influence of Fickian and Knudsen transport is unclear in the transition regions but [Welty *et al* 2001] has proposed a combined vapor diffusivity:

$$D_{v,eff} = (1/D_v + 1/D_{v,K})^{-1} \qquad \text{(Eq. 8.12)}$$

where D_v is the Fickian vapor diffusivity.

In most cases, researchers choose to use experimental data which includes both mechanisms rather than using explicit and separate considerations of diffusion and effusion. Separation of the two mechanisms in the study of building enclosures would only be considered because the temperature dependence of the two mechanisms can then be accounted for: Knudsen flow varies as $T^{0.5}$ and Fickian flow varies with $T^{3/2}$. A second reason is the variable effect of RH and vapor pressure on Knudsen diffusion. In most situations, it is acceptable to ignore Knudsen diffusive transport, but this is not necessarily true in all cases.

Fickian diffusion

As stated in Chapter 6, the mass rate of Fickian vapor (q_v) diffusion flow through still (no convection) air can be found from:

$$q_v = -\delta_a \cdot \nabla P_w \quad \text{(Eq. 8.13)}$$

where δ_a is the water vapor permeability of air (kg/(m · Pa· s)),

∇ is the divergent operator, $\nabla = \left(\frac{\partial}{\partial x}, \frac{\partial}{\partial y}, \frac{\partial}{\partial z}\right)$ and

P_w is the water vapor pressure (Pa).

This equation is valid provided there are small or no changes in total pressure and temperature. The water vapor permeability of still air, δ_a, can be found from Shirmer's widely used equation [Shirmer 1938]:

$$\delta_a = 2.0 \times 10^{-7} \cdot T^{0.81} / Pa \quad \text{(Eq. 8.14)}$$

where P_a is the ambient total pressure (Pa), and

T is the absolute temperature (K)

For an idealized porous body, the ratio of stagnant air permeability to the measured vapor permeability of a dry sample of a porous material (δ_p) should give:

$$\delta_p = \psi_0 \cdot \delta_a \quad \text{(Eq. 8.15)}$$

where ψ_0 is the open porosity for vapor flow (m^3/m^3)

Equation 8.15 assumes that no vapor can be transported through the solid fraction of a porous material and that vapor flow through pores is exactly the same as vapor flow through still air. Fickian diffusion through porous materials is, however, not exactly the same as diffusion through air. Experimental values of vapor permeability are lower than that predicted by Equations 8.13 and 8.15. As described earlier, the open porosity theory is only valid for a material with pores large enough that vapor molecule collisions with the pore walls do not retard flow (e.g., pure Fickian diffusion). Real materials have a range of pore sizes and complicated interconnectedness, and therefore the actual vapor flow path is considerably longer than the thickness of the sample.

A tortuosity factor, τ, is defined which accounts for the longer flow path, acknowledges that effective porosity, ψ_{eff}, is less than the open porosity, and provides a better match between theory and experimental results.

$$\delta_p = \psi_0 \cdot \delta_a \cdot \tau \quad \text{(Eq. 8.16)}$$

Pennman originally defined tortuosity as the ratio of the actual vapor molecule path length to specimen thickness [Hartley 1987, Handbook of Drying 1996] and suggested a constant value of about 0.66, although many references appear to use the constant as an empirical fitting coefficient. The tortuosity factor is not likely to be a constant for a given material (if it has any physical meaning at all), since as the moisture content increases, the volume filled with air (and thus available for vapor transport) decreases. (Note: In different literature sources the tortuosity occurs in either the numerator or denominator of Equation 8.15. It is important to be aware of the different definitions.)

Millington and Quirk [1961] recommended a tortuosity equation widely used in soil sciences:

$$\tau(\psi) = \frac{(\psi_0-\psi)^{10/3}}{\psi_0{}^3} \qquad \text{(Eq. 8.17)}$$

which reflects the fact that pores filled with liquid water do not participate in vapor flow. More recent results suggest that the geometrically derived exponent in the numerator of Equation 8.17 may be too high and a value of just over three (e.g., 3.1 to 3.2) maybe more generally applicable [Sallam *et al* 1984]. The fact that open porosity is used in Equation 8.16 instead of effective porosity (which is difficult to measure but more realistic) might be one reason why a lower exponent provides a better match to experimental values

It is expected that large pores will obstruct vapor diffusion less because more area will be available for the water molecules to move freely. In small pores, the addition of several layers of adsorbed moisture will have a proportionately larger influence because of the larger reduction in the free area available for vapor molecules to move. As the RH (moisture content) increases, small pores will fill with water from capillary condensation and will also become effectively smaller because of the increased thickness of the adsorbed layers on the pore walls. This should result in decreasing vapor permeability with increasing RH, i.e., theory would suggest a vapor permeability plot that steadily decreases with increasing moisture content (and RH).

While the concepts of porosity and tortuosity provide a theoretical background for the prediction of vapor permeability through porous materials, this theory has had limited success in the prediction of the vapor permeability of most porous building materials. Vapor permeability test results often have little relationship to porosity and may exhibit a strong moisture content or RH dependency especially for hygroscopic materials (e.g., see [Burch *et al* 1992, Kumaran 1996]). For example, wood-based products exhibit a powerful change in apparent "vapor" permeability with RH. This effect is due to liquid and absorbate transport. Figure 8.17 presents the results of standard vapor permeance tests on Oriented Strand Board (OSB) and plywood. The permeance at 90% RH can be seen to be 30 times greater than at 20%RH. Mechanisms other than vapor diffusion through pores is clearly at work.

Various vapor flow enhancement mechanisms have been suggested to explain the observed higher vapor permeability. The "water island" or condensation-evaporation theory proposed by Philip and DeVries [1957] acknowledges that some capillary condensation has occurred in the smaller pores or in the narrow necks of larger pores, and that this will accelerate the transfer of vapor. This hypothesis suggests that vapor is transported to the meniscus on one side of a pore, where it condenses, while at the same time a water molecule on the meniscus on the other side of the pore would then break free and begin moving along the vapor gradient (Figure 8.28). To reflect the fact that the actual length of vapor diffusion becomes less as more water islands form, this mechanism's effect might be modeled by an enhancement factor that increases with increasing RH. Once a continuous phase of water forms (the critical moisture content), the enhancement factor should drop, until it reaches zero at capillary saturation (when there should be practically no transport by vapor diffusion).

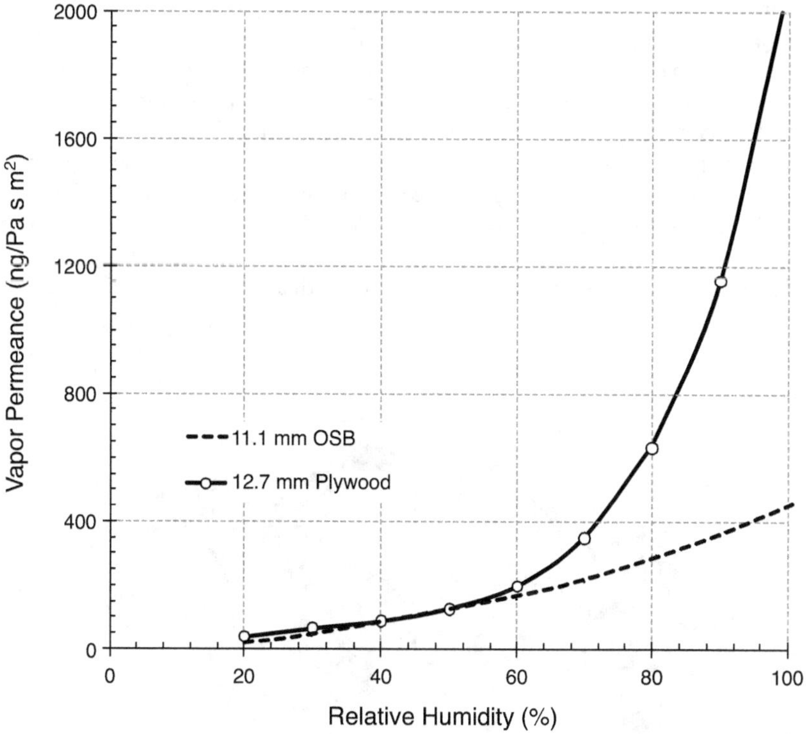

Figure 8.26: Vapor permeability test results for wood-based products as a function of RH [Kumaran *et al* 2002]

A related explanation for increased vapor permeability with increasing RH is capillary transport in small pores (micropores) supplementing the mass flow of vapor. This could be termed a "water canal" effect. This mechanism is merely an extension of the evaporation-condensation theory but with liquid flow over larger distances.

Finally, surface diffusion (discussed in the next section) will always be occurring in those pores not small enough to precipitate condensation, but large enough to contain thick adsorbed layers. Some researchers (e.g., [Kuenzel 1994, Krus 1996, Daian 1990a, Daian 1990b]) maintain that surface diffusion may be sufficient to explain all of the observed increased vapor flow with increasing RH.

It is questionable if any of these enhancement phenomena should be labeled as vapor flow mechanisms (as Philip and de Vries did) or liquid flow mechanisms, since the increased flow is occurring in liquid, vapor and/or the adsorbed states. In reality a combination of series flow (e.g., diffusive vapor transport - condensation - liquid transport - evaporation - diffusive vapor transport) and parallel flow (e.g., surface diffusion in larger pores and capillary flow in smaller pores) is likely occurring. The relative importance of each proposed mechanism will theoretically vary with the pore topology, the pore size distribution, and the actual pore size distributions. In fine- pored materials, such as cement paste, micropore capillary transport will be very important, whereas in materials with mid-sized pores, such as brick, adsorbed flow may be more important. In all cases there is no reason to believe that the three moisture phases are not interacting and transporting moisture over much of the moisture content range. In any case, in materials that exhibit a strong RH dependency, it is clear that mechanisms other than Fickian diffusion are acting.

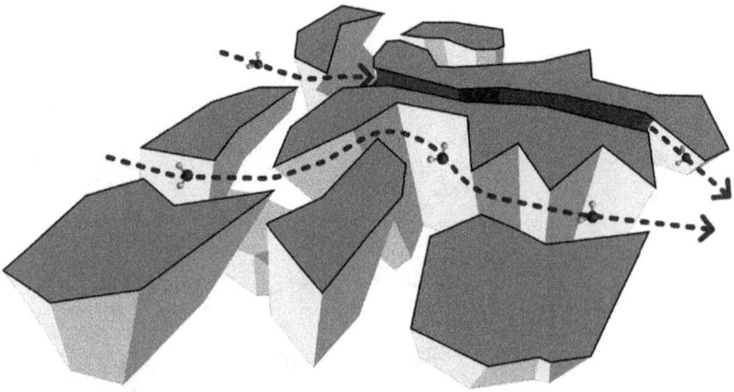

Figure 8.27: Water island / water canal vapor flow enhancement mechanism demonstrating parallel and series moisture flow

Because the prediction of the vapor permeability of building materials has not received the same amount of attention as it has in other materials (e.g., soils, food, etc.) theory is unable to provide even reasonably useful values for vapor permeability. Therefore, standardized tests must be conducted to find empirical material-specific isothermal vapor permeabilities, δ_p.

Because this permeability factor accounts for both Fickian and Knudsen transport, tortuosity, surface diffusion, micropore capillary conduction, etc., the measured permeability is only strictly valid for the conditions of the test.

It is essentially an empirical fitting coefficient. Practical approximate values of vapor permeability have been presented in Chapter 6.

A comprehensive description of vapor flow would have the form:

$$q_v = -D_v(w) \cdot \nabla\rho_v + D_{T,v}(w) \cdot \nabla T \qquad \text{(Eq. 8.18)}$$

$D_v(w)$ is the moisture content dependent water vapor diffusivity (m^2/s),

$D_{T,v}(w)$ is the moisture content dependent thermal vapor diffusivity ($m^2/K \cdot s$),

w is the material moisture content (kg/m^3), and

ρ_v is the water vapor concentration (kg water vapor/m^3).

The second term in Equation 8.18 describes thermal diffusion, or diffusive flow driven along a thermal gradient. This is also called the Soret effect and can be ignored as negligible in building applications since its value is usually several orders of magnitude less than the vapor (or total moisture) diffusivity [Welty *et al* 2001, Kumaran 1996, Kuenzel 1996, Daian 1990a]. Thermal diffusion is not to be confused with surface diffusion and vapor diffusion, which can be greatly enhanced by thermal gradients because of relative humidity changes with temperature.

Measuring vapor permeability

The most common North American vapor permeability test method for building materials is the ASTM E96 standard. Another common standard used to measure the permeance of coatings and films is ASTM D1653[1].

All test methods tightly seal the sample over the mouth of a vapor and airtight container which holds a salt solution or desiccant with a known relative humidity (Figure 8.28). The container and sample are then placed within a larger container that also has a controlled relative humidity. The conditions are maintained for a period of days to weeks (depending on the permeability of the sample and the size of the vapor pressure difference). The rate at which the interior container gains or loses mass is measured at intervals. When the rate of moisture gain has reached a steady-state the slope (mass loss per unit time) is divided by the vapor pressure and the exposed area of the sample (equal to the size of the opening of the container) to calculate vapor permeance.

Several RH and temperature values can be used inside and outside the container. It is important to understand which conditions were used. In all ASTM E96 tests (Figure 8.29), the sample is sealed over a dish which contains either a desiccant (0% RH, and thus termed the dry cup or desiccant method) or water (100% RH and thus the wet cup or water method). The most common

[1] The latter must be used with care as coating and films often interact with their substrate and hence have different characteristics when tested to ASTM D1653 than they do in service.

method is Procedure A, a dry-cup test in which the cabinet is maintained at 23 °C and 50% RH (Figure 8.29). Procedure B is the wet-cup version of Procedure A. Procedure C and D tests are the Procedure A and B tests conducted at 32.2 °C (90°F). Procedure E, is a dry-cup method in which the cabinet is maintained at 37.8 °C (100 °F). Type BW is termed the inverted wet-cup test, because the cup is an inverted version to Procedure B in which the water in the cup contacts the sample. Clearly, the Type BW test is a crude test of a material's liquid moisture transport properties.

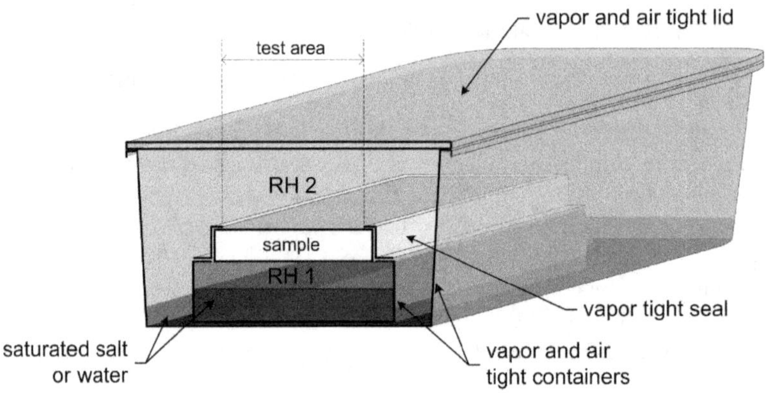

Figure 8.28: Test apparatus for measuring vapor permeance

Figure 8.29: ASTM E96 Wet cup and dry cup vapor permeance tests

A comparison of the wet-cup and dry-cup vapor permeability values from the ASTM E96 test standard is useful for assessing the sensitivity of a material's vapor permeability to changing RH. If the wet-cup and dry-cup values are significantly different, surface diffusion, or some other flow enhancement mechanism, is likely at work. The use of a constant vapor permeability for such materials is unlikely to yield accurate results in analysis.

8.4.2 Adsorbed moisture surface diffusion

As described earlier, the thickness of the adsorbed layer of water molecules as a function of water content can be estimated by the modified BET equation. Adsorbed multi-layers are not as strongly tied to the pore walls as the first mono-layer and, of course, the molecules continue to vibrate (with slightly less energy than vapor molecules at the same temperature). If an area with several adsorbed layers adjoins an area with fewer adsorbed layers, there is a likelihood that the more loosely bound molecules from the thicker layer will jump free and settle on the more attractive site with fewer layers. The mass transfer of molecules by this mechanism is commonly termed surface diffusion, but is sometimes more accurately labeled surface flow. It is a diffusion process since the flow is driven by the mass density gradient of adsorbed water molecules.

Okazaki *et al* [1981, 1983] proposed the random walk process or "hopping model" as described in the previous paragraph. By accounting for the reduced attraction that holds each subsequent layer of adsorbed molecules to the surface they were able to improve the predictive ability of previous similar models by Hill [1956].

Adsorbed moisture will clearly be driven along gradients of mass density. However, the magnitude of the transport function will vary with moisture content. As the RH increases, additional layers of adsorbed molecules will line the capillary walls. Thicker layers of adsorbate are attached less strongly than thin layers and so are more likely to move along the concentration gradient of adsorbed moisture along the pore walls. If a mass density gradient exists, adsorbed flow will increase. Although the mass density gradient of adsorbed moisture is the true fundamental transport potential of adsorbed moisture, it is easy and accurate to use the RH or moisture content as a transport potential since the adsorbed layer thickness can be approximately related to the RH and/or moisture content via the sorption isotherm.

The relative significance of adsorbed surface flow can be judged from the internal surface area of a porous material. For materials with a high proportion of small pores (e.g. cement paste, calcium silicate), adsorption tends to play a very significant role in moisture storage and transport. Confusing the issue is the fact that capillary condensation is also more likely in these materials at mid-range RH, and so coincident micropore liquid transport and "water island" effects are also possible.

As discussed in the previous section, the vapor permeability of many materials exhibits a strong moisture content or RH dependency (Figure 8.26). One of the several plausible explanations for this dependency is the surface diffusion of adsorbed water molecules. Hence, while vapor diffusion continues to move water vapor as the moisture content increases, the less-tightly bound adsorbed molecules also begin to move from thick layers to thin. This process is shown graphically in Figure 8.30.

Surface diffusion would be expected to increase with moisture content. However, a non-linear dependence might be expected for each additional increase in moisture content because each additional layer of adsorbed molecules is held less firmly than the previous. Also, as the RH and the moisture content increase, capillary condensation will occur in the smallest pores. This capillary condensate may aid vapor transport by the water-island effect and can also move by capillary suction driven transport. Surface diffusion very likely no longer occurs during liquid conduction, since no adsorbed film thickness gradients will exist because the pore is filled with water molecules. As the proportion of condensate-filled pores increases, the rate of surface diffusion will continue to increase in the remaining open pores but the proportion of pores in which surface flow is possible will decrease. As the moisture content increases, capillary condensation will fill pores and counter-act the increasing ease of surface diffusion. The transition between adsorbed surface diffusion and liquid capillary transport likely occurs over a large range of RH.

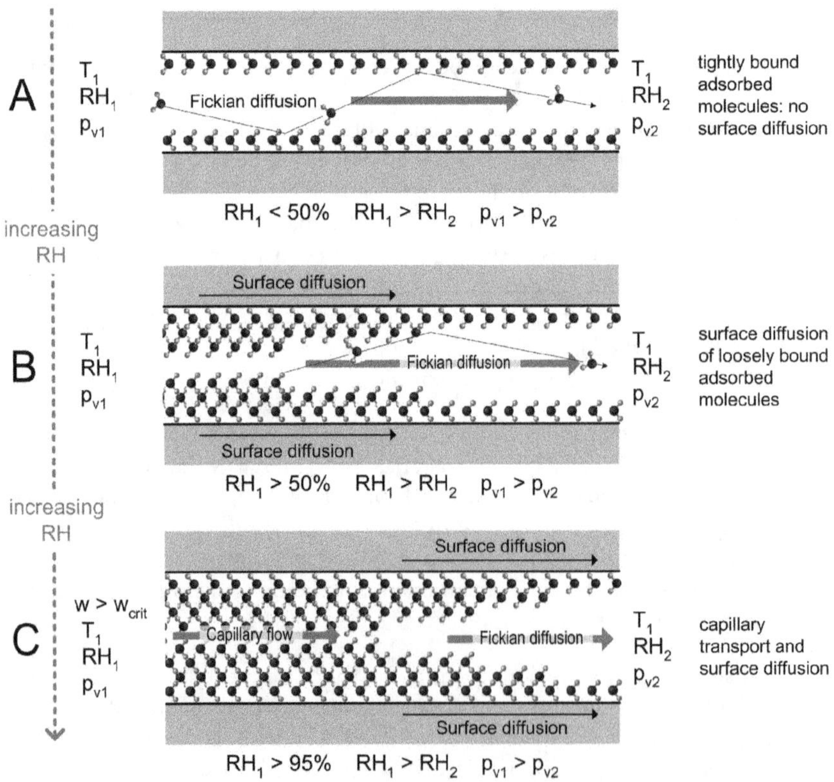

Figure 8.30: Isothermal vapor and surface diffusion as a function of material RH

The behavior of some materials under non-isothermal conditions provides strong evidence for the theory presented in Figure 8.30. Under a temperature gradient, the vapor pressure gradient can be sloping in a different direction than the relative humidity gradient (Figure 8.32). Since vapor pressure drives diffusion and relative humidity indirectly drives surface diffusion, complex moisture transport scenarios are possible. As Kuenzel [1994] pointed out, this is of practical concern because most building enclosures in cold climates during winter separate indoor environments of low RH (30%) and moderate vapor pressure (e.g., 700 Pa) from outdoor environments of high RH (>80%) and low vapor pressure (e.g., < 200 Pa). In this situation surface diffusion is expected to drive flow in the opposite direction as vapor diffusion.

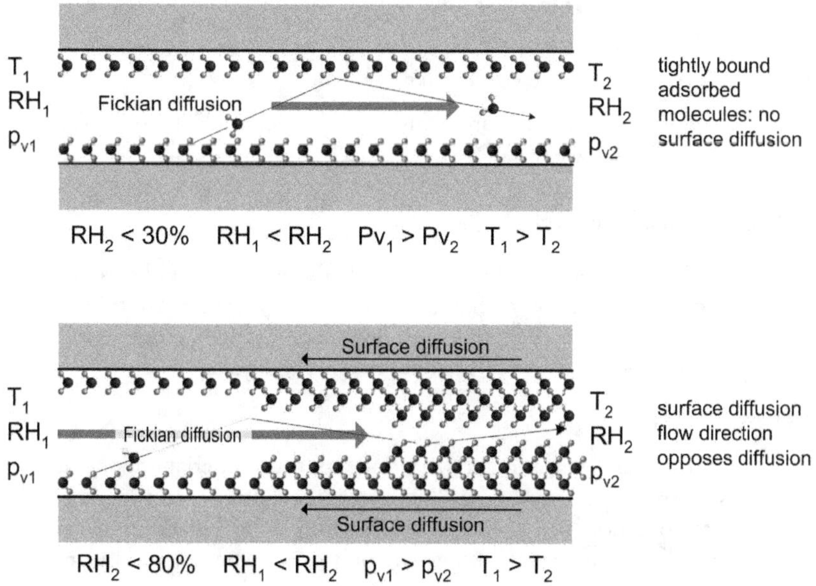

Figure 8.31: Vapor and surface diffusion under a temperature gradient

Under non-isothermal conditions (i.e., most building science situations), serious over- and under-estimations of moisture flow can result from the use of vapor diffusion model to predict surface diffusion. The theory behind Figure 8.32 strongly suggests that surface and vapor diffusion moisture flow should be dealt with as separate mechanisms when predicting moisture flow through porous media with a thermal gradient, especially those materials with a significant proportion of small pores. Moisture transport through materials composed of fine fibers (e.g., paper, cloth, mineral fiber scrims, etc.) may also be strongly influenced by adsorbed surface flow. Confusion is often caused because the prediction of isothermal flow, (that which is measured in laboratories most of the time) will not be affected by lumping surface diffusion and Fickian vapor diffusion into one mechanism. Under isothermal conditions, a relationship with the same form as vapor diffusion transport can provide accurate predictions, because the relative humidity gradient is directly related to the vapor concentration (or pressure) gradient.

It is preferable to include surface flow as part of liquid flow because the driving potentials can both be related to moisture content. In any case, the RH dependent nature of the surface diffusion should be taken into account.

For surface diffusion flow, i.e., transport of moisture in the adsorbed phase, the following relationship is recommended:

$$q_a = - D_a(w) \cdot \nabla w \qquad \text{(Eq. 8.19)}$$

where q_a is the mass flux density of adsorbed moisture flow ($kg/m^2 \cdot s$),

w is the moisture content (kg/m^3), and

$D_a(w)$ is the moisture content dependent adsorbed moisture diffusivity (m^2/s).

8.4.3 Liquid moisture transport

This section will consider liquid transport, although it should be clear by this point that the separation of liquid, adsorbed, and vapor phase transport is not clearly delineated, i.e., thick adsorbed layers and capillary condensate in micropores can both behave in a similar manner to liquid.

Saturated flow

Flow through an oversaturated (Region E) porous material under external forces is termed saturated flow. Since internal capillary forces are in equilibrium in this region, moisture flow will not be driven along a gradient in moisture content. Therefore, saturated flow will only occur when driven by gradients of air pressure or hydraulic head. Although saturated flow is not often of practical importance in building science, an understanding may be necessary to describe gravity drainage from saturated porous materials.

Water flow through a fully saturated porous media is usually modeled using Darcy's empirical law [Hall and Huff 2021]:

$$q_l = k_l \cdot \nabla P \qquad \text{(Eq. 8.20)}$$

where q_l is the mass flux density of liquid water flow ($kg/m^2\ s$),

k_l is the hydraulic conductivity ($kg/m \cdot s \cdot Pa$) and

P is the pressure (or hydraulic potential) (Pa).

The term hydraulic potential (in units of m of head or Pa) is borrowed from soil science. It can include components of gravity head, air pressure, and differential water pressure.

In general, Darcy's law applies to laminar flow (i.e., slow, percolating flow) defined as a Reynolds number (based on the particle diameter) of less than about 10 [Bear 1972]. In very course-grained materials, Darcy's law may not apply for fully saturated flow, and recourse to the more complete Navier-Stokes equation may be required.

The hydraulic conductivity or Darcy coefficient is affected by temperature through the temperature dependency of the viscosity of water. A standard means of correcting for temperature in soil science is [Craig 1987]:

$$k_l(T) = \frac{\rho g}{\mu(T)} \cdot K \qquad \text{(Eq. 8.21)}$$

where K is the intrinsic hydraulic conductivity or permeability coefficient (m^2), a function of the nature of the porous body only

$\mu(T)$ is the temperature dependent dynamic viscosity of water (1000 Pa·s at 20°C),

ρ is the density of water (kg/m^3), and

g is the acceleration due to gravity (m/s^2).

The temperature effect is not insignificant: the hydraulic conductivity at 0°C is only 56% of the conductivity at 20 °C. Below 0 °C, the hydraulic conductivity is typically assumed to be zero.

There have been numerous attempts to predict the hydraulic conductivity from the characteristics of the pore space. For soils, Hazen [Bear 1972] showed that an approximate value for k is given by:

$$k_l = 10^{-8} \cdot D^2_{eff} \qquad \text{(Eq. 8.22)}$$

where D_{eff} is the effective particle size (m).

The value of k_l is in units of m/s for use in Darcy's equation with hydraulic potential expressed in m of head. While predictions can be relatively accurate for materials with unfused, well-graded, spherical particles, (sometimes a practical assumption in soil sciences), the prediction of porous building materials' properties is more difficult. Nevertheless, equation 8.22 is sometimes of practical use. For example, the flow rate through a crushed stone or coarse sand drainage layer under gravity forces can be predicted. Drainage through materials sold as drainage layers can also be checked with the above information.

The Darcy coefficient in Equation 8.19 is reasonably constant so long as the material is fully oversaturated, i.e., so long as all pores remain full. In fact, the Darcy coefficient used in soil science is sensitive to the amount of air in the water: deaired water must be used to ensure that two-phase unsaturated flow permeability is not measured. Depending on the size of the pressure driving flow, small capillaries may not participate in flow if capillary suction is stronger than the applied hydraulic gradient. Water in these small capillaries will be retained by capillarity.

Unsaturated two-phase flow

As noted above, in most building science applications unsaturated flow, i.e., two phase (air and water) moisture flow through an unsaturated porous body is of much more practical interest than saturated flow. Saturated flow is driven primarily by gradients in capillary suction pressures caused by moisture content variations. Relative to capillary pressures, wind-induced air pressures, and hydraulic head are rarely large enough to influence unsaturated flow.

Many authors have attempted to model the flow of liquid water through unsaturated media by assuming the pore space is a bundle of parallel tubes, notably Krischer's analysis of the drying process [Keey 1972, Krischer 1963]. The advantage of the parallel-tube approach is that laminar fluid flow through a single pipe can be predicted using the simple Hagen-Poiseuille equation [Welty *et al* 2001]:

$$Q = \frac{\pi \cdot P \cdot r^4}{8 \cdot \mu \cdot L} \qquad \text{(Eq. 8.23)}$$

where Q is the volumetric flow rate of liquid water (m^3/s),

P is the pressure difference (Pa) along the flow path,

r is the equivalent radius of the pore or crack (m),

μ is the dynamic viscosity of water (1000 Pa· s at 20°C), and

L is the length of the flow path (m).

This equation can be re-arranged in terms of viscous friction-induced pressure. The fluid friction pressure induced by a flow is:

$$P = \frac{8 \cdot Q \cdot \mu \cdot L}{\pi \cdot r^4} \qquad \text{(Eq. 8.24)}$$

An examination of both forms of the Hagen-Poiseuille equation shows that:

1. the resistance to flow decreases as the fourth power of pore radius,
2. the flow quantity increases linearly with pressure (or capillary suction pressure) and,
3. the flow quantity decreases as the inverse of flow path length.

The absorption of water into a porous material is driven by capillary suction and is slowed by viscous drag with the pore walls. Water absorption cannot, however, move through all pores because of trapped air bubbles in dead end pores; the only means for this air to escape is by gas diffusion into the water, a very slow process.

If a dry material is placed in contact with water, the effective Darcy permeability coefficient will change drastically with moisture content as the smaller pores are filled first (capillary suction draws the water into the smaller pores first) at the expense of the larger pores. However, as Equation 8.24 shows, the resistance to flow increase as the fourth power of pore radius. Therefore, as a material dries and the water content of the material decreases, the largest pores no longer participate in the flow (since they empty first) and the resistance to flow increases.

Water flow through a capillary porous body could be estimated by calculating the flow through each small pore or crack using the Hagen-Poiseulle relation and then summing the results. Unfortunately, the pores in real materials have short lengths (perhaps 3 to 100 times the pore radius), are highly interconnected, and are variable in cross-sectional area.

Series models, which string several different pore sizes and lengths together in series are simple extensions of the parallel tube model which attempt to incorporate some of the variations of the pore space. A major flaw with both the series and parallel-tube approach is that these models ignore the interconnectivity of pores in the direction normal to the flow direction.

Various network models have also been employed using electrical analogies, etc. These models attempt to include the effect of the pore space properties perpendicular to the direction of flow. This approach was first used by Fatt [1956], but has been revisited by many other researchers. The fundamental difficulty with these approaches is the proper modeling of the pore space: the distribution of size, length, number, and nature of pores are major variables. Although these variables can be manipulated so that a model predicts the measured permeability of a porous material, there is no guarantee that the chosen mix of variables will also predict the behavior of materials slightly different than the test specimen that was used for calibration.

Percolation theory attempts to more accurately model the behavior of water flow through a porous body. The advantage of percolation theory is that it accepts that the pore properties are randomly distributed throughout the material. Descamps [1997] has provided a recent summary of network theory and advanced the use of percolation theory for the prediction of unsaturated liquid moisture flow.

Because a rigorous and reliable theory has yet to be developed, unsaturated flow is often modeled using a phenomenological approach using a moisture content dependent moisture diffusivity, i.e.:

$$m_l = -D_l\,(w) \cdot \nabla w + D_{T,l}\,(w) \cdot \nabla T \qquad \text{(Eq. 8.25)}$$

where m_l is the liquid moisture mass flux density (kg/m$^2\cdot$ s)

D_l (w) is the moisture content dependent liquid moisture diffusivity (m^2/s)

$D_{T,l}$ (w) is the moisture content dependent thermal liquid diffusivity (m^2/ (K$\cdot$ s)) and w is the volumetric moisture content (kg/m^3)

As for pure Fickian diffusion, the second term (called thermal diffusion or Soret effect) in Equation 8.25 is usually ignored because its effect is one to several orders of magnitude smaller than the isothermal liquid diffusivity [Welty *et al* 2001, Hartley 1987]. The thermal diffusivity should not be confused with the very significant effects of temperature on vapor and adsorbed moisture flow and the somewhat important impacts of temperature on viscosity and surface tension.

Measuring liquid diffusivity

The liquid moisture diffusivity in Equation 8.25 is usually derived from experimentally measured isothermal moisture fluxes.

The advent of specialized experimental equipment, such as Nuclear Magnetic Resonance (NMR), gamma ray scanning, and neutron transmission, allows highly accurate (e.g., 1-2%), high speed (10 to 60 s per reading) non- intrusive measurements of moisture content with a spatial resolution of 1 to 3 mm. Although these measurements are normally taken under isothermal conditions, the results they provide are exceptionally useful. Moisture content profiles measured during wetting, redistribution, and drying can be used to calculate a "best-fit" total moisture diffusivity assuming an equation of the form:

$$q_m = D_m(w) \cdot \nabla w \qquad \text{(Eq. 8.26)}$$

Since the measured coefficient, D_m (w), is an aggregate value which describes the transport of moisture by all mechanisms (e.g. Knudsen and Fickian diffusion, surface diffusion, capillary transport), it is in fact a purely empirical phenomenological factor that is only valid for the conditions of the test. The driving potential shown in Equation 8.26 is the moisture content, but the different driving forces (vapor pressure, RH, and suction) will be represented in the total diffusivity calculated from the measured profiles. In isothermal tests, this is valid since vapor pressure is linearly related to RH, and RH and suction stress are essentially the same measure (via Kelvin's equation and the sorption isotherm).

Keey [1972] was apparently the first to point out what the more detailed measurements of Krus [1995a, 1992], Kuenzel [1994], Winter [1997], and Daian [1990b] have all shown: the wetting (suction), redistribution and drying moisture transport functions are often different, often by a large margin. Adan's measurements of the moisture diffusivity of gypsum plaster [Adan 1996] has shown that the moisture diffusivity of transient RH variations was significantly different (two or three orders of magnitude) than either the wetting or drying values. Typically, the diffusivity during wetting is an order of magnitude greater at low moisture contents than it is for redistribution. Krus [1995a, 1995b] has found that some materials exhibited a far greater difference (more than two orders of magnitude) while other materials showed almost no difference. The behavior of wood is especially difficult to understand. Although Pel [1996] conducted both wetting and drying measurements, he combined the results of the two processes to generate his total diffusivity results and made no comments regarding the need for different diffusivity functions.

Exactly why there is a need for different functions is neither fully understood nor predictable. The different behavior of porous media to capillary suction and redistribution are, however, qualitatively predicted by the percolation theory discussed earlier. Sorption isotherms and capillary pressure curves also both exhibit hysteresis and various theories to explain this hysteresis have also been presented earlier. Kuenzel [1994] provides one possible explanation: wetting by contact with liquid water is expected to occur quickly because the large pores offer less resistance to flow, while redistribution occurs more slowly through the smaller pores which have stronger capillary suction, but more flow resistance. Hence wetting initiates through the large pores, but when the source of water is removed redistribution takes place via the smaller pores at a slower rate.

Several researchers have found that an exponential curve fit matches experimentally derived liquid diffusivity data. A popular form is:

$$D_l(\psi) = D_0 \cdot e^{(\beta \cdot \psi)} \quad \text{(Eq. 8.27)}$$

where D_0 is a constant (m^2/s), and

β is a fitting exponent.

Pel [1996], who used this form, found that the product $\beta \cdot \psi$ was, as a rough approximation, equal to 7.5 for several inorganic porous building materials, and quotes Kalimeris [1984], who found the same. This information may be useful for a simple analysis when no other information is available. However, some materials, like gypsum have been found to violate the exponential form.

Many measurements find that liquid water moving into an initially dry media generally follows a simplified form of Darcy's equation [Hall 1977]:

$$m_w = A \cdot \sqrt{t} \quad \text{(Eq. 8.28)}$$

where m_w is the mass of the water absorbed per unit area (kg/m^2),

A is the water absorption coefficient ($kg/m^2\,s^{0.5}$ or $kg/m^2\,hr^{0.5}$) and
t is time (s or hr)

Equation 8.28 is, in fact, a special case of Equation 8.26 that presumes that the liquid moisture diffusivity changes exponentially with moisture content. The value of A is therefore related to the liquid moisture diffusivity during suction (see also Equation 8.27). Table 8.5 provides the moisture absorption coefficient for several common building materials from a number of different sources.

It is rather simple to measure the water absorption coefficient, A. Samples are placed in contact with water to a depth of 1 to 2 mm (1/16") to ensure a steady supply of water without generating significant hydrostatic pressures (Figure 8.32). Standard tests include ASTM C1794, EN ISO 15148 (the replacement for DIN 52617). The total weight gain (measured weight less dry weight) per unit area is plotted versus the square root of time.

Table 8.5: Water absorption coefficients for various materials

Compiled from various sources, including [Kumaran 1996, Pel 1996, Kuenzel 1994, Lohmeyer 1996, Kuenzel 1993, Kuenzel 1986, Kumaran et al 2002, Schumacher et al 2020]

Material	**A ($kg/m^2/s^{0.5}$)**
Building Paper	0.0005 - 0.001
Water-repellent stucco (per DIN 18550)	< 0.0083
Good quality concrete	< 0.02
Cement stucco	0.03 – 0.06
Face brick	0.005 – 1.0
Cement mortar	0.04 – 0.80
Fiber cement sheet	0.005 – 0.025
Plywood: face	0.001 – 0.006
Plywood: along plies	0.10
Pine: transverse	0.004
Pine: longitudinal	0.016

The resulting plot (Figure 8.33), usually exhibits a straight line, and the water absorption coefficient (or "A-value") is defined as the slope of this line in units of $kg/(m^2\,hr^{1/2})$ or $kg/(m^2\,s^{1/2})$. If the line exhibits an initial slope that is different from the final slope, the initial straight portion is used. If there is no significant change in slope, the total water uptake at 24 hours is used to define the water absorption coefficient, A.

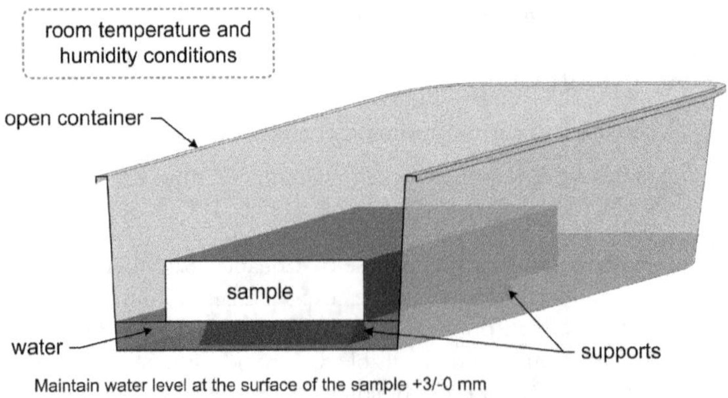

Figure 8.32: Test set-up for water uptake test

Kuenzel [1994] and Krus [1996] have found that the absorption coefficient can be used to calculate an approximate liquid diffusivity for many inorganic porous and wood-based materials. This approximation takes the form:

$$D_l(w)= 3.8\cdot (A / w_{cap}) \cdot 1000(w/w_{cap})^{-1} \qquad \text{(Eq. 8.29)}$$

where A is the absorption coefficient ($kg/s^{0.5}/m^2$)

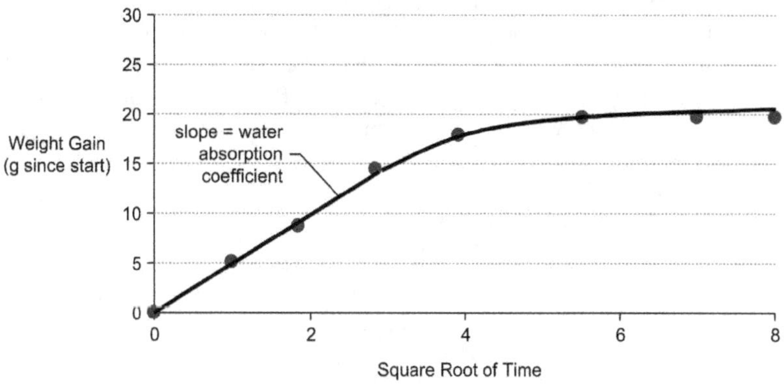

Figure 8.33: Typical test data from water uptake test

The use of the absorption coefficient, A, and the saturation moisture content, w_{cap}, allows simple lab tests to be used to generate liquid moisture diffusivity. Kuenzel also suggests that the wetting diffusivity be reduced by an order of magnitude for redistribution or drying if mineral-based (i.e., not wood) materials are involved.

8.4.4 The moisture transport function

Equations 8.18, 8.19, and 8.25 provide general transport functions for unsaturated flow of each water phases in a porous material: vapor, adsorbed, and liquid. (Saturated flow is not generally important in building enclosures). The rate of moisture flow for each phase is governed by material- specific diffusivities. It will be noted that the driving potential for vapor flow and adsorbed and capillary transport are different. Vapor pressure gradients drive vapor diffusion, relative humidity gradients drive surface diffusion, and suction pressure gradients drive liquid flow. The temperature dependencies of each of the diffusivities also vary at a different rate. For all of these reasons, the development of a general moisture transport function for a given material under gradients of moisture content, temperature and vapor pressure is very difficult.

Figure 8.34 shows the magnitude and variation of the various isothermal moisture diffusivities as a function of moisture content for a hypothetical hygroscopic porous material at a certain temperature. At the lowest moisture contents, Knudsen diffusion controls vapor flow and tightly bound adsorbed molecules cannot easily move. As the moisture content increases, Fickian vapor diffusion steadily decreases as the smallest pores fill with water. At moisture levels over about 30% RH, surface diffusion of adsorbed moisture begins, and becomes important in the 50 to 90% RH range. Surface diffusion may continue up to full saturation in unfilled pores, but will slow dramatically as the moisture content exceeds the hygroscopic region. Liquid transport of condensed water may begin in the smallest pores at moisture contents as low as 30%, but only becomes significant as the RH exceeds 90% for most materials. At around 95%, liquid transport becomes the dominant mechanism and increases as capillary saturation is approached. By necessity, the liquid moisture diffusivity becomes equal to the hydraulic conductivity when the moisture content reaches the saturation level (e.g. when $\psi = \psi_{sat}$).

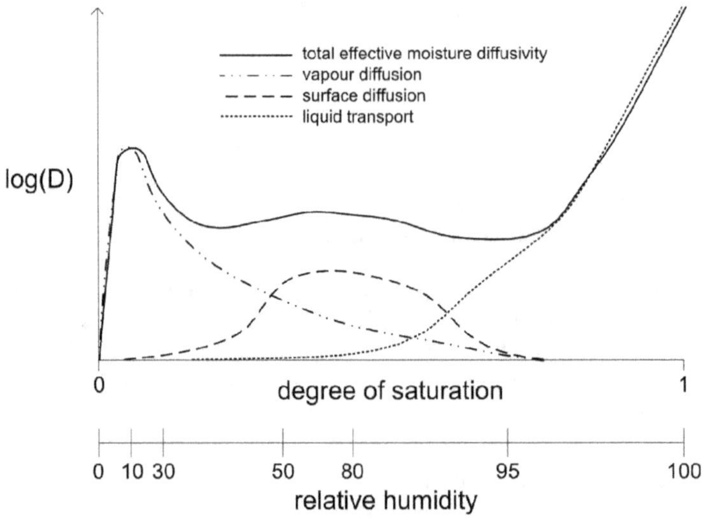

Figure 8.34: Hypothetical total isothermal moisture transport function
Note: The saturation and RH scales are not linearly related because the moisture storage function is not a linear function.

The above is a simplified description for a hypothetical material. Depending on the material characteristics (primarily the topology of the pore space) each of the three transport functions is expected to have different shapes and magnitudes.

The transport functions for each of the moisture phases are shown for isothermal conditions. Some temperature dependencies of the transport properties is likely, although based on most available evidence, thermal moisture diffusivity is negligible and can be ignored. A total moisture diffusivity might be chosen as:

$$D_m(w) = D_v(w,T) + D_a(w,T) + D_l(w,T) \qquad \text{(Eq. 8.30)}$$

The total moisture transport function might have the form:

$$m_m(m,T) = -D_v(w,T) \cdot \nabla \rho_v - (D_a(w,T) + D_l(w,T)) \cdot \nabla w \qquad \text{(Eq. 8.31)}$$

If Equation 8.31 is used in a finite-volume computer program, the important temperature and moisture content dependencies are already incorporated in the choice of driving potentials and diffusivity functions. Some models use the moisture content (in various forms, e.g., mass or volumetric measures) directly or employ the RH and suction isotherm to couple moisture content with vapor pressure and capillary suction.

Supported by experimental evidence of simple well-characterized materials, Imakoma *et al* [1992] have shown that the separation of the three moisture transport functions is viable. In fact, using the theory presented above (with detailed quantification of the constants), they have predicted both the moisture storage and moisture transport function from pore size distribution information. Their model accurately predicted experimental results of the drying process. Although much more work is necessary, this is a promising research direction.

8.4.5 Moisture transport across interfaces

Mass transfer (wetting, capillary transport, evaporative drying and diffusion) at exterior surfaces and inter-layer interfaces is another complex aspect of moisture transport through building enclosures. Although the moisture transport function can be used across interfaces, this approach is in fact an approximation. Surface resistance to vapor flow and interface or contact resistance to capillary and adsorbed moisture transport exist and may be significant in some cases. Moisture transport across interfaces is a relatively unexplored, yet important area of moisture transport research.

If two material layers are in direct physical contact on a pore-level scale, the normal liquid transport function is theoretically valid. However, since moisture transport primarily occurs through very small capillaries and pores, direct contact is exceptionally unlikely to occur in real building assemblies. Where such contact is desired, special material layers, like bentonite (a very fine particle clay), might be used as they are in pressure plate apparatus tests to ensure capillary contact.

When materials are not in direct contact, only vapor flow is possible. However, vapor flow can also be interrupted at interfaces with some incomplete contact. Small interruptions in material contact or changes in thermal resistance may also precipitate condensation at interfaces because of small temperature changes across the interface. Many building materials also have higher vapor resistances at their exterior surfaces because of their manufacturing process (e.g., extruded polystyrene exhibits a surface skin, and the outer layers of concrete, plaster and extruded bricks are all typically more dense and less vapor and liquid permeable than the body of the material).

Because dissolved salts can only be transported in liquid form, not vapor, understanding liquid transport across interfaces is very important for the prevention and understanding of expansive salt recrystallization damage (subflorescence).

Liquid water transport across interfaces

Holm *et al* [1996] found that the nature of the bond between plaster and brick results in a significant "contact resistance" (to borrow heat flow terminology) that could result in significant errors in modeling moisture behavior. Krus [1996a] showed that one can model the capillary transport between two material layers (natural stones) if hydraulic contact was ensured through the use of kaolin (a clay with very fine pores and good capillary conductivity). In related research, Groot [1990] and Brocken *et al* [1997] have also found that there are hydraulic discontinuities in the mortar to brick interface.

Bomberg [1974] investigated the importance of very small cracks and separations in his work on moisture transport. He showed that for some materials simple contact ensured dramatically faster moisture transport than across a 1 mm space, while for other materials the 1 mm space affected moisture transport very little.

Clearly, the importance of interface transport depends on the relative magnitude of the test materials' vapor and liquid diffusivities. If two highly capillary active but reasonably vapor impermeable materials (e.g., brick and cement plaster) are involved, the nature of the interface is critical to moisture transport. Results from the research referenced above suggest that liquid transport will be an order of magnitude less at the almost invisible masonry-mortar interfaces. For materials that are capillary inactive or hydrophobic but vapor permeable (e.g., mineral wool insulation and acrylic stucco), even a gap visible to the naked eye will not significantly reduce the moisture flux between layers because vapor diffusion governs the flow of moisture, and no additional contact or interface resistance exists.

Water vapor transport across interfaces

Vapor diffusion transport within enclosures is rarely influenced by imperfect inter-layer contact since still air is far more vapor permeable than most building materials. However, vapor flow at exterior surfaces is affected by equivalent surface resistances and a combination of convective and diffusive transport effects. Moisture transport from a free surface by pure vapor diffusion is unlikely to occur in most situations since some airflow is always present, either because of driving forces from outside the system or because evaporation induces temperature depressions and vapor gradients that result in airflow.

Water vapor transport from surfaces

Moisture within a porous material or solid assembly must eventually leave, and can only do so by drainage in the oversaturated region and by diffusion, convection, and evaporation in the hygroscopic region. Drainage is well understood and rarely plays an important role in drying materials (although it can be very important for the removal of water from cracks and voids large enough that capillary suction does not play a role (in the order of 1 mm.) Drying of porous materials, such as rain-wetted cladding and sheathing wetted by condensation, is an important consideration for the good performance of most building enclosures. In these cases the material is drying down from the over hygroscopic region to the hygroscopic region. Both diffusion and unsaturated flow are occurring. An understanding of drying behavior under these conditions requires information about mass (specifically water vapor) transfer coefficients.

Lacy [1977] measured evaporation rates from brickwork in the field and reported maximum rates of 68 $g/m^2/hr$ for a few hours, 20 $g/m^2/hr$ for a few days, and 1- 7 $g/m^2/hr$ for most of the time (when the walls were likely not saturated). Schwarz [1978] measured rates of between 50 and 400 $g/m^2/hr$ (depending on the influence of solar heating and wind speed) from an artificially saturated surface. Because of the nature of evaporative drying described in the previous section, drying rates of fully saturated materials are likely to be similar for wood, concrete, and stone claddings. Unfortunately, significant evaporative drying can only occur from the outside face and the inner side of the cladding will remain wet for a relatively longer time unless a well ventilated airspace is provided behind the cladding. Theoretical calculations [Straube and Burnett 1995] and laboratory experiments [Hall 1984] suggest that rates as high as 200 to 400 $g/m^2/hr$ are possible.

One of the reasons that the flow of air across the surface of a material can accelerate drying is the reduction or elimination of the resistance of stagnant air to diffusion. Another reason is the delivery of the energy required to evaporate the moisture. Exchanging the air adjoining a wet material with dry air ensures that the vapor pressure difference between the air and the material surface (and hence the drying rate) will remain the same. Perfect mixing is a general assumption in building sciences, but the validity of this assumption should be investigated since it actually depends on vapor pressure difference, flow rate, and flow path.

The combined effect of diffusion and convection in the boundary layer next to a surface is normally accounted for in the same way as for heat flow at surfaces: by the use of a fictitious equivalent coefficient. Just as for convective heat transfer, one can write a linear equation which relates the mass flux at a surface (either the exterior surface or surfaces within a building cavity) to a difference in driving potentials, in this case the vapor pressures:

$$q_v = h_w \cdot (P_{ws} - P_\infty) \qquad \text{(Eq. 8.32)}$$

where q_v is the rate of water vapor convection (g/s),

h_W is the convective water vapor transfer coefficient ($g/m^2 \cdot Pa \cdot s$),

P_{WS} is the water vapor pressure at the surface (Pa), and

P_∞ is the vapor pressure far from the surface (Pa).

The magnitude of this surface film coefficient is not well known for the pure diffusive case. In laboratory vapor permeability tests, Burch *et al* [1989] found a water vapor transfer coefficient of 2.87 x 10^{-6} $g/(Pa \cdot m^2 \cdot s)$. This value was arrived at by using a special experimental technique to eliminate all convective effects.

In real building science situations thermal- and moisture-induced convective airflows over surfaces are inevitable. Therefore, most coefficients measured in the field are an order of magnitude larger than the purely diffusive case [Worch 2004]. For example, the mixed diffusive-convective mass transfer coefficient has been reported as ranging from 13.4x10^{-6} $g/(Pa \cdot m^2 \cdot s)$ for saturated earth in a subfloor crawl space, 25 x10^{-6} $g/(Pa \cdot m^2 \cdot s)$ for swimming pools during calms [Cuningham 1984] and between 17 and 102 x10^{-6} $g/(Pa \cdot m^2 \cdot s)$ for saturated brick wall samples exposed to wind speeds of 1 to 8 m/s [Schwarz 1978]. Kuenzel [1994] has found that an exterior surface mass transfer coefficient of 75 x 10^{-6} $g/(Pa \cdot m^2 \cdot s)$ allowed for good computer model approximations of measured results from wall drying field exposure experiments.

Convective mass transfer is analogous to convective heat transfer for the same geometry and flow conditions. The accelerated drying caused by air- flowing over a surface can be calculated for many conditions by the use of the very powerful Lewis correlation between heat and mass transfer [Luikov 1965, Kays and Crawford 1980, Incropera and DeWitt 2002]. This correlation allows any convective-conductive heat transfer coefficient, h_c, to be converted to a mass transfer coefficient for the same geometry and conditions:

$$h_W = \frac{h_c}{R_{wv} \cdot T \cdot \rho \cdot c_o} \quad \text{(Eq. 8.33)}$$

where R_{wv} is the gas constant for water vapor under ($J/g \cdot K$),

T is the absolute temperature (K),

ρ is the air density (g/m^3), and

c_o is the specific heat capacity of air ($J/g \cdot K$).

This relationship is very useful because h_c has been defined theoretically and empirically for a wide variety of flow conditions and geometries, whereas the mass transfer coefficient is known for only a few cases.

Substituting typical values for air and water vapor into the Lewis correlation, the mass transfer coefficient for water vapor, h_w, can be found at room temperatures from:

$$h_w = \frac{h_c}{R_{wv} \cdot T \cdot \rho \cdot c_o} = \frac{h_c}{0.4615 \cdot 293 \cdot 1200 \cdot 1.005} \qquad \text{(Eq. 8.34)}$$

$= 6.1 \times 10^{-6}\, h_c$

where h_w is in units of $g/m^2 \cdot s \cdot Pa$ and h_c is in $W/m^2 \cdot K$

For exterior and interior surfaces under average conditions, typical average heat transfer coefficients are 20 and 8 $W/m^2 \cdot K$ respectively, of which the combined effect of convection and conduction comprises about 8-12 and 2-3 $W/m^2 \cdot K$ respectively (see Chapter 4).

By substituting the convective heat transfer coefficients into the Lewis correlation it can be seen that the water vapor transfer coefficient will typically take on values ranging from 12-18 and 50-75 x 10^{-6} $g/(Pa \cdot m^2 s)$ respectively. Higher values near corners and projections and rougher surfaces can be expected and lower values are likely to occur in small pockets behind plantings, inside of architectural reveals, etc.

As a surface dries the latent energy required for the evaporation of liquid water will depress the temperature at the surface and hence reduce the vapor pressure, resulting in an apparent mass transfer coefficient that is somewhat lower. Therefore, to predict surface drying, either an accurate assessment of surface temperatures which includes latent heat effects must be conducted or a lower equivalent mass transfer coefficient should be used.

The vapor pressure at a wet surface is the saturation vapor pressure for that temperature (since the relative humidity is 100% for the over-hygroscopic range). The use of the saturation vapor pressure at the cladding temperature is a reasonably accurate assumption for moisture contents considerably less than capillary saturation because the shape of the sorption isotherm for most cladding materials is steep in the 90% to 100% RH range (consider the isotherms shown earlier for wood, concrete and clay brick). Wood siding is more hygroscopic than brick, concrete or stucco, and therefore this assumption is not valid over as large a range.

Example: Consider a rain saturated masonry wall at 10 °C and outdoor air at 10 °C and 80% RH (860 Pa). Calculate the drying rate.

Solution

Assuming the masonry is at 100% RH (reasonable since it is saturated), the vapor pressure of the wall and the air (P_v= 1230 Pa and 985 Pa respectively) can be found from psychrometrics. Using 75 x 10^{-6} $g/Pa \cdot m^2 \cdot s$ as the mass transfer coefficient, the rate of surface drying rate can be calculated as:

$q_v = h_m \times A \times \Delta P_w \times \Delta t$

$= 75 \times 10^{-6} \times 1\ m^2 \times (1230\text{-}985)\ Pa \times 3600\ s/hr$

$= 66\ g/(m^2 \cdot hr)$

Which is close to both Lacy's and Schwarz's field measurements of the drying rate for saturated brickwork.

Under real conditions the latent energy of evaporation is required to maintain the temperature of the evaporating surface above the ambient temperature. Given that the latent heat of evaporation of water is 2300 J/g, a total of 2300 times 66 g = 151,800 J would be required to maintain the surface temperature of the masonry in the previous example. An absorbed solar energy intensity of only 151,800 /3600 = 42 W/m^2 would be sufficient to provide the latent heat of evaporation. As discussed in Chapter 3, normal daytime solar exposure is in the order of several hundred W/m^2.

The surface of cladding, especially brickwork, will on average be several degrees above the ambient air temperature. Solar radiation typically increases the temperature of claddings from 20 to 40 °C above ambient for several hours on all sunny days. While this solar heating only occurs for orientations exposed to the sun and for a short time, the effect on drying can be very significant because of the very significant increase in surface vapor pressures and energy delivered to evaporate the moisture.

Consider the effect on the drying rate of the previous example if a cladding temperature of 40 °C (30 °C above exterior air) were assumed. The saturation vapor pressure at 40 °C is about 7400 Pa. Substitution shows that this would increase the drying rate by a factor of more than 10, to 1730 $g/m^2 \cdot hr$! The moisture within the masonry is unlikely to be transported to the surface at this rate (the drying front would recede below the surface), and 1100 W/m^2 of energy would need to be provided (more than the sun can provide). For a given scenario one can iterate toward a solution that balances evaporation rate, temperature and latent energy requirements.

The inability of capillary flow to transport liquid water to the surface fast enough will almost immediately reduce the drying rate to below that predicted above. The result is that the outside few mm of the surface will dry very quickly, but depending on the vapor permeance and the speed of capillary redistribution within the material, the material will remain almost saturated some distance inside.

The rate of drying of all but the surface of the cladding is therefore likely to be governed by the ability of the material to transport sufficient water (by capillary conduction) to the surface for evaporation, or the vapor resistance of the material and/or coatings, not by the effective mass transfer coefficient at the surface.

8.4.6 Convective water vapor transport

As described in Chapter 7, the convective flow of air can transport significant quantities of water vapor. In most practical applications, the airflow is driven through openings and cracks, not porous materials. However, there are some very important applications that involve the flow of air through porous building materials. The most important of these is forced and natural convection through low-density, highly porous insulations described in Chapter 7.

The flow of air through porous materials is governed by the Navier-Stokes equation. For the laminar flow regime encountered in most porous building materials, Darcy's simplified equation is usually an acceptable approximation:

$$m_a = k_a \cdot \nabla P \qquad \text{(Eq. 8.35)}$$

where m_a is the mass flow rate of air (kg/s),

k_a is the air permeability (kg/m· Pa· s), and

P is the total air pressure (Pa).

The air permeability of most building materials is less than 10^{-8} kg/m· Pa· s [Kumaran 1996, Kronvall 1980, Bumbaru *et al* 1988], which means that airflow through even small cracks, i.e., of the order of 1 mm, are much more significant to airflow.

As shown in Chapter 6, it is simple to describe the convective water vapor flow rate if the mass flow rate of air is known:

$$m_{v,conv} = m_a \cdot W_a \qquad \text{(Eq. 8.36)}$$

where W_a is the humidity ratio of the air (kg/kg) and

m_a is the mass flow rate of dry air (kg/s).

Very detailed modeling of the physical system, material properties, and fluid dynamics is required to accurately and usefully assess convective vapor (and heat) transfer. The precise geometry of the flow path is very difficult to define for many building science problems. The extreme sensitivity of the calculated airflow (and thus moisture and heat flow) to the assumptions of how the building system is modeled (i.e., the size and shape of every crack and wrinkle) severely limits the accuracy, and even usefulness, of modeling airflow. In many cases, detail computational fluid dynamics (CFD) approaches are limited to computerized research models and generate relative, not absolute results. Modeling the flows as if it were through a series of cracks, as Kronvall [1980] has done in detail may be just as useful as more complex approaches.

8.5 Conclusions

This chapter has presented a relatively detailed review of moisture, moisture storage, and moisture transport. There is clearly limited merit (except, perhaps, for research) in attempting to model all moisture wetting, drying, and storage in the detail presented here because:

1. The effort required is considerable and most projects/problems do not allow or justify the expenditure of the necessary resources,
2. Knowledge of the actual geometry of the enclosure is typically incomplete, i.e., the location of every crack, twist and opening is rarely known,
3. The material data necessary for detailed computations are rarely available or sufficiently accurate, and
4. The boundary conditions (i.e., interior and exterior environmental conditions) are rarely available with sufficient detail and accuracy (e.g., driving rain deposition spatial distribution).

Simple methods of comparison, or safe methods of prediction are typically much more useful than detailed methods for most designers and analysts. However, there are certain problems that require detailed analysis using some of the information covered here. Nevertheless, a solid understanding of the moisture physics underlying the phenomenon observed in the lab and field, and the complexities and variables involved in computer models is important, perhaps critical for good building science research and practice.

8.6 References

Adan, O.C.G., 1996. *On the Fungal Defacement of Interior Finishes.* Doctoral Dissertation, Eindhoven University of Technology.

Andersen, N.E., 1987. "Summer Condensation in an Unheated Building," *Proc. of Symposium and Day of Building Physics.* Lund University, Swedish Council for Building Research, August 24-27, pp. 164-166.

Arnott, M.R., Litvan, G.G., 1988. "Quality-Control Test for Clay Brick Based on Air Permeability," *Amer. Ceramic Soc. Bull.*, Vol 67, No. 8, Aug, pp. 1412-1417.

Arnott, M.R., 1990. *Investigation of Freeze-Thaw Durability.* CBAC/IRC Industrial Research Fellowship Final Report, National Research Council of Canada, Ottawa, Dec..

Ashton, H.E., 1970. Canadian Building Digest 124 *Biological Attack on Organic Materials.* NRCC, Ottawa, April.

ASTM C1498. *Standard Test Method for Hygroscopic Sorption Isotherms of Building Materials.* American Society of Testing and Materials, West Conshohocken, Pennsylvania.

ASTM C1699. *Standard Test Method for Moisture Retention Curves of Porous Building Materials Using Pressure Plates.* American Society of Testing and Materials, West Conshohocken, Pennsylvania.

ASTM C1794 -19 *Standard Test Methods for Determination of the Water Absorption Coefficient by Partial Immersion.* American Society of Testing and Materials, West Conshohocken,

Pennsylvania, 2019.

ASTM D1653 (Reapproved 2021) *Standard Test Methods for Water Vapor Transmission of Organic Coating Films.* American Society of Testing and Materials, West Conshohocken, Pennsylvania.

ASTM D2325-68 (withdrawn 2007) *Standard Test Method for Capillary-Moisture Relationships for Coarse- and Medium-Textured Soils by the Porous-Plate Apparatus.* American Society of Testing and Materials. West Conshohocken, Pennsylvania.

ASTM D3152-72 (withdrawn 2007) *Standard Test Method for Capillary- Moisture Relationships for Fine-Textured Soils by the Pressure-Membrane Apparatus.* American Society of Testing and Materials, West Conshohocken, Pennsylvania.

ASTM D6836-16 *Standard Test Methods for Determination of the Soil Water Characteristic Curve for Desorption Using Hanging Column, Pressure Extractor, Chilled Mirror Hygrometer, or Centrifuge.* American Society of Testing and Materials, West Conshohocken, Pennsylvania.

ASTM E96. *Standard Test Methods for Water Vapor Transmission of Materials.* American Society of Testing and Materials. West Conshohocken, Pennsylvania.

Baker, M.C., 1969. Canadian Building Digest 111 *The Decay of Wood.* NRCC, Ottawa, March.

Bao, Jiuwen, & Licheng Wang. 2017. "Capillary imbibition of water in discrete planar cracks." *Construction and Building Materials* 146, pp. 381-392.

Beaglehole, D. and Christenson, H.K., "Vapor Adsorption on Mica and Silicon: Entropy Effects, Layering, and Surface Forces," *J. Phys. Chem.* Vol 96, p 3395, 1992.

Bear, J. *Dynamics of Fluids in Porous Media.* Dover Publications, New York, 1972.

Bomberg, M., & Shirtcliffe, C.J., "Influence of moisture and moisture gradients on heat transfer through porous building materials," *Thermal Transmission Measurements of Insulation,* ASTM STP 660, ed. R.P. Tyre, ASTM, Philadelphia, 1978, pp. 211-233.

Bomberg. M., *Moisture Flow Through Porous Building Materials.* Report 52, Lund Institute of Technology, Lund Sweden, 1974.

Brocken, H.J.P., Adan, O.C.G., Pel, L., "Moisture Transport Properties of Mortar and Mortar Joint: a NMR Study," *Heron,* Vol. 42, No. 1, pp. 55-69, 1997.

Brunauer, P.H., Emmet, P.H., Teller, E., "Adsorption of Gases in Multimolecular Layers," *J. of Amer. Chem. Soc.*, Vol. 60, pp. 309- 319, 1968.

Bumbaru, D., Jutras, R., Patenaude, A., *Air Permeance of Building Materials.* Report by AIR-INS Inc. for Canada Mortgage and Housing Corp., Ottawa, 1988.

Burch, D., Thomas, W.C., Mathena, L.R., Licitra, B.A., Ward, D.B., "Transient Moisture and Heat transfer in Multi-Layer Non-Isothermal Walls - Comparison of Predicted and Measured Results," *Proc. of Thermal Performance of the Exterior Envelopes of Buildings IV*, Orlando, Florida, Dec 4-7, pp. 513 - 531, 1989.

Burch, D.M., Thomas, W.C., Fanney, A.H., "Water Vapor Permeability Measurements of Common Building Materials," *ASHRAE Transactions*: Symposia, Atlanta, 1992, pp. 486-494.

CEB Guide to Durable Concrete Structures, Bulletin d'Information No. 166, Comite Euro-International d'Beton, 1986.

Cunningham, M.J., "A New Analytical Approach to the Long Term Behavior of Moisture Concentrations in Building Cavities - I. Non-Condensing Cavity," *Bldg & Environment*, Vol. 18, No. 3, pp. 109-116, 1983.

Cunningham, M.J., "A New Analytical Approach to the Long Term Behavior of Moisture Concentrations in Building Cavities - II. Condensing Cavity," *Building & Environ*, Vol. 18, No. 3, pp. 117-124, 1983.

Cunningham, M.J., "Further Analytical Studies of Building Cavity Moisture Concentrations," *Building & Environ*, Vol. 19, No. 1, pp. 21-29, 1984.

Cunningham, M.J., Tsongas, G.A., McQuade, D., "Solar-Driven Moisture Transfer Through Absorbent Roofing Materials," *ASHRAE Transactions* 96, Part 1, 1990, pp. 465-471.

Daian, J.F., "Experimental and Numerical Study of Drying. Compared Behavior of Two Building Materials," *Drying '89*, ed. Arun S. Mujumdar and Michel Roques, Hemisphere Publishing Corp., New York, 1990, pp. 76-82.

Daian, J.F., "The Interaction Between Condensed and Gaseous Water During Migration Processes in Porous Media," Drying '89, ed. Arun S. Mujumdar and Michel Roques, Hemisphere Publishing Corp., New York, 1990, pp. 113-118.

Descamps, Filip, *Continuum and Discrete Modelling of Isothermal Water and Air Transfer in Porous Media.* Doctoral Dissertation, Katholieke Universiteit Leuven, Leuven, Belgium, 1997.

Duffie, J.A., and Beckman, W.A., *Solar Engineering of Thermal Processes.* John Wiley & Sons, New York, 1980.

El-Sabaawi, M., and Pei, D.C., "Moisture Isotherms of Hygroscopic Porous Solids," *Ind. Eng. Chem. Fundam.,* Vol. 16, pp. 321-326, 1977.

Enevoldsen, J.M., Hansson, C.M., Hope, B.B., "The Influence of Internal Relative Humidity on the Rate of Corrosion of Steel Embedded in Concrete and Mortar," *Cem. and Conc. Res.*, Vol. 24, pp. 1373-1382, 1994.

Fatt, I., "The Network Model of Porous Media," *AIME Petrol. Trans.*, Vol. 207, pp. 144-177, 1956.

Fisher, L.R., Gamble, R.A., and Middlehurst, J., "The Kelvin Equation and the Capillary Condensation of Water," *Nature*, Vol. 290, 16 April, 1981, pp. 575-576.

FPL, *Wood Handbook: Wood as an engineering material.* Forest Products Laboratory, Forest Service, U.S. Dept. Of Agriculture, 1999.

Freund J., Halbritter, J., Hoerber, J., "How Dry Are Dried Samples? Water Adsorption Measured by STM," *Microsc. Res. Tech.* Vol. 44, pp. 327–338, 1999.

Geving, S. Thue, J.V., 2002. *Fukt i bygninger.* Handbok 50. Norges byggforskningsinstitutt, Oslo.

Groot, C.J., "Effects of brick suction on the water distribution in joints investigated with thermal neutrons," *Proc. 5th N.A. Masonry Conf.*, Urbana Champaign, pp. 723-734, 1990.

Hall, C., "Water Movement in Porous Building Materials - I. Unsaturated Flow Theory and its Applications," *Bldg & Environ*, Vol. 12, pp. 177-125, 1977.

Hall, C., Hoff, W.D., Nixon, M.R., "Water Movement in Porous Building Materials - VI. Evaporation and Drying in Brick and Block Materials," *Bldg & Environ*, Vol. 19, No. 1, pp. 13-20, 1984.

Hall, C. and Hoff, W.D. *Water transport in brick, stone and concrete.* 3rd edition. CRC Press, 2021.

Handbook of Industrial Drying, pg. 115, 2nd ed. Vol 1. Hemisphere Publishing, Washington D.C.,1996.

Hartley, J.G., "Coupled Heat and Moisture Transfer in Soils: A Review," in *Advances in Drying*, vol. 4., ed. A. Mujumdar, Hemisphere Publishing, Washington, D.C., 1987, pp. 199-248.

Haynes, J.M., "Determination of pore properties of constructional and other materials," *Materials and Structures*, Vol 6, No. 33, May-June 1973, pp. 169-179. Also refer to the 11 other papers in this issue for specific details of each measuring approach.

Hedlin, C.P., "Heat Flow Through a Roof Insulation Having Moisture Contents Between 0 and 1% by Volume, in Summer," *ASHRAE Transactions* 94, 1988, pp. 1579-1594.

Hiemenz, P. C., and Rajagopalan, R., *Principles of Colloid and Surface Chemistry*, 3rd. ed., Marcel Dekker, New York, 1997, pp. 257-270.

Hill, T.L., "Surface Diffusion and Thermal Transpiration in Fine Tubes and Pores," *J. Chem. Phys.*, Vol 25, pp. 730, 1956.

Holm, A., Krus, M., Kenzel, H., "Feuchtetransport ueber Materialgrenzen im Mauerwerk," *Bauinstandsetzen*, Vol. 2, Heft 5, pp. 375-396, 1996.

Imakoma, H., Okazaki, M., Toei, R., "Drying Mechanism of Adsorptive Porous Soids," in *Advances in Drying*, vol. 6., ed. Arun Mujumdar, Hemisphere Publishing, Washington, D.C., 1992, pp. 77-107.

Incropera, F.P. and DeWitt, D.P., *Fundamentals of Heat and Mass Transfer.* 5th ed., John Wiley & Sons, New York, 2002.

ISO 9223. International Standard 1992. *Corrosion of Metals and Alloys.* Classification of Corrosivity of Atmospheres. International Organization for Stan- dardization.

Janz, M., and Johannesson, B.J., "Measurement of the Moisture Storage Capacity Using Sorption Balance and Pressure Extractors," *J. of Thermal Env. and Bldg. Sci.*, Vol. 24, April 2001, pp. 316-334.

Kalimeris, A.N., *Water flow processes in porous building materials*, Ph.D. thesis, University of Manchester, U.K., 1984.

Kays, W.M.., and Crawford, M.E., *Convective Heat and Mass Transfer.* Mc- Graw-Hill, New

York, 1980.

Keey, R.B., *Drying: Principles and Practice*, Pergamon Press, Oxford, 1972.

Kohonen, M., and Christenson, H.K., "Capillary Condensation of Water between Rinsed Mica Surfaces," *Langmuir*, Vol 16, pp. 7285-7288, 2000.

Krischer, O., *Die wissenschaftlichen Grundlagen der Trocknungstecknik*, Springer Verlag, Berlin, 1963.

Kronvall, J., *Airflows in Building Components*. Doctoral Dissertation, Lund Institute of Technology, Lund, Sweden, 1980.

Krus, M, Kuenzel, Hartwig, *Fluessigtransport im Uebersaettigungsbereich*, IBP Mitteilung 270, Fraunhofer Institut fuer Bauphysik, Stuttgart, 1996.

Krus, M., "Effect of Different Liquid Water Diffusivities for Wetting and Drying on the Moisture Behaviour," Annex 24 Paper T3-D-92/01, International Energy Agency, 1992.

Krus, M., *Feuchtetransport- und Speicherkoeffizienten poroeser mineralischer Baustoffe. Theoretische Grundlagen und neue Messtecchniken*. Doctoral Dissertation, Universität Stuttgart, Germany, 1996.

Krus, M., Sedlbauer, K., Zillig, W., Kuenzel, H., "A New Model for Mould Prediction and its Application on a Test Roof," *Proc. Of Current Problems of Building Physics in the Rural Building*, Cracow, Poland, Nov. 2001.

Kuenzel, Hartwig. 1994. *Verfahren zur ein- und zweidimensionalen Berechnung des gekoppelten Waerme- und Feuchtetransports in Bauteilen mit einfachen Kennwerten*, Doctoral Dissertation, Universität Stuttgart, Germany.

Kuenzel, Helmut, "Der Regenschutz von wasserabweisenden Aussenputz," *Bauphysik*, Vol. 5, 1986.

Kuenzel, Helmut, "Risse in Bewittertem Holz," *bauen mit holz*, Vol. 95, Heft 12, pp. 1018-1025, 1993.

Kuenzel, Helmut, "Trocknungsblockade durch Mauerversalzung," *Bautenschutz + Bausanierung*, 1990.

Kumaran, M.K., Task 3: *Hygrothermal Properties of Building Materials*, Annex 24, Final Report, International Energy Agency, 1996.

Kumaran, K., John Lackey, Nicole Normandin, David van Reenen, Fitsum Tariku. *MEWS Report 3: Material Properties*. NRCC/IRC, Ottawa 2002.

Kumaran, M.K., 2006. "A thermal and moisture property database for common building and insulation materials". *ASHRAE transactions, 112*(2).

Lacy, R.E., *Climate and Building in Britain*. HMSO, London, 1977.

Litvan, G.G., "Testing for the Frost-Susceptibility of Bricks," Special Technical Publication 589, American Society of Testing and Materials, 1975, pp. 123-132.

Litvan, G. G., 1980. "Freeze-Thaw Durability of Porous Building Materials," *Durability of Building Materials and Components*. ASTM STP691. P. J. Sereda and G. G . Litvan, Eds., American Society for Testing and Materials, 1980, pp. 455-463.

Lohmeyer, G. *Praktische Bauphysik*, B.G. Teubner, Stuttgart, 1996.

Lopez, W., Gonzalez, J.A., "Influence of the Degree of Pore Saturation on the Resistivity of Concrete and the Corrosion Rate in the Presence of Chlorides," *Cem. and Conc. Res.*, Vol. 23, pp. 368-376, 1993.

Luikov, A.V., *Heat and Mass Transfer in Capillary Porous Bodies*. Permagon Press, Oxford, 1966.

Mehta, P.K., *Concrete: Structure, properties, and materials*. Prentice-Hall, 1986.

Mensinga, P., 2009. *Determining the critical degree of saturation of brick using frost dilatometry* (Master's thesis, Faculty of Engineering, University of Waterloo.

Millington, R.J., and Quirk, J.P., "Permeability of Porous Solids," *Trans. Faraday Society*. Vol 57, pp. 1200-1207, 1961.

Nevander, L.E., Elmarsson, B., *Fukthandbok: praktik och teori (Moisture handbook: practice and theory, in Swedish)*. Svensk byggtjanst, Solna, Sverige, 1994.

Okazaki, M., Tamon, H., and Toei, R., "Interpretation of Surface Flow Phenomenon of Adsorbed Gases by Hopping Model," *Amer. Inst. Chem. Eng. Journ.*, Vol 27, No. 2, pp. 262-270, 1981.

Panella, V., Chiarello R., Krim, J. "Adequacy of the Lifshitz Theory for Certain Thin Adsorbed Films," *Phys. Rev. Letters*, Vol. 76, No.19, pp. 3606-3610, 1996.

Pauling, L. *General Chemistry*. Dover Publications, New York, 1970.

Peake, F. and Ford, R.W., 1882. "A Comparison of the Vacuum and Boiling Methods for Measuring the Water Absorption of Bricks," *Trans. J. Br. Ceram. Soc.*, Vol 81, pp. 160-162.

Pel, Leo, 1996. *Moisture Transport in Porous Building Materials*, Doctoral Dissertation, Eindhoven University of Technology, Eindhoven.

Penner, E., 1963. "Suction and its use as a measure of moisture contents and potentials in porous materials," Vol. 4, *Proc. Of Int. Symp. On Humidity and Moisture*, Washington D.C., Chap 29, pp. 245-252.

Philip, J.R., and de Vries, D.A., "Moisture Movement in Porous Materials under Temperature Gradients," *Trans. Amer. Geophys. Union*, Vol 38, 1957 , pp. 222-232.

Quenard, D., Sallee, H., "The Transport of Condensible Water Vapor Through Microporous Building Materials," *Drying '89*, ed. By Arun S. Mujumdar and Michel Roques, Hemisphere Publishing Corp., New York, 1990, pp. 106-112.

Rounsley,R.R., "Multimolecular Adsorption Equation," *J. of Amer. Chem. Soc.*, Vol 7, No. 2, pp. 308 -311, 1961.

Sallam, A., Jury, W.A., Letey, J., "Measurement of gas diffusion under relatively low air-filled porosity," *Soil Sci. Soc. Amer.* J., vol 48., pp. 3-6, 1984.

Sanders, C., *Task 2: Environmental Conditions*, Annex 24, Final Report, International Energy Agency, 1996.

Schumacher, C.J., Lepine, C., Straube, J.F., 2020. *Thermal, Moisture and Air Transport*

Property Values for New Building and Insulating Materials. Research Project RP-1696, ASHRAE, Atlanta, Ga.

Schwarz, B., 1978. *Die Wärme und Stoffübertragung an Außenwandoberflächen*. Doctoral Dissertation, Universität Stuttgart, Germany.

Sereda, P.J., and Feldman, R.F., 1970. CBD 130 *Wetting and Drying of Porous Materials*. NRCC, Ottawa, October.

Sereda, P.J., 1970. Canadian Building Digest 127 *The Structure of Porous Building Materials*, NRCC, Ottawa, July.

Sereda, P.J., 1976. Canadian Building Digest 170 *Atmospheric Corrosion of Metals*, NRCC, Ottawa.

Shirmer, R., 1938. "Diffusionszahl von Wasserdampf-Luft-Gemischen und die Verdampfungeschwindigkeit," *Verfahrenstechnik*, Vol. 7, pp. 170-177.

Straube, J.F., and Burnett, E.F.P., 1996. *Vents, Ventilation, and Pressure Moderation*. External Research Program Report for CMHC, Ottawa, Dec.

Tamon, H., Okazaki, M., and Toei, R.,1981. "Flow Mechanism of Adsorbate Through Porous Media in Presence of Capillary Condensation," *Amer. Inst. Chem. Eng. Journ.*, Vol 27, No. 2, pp. 271-277.

TenWolde, A., Mei, M.T., "Moisture Movement in Walls in a Warm Humid Climate," *Proceedings of Thermal Performance of the Exterior Envelopes of Buildings III*, Clearwater Beach, Florida, Dec 2-5, 1985, pp. 570 - 582.

Toei, R., "Drying mechanism of capillary porous bodies," in Advances in Drying, vol. 2., ed. A. Mujumdar, Hemisphere Publishing, Washington, D.C., 1983.

Washburn, E. W. "The dynamics of capillary flow". *Physical review*, *17*(3), 273, 2001.

Welty, J.R., Wicks, C.E., Wilson, R.E., Rorrer, G.L. *Fundamentals of Momentum, Heat and Mass Transfer*. John Wiley & Sons, New York, 2001.

Whiteley, P., Russman, H.D., Bishop, J.D., *Porosity of Building Materials - a Collection of Published Results*, BRE Current Paper 21/77, HMSO, London, 1977.

Wilson, A.G., "Condensation in Insulated Masonry Walls in Summer," *RILEM/CIB Symposium*, Helsinki, 1965, pp. 2-7.

Winkler, E.M. and Wilhelm, E.J., 1970. "Salt burst by hydration pressures in architectural stone in urban atmosphere". *Geological Society of America Bulletin*, *81*(2), pp.567-572.

Winter, K., Krus, M., Kuenzel, H., "Feuchtetransport in feinporigen Materialien," *Bauinstandsetzen*, Vol. 3, Heft 1, 1997, pp. 15-28.

Worch, Anatol. "The Behaviour of Vapour Transfer on Building Material Surfaces: The Vapour Transfer Resistance." *Journal of Thermal Envelope and Building Science*, vol. 28, no. 2, Oct. 2004, pp. 187–200.

Chapter 9

Moisture management

The previous chapters have explored the physics of heat, air and moisture as well as the wetting, storage and drying of moisture in enclosure assemblies. This chapter and the next three introduce the philosophy and concepts used to apply building physics to manage moisture in enclosure design, construction, repair, and retrofit.

Moisture control or management is acknowledged as a very important consideration because moisture is a common factor in many problems that continue to plague the building industry. Heat and air movement are also important enclosure control functions in their own right and are often highly coupled to managing moisture and limiting moisture-related problems.

9.1 Moisture problems

Moisture is involved in many performance problems or deterioration processes that afflict the building enclosure (see Chapter 8 for example), such as:

- leakage of water into the building;
- freeze-thaw deterioration of concrete, stone, and masonry;
- electrochemical corrosion of metal components such as structural framing, reinforcing bars, masonry anchors, ties, flashing, etc.;
- biological, especially fungal (mold, rot, decay) growth, which can have a major effect on occupant health, structural capacity, and appearance;
- the chemical deterioration and dissolution of materials such as gypsum sheathing, wood products, and damaging chemical processes such as carbonation and alkali-aggregate reaction;
- volume changes (expansion, shrinkage) that can cause structural failure, cracking, degradation of appearance, etc.; and

- discoloration (staining, "dusting," irregular wetting, etc.) of building finishes.

For a moisture-related problem to occur, at least four conditions must be satisfied:

1. a *moisture* source must be available,
2. there must be a *route* or means for this moisture to travel,
3. there must be some *driving force* to cause moisture movement, and
4. the material(s) involved must be *susceptible* to moisture damage.

To avoid a moisture problem one could, in theory, choose to eliminate any one of the four conditions listed above. In reality, it is practically impossible to remove all moisture sources, to build walls with no imperfections, or to remove all forces driving moisture movement. It is also not economical to use only those materials that are not susceptible to moisture damage. Therefore, in practice, it is often advantageous to address two or more of these prerequisites so as to reduce the probability of having a problem.

Accordingly, controlling moisture and reducing the risk of failure by judicious design, choice of materials, assembly and operation must be the approach taken in the design of durable building enclosures. This approach requires not only an understanding of moisture and moisture control in enclosure systems but also some background in material science and micro-meteorology.

9.1.1 Susceptibility and vulnerability

Different materials and assemblies vary in their susceptibility to different kinds of moisture-related damage. Standards, codes, and industry criteria are available to assess the susceptibility of materials or assemblies to some of the moisture-related concerns listed earlier.

Whether a moisture problem occurs depends on whether a susceptible material or assembly is placed in a vulnerable environment. This vulnerability — or more generally, the level of risk associated with a moisture-related problem — should be considered together with the consequences of the problem. While the susceptibility of a material to a particular problem (e.g., freeze-thaw deterioration) may be measurable (e.g., by means of freeze thaw tests and target criteria for acceptable performance), designers and builders are responsible for ensuring that a material or assembly is not used in a manner that renders it disproportionately likely to incur a problem.

This moisture vulnerability, or degree of risk, affects the probability of a problem occurring and may be considered to be a function of three potentials, that is:

- wetting,
- storage, and
- drying.

Materials, assemblies, and environments may all be assessed by considering these potentials. The environments (Chapter 3) that the enclosure separates provide both the source of moisture and a sink for moisture that is removed. Therefore, the wetting and drying potentials of the interior and exterior environments are just as important for assessing vulnerability as the wetting and drying potentials of an assembly. Location is the primary determinant of exposure: the location of the material in the wall, of the wall on the building, of the building on the site, and the geographical location of the site.

The relevant characteristics of an environment can be grouped together with the characteristics of an enclosure to define an overall (i.e., enclosure plus environment) wetting potential and an overall drying potential. Such potentials are necessarily climate- as well as assembly-specific. Therefore, any such statement as "This is a durable, high-performance wall assembly" cannot be taken seriously without the context of climate. For example, is the enclosure expected to separate an operating room from the Antarctic winter or a warehouse from the Saharan desert?

Using a material that is not supposedly susceptible to moisture damage (e.g., good quality face brick) in locations with high wetting potential (such as at windowsills or as garden wall copings) often leads to a problem (freeze-thaw damage in cold climates). In traditional multi-wythe brick walls, the moisture storage capacity is so high that the moisture content of the masonry often never exceeds a performance threshold. In a drier climate, such as the Canadian prairies or US Southwest, the potential for drying is higher than the potential for wetting and, again, the moisture content of the brickwork remains below its performance threshold. Just as for brick, understanding the moisture content of materials such as wood and the time-of-wetness (TOW) of steel can be used as a measure of their vulnerability to premature deterioration.

Complicating the assessment outlined above is the fact that the performance thresholds, and even what physical measures one should use to define a threshold, are unknown or poorly defined for most materials and assemblies. Some of these performance thresholds are briefly described in Chapter 8.

The ability of a wall assembly to dry and store moisture as well as its resistance to wetting are of great importance in assessing its vulnerability to moisture-related damage and deterioration. To develop some understanding of how and how much moisture can be removed from a wall by drainage, evaporation, diffusion, and ventilation, it is necessary to understand the fundamental behavior of moisture in enclosures and the physics of the wetting and drying processes in general. The sources of moisture, the available storage, and the mechanisms of moisture removal in building enclosures are considered briefly below. A predictive and quantitative understanding of the wetting and drying mechanisms of common building materials and assemblies in various climates is necessary for the quantification of vulnerability. This knowledge can in turn be used to understand the causes of failure and to design against moisture damage.

9.2 Moisture control

If a balance between wetting and drying is maintained, moisture will not accumulate over time, and moisture-related problems are unlikely (Figure 9.1). The extent and duration of wetting, storage, and drying must, however, always be considered when assessing the risk of moisture damage [Straube 2002].

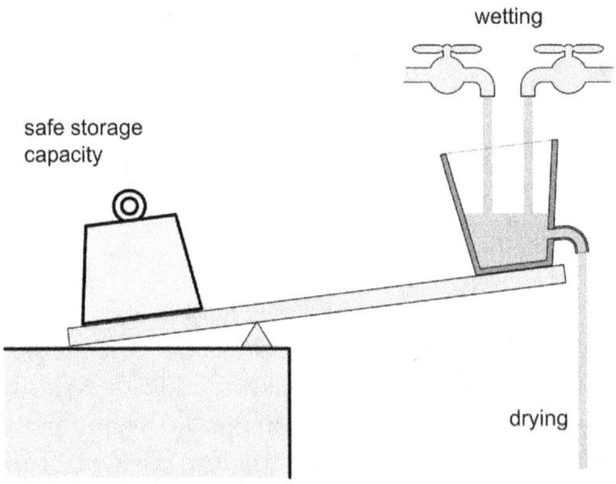

Figure 9.1: The moisture balance

Most moisture control strategies tend reduce the amount of wetting by, for example, increasing air tightness and vapor resistance, reducing the volume of rain water penetration and absorption, etc. However, it has become generally accepted that most building construction will not be perfect, and thus wetting will occur (even if only during shipping and construction). Therefore, the provision of greater drying potential and storage capacity has begun to receive more attention. These are powerful but often-overlooked design strategies for controlling moisture. Finally, reducing vulnerability through intelligent design choices relating to building location, orientation, geometry, etc., is often the least expensive and best approach (although it must be considered very early in the concept stage).

The major wetting and drying processes and the moisture transport mechanisms involved in the movement of moisture into and out of the enclosure are summarized in Figure 9.2.

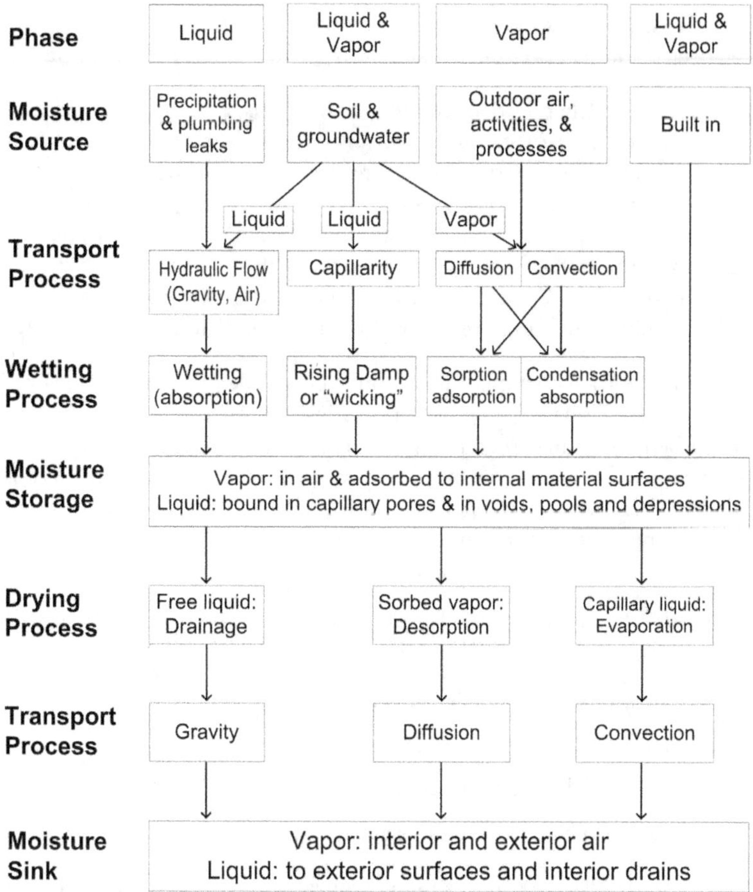

Figure 9.2: Wetting, storage, and drying processes in building enclosures

9.3 The wetting process

Wetting occurs when water vapor is adsorbed, liquid water is drawn into the pores by capillary forces, or external forces drive liquid water into an already saturated material. Moisture is redistributed within the materials and layers of an enclosure assembly primarily by liquid transport (capillary and gravity), adsorbed flow, vapor diffusion and vapor convection. Either the transport process or the source of moisture (e.g., condensation, rain penetration or absorption) is used to define the type of wetting.

It should be noted that if the rate of water accumulation on the surface of a material (i.e., the rate of wetting) is greater than the product of the water absorption coefficient, A and the root of the time since wetting began, water will no longer be absorbed. Any water that is not absorbed will form a film on the surface and will eventually drain away, if drainage is provided for.

As discussed in Chapter 8, once the wetting stops, e.g., rain or condensation, the nature of the absorption changes (and the liquid moisture diffusivity drops). During wetting, the larger pores transport the largest volume of water, because their resistance to flow is the least. However, once wetting of the surface stops, the larger capillary suction of the smaller capillaries begin to redistribute the water from the larger pores to the smaller ones. Redistribution takes a longer time because the flow velocity within the small pores is reduced by the large viscous forces resisting flows into and through these pores.

The moisture transport mechanisms may be simple or a complex chain. For example, rain deposited on an exterior brick veneer will cause wetting when it is absorbed, but it may then be evaporated by solar heating (a phase change) followed by the parallel transport by diffusion and convection of the vapor inward. There it may condense (a phase change) on an interior layer of polyethylene, whereupon it may drain down to the bottom and be absorbed by the wood plate and evaporate and/or be adsorbed by the wood studs.

The four major sources of moisture for the above-grade building enclosure are (Figure 9.3):

1. precipitation, especially driving rain,
2. water vapor in the air (either the interior or exterior),
3. built-in and stored moisture, and
4. liquid and bound ground water.

Rain wetting by wind driven rain is usually the largest moisture source for above-grade walls. Rain deposition on one or more faces of an exposed low-rise building can easily exceed 100 kg/m^2/yr in many climates.

Water vapor transported to cold surfaces inside the enclosure by convection or diffusion can cause condensation during cold and hot weather. Cold weather exfiltration-induced condensation on the inner face of the wall sheathings, roof membranes, and wall cladding is a relatively common problem which results in slow wetting of surfaces within enclosure assemblies. Water vapor adsorption will increase the moisture content of hygroscopic materials but only rarely to a dangerous level.

While diffusion wetting is usually not a powerful wetting mechanism, diffusion is an important means of moisture movement and redistribution between layers and materials within a wall. Solar heating of absorbent exterior layers (especially cladding) will often cause evaporation from the interior face of capillary active rain-wetted claddings (such as brick, stucco, and wood products) and condensation further inside the enclosure assembly, especially on air-conditioned impermeable surfaces near the inside (such as polyethylene sheets, vinyl wallpaper, epoxy-paint, mirrors). [Hutcheon 1953, Wilson 1965, TenWolde & Mei 1985, Andersen 1987, Cunningham *et al* 1990, Straube & Burnett 1998]. These inward vapor drives are intense (several thousand Pascals) and of short duration (a few hours per day). Although a redistribution mechanism that can dry outer layers it can also cause dangerous wetting of inner enclosure layers. Inward drives can be managed simply in design by ventilation [Straube et al 2004] or semi-permeable exterior insulation [Smegal et al, 2015]. This mechanism, while not often considered in design, can also be significant to heat transfer [Bomberg and Shirtcliffe 1978, Hedlin 1988], especially for exposed membrane roofs with rain penetration and air leakage condensation problems.

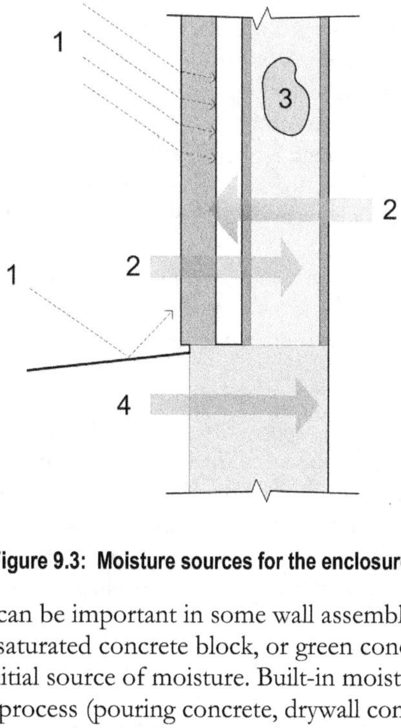

Figure 9.3: Moisture sources for the enclosure

Built-in moisture can be important in some wall assemblies. The use of wet framing lumber, saturated concrete block, or green concrete within a wall may provide a large initial source of moisture. Built-in moisture may be a result of the construction process (pouring concrete, drywall compound), the manufacturing process (wet lumber, aerated concrete masonry), or construction site exposure (rain wetting of wood, masonry, gypsum).

Wetting by the capillary movement of water from other moisture sources (i.e., rising damp from below-grade sources) is generally not a significant problem in North American buildings (capillarity is, however, very important as a redistribution mechanism). Nevertheless, capillary breaks in the form of air gaps or capillary inactive materials should be provided to decouple the enclosure from wet soil.

9.4 The drying process

Water leaves hygroscopic materials in building enclosures by the processes of:

- drainage of liquid (from oversaturated materials)
- evaporation/sublimation of liquid/ice (from wet materials), and
- desorption of adsorbed vapor.

An initially oversaturated material loses moisture first by gravity drainage until the capillary saturation moisture content is reached. The remaining liquid moisture in the material is then removed by evaporation at the surface of the material. Liquid water evaporated from the surface is replaced by water that moves through the material toward the relatively drier surface by capillary conduction, driven by the suction gradient. Once the material is no longer saturated, unsaturated flow and desorption begin. The rate of free water evaporation is usually much higher than the rate of desorption. The two processes and their different rates result in a moisture flux (or drying rate) with two reasonably distinct stages (Figure 9.4) [Hall *et al* 1984]:

- Stage I is the surface-saturated rate of drying. This rate is constant and approximately the same as free-water evaporation under the same conditions; hence it does not vary significantly from material to material.
- Stage II is the desorption drying rate. This rate is controlled by material properties (total moisture diffusivity, sorption isotherm), geometry, and the surrounding air conditions. The rate decreases in an approximately exponential manner as the material approaches equilibrium with its environment. The variation of moisture diffusivity with moisture content can dramatically change the shape of this curve.

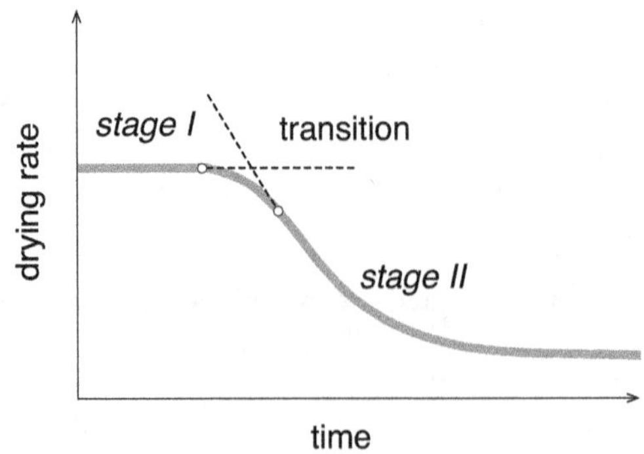

Figure 9.4: Drying stages for a saturated hygroscopic material

Although the rate of Stage I drying is essentially the same as that for free water under the same conditions, evaluating the evaporation rate of free water is also difficult. For some materials, geometries, and conditions of surrounding air, the transport of water through the saturated material may not replace the surface water as quickly as it is removed by evaporation. In such (rare) cases, the flow of water through the material controls the Stage I drying. Provided that free water is available over a constant surface area, the Stage I drying rate remains approximately constant with constant air temperature and vapor pressure.

Stage I ends when insufficient liquid water is available at the surface of the porous material. When Stage I begins, all pores are typically full or almost full. As liquid water evaporates from the drying surface, the moisture content at this layer drops and a potential for liquid water flow develops. Water will flow toward the drying surface from the interior of the material at a rate dependent on liquid water diffusivity (D_l) of the material. If the drying rate at the surface exceeds the ability of water to move towards it, drying is restricted and Stage II drying begins. The flow of water towards the drying surface then proceeds by a combination of vapor diffusion, surface diffusion, and unsaturated liquid flow.

As Stage I comes to an end, the plane where liquid water evaporates into the air, often called the drying front, moves below the surface. Stage II drying occurs at a slower rate for two reasons:

1. Since most porous materials have some internal resistance to water vapor flow, water vapor removal is hindered by the vapor resistance of the material between the exterior surface and the drying front, and

2. The vapor pressure immediately over the drying front will be higher because convective effects cannot produce a strong vapor gradient. Since the gradient of vapor pressure from the drying plane is lower than if it were at the surface, evaporation from below the surface occurs at a lower rate.

Thus, the Stage II drying rate is expected to follow an approximately exponential decay for constant ambient air conditions since the vapor resistance of the material increases inversely as the distance of the drying front from the surface increases. The actual shape of the drying versus time curve depends on the moisture content dependent nature of the material's moisture diffusivity.

Except in cases of very high surface evaporation rates and/or thick material layers with low moisture diffusivities, Stage I is approximately the time taken until the material is no longer capillary-saturated ($w<w_{cap}$). This is usually a short period of time and can often be ignored. Stage II drying then becomes a drying period during which vapor flow and unsaturated liquid flow dominate.

Consideration of the physics of drying leads to the expectation that material layers within an enclosure that are initially saturated will exhibit an initially high and constant rate of drying (as all free and surface water is evaporated) and an exponentially decreasing drying rate as all porous materials approach their equilibrium moisture content. Between these two extremes there will be a transition in which drying is due to a combination of the two stages.

Constant air temperature and vapor pressure conditions are not common in the natural environment, and conditions within an enclosure might be characterized as "constant" for a few hours at most. During the day, solar radiation may change the temperature of the surface so quickly that even an hourly average is not sufficiently accurate to capture the full range of behavior. The expected environmental variations can be superimposed on the plot of expected drying to qualitatively gauge actual performance.

It is important to note that all of the water stored in the materials that constitute an enclosure can be removed only by vapor diffusion, convection, or drainage[1]. Of course, water can be delivered to a surface by capillary flow followed by evaporative drying (i.e., diffusion and convection). Supersaturated materials will drain until they reach capillary saturation if a drainage path is provided. Gravity drainage will not remove any water stored by absorption or adsorption (moisture storage regions A through D in Chapter 8) which in practise means drainage will not lower the equilibrium RH of a material below approximately 100%. An example is the drainage of gravel layers below concrete slabs – water will drain through the gravel, but if the sand content is high a significant quantity of water will remain absorbed in the smaller pore spaces.

9.4.1 Drying of enclosure assemblies

An assembly's drying potential is an important factor in assessing its vulnerability to moisture problems. Moisture is usually removed from an enclosure by (Figure 9.5):

1. evaporation of liquid water transported by capillarity to the inside or outside surfaces;
2. vapor transport by diffusion, air leakage, or both, either outward or inward;
3. drainage, driven by gravity; and
4. ventilation (ventilation drying).

[1] liquid transport through capillary active materials connected to drier capillary active materials on the exterior or interior can theoretically remove moisture, but it is difficult to imagine a practical example.

Drainage is capable of removing the greatest volume of water in the shortest period of time. Hence it is a very important mechanism for moisture control. Provided a clear drainage path exists (e.g., cavities, slopes, drainage openings), a large proportion of rain water penetration or condensation can flow out of a wall. The gap required for drainage is usually in the order of a millimeter (less than 1/16") [Smegal 2006, Straube & Smegal 2007, van den Linden 2018] and in fact some highly porous materials allow drainage through them. However, even in systems with excellent drainage, a small but significant amount of water will remain in the enclosure, attached to surfaces by surface tension, and be held in materials by capillary forces. Cladding or sheathing must be almost saturated by condensation (which deposits water very slowly) before sufficient volumes of water will bead on the surface for drainage to occur. Therefore, it must be assumed that water that cannot be removed by drainage will be stored.

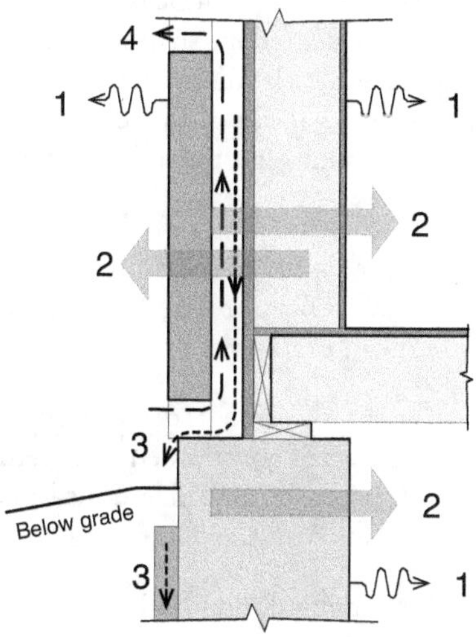

Figure 9.5: Moisture removal mechanisms for enclosures

Under the right conditions, moisture not drained from an assembly will dry by evaporation or desorption. The resulting water vapor can be transported out of and within the wall system either by diffusion or advection (i.e., combined diffusive and mass flow).

Diffusive drying will move vapor in all directions driven by vapor pressure differences and resisted by the vapor resistance of enclosure materials. Diffusion outward, especially when driven by large vapor gradients in the winter, can remove a significant amount of moisture and be an effective drying mechanism. Inward diffusive drying can occur in many climates, even cold ones, if the inner layers are sufficiently vapor permeable. The warmer and sunnier the climate, the more powerful inward diffusive drying is.

Air movement (or leakage) through the enclosure can, under the proper conditions, move a large quantity of moisture. While air leakage usually leads to condensation wetting under many winter conditions, it can also remove moisture. Periodic reversals of air flow from exfiltration to infiltration (when the wind changes direction, for example) can allow drying even under winter conditions. Outward air leakage during warmer weather will dry moisture that condensed during colder weather. Similarly, winter-time, stack-effect-driven infiltration and summer exfiltration can cause drying.

Ventilation, or air flow through a space, uses the drier outdoor air to transport water vapor out of the assembly. Ventilation drying can be a very useful, and sometime critical, design strategy for some enclosure types. Ventilation drying has been measured in many field studies [Popp et al 1980, Straube et al 2004] and predictive methods developed [Straube & Burnett 1995, Burnett et al 2004] some of which have been presented in Chapters 6 and 7. For example, ventilation around impermeable cladding (metal cladding, asphalt shingles) is often required to bypass their vapor resistance especially in situations with solar heating, small amounts of drying, and relatively impermeable materials [Finch & Straube 2007].

Capillary transport acts to redistribute moisture within a system. For example, water on the back of a brick veneer or wood siding can move by capillarity to the exterior face where it can evaporate.

Diffusion of water vapor, air movement, desorption, capillarity, evaporation, and of course gravity drainage all contribute to the removal of moisture from an enclosure assembly. Obviously, considerations of configuration, location and exposure have an influence on the nature and rate of moisture removal.

In an enclosure assembly, drying during both stages is affected by:

- the volume, various degrees of wetting/saturation, and the distribution of different materials and different parts of the assembly,
- the amount and nature of moisture flow into the wall (by airflow, diffusion, or rainfall),
- the temperature and vapor pressure conditions and gradients,
- the different sorption isotherms of the hygroscopic materials,
- the vapor resistance of different layers in the wall assembly,
- ventilation air flow, and
- interior and exterior ambient air conditions.

When most non-adsorbed or weakly adsorbed water has been removed, a continuous and mobile internal water layer no longer exists, and drying proceeds by vapor diffusion and convection only. The material eventually comes to equilibrium with its environment as predicted by the sorption isotherm.

The moisture in saturated exterior surfaces can usually easily evaporate to the exterior. Materials below the surface must be capillary active and in contact with wet materials further in the wall for drying to occur. Typical cladding materials such as brick veneer, vinyl siding, precast concrete, wood, natural stone, metal siding, polymer-modified stuccos, etc., have relatively high vapor resistances (for metals and plastics, essentially zero vapor permeance, for brick veneers and wood, between about 25 and 200 ng/s·Pa·m^2). Highly vapor-resistance claddings can greatly retard drying of the inner wythe by evaporation and diffusion. Ventilation drying is therefore a potentially important drying mechanism because it is able to remove the moisture that remains behind claddings with a high water-vapor resistance.

9.5 Storage

It has become generally accepted that most building construction will not be perfect, and thus wetting will occur. Therefore, more attention is being paid in design practise to approaches that also consider the drying ability of the enclosure. However, moisture removal occurs at times other than the time(s) at which wetting occurs, and thus moisture storage must be provided to bridge the gap in time between wetting events (often brief and relatively intense) and drying conditions (often long periods of slow moisture removal).

The ability of a wall assembly to store moisture may therefore be an important measure of its durability because storage acts as the vital buffer, or capacitor, between the deposition and removal of moisture. However, if the volume of stored water exceeds the safe level for a material and is present for long enough, deterioration can occur, i.e., rotting of wood, freeze-thaw damage of masonry, and corrosion of metal. Therefore, the two most important characteristics of moisture storage are: how *much* moisture can be stored and for what *duration* without crossing a performance threshold.

Moisture can be stored in a variety of ways in enclosure assemblies — as vapor, water, or solid (frost, ice). All mechanisms have practical importance: adsorption and absorption of water in hygroscopic materials, free liquid water stored in small depressions or in undrained portions of an assembly, and even as droplets or frost on surfaces.

The volume of water that is stored in an enclosure can be large, in the order of a few to tens of kg per square meter (from ounces to pounds per square foot). This moisture can be stored as vapor, liquid, or solid in a variety of ways (Figure 9.6):

1. trapped in small depressions or in poorly drained portions of assemblies,
2. adhered by surface tension as droplets (or frost, even ice) to materials and surfaces,
3. adsorbed in or on hygroscopic building materials (esp. brick, wood, fibrous insulation, paper),
4. retained by capillarity (absorbed) in porous materials, and
5. stored in the air as vapor.

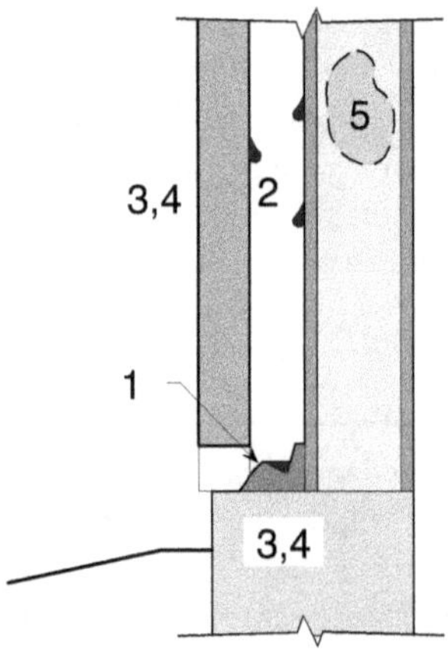

Figure 9.6: Moisture storage in building enclosures

Sorption of water vapor by hygroscopic materials is an important storage mechanism. A significant amount of moisture can be stored within a porous material as water vapor molecules adsorb to the large internal surface areas of materials such as wood, concrete and masonry (e.g., see the sorption isotherm in Chapter 8). When a porous material has adsorbed all the moisture it can, further moisture will be stored in the pores and cracks within the material by capillary suction, or by absorption.

Only when all pores are filled with water is a material saturated. For example, wood will adsorb moisture up to approximately 30% of its mass in the adsorbed state, and fully saturated wood can hold two to four times this amount of water. When all easily filled pores of a material are filled with water, the material is capillary-saturated. A capillary-saturated material is generally unable to store any more moisture. Water stored in pores must usually leave in vapor form. (Chapter 8 deals with the physics of moisture storage in much greater detail.)

Table 9.1 uses the average values to calculate the amount of moisture that might be stored in the materials that constitute a representative wood framed house wall.

The representative wall comprises an 85 mm clay brick veneer with cement mortar (approximately 10 mm wide joints), an airspace, 12.5 mm OSB sheathing, on a 38x140 mm wood frame at 610 mm centers. The batt insulation, the sheathing paper, and the air in the air space have not been accounted for, since their contribution is small. The interior drywall finish is assumed to be well separated from the exterior by a 6 mil (0.15 mm) polyethylene vapor barrier. Table 9.1 considers only the moisture stored through absorption and adsorption. Moisture may also be stored as trapped liquid water, although this amount is often less than one kg per square meter.

Table 9.1: Moisture contents of representative wood-framed wall assembly

Material	Dry density (kg/m^3)	Volume (m^3/m^2)	At 95%RH (kg/m^2)	At capillary saturation (kg/m^2)
Brick veneer (85 mm)	1700	0.0765	6.5	13.0
Mortar (10% area)	1900	0.0085	1.0	2.4
OSB (12.5 mm)	700	0.0125	1.2	8.7
Wood studs (38/140@610) †	500	0.0167	2.2	8.5
Total		**0.189**	**10.9**	**32.6**
EPS sheathing (38)	16	0.038	<0.1	<0.1
Wood studs (38/89@410)	500	0.0141	1.9	7.0

†Includes two top and one bottom plate and a 38/185 rim joist per 2.64 m height

It is clear from Table 9.1 that a significant amount of water can be stored in such a representative wall. While it is unlikely that the framing will be at 95% RH (in fact, fungal growth would be expected at this moisture content), both the OSB and the wood framing can safely act as short-term accumulators of moisture, especially in cold weather[2]. Field measurements have shown that exposed masonry veneers will often be at between 95% RH and saturation in service. The volume of water that can safely be stored in most typical enclosures will therefore be of the order of several to tens of kg per square meter.

One of the important changes in enclosure design over the last 50 or more years [Lstiburek 2009], for purposes of saving costs, resources, and weight

[2] As most damage mechanisms act more quickly at higher temperatures, the time-scale of moisture storage is usefully extended for parts of the enclosure that are cold during cold weather.

(important for high rise buildings), has been the reduction in material content in building enclosures. This reduction in material use has reduced, sometimes by orders of magnitude, the safe storage capacity of enclosures. Combined with the change to materials that are more susceptible to moisture damage (e.g., the switch from concrete to wood, or to paper-faced gypsum from mineral based plasters) most enclosures can tolerate less wetting without performance problems. Enclosures that can tolerate fewer errors, less construction wetting, and no floods demand more care in design, specification, construction and operation.

9.6 Designing enclosures for moisture control

It has been argued above that moisture-related enclosure design must be a balance of wetting, drying, and storage potentials in conjunction with a judicious assessment of the susceptibility and vulnerability of the enclosure materials and sub-assemblies.

The physics of heat, air and moisture flow presented in earlier chapters can be applied, with the appropriate material properties, environmental data and assumptions, to predict and understand the wetting, drying and storage potentials. A quantitatively accurate value cannot be calculated for most cases. However, it is important to realize that

> **building enclosure design usually involves the assessment of relative performance, pass-fail comparisons, or the ranking of competing design choices, not absolute values, and therefore there is rarely a need for precision.**

The results generated by an understanding of the physics involved and the approximate calculation techniques presented so far should therefore be considered in this context and can be applied to arrive at reliable relative assessments of performance useful for design rather than precise quantities. The limited benefit of precision offered by computer simulations also needs to be recognized (see Chapter 14). The moisture management approach described above is useful to guide a designer and analyst. However, calibrating predictions and reasoning with real field experience over at least several years is fundamental to good building science. Experienced-based recommendations are extremely useful, particularly if they are supported by an understanding of building physics: for example, books such as Lstiburek [2020] should be referenced to confirm one's application of building science to real buildings.

9.6.1 Material choices

The enclosure designer has an almost infinite range of choices if one considers the possible combinations of materials, layers, shape, orientation, etc. With respect to durable or high-performance enclosures, there are no "good" or durable materials. As described in Chapter 4, all materials can be used in a durable way, and the application of building science allows one to design effective durable enclosures from materials such as straw, paper, and wood fiber [Straube 2006].

Almost by definition, a durable enclosure design will expose the materials selected to environments conducive to their good performance, i.e., a combination of temperature, moisture, solar radiation, freezing, and time that will not cause problems. This means that the enclosure design often is driven by the need to allow a certain material to be used at a certain location in the enclosure given the environments expected on either side.

For example, if one uses materials that are sensitive to moisture (e.g., wood), the design of the enclosure and the building must ensure that the probability of wetting is low, and the probability of drying is high. If the potential for wetting is high, frequent, and intense, with little chance of drying then a material such as concrete would be more appropriate. If the use results in repeated and severe freezing as well, air-entrained concrete would be a better material choice. Polymers, increasingly part of the building enclosure material palette, are often located to avoid direct exposure to ultraviolet radiation and high temperatures to enhance their durability.

Of course, as for all engineering work, the risks and consequences of failure and the desired performance must be weighed together with the cost, the client's expectations, government regulation, local availability, and many other factors.

9.7 Enclosure design principles

The starting point in the design of a building enclosure should be a consideration of what climate it is intended to exclude (e.g., climatic region: cold-dry, hot-wet, arid) what micro-climatic conditions are expected (due to site, orientation, and height), and what type of climate it is intended to enclose (e.g., museum, apartment, office, factory). Together with the client's expectations and constraints, these lead naturally to the specification of the required level of enclosure performance and durability (e.g. service life). The designer(s) must then balance the design with the available resources (money, labor, materials, time, etc.).

Regardless of the durability and performance requirements, or of the available resources, certain principles of design [Straube 2012] should be followed.

Be sure to have defined a complete load transfer path, including all connections, to the ground for all loads. This means designed connections of roofs to foundations, from floors to walls, walls to foundations and windows or canopies to walls. The most common problem affecting enclosure support noted in practise is the lack of design documentation for the connection of secondary components to the structure (e.g. windows to wall, steel studs to concrete slab). The most common primary superstructure performance problem experienced is that of too much flexibility caused by excessive slenderness.

Control rain penetration by proper siting, building shape, surface features and enclosure rain penetration control strategies. Deflection, drainage and drying are the proven approaches to rain control. Orient and site the building to deflect rain if possible use surface features, drips, and flashing to redirect water off the building and away from critical joints. Identify the rain control strategies (drained, storage, or exclusion) intended and *label the rain/water control layer with connections between all components and around all penetrations on the drawings*!

Where possible, use a drained system to control rain penetration. The ability of the design to drain water in the likely as-built condition should be a high priority. Remember that the enclosure system includes joints and elements. *Drawings should explicitly label the intended water control layer, all flashings, and laps.* Slopes and drip edges should be dimensioned and labelled.

If a mass system is chosen (e.g., solid masonry, adobe) ensure that the amount of mass, both storage and drying capacity, and its absorptance is matched to its exposure and the climate. Deflect as much rain from the building as possible with good details and consider the buildability of the system.

If perfect barrier elements are used (e.g., insulated glazing units, low-slope roof membrane, precast panels, insulated metal panels) pay extra attention to surface drainage, joints, and buildable details. Joints between elements should be of a drained design. Provide as much rain shedding and solar protection as possible to polymer-based sealants. Quality control is essential both in the factory and on site, and a maintenance plan may be needed.

Control air flow by using an effective air barrier system. A strong, stiff, durable and, above all, continuous air barrier can be placed anywhere within the assembly. Consider inspection, repair, and replacement. The buildability of the air barrier is critical; all joints and interfaces in three dimensions must be tight and remain so. *The air barrier system should be labeled (i.e., its function and the specific materials) on drawings and connections between all components and around all penetrations detailed.*

Insulate the structural components of the assembly. While not as critical for some structural materials and systems as for others, positioning the structural support components of the enclosure in a location with stable temperatures close to the interior will always increase the durability. *A continuous layer of insulation should be provided around the support and super structure, including studs, columns, floor slabs, walls, etc.* Always consider the size and effect of thermal bridges. The durability of most polymeric materials (e.g. common air and water barriers) will benefit from being placed inside some or all of the insulation. Thermal mass located inside of the primary insulation can usefully stabilize temperature in almost all permanently occupied buildings.

Design to accommodate construction tolerances and movements. Include realistic construction tolerances and provide a design that allows for them. This will often require an understanding of realistic expectations for workmanship (e.g., footings poured on soil will always have significant variations) and tolerances of materials (sawn timber has implicit variations as do fired ceramic plumbing fixtures).

Movement joints (not just control, shrinkage, or expansion joints) or flexible materials should be designed and provided wherever movements could be expected. *Movement joints must be indicated and dimensioned on drawings.*

Provide a moisture-tolerant design by balancing wetting, drying, and storage potentials. Despite the previous measures, always presume that some wetting will occur. Assess the susceptibility of each material to moisture damage relative to its vulnerability (i.e., its potential for getting wet). Provide as much drying capacity as practical and always consider the impact of safe moisture storage.

Control diffusion by providing appropriate layers of vapor resistance throughout the assembly. A specific vapor barrier plane is often not required but if provided, *the vapor control layer should be explicitly labelled as to its function on the drawings.* Vapor barriers rarely solve problems on their own, and can cause them [Straube & Smith 2009, Wilkinson et al 2007]. Reduce the risk of condensation within walls by making it difficult for vapor to enter the assembly and relatively easy to leave. Vapor barriers should be located close to the "warm side" of the wall, although in many climates, the warm side changes with seasons. Beware of inward vapor drives in air-conditioned buildings and warm weather -- some vapor resistance inside of absorbent exterior cladding is beneficial.

Control unwanted solar gains. Provide exterior shade and solar control (low SHGC) glazing to control glare and excessive solar heating. Interior shading does not control solar heat gain well. Too much west-facing glass causes overheating and the worst comfort complaints. Almost twice as much energy falls on a south wall in the winter than in the summer.

Consider the enclosure as part of the building system. The enclosure and the HVAC system should not be considered in isolation. Understand how the building will provide heating, air conditioning, humidity control, fresh air, and filtration. Resources expended on the enclosure will often displace resources expended on HVAC. Ensure that interior conditions are controlled to levels that the chosen enclosure design can tolerate (e.g., ventilate to control RH, ensure building spaces are not over-pressurized or de-pressurized).

Consider the future. Attempt to design an enclosure that allows for maintenance, repairs, improvements, changes in use, additions, and deletions. What is the likely life-cycle of the building? How will it be maintained, repaired, retrofit, converted, demolished, and recycled? Can the windows be replaced in 35 years and the roofing membrane parapet in 25 yrs?

In addition to the guidance described above, ensuring continuity, in design and construction, of the primary control layers remains a practical matter of great importance. In fact, this is such a basic need, methods to help ensure designers ensure control layer continuity have been developed. The 'pen test' is one such technique described in detail in Appendix A.

9.8 References

Andersen, N.E., "Summer Condensation in an Unheated Building," *Proc. of Symposium and Day of Building Physics*, Lund University, Swedish Council for Building Research, August 24-27, 1987, pp. 164-166.

Bomberg, M., and Shirtcliffe, C.J., "Influence of moisture and moisture gradients on heat transfer through porous building materials," *Thermal Transmission Measurements of Insulation*, ASTM STP 660, ed. R.P. Tyre, ASTM, Philadelphia, 1978, pp. 211-233.

Burnett, E.F.P., Straube, J.F., Karagiozis, A.N., *Development of Design Strategies for Rainscreen Sheathing Membrane Performance in Wood Frame Walls*, ASHRAE RP1091, 2004.

Cunningham, M.J., Tsongas, G.A., McQuade, D., "Solar-Driven Moisture Transfer through Absorbent Roofing Materials," *ASHRAE Transactions* 96, Part 1, 1990, pp. 465-471.

Finch, G, and Straube, J.F., "Ventilated Wall Claddings: Review, Field Performance, and Hygrothermal Modeling", *Thermal Performance of the Exterior Envelopes of Whole Buildings X International Conference*, Clearwater, Dec., 2007.

Hall, C., Hoff, W.D., Nixon, M.R., "Water Movement in Porous Building Materials - VI. Evaporation and Drying in Brick and Block Materials," *Bldg & Environ*, Vol. 19, No. 1, pp. 13-20, 1984.

Hedlin, C.P., "Heat Flow Through a Roof Insulation Having Moisture Contents Between 0 and 1% by Volume, in Summer," *ASHRAE Transactions* 94, 1988, pp. 1579-1594.

Hutcheon, N.B., "Fundamental Considerations in the Design of Exterior Walls for Buildings", *Proc. 47th Annual Meeting of the Engineering Institute of Canada*, Halifax, May 1953.

Lstiburek J.W. "Five fundamental changes in the last 50 years". *ASHRAE Journal*. July 2009; Vol 51(No. 7), pp.52-56.

Lstiburek, J.W. 2020. *Moisture Control for Residential Buildings: Principles and Practices.* Building Science Press, Westford, MA.

Popp, W., Mayer, E., Künzel, H., 1980. *Untersuchungen über die Belüftung des Luftraumes hinter vorgesetzten Fassadenbekleidung aus kleinformatigen Elementen*, Fraunhofer Institut für Bauphysik, Forschungsbericht B Ho 22/80, April.

Smegal, J., Grin, A., Straube, J. "Controlling Inward Vapor Drives Behind Adhered Veneers Using Continuous Insulation", Fourth *NIBS Building Enclosure Science Technology Conference*, Kansas City, April 13-15, 2015.

Smegal, J., *Drainage and Drying of Small Gaps in Wall Systems.* MASc Dissertation, Civil Engineering Department, University of Waterloo, 2006.

Straube, J.F., "Chapter 5: Moisture", in *Strawbale Design and Construction*, Bruce King, ed., Green Building Press, San Francisco, 2006.

Straube, J. F., & Burnett, E. F. "Drainage, ventilation drying, and enclosure performance". *Proc of Thermal Performance of Building Envelopes VIII*, Clearwater Beach Florida, December,1998, pp. 189-198.

Straube, J.F., "Moisture in Buildings". *ASHRAE Journal*, Jan. 2002, pp. 15-19.

Straube, J.F., VanStraaten, R., Burnett, E., "Field Studies of Ventilation Drying", *Proc. of Performance of Exterior Envelopes of Whole Building IX*, Clearwater Beach Florida, December, 2004.

Straube, J.F., and Smegal, J. "Modeled and Measured Drainage, Storage, and Drying Behind Cladding Systems", *Thermal Performance of the Exterior Envelopes of Whole Buildings X* International *Conference*, Clearwater, Dec., 2007.

Straube, J.F., and Smegal, J., "The Role of Small Gaps Behind Wall Claddings on Drainage and Drying", *Proc. of 11th Canadian Building Science & Technology Conference*, Banff, March, 2007.

Straube, J.F., *High Performance Enclosures: Design Guide for Institutional Commercial and Industrial Buildings in Cold Climates.* Building Science Press, Somerville, Massachusetts, 2012. 320 pp.

Straube, J.F., Smith, R., "Need for Vapor Barriers in Above-Grade Walls with Spray Polyurethane Faom Cavity Insulation", *Proc. of Can. Bldg. Sci. & Tech. Conference*, Montreal, May 6-8, 2009, pp. 473-481.

TenWolde, A., Mei, M.T., "Moisture Movement in Walls in a Warm Humid Climate," *Proceedings of Thermal Performance of the Exterior Envelopes of Buildings III*, Clearwater Beach, Florida, Dec 2-5, 1985, pp. 570 - 582.

Van Linden, S., Lacasse, M., & Van Den Bossche, N., 2018. "Drainage and retention of water in small drainage cavities: Experimental assessment." *7th International Building Physics Conference*, Syracuse, NY.

Wilkinson, J., Ueno, K., DeRose, D., Straube, J.F., Fugler, D., "Understanding Vapour Permeance and Condensation in Wall Assemblies", *Proc. of 11th Canadian Building Science & Technology Conference*, Banff, March, 2007, 12 pp.

Wilson, A.G., "Condensation in Insulated Masonry Walls in Summer," *RILEM/CIB Symposium*, Helsinki, 1965, pp. 2-7.

Chapter 10

Condensation Control

10.1 Introduction

Controlling the occurrence and severity of condensation is a major requirement of most enclosures. Condensation control is just as important in hot-humid climates as it is in cold climates and it can be a critical consideration for buildings with demanding interior environments, such as enclosed pools, freezer warehouses, humidity-controlled factories, etc.

Condensation has long been a concern in buildings, and science has been applied to understanding and controlling it over multiple generations [Teesdale 1939, Wilson & Garden 1965, Andersen 1972, Bomberg & Onysko 2002]. Current codes, standards and guides continue to focus on condensation control [BS 5250 2021, EPA 2013].

There are two broad types of condensation that must be considered:

- interior surface condensation, and
- interstitial (within enclosure) condensation.

The following sections will consider the processes that result in surface and interstitial condensation, and then explore the means to assess, predict, and control condensation occurrence and intensity.

10.2 Surface condensation

Condensation on interior building surfaces is undesirable for a number of reasons other than just inconvenience. First, the moisture provided by condensation will often damage moisture-sensitive interior finishes such as wallpaper, paint, wood, and gypsum wallboard. Secondly, the condensation provides moisture for mold growth. Mold growth not only results in material damage and staining it can also be a health hazard. To avoid mold growth, the relative humidity at a material surface should be kept below about 80%, not just below the 100% RH required for surface condensation (see Chapter 8).

Surface condensation is often the result of dynamic and short-term variations in temperature, (e.g., a cold and windy night, or a cool morning), or absolute humidity, (e.g. after a shower, during cooking, etc.) [Adan 1994]. These short-term and dynamic effects should be considered when conducting a detailed hygrothermal analysis. Some materials are able to adsorb/absorb moisture quickly, e.g., porous lime stuccos, uncoated concrete block, porous brick, cotton, etc. and thus reduce the potential for surface condensation during short-term events. This hygric buffering capacity can be useful for reducing variations in interior relative humidity despite rapid changes in temperature and moisture production [Straube and deGraauw 2001, Padfield 1999, Rode 2004]. Materials such as plastic wall coverings and glazed tile cannot absorb moisture and will experience surface condensation almost instantly after exposure to the proper conditions.

10.2.1 Cold-weather surface condensation

Condensation will only occur on a surface that is at or below the dewpoint of the adjacent air. Hence, understanding and predicting surface temperatures are the key to the understanding and prediction of surface condensation.

The methods developed in Chapter 5 can be used to predict surface temperatures:

$$t_{surf} = t_{in} - \frac{R_{film}}{R_{total}} \cdot (t_{out} - t_{in}) \qquad \text{(Eq. 10.1)}$$

where t_{in} and t_{out} are the interior and exterior air temperature respectively, and

R_{film} and R_{total} are the thermal resistance of the surface film and enclosure assembly respectively.

Any situation which results in interior surface temperatures significantly below interior air temperature (i.e., low insulation levels, low surface film conductances, wind washing, and thermal bridges) can result in surface condensation and its related problems. The drop in temperature at an interior enclosure surface (Figure 10.1) during cold weather results in a rise in RH (Figure 10.2). The accuracy of these results is heavily impacted by the estimate of the surface transfer coefficients: computer models that have detailed radiation exchange physics can often improve the accuracy of the surface film. In particularly demanding applications, computational fluid dynamics can improve the convective component.

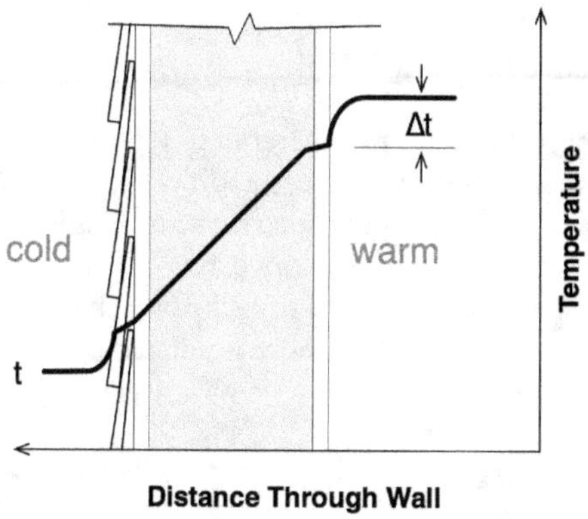

Figure 10.1: Temperature drop at enclosure surface

Almost all areas of modern enclosure walls and roofs are so well insulated that surface condensation during cold weather should not be a problem. This is so because the interior surface film (with a conductance of about 8 W/m²°C, and a resistance of approximately RSI0.12/R-0.7) provides only a small proportion of the total enclosure thermal resistance and thus the temperature drop across the interior surface film is small. Under cold exterior conditions, the temperature of the interior surface is therefore high in well-insulated assemblies, and hence the RH at this surface is only slightly higher than the interior RH.

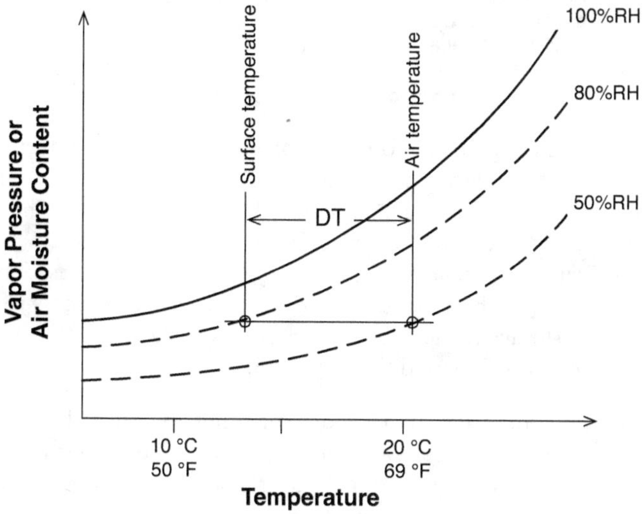

Figure 10.2: Relative humidity change due to surface temperature drop

Surface condensation tends to become a problem in practice when:

1. the thermal resistance of the enclosure is low (as it is at thermal bridges),

2. the surface film has an unusually high resistance value, or
3. interior relative humidity is high.

Thermal bridging, especially at steel framing (Figure 10.3, Figure 10.4), or at the intersection of wall corners with roofs and floors, can cause low surface temperatures and thus allow condensation to occur. This is one of many reasons to avoid thermal bridging in design. Convective cooling, or wind-washing, should be controlled, especially at corners, since this mechanism can result in interior surface temperatures much below those predicted by simple conductive heat transfer.

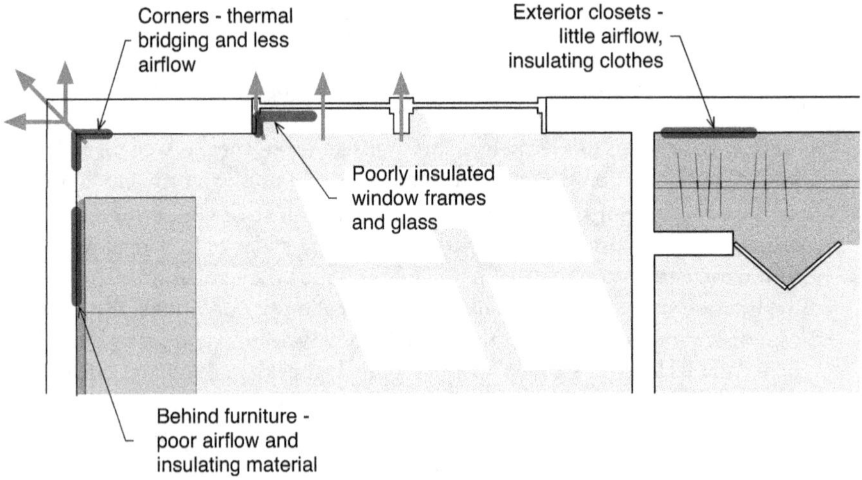

Figure 10.3: Common locations of surface condensation

Surface condensation often forms first in corners and behind furniture. This occurs because the surface film conductance for these situations is much lower than over the face of the wall, perhaps only 2 to 4 $W/m^2 \cdot °C$. The lower convective and radiative heat transfer results in a lower surface temperature and hence a greater probability of, and severity of, surface condensation. In other situations, radiative heat loss (to a cold surface, such as the ice surface of an ice rink) from a surface reduces its temperature so far below the adjacent air that condensation or high humidities can occur.

Windows and doors are often a source of condensation problems in modern buildings because they are not as well insulated as the opaque enclosure (they could be considered large thermal bridges). Condensation on window glazing is not directly a problem (other than by reducing visibility) but the moisture may collect, run down and cause staining, mold, and rot. The coldest parts of modern double-glazed windows are typically the lower corners for two reasons. First, cold, dense air falls along the surface of the glazing and transports cold air to the bottom of the window. Secondly, corners are more susceptible because both the window frame and the glazing spacers (often aluminum) tend to be much more conductive than the glazing unit itself. Figure 10.5 presents carefully measured air velocity and heat transfer coefficients for a modern curtainwall system during very cold weather [Ge & Fazio 2003][1].

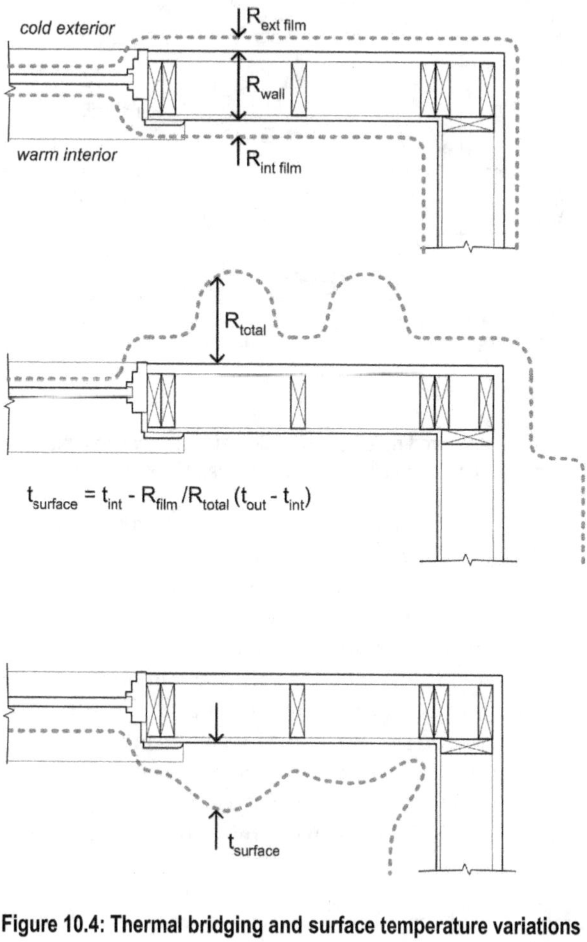

Figure 10.4: Thermal bridging and surface temperature variations

Very tall and narrow windows (and patio doors) with closely-spaced mullions are therefore often the most susceptible to condensation problems. Well-insulated frames and insulated spacers combined with proper HVAC design and operation can alleviate most window surface condensation problems.

It should be noted that field experience has shown that deficient design or construction can also allow exterior airflow within hollow window frames and glazing pockets. This airflow can cause cold interior surface temperatures, much colder than predicted by conductive heat flow calculations.

[1] For comparison, in Chapter 5, the radiation film coefficient was to often have a value of 2 to 4 W/m^2 K.

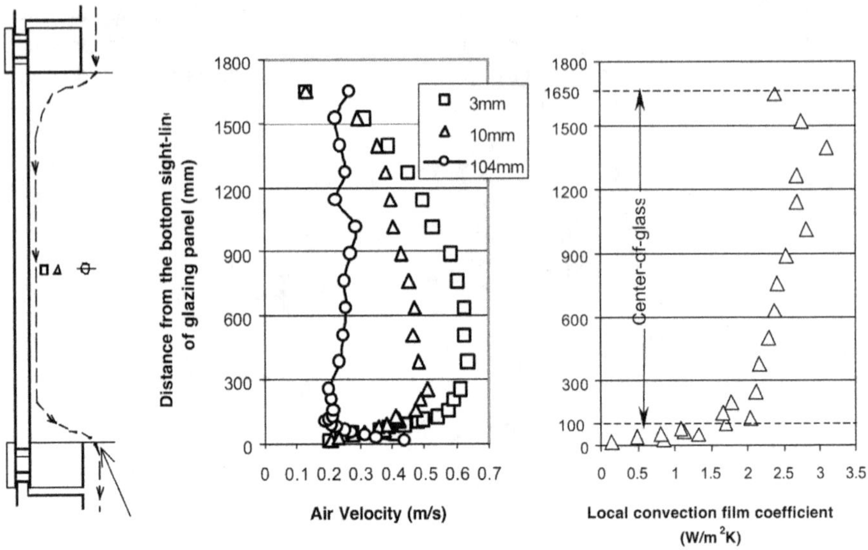

Figure 10.5: Measured natural convection air velocity and heat transfer coefficient for an aluminum curtainwall [Ge & Fazio 2003]

Because of their low thermal resistance, natural convection is more important for understanding surface condensation on windows and doors than it is for walls. For example, windows located close to or flush with the interior of a wall surface will be kept warmer than windows located close to the exterior because natural convection will increase the convective film conductance (Figure 10.6). Similarly, relatively small windows installed within deep openings will have low surface conductances on both the interior and exterior surfaces, resulting in higher overall thermal resistance and a greater chance for surface condensation. Curtains and blinds drawn over the interior surfaces of windows can dramatically reduce outward heat transfer during cold weather and thus cause surface condensation. Special blinds can avoid this if they are sufficiently hygroscopic or air and vapor tight.

The natural heat transfer coefficient varies over the width and height of a window and wall. Near corners, bouyancy flows are retarded more by friction, and cold dense falling air will increase in velocity with height. For most design purposes standard estimates are sufficient, but if more accuracy is required CFD modeling may be useful (if properly benchmarked to physical measurements).

One common method of avoiding condensation on windows is to enhance convective heat transfer by placing a source of heat directly below the window: although natural convection from a hot source will help, in challenging applications like swimming pools forced convection is utilized.

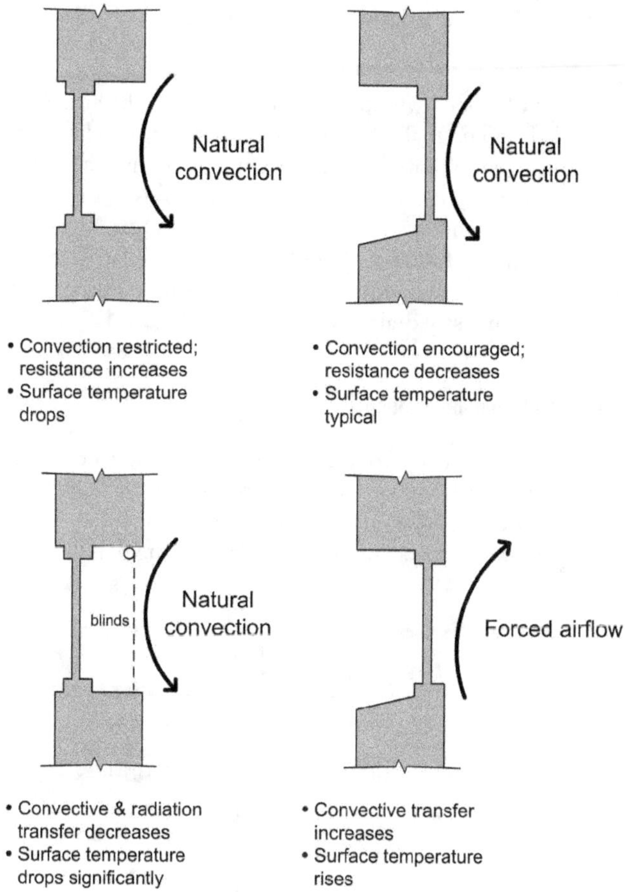

Figure 10.6: Natural convection effects on windows

Radiantly heated panels above a window can also be effective and do not use valuable floor space, are much harder to block by users, and can be visually unobtrusive. Another effective and more energy intensive technique is to conductively heat the window frames with heat tracing. Finally, it is often recommended that one open a problem window slightly: this reduces the moisture content of the air in the near vicinity (by dilution ventilation) and thereby can reduce condensation problems. Naturally, the use of highly insulated window frames and glazing with insulated spacers should be encouraged, along with the control of interior RH (if possible). Modern well-insulated glazing systems can even obviate the need for the below-window heating vents or convectors.

Skylights are particularly problematic for a number of reasons. First, night- sky radiation reduces exterior surface temperatures more than for vertical windows. Secondly, skylights are often located in a "tunnel." This tunnel may be as much as a meter deep and dramatically lowers convective and radiative heat transfer from the interior.

Large expanses of sloped windows are often used in the atria of commercial buildings and shopping malls. Surface condensation is so likely in poorly insulated (i.e., typical) aluminum-framed enclosure systems that condensation gutters are provided along the sides of framing members in many sloped glazing systems (and some windows). Although providing more highly insulated glazing systems (especially frames) and reduced interior relative humidity is the preferred solution to surface condensation, condensation gutters can also be used to capture and harmlessly drain the water away (Figure 10.7).

10.2.2 Warm-weather surface condensation

In the summer the interior surfaces of enclosures tend to be warmer than the interior air temperature and surface condensation is unlikely. Interior surfaces can be cooled by exposure to streams of air from an air conditioning air supply or radiatively cooled by exposure to a radiant cooling panel.

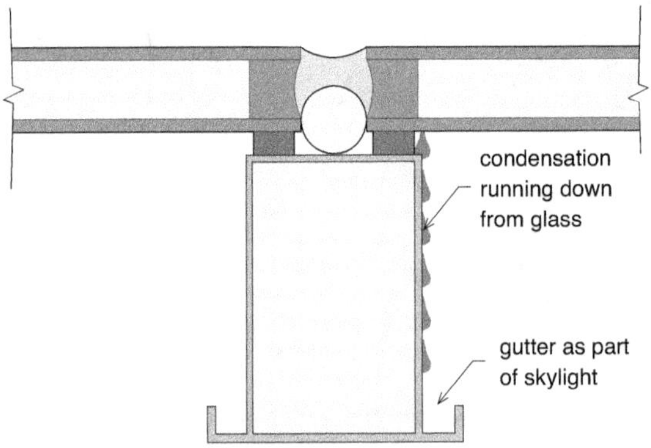

Figure 10.7: Sloped skylight frame with integral condensation gutter

Warm-weather surface condensation on concrete floors or other high thermal mass elements can often be a problem in both warm and cold climates. The thermal storage capacity of the concrete causes its temperature to lag behind the air temperature. Also, if the concrete floor is a slab on grade, the cool soil (cool because of its large thermal mass) will further reduce the temperature. It is not uncommon for the combined effect of seasonal soil temperature lag and diurnal concrete slab temperature lag to cause as much as a 10 °C (20 °F) temperature difference between the air and the slab surface. In humid weather these depressed temperatures are likely to result in condensation.

Compounding the problem of thermally massive slabs is the poor heat transfer coefficient from the air to the slab (approximately 4 to 6 W/m2· °C or as little as half that of walls) and the insulating effect of vapor permeable floor coverings, especially carpet (thick carpet can have a thermal resistance exceed RSI0.2 / R1). Carpet covered, uninsulated concrete slab-on-grade floors are very likely to experience significant surface condensation in humid climates, and the carpet is an ideal mold growing substrate, unless interior RH is strictly controlled. Concrete slabs on grade should be placed on insulation in climates with humid springs and summers (both in cold and warm climates). This both reduces the thermal lag and increases the temperature of the concrete. Excessive ventilation with untreated outdoor air should be avoided during humid periods (when the soil and concrete temperatures are likely to be at a minimum).

Surface condensation problems during warm weather can also be associated with cold water plumbing. Cold water piping within interior partition walls and cold water storage tanks such a toilets, will often be sufficiently cool to allow condensation of humid interior air. This surface condensation may become a serious problem if the moisture collects within hidden service spaces where it can support mold, rot, and corrosion. The usual solution is either to apply a small amount of insulation, either inside or outside, the tank or pipe or better manage interior humidity via the mechanical system.

10.2.3 Surface condensation resistance ratings

It is customary to rate the “condensation resistance” of window products because their typically low thermal resistance means they are the most susceptible to cold-weather condensation. Condensation resistance is, of course, a function of the interior humidity and temperature, exterior temperature, thermal mass and hygric buffer capacity.

The most fundamental measure is the surface temperature index, TI, defined as the temperature drop from the inside air to surface divided by the total temperature difference between inside and outside) (Figure 10.8):

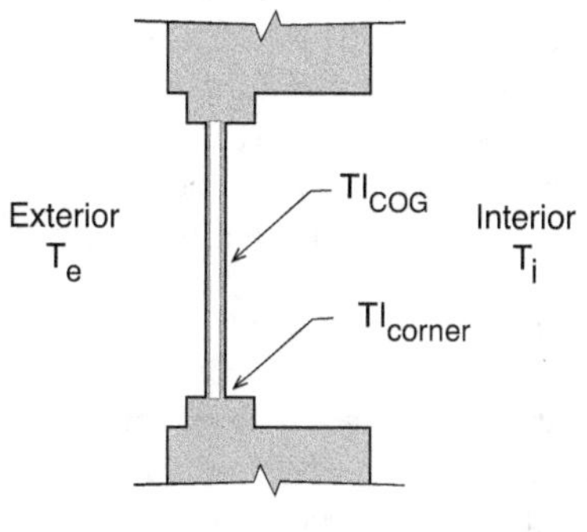

$$TI_x = \frac{(T_x - T_e)}{(T_i - T_e)}$$

Figure 10.8: Temperature Index Definition for an Arbitrary Location, X

$$TI_{surface} = \frac{t_{int} - t_{surface}}{t_{in} - t_{out}} = \frac{R_{surfacefilm}}{R_{total}} \quad \text{(Eq. 10.2)}$$

Interior surface condensation is therefore expected, given TI, when

$$T_{dewpoint,in} > t_{in} - TI \bullet (t_{in} - t_{out}). \qquad \text{(Eq. 10.3)}$$

As the surface film coefficient varies over a window and the thermal resistance varies as well, different locations will have different temperature indices: the center of glass (COG) and lower corner will each have different values.

Attempts to standardize condensation ratings have been on-going for at least half a century [Sasaki 1971]. Some of the terms used include the Condensation Index (CI) [e.g., NFRC 500-2017], Temperature Index (TI), and condensation resistance factor (CRF)[Lacasse et al 2009]. These indices vary in how many points are measured over the surface, how they are spatially averaged, and under what environmental conditions the tested or measured.

10.3 Interstitial condensation

Interstitial condensation of water vapor is one of the two most important sources of moisture in the above-grade enclosure (driving rain is the other). One means of controlling interstitial condensation is the provision of appropriate vapor and air control layers (aka barriers). However, other complementary approaches can, and usually should, be used to control interstitial condensation, e.g., the control of driving forces and the control of potential condensation surface temperatures.

Condensation will occur whenever air contacts a surface with a temperature below its dewpoint. Water vapor in air can move to an interstitial surface by two mechanisms: diffusion and convection. Figure 10.9, calculated using the methods of Chapter 6, shows how water vapor moving by convection[2] and/or diffusion can condense on a cold surface within an enclosure. The water vapor must contact a colder surface for condensation to occur. Therefore, if the direction of vapor flow is from the cold side to the warm side of an enclosure, condensation will not occur. Vapor flow from the cold side to the hot side can only occur by diffusion if the absolute humidity is higher on the cold side than on the warm side. This is not common since it requires a very significant relative humidity difference or a small temperature difference.

Air passing through an enclosure will not change its vapor content unless a) it becomes saturated and condenses water onto a surface below its dewpoint temperature or b) it gains water as it passes over or through a material with a higher vapor pressure. Therefore, depending on the flow path and the temperature and moisture content of the air and enclosure layers, airflow may cause wetting (condensation and adsorption) or drying (evaporation and desorption).

[2] Air leaking outward through an insulated enclosure will of course often follow a tortuous path.

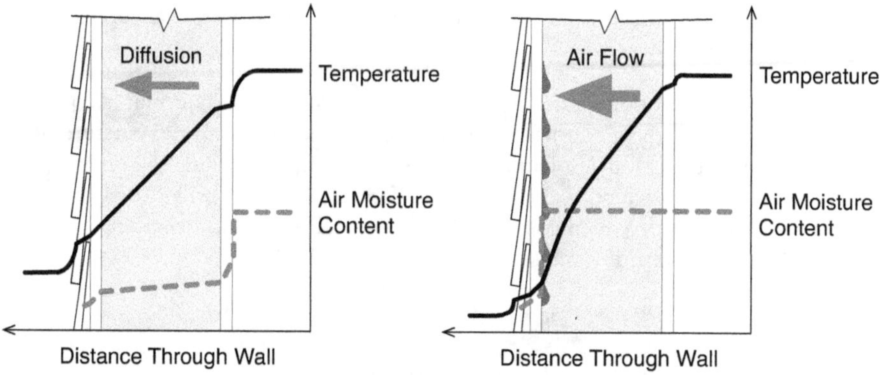

Figure 10.9: Water vapor transported by air leakage and diffusion can cause interstitial condensation on cool surfaces within walls

The process of cold weather condensation at a building system level is illustrated in Figure 10.10 and Figure 10.11.

1. In the first stage, exterior air is heated when it is brought in from outdoors (shown as about -7 °C and 70%RH). The absolute moisture content remains the same, but the relative humidity drops significantly, often to as low as 10 or 20% RH during very cold weather.

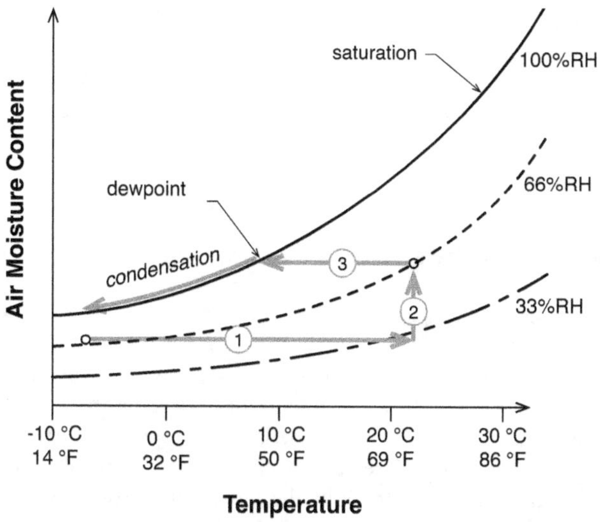

Figure 10.10: Cold weather air leakage condensation process

2. Mechanical humidification or, more commonly, moisture released by activity within the building increases the absolute and relative humidity. This is a critically important variable in the process. If little moisture is released, or a large amount of air dilutes the moisture that is released,

the interior moisture content will rise little, and interstitial condensation is unlikely.

3. As air is cooled, the RH rises. Condensation will occur if air is cooled to, or if the air contacts surfaces below, the dewpoint temperature of the interior air. Potential condensation surfaces include the interior surface of a window, the interior surface of a thermal bridge, or a cold surface inside the enclosure, e.g., the underside of attic sheathing, the back of wall cladding, etc.

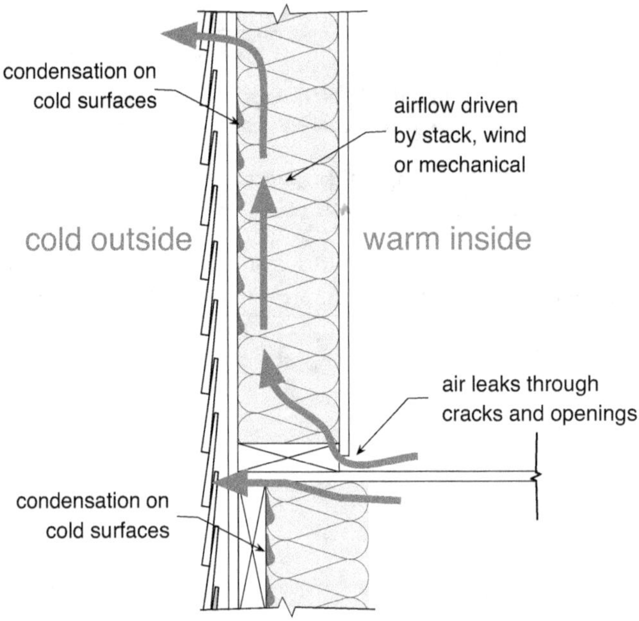

Figure 10.11: Air leakage path resulting in cold-weather condensation

10.3.1 Warm-weather diffusion condensation

The literature abounds with reports of cold-climate, cold weather interstitial condensation problems. This is likely because of the airtight construction used to save energy in such climates and the large temperature difference between interior and exterior (often greater than 30 °C / 55 °F for extended periods). However, both surface and interstitial condensation in warm weather and warm climates can be a problem because of the often high exterior relative humidity levels and the increasing use of air conditioning (although temperature differences are usually only 10 °C / 20 °F for most of the time).

Warm weather interstitial condensation can occur if warm, humid air contacts cool surfaces inside an assembly. This mechanism of course requires inward air leakage and outdoor dewpoint temperatures of near interior air temperature or higher (Figure 10.12). The growing use of air conditioning means that the interior surfaces are often kept cool even during hot weather.

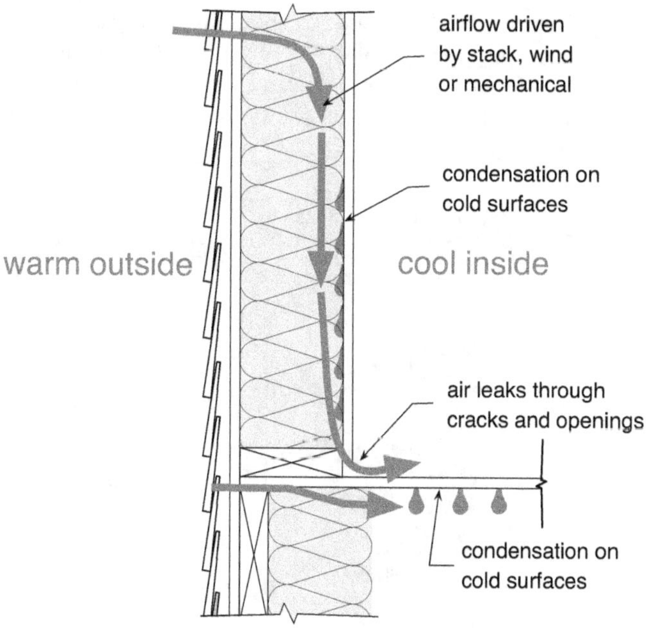

Figure 10.12: Warm-weather air leakage condensation

Inspection of the slope of a saturation vapor pressure versus temperature plot (i.e., the psychrometric chart) will show that the size of the vapor pressure difference between indoors and outdoors can be as large during hot weather as it is during cold weather even if temperature differences experienced are less.

Inward-flowing air is commonly the source of damage warm weather condensation, but an important category of "inward vapor drive" or "solar driven vapor" field performance problems have long been reported [Wilson 1965, Andersen 1987, Southern 1986, Sandin 1993, Straube 1997]. Ironically mostly of this early research was in cold climate countries with warm humid summers.

Warm-weather condensation can also occur because of solar heating driving water vapor from damp materials near the exterior of the assembly towards the cooler interior layers. Although solar heating of above-grade enclosure elements increases the temperatures of all layers of an enclosure and thereby reduces the risk and magnitude of cold- and warm-weather condensation, solar heating can cause interstitial condensation in some unique situations.

As described in Chapter 8, the vapor pressure at the surface of a material is a function of its moisture content and temperature (via the sorption isotherm). Absorbent claddings, e.g. brick, stucco, stone, concrete, wood, will have very high vapor pressures when heated.

For example, the sorption isotherm for wood indicates that at 12% MC, an average moisture content for exposed wood, the RH at the surface would be about 60%. If this wood were used as cladding on the east, west, or south exposure, it is likely that it would be heated to about 40 °C on many sunny summer days. A surface vapor pressure of 0.60 · 7380 = 4430 Pa would be generated by this heating. If a brick veneer had been exposed to rain the day earlier, the brick would be close to saturation and vapor pressures of about 7000 Pa would be generated (Figure 10.13).

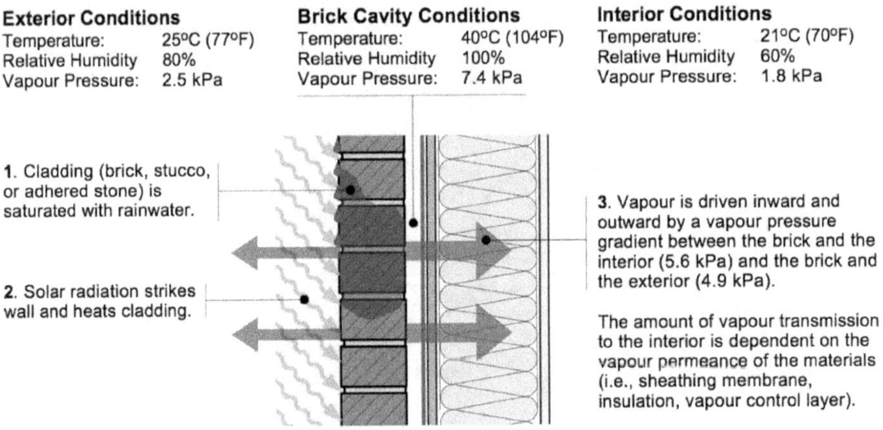

Figure 10.13: Warm weather solar-driven inward vapor diffusion condensation process

These high vapor pressures both promote drying to the outside and vapor flow into the assembly. If the interior is air conditioned, the temperature of the innermost layer could be expected to be about 20 - 25 °C / 68 – 80°F. The saturation vapor pressure at these temperatures is in the order of 2500 to 3000 Pa. The vapor pressure difference acting inward would therefore be in the order of one to five thousand Pascals. Clearly, these vapor pressures are much higher than those that will be experienced under cold weather conditions.

If little vapor resistance is provided between the back of the cladding and an interior vapor resistant layer is provided (e.g. polyethylene, vinyl wallpaper, oil paint) large inward vapor flow rates will feed condensation on these inner layers. This moisture can quickly cause damage because 1) it is deposited on the inner, usually more moisture sensitive, part of the enclosure and, 2) corrosion, mold growth and rot all occur much more quickly at elevated temperatures.

Fortunately, this situation does not normally occur often or for a long duration. The requirements for damaging inward vapor drives, e.g., a relatively wet and porous cladding (typically after rain), solar heating, and vapor permeable outer layers, are not often coincident. The storage capacity of masonry inner wythes can easily deal with the moisture from inward drives. Wood-framing, especially unsheathed framing is quite susceptible if no vapor retarding sheathing is used.

Because of its limited moisture storage capacity, light-gauge steel framing is perhaps the most susceptible to damage from inward drives.

Insulated sheathing in the form of EXPS is sufficiently vapor impermeable (usually 35 to 70 ng/Pa· s· m²) to control inward flows during warm weather. Polymer- based housewraps, exterior gypsum sheathing, and modern "breathing" building papers all provide too little control (e.g., vapor permeance of 1000 to 4000 ng/Pa· s· m² at high RH levels). Moderate levels of ventilation behind the cladding also control the problem [Burnett et al 2004] by allowing the vapor to leave through convective flow -- good ventilation is, however, difficult to achieve in some systems.

Stucco or EIFS-clad walls in warmer climates may experience inward-vapor-drive related problems if the finish is less than perfectly constructed. Small amounts of rainwater may penetrate behind the cladding but cannot dry quickly if 1) the EIFS finish/stucco and foam insulation is relatively vapor impermeable and 2) a vapor barrier is placed on the interior. Water trapped in the expanded polystyrene sheathing/substrate can be driven inward to condense on the air-conditioned polyethylene vapor barrier or vinyl wall paper, causing deterioration of the wood or steel framing and drywall commonly used in such walls.

10.4 Condensation control

It must be recognized that condensation of moisture within enclosure assemblies is not necessarily a problem. The quantity of condensation, its location, and duration must be considered in the context of the moisture balance.

If the condensate is in liquid form, it can be absorbed by porous materials (e.g., wood, concrete) and wet the surfaces of non-porous or hydrophobic materials (e.g. steel, plastic). The moisture will cause a problem if, and only if:

1. sufficient drying, by diffusion or convection, does not occur,
2. the safe storage capacity of the material is exceeded, and
3. the material is susceptible to moisture damage.

If the condensing surface is below freezing, any condensate will form as frost. This frost is unlikely to cause any expansive damage, but may, in extreme cases, completely fill a space. Corrosion, rot, mold growth and most other moisture-related deterioration mechanisms will not occur below freezing temperatures. However, when the surface temperature rises above freezing, a large amount of liquid water will be released in a short period of time. When this occurs, drainage may remove a large proportion of any water before it is absorbed.

Thawing will occur daily in many cold climate locations because of solar heating and heat flow from the interior. Even in very cold climates such as

Bismark, ND, and Edmonton, AB, it is unlikely that exterior layers of an *unshaded* roof or wall will accumulate moisture as frost for more than a few weeks at a time. However, in assemblies that are not exposed to the sun, frost can accumulate for several weeks or months before near freezing air temperatures and diffuse solar radiation allow for thawing.

Solar heating, even on the north side of a building, will often maintain the temperature higher than air temperatures alone would suggest. For example, field monitoring has shown that the average monthly temperature of a dark colored south-facing brick wall can be as much as 10 °C above air temperature during winter in high latitudes [Straube & Burnett 1998]. Therefore, the risk of condensation is lower (higher surface temperatures) and the drying potential is greater (greater material vapor pressure differences). As described earlier, solar heating in the summer can, however, cause performance problems by reversing vapor flow.

Cold-weather condensation is likely to occur on the back of many cladding materials (e.g. steel, concrete, brick, vinyl) because this cladding is both cold and relatively vapor impermeable. In these cases, ventilation can be a useful means of reducing the effective vapor permeance of the cladding.

Absorbent claddings like brick will also transport liquid condensate to the exterior surface by capillarity where it can evaporate to the exterior air. This capillary movement of moisture can transport salts with it. When the water reaches the surface and evaporates, the salts are left behind and the whitish residue is called efflorescence.

If the cladding becomes saturated (almost certainly the case in practice, since capillary transport through partly saturated materials is a slow process), damage may occur, for example, freeze-thaw damage and rot. To avoid this possibility, moisture absorbing cladding (especially sawn wood siding) should be back primed with a hydrophobic coating or paint that resists absorption. Condensate will then be forced to harmlessly drain and/or ventilate and diffuse through the cladding without saturating it.

Wetting, drying, and storage can all be assessed with the aid of the physics presented earlier. That the assumptions, material properties, and physics are approximate should always be borne in mind during any analysis. The basic moisture control approach should be used for condensation moisture control: balance wetting with drying accounting for the safe storage capacity available.

Drainage cannot often be considered as the sole means of removing condensation because, in almost all practical cases, materials need to be saturated before drainage can occur. And most materials can be damaged by long-term exposure to liquid water. Nevertheless, drainage is a very effective "emergency relief valve" which allows the removal of excess moisture because of short-term extreme wetting. Diffusive and convective drying are necessary to subsequently lower the moisture content to safe levels.

10.4.1 Diffusion control

Vapor diffusion can be controlled by locating vapor retardent layers in the appropriate locations. This is not always easy to achieve because vapor flow can occur in either direction, and because one wishes to retard vapor movement at one time (when wetting is possible) while at other times it is often best allow vapor to flow through (when drying is desired). Therefore, the vapor permeance of each layer in the enclosure must be carefully chosen to achieve good performance. Although some building codes require the placement of highly vapor resistant layers on the inside of an assembly, this is now known to be a far too simplistic solution, and in many cases may cause, not avoid, damage.

The simple approach described in Chapter 6 can be used to predict diffusion-related condensation and diffusive drying [Straube 2001]. If a condensation plane is identified, the rate of accumulation must be calculated to provide meaningful information. In many cases, the amount of accumulation due to vapor diffusion is so small (e.g. a few g/m^2/day) it can be ignored. When this is not the case, vapor resisting layers should be added or the temperature of the surface changed by rearranging the relative insulation values of the layer. Alternatively, the interior humidity within the building may be too high for the enclosure under consideration. (Note: Large amounts of condensation will cause a temperature rise at a surface due to the heat of condensation released. This normally does not need to be considered in simple analysis since the volumes of condensation should be relatively small.)

Diffusive drying of a material or layer saturated by rain, condensation, built-in moisture or rising damp can be calculated by assuming that the relative humidity of the saturated material is 100% (or some more accurate assessment from that material's sorption isotherm). Vapor flow can then be calculated using Glaser's approach. Calculating the drying of cladding with this method is rather inaccurate because the only vapor resistance is provided by the surface film and the surface film for vapor diffusion is both highly variable and not accurately known. Example diffusive drying calculations are provided in Chapter 6.

It can be shown from this type of drying assessment that inward drying is often possible in common enclosure assemblies if the vapor resistance of the interior layers is not too high. Basement and roof enclosures with vapor impermeable dampproofing/waterproofing layers on their exterior can only dry to the inside. Many walls in warm climates should be presumed to dry predominately to the interior.

It can also be shown that materials placed between two vapor resistant layers will dry very slowly once wetted. It is for this reason that two low-permeance vapor barriers should not be provided in an enclosure unless there is an exceptionally small likelihood of wetting. This is the approach taken in modern sealed Insulated Glazing Units and most low-slope exposed membrane roofs.

It is often stated that for climates where the interior temperature is higher on average than the exterior, the vapor permeance should increase from the inside to outside of an assembly. The reverse is true for cooling climates. The are however many examples of successful enclosure assemblies. There are, however, many examples of successful enclosures that violate these rules. Analysis that is more complex, and may involve storage, air leakage, and ventilation is often needed to explain the successful field performance and the limitations of the simple rules.

A common design question raised is how to limit outward flowing vapor diffusion in cold weather without trapping inward-flowing diffusion in warm weather. Smart, or "moisture adaptive" vapor control membranes is a product which has been developed to mitigate such problems. Such membranes changes their vapor permeance by as much as order of magnitude as a function of RH: when interior RH is low in winter outward vapor flows is resisted, but during warm weather if the RH on the outer side of the membrane rises, or condensation occurs, the permeance increases dramatically allowing vapor to pass inward.

Inward vapor drives can be a very important drying mechanism in some assemblies and situations if interstitial condensation is avoided. In warm weather, especially sunny warm weather, inward vapor drives can be the predominant direction of drying if the enclosure design allows this. In hot-humid climates it is actually difficult to design a proper enclosure with vapor tight interior surfaces because the vapor drive is almost always from the exterior to the interior. Condensation behind cabinetry, white boards, trim, and other vapor impermeable components on the inside routinely cause damage in hot-humid climates.

Water that has leaked through waterproof (and vapor impermeable) roofing membranes of low-slope roofs can also be redistributed by solar heating. In fact, many low-slope exposed membrane roofs without interior vapor retarders can dry quickly and effectively during warm weather because of the inward vapor drive phenomenon. Because most exposed membrane low-slope roofs place highly vapor impermeable layers (i.e., far less than 0.5 US perms or 30 $ng/Pa \cdot s \cdot m^2$) on their exterior-most surface, if moisture is present (from condensation, rain penetration, or built-in) it can diffuse outward only very slowly. The significant solar heating of a roof enclosure in the summer will evaporate moisture and drive it inward. If highly vapor-resistant layers are located close to the interior, water will condense here. Ventilation of such roofs may increase the drying potential but is rarely the best solution. In many moderate climates, avoiding an inner poly vapor barrier and providing some moisture storage in the roof assembly (often in the form of fibreboard) is an effective moisture control strategy that acknowledges and takes advantage of inward vapor drives.

This strategy has been labeled the "self-drying roof " by Desjarlais at the Oak Ridge National Labs [Desjarlais 1995, Smegal et al 2018]. In colder climates, an adaptive vapor control layer would be a possible solution.

10.4.2 Vapor convection control

Except for warm-weather condensation and atypical assemblies or environments, diffusion does not usually transport sufficient amounts of water to cause damage. Airflow, outward in winter, and inward in summer, will often cause significantly greater volumes of interstitial condensation. It is for this reason that the provision of air barriers in enclosure assemblies is so much more important than providing vapor barriers.

The process of air leakage condensation has been described above and condensation during both cold and warm weather. Control of air leakage condensation "simply" requires a perfect barrier to air movement (Chapter 11). However, a perfect air barrier is very difficult to achieve in practise. Rather than require perfection, it is often more practical and reliable to control the *direction* of airflow by controlling pressures using mechanical means (see below).

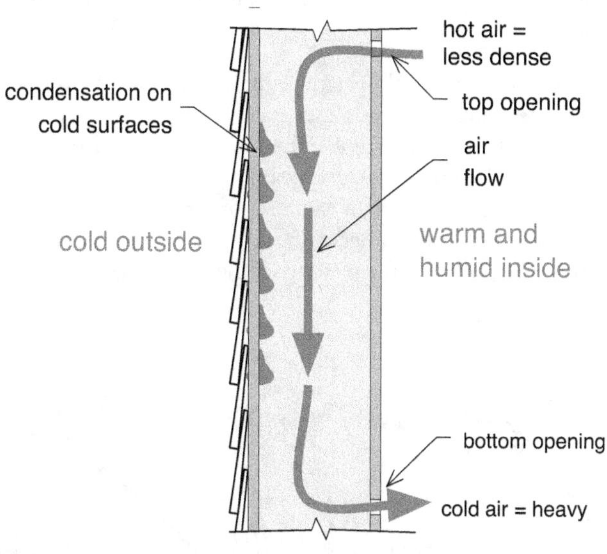

Figure 10.14: Convective loops can cause interstitial cold weather condensation in some situations

Controlling convective loops (Figure 10.14) within and through the enclosure is also important [Bankvall 1978, Huttunen & Vinha 2013]. This is achieved by providing an appropriate level of air permeability for the materials within the enclosure, avoid vertically connected air space, and maintaining a minimum level of airflow resistance on both the exterior and interior surfaces. The most common problem occurs when insulation is not pressed tight to the face of the wall to which it is attached or the stud cavity in which it is installed. Low-density batt insulation is rather air permeable and is the most likely to allow convective loops. Higher-density batts and dense-pack or damp spray cellulose tend to minimize the problem, but proper installation is required.

Chapter 7 and 11 present the physics of airflow and some details of its control and therefore will not be further discussed here.

10.4.3 General condensation control strategies

Three basic strategies can be used to control interstitial condensation:

1. control the flow of water vapor
2. control the forces driving vapor movement, and
3. control the temperatures of potential condensation surfaces.

Note, however, that the tolerance of enclosure assemblies to errors in assembly and workmanship should also be considered in design and therefore no one strategy should be relied upon exclusively. As discussed in the previous sections, practical enclosure designs include means of controlling both convective and diffusive vapor flows (strategy 1), but air barriers are usually built far from perfectly tight.

One of the most effective cold weather interstitial condensation control strategies is to provide a layer of continuous insulation ("ci" or "insulated sheathing") outside of the main structure. When properly implemented, this strategy maintains most of the moisture-susceptible materials at a temperature higher than the dewpoint temperature of the interior air in winter and thereby precludes any condensation from occurring, even if air leakage does occur (Figure 10.15). Hence, the potential for wetting of these materials has been greatly reduced.

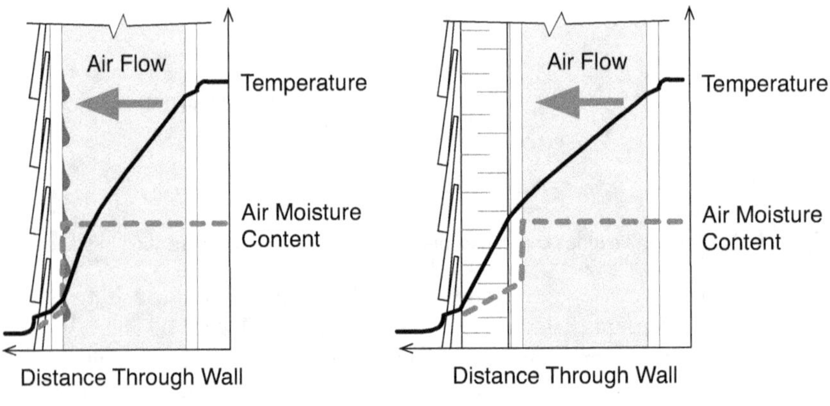

Figure 10.15: Control of interstitial cold weather condensation with exterior insulation

If vapor semi-permeable EXPS or EPS is used as this sheathing, inward vapor drives will also be controlled. Cold-weather diffusion condensation is also easy to eliminate with exterior insulation. It is for these reasons that, for example, 2x6 framed walls are hygrothermally inferior to 2x4 walls with insulating sheathing [Timusk & Lischkoff 1983, Timusk & Doshi 1986, Straube 2011, Smegal et al 2013, Trainor et al 2016, Ge et al 2019, Viljanen 2022].

Minimizing the pressure differences acting across the enclosure will reduce air leakage and thus is another important means of reducing the severity of condensation. Stack effect can be controlled by compartmentalization between floors. This approach has been shown to be quite effective in high-rise buildings but requires lobbies with airtight doors around the elevator shafts [CMHC 1996].

HVAC pressurization can be almost eliminated by proper balancing and operation, but this may be difficult to achieve in large multi-zone buildings or in tall buildings.

Reducing the amount of moisture in the air is often a practical condensation control measure, although primarily useful for cold weather conditions. For example, in well-built airtight homes the control of air leakage is so good that interior air moisture levels can build up to 50% RH or more in the winter. To reduce the RH to "safer" but still comfortable and healthy levels (e.g., 25-40% is often recommended), ventilation should be used.

This may simply be a continuously operated exhaust fan, but heat recovery can be added to the ventilation system to reduce the energy penalty of this approach. If energy savings are not a priority, direct mechanical exhaust ventilation of rooms with large moisture production (e.g., kitchens and baths), can be used to effectively reduce interior moisture levels during cool weather. In fact, the motto "build tight, then ventilate right" is commonly used to explain the modern philosophy of controlling airflow through buildings, not just eliminating it. The role of ventilation on interior humidity levels in cool weather is further explored in Chapter 3.

Ventilation will NOT reduce interior RH levels in weather with high exterior moisture content. This situation occurs for a large proportion of the year in the American Southeast and is somewhat of a problem in most of the eastern part of North America during the summer months. In these situations, mechanical dehumidification is the most practical means of controlling interior humidity levels. Note that controlling interior humidity during hot humid weather only controls surface condensation. Interstitial condensation during hot humid weather is governed by the temperature of the interior.

10.5 Advanced techniques

Surface condensation is avoided by ensuring that the surface is close to the interior air temperature. This is best achieved by providing a well-insulated enclosure. As described earlier, energy intensive methods of achieving warm interior surfaces include heat tracing and directing streams of warm air onto the cold surface. Radiative heating of the interior enclosure can also be used to deal with problems. Energy-saving approaches, such as better insulated enclosures, avoidance of serious thermal bridges, and reduced interior humidity during cold weather, are recommended for the design of new buildings.

Various advanced techniques involving airflow can be applied to control interstitial condensation in specific cases. Most rely on mechanical equipment to control the quality of the air or the direction of pressure.

A small amount (pressure of a few Pascals) of mechanical depressurization in the winter and pressurization in the summer can almost eliminate the potential for interstitial air-leakage condensation. This strategy must be used carefully: large air pressure differences cause other enclosure problems, such as over cooling interior surface, excess energy use, poor air quality, etc.

The Dynamic Buffer Zone (or DBZ) is a term applied by some consultants to a range of strategies involving air forced into enclosures originally proposed and implemented by researchers at the Canadian Division of Building Research [Tamura et al 1974]. A broader and less proprietary term is preferred. Installing fans to force outdoor air into problem attic spaces in the winter is an example of active pressure control used as a retrofit measure. For example, instead of repairing a leaky air barrier system between the ceiling and attic of a building, a fan is used to induce a positive pressure in the attic to reduce, eliminate, or even reverse the flow of air [Lstiburek, 1993]. Another application has been the management of condensation problems in ice arenas [Straube 2006]. These approaches, of course, requires constant energy input and proper maintenance and operation. See Section 11.8 for more coverage of this topic.

Provided that the enclosure system is reasonably airtight, the energy cost is small. To ensure that exterior air is delivered only when needed, a dewpoint controller can be used. This electronic control unit senses the dewpoint temperature of the interior air and the exterior enclosure surface temperature. When condensation could occur (when $T_{ext} < T_{int,dewpoint}$) the controller operates the fan. In other cases (often a large fraction of annual hours), the fan does not operate.

10.6 References

Adan, O. 1994. *On the Fungal Defacement of Interior Finishes*. Ph.D Thesis, Eindhoven University of Technology, Eindhoven, Netherlands.

Andersen, N.E., 1987. "Summer condensation in an unheated building". In *Nordic Symposium on Building Physics*, August.

Anderson, L.O., 1972 *Condensation Problems: Their Prevention and Solution*. Research paper 132, US Dept of Agriculture, Forest Product Labs, Madison, WI.

BS 5250:2021, 2021. *Management of moisture in buildings - code of practice*. British Standards Institute, UK.

Bankvall C.G. 1978. "Forced Convection: Practical Thermal Conductivity in an Insulated Structure under the Influence of Workmanship and Wind," *Thermal Transmission Measurements of Insulation*, ASTM STP 660, R. P. Tye, ed., American Society for Testing and Materials, pp. 409-425.

Bomberg, M. and Onysko, D.M. 2002. "Heat, Air and Moisture Control in Walls of Canadian Houses: A Review of the Historic Basis for Current Practices", *J. Therm. Envelope Build. Sci.*, 26: 3–31.

CMHC, 1996. *Controlling Stack Pressure in High-Rise Buildings by Compartmenting the Building*. Report by Morrison-Hershfield for Canada Mortgage and Housing Corporation, Ottawa.

Desjarlais, A., "Self-Drying Roofs: What?! No Dripping!", *Proc. Of Thermal Performance of Building Envelopes VI*, Clearwater, Florida, December 4-8, 1995, pp. 763-773.

EPA, 2013. *Moisture control guidance for building design, construction and maintenance*. U.S. Environmental Protection Agency, document 402-F-13053.

Ge, H. and Fazio, P., 2003. "Experimentally establishing local convection film coefficients for a standard metal curtain wall system", *Proceedings of 2nd Int Conf on Research in Building Physics*, Leuven, Belgium, Sept, pp. 225-232.

Ge, H., Straube, J.F, Wang, L., Fox,M.J. 2019. "Field study of hygrothermal performance of highly insulated wood-frame walls under simulated air leakage". *Bldg & Environ.*, Vol 160.

Huttunen, P & Vinha, J, 2103. "Effect of Natural Convection on the Thermal Transmittance of Exterior Walls with Air-Permeable Insulation," in *Thermal Performance of Exterior Envelopes of Whole Buildings XII Conference* , December 1-5, 2013, Clearwater, Florida.

Lacasse, M.A., Maref, W., Elmahdy, A.H. and Armstrong, M.M., 2009. "Evaluating window-wall interface details for risk of condensation on windows". *Proceedings Symposium on Building Envelope Sustainability*. pp 1-12.

Lstiburek, J., Carmody, J.1993. *Moisture Control Handbook*. Van Nostrand Reinhold, New York.

Manual on moisture control in buildings. Heinz R Trechsel, editor. ASTM manual series. MNL 18, American Society of Testing and Materials, West Conshocken, Pa., 1994.

NRFC 500-2017. *Procedure for Determining Fenestration Product Condensation Resistance Values*. National Fenestration Rating Council, 2017.

Padfield, T.,1999. *The role of absorbent building materials in moderating changes of relative humidity*. Ph.D. thesis, Dep. of Struct. Eng and Materials, Danish Technical University.

Rode, C., Holm, A., Padfield, T. 2004. "A Review of Humidity Buffering in the Interior Spaces", *J. of Therm. Env. and Bldg. Sci*, Vol. 27, No. 3, Jan, pp. 221-226.

Sandin, K., 1993. "Moisture conditions in cavity walls with wooden framework". *Building Research and Information, 21*(4), pp. 235-238.

Sasaki, J.R. 1971. "Developing a standard test for window condensation performance," *Materials Research and Standards*, 11, (10), pp. 17-20, October.

Smegal, J., Straube, J.F., Lstiburek,J., 2013. "Moisture-Related Durability of Walls with Exterior Insulation in the Pacific Northwest", *Proc. Of Buildings XII*, Clearwater Beach, FL., Dec.

Southern, J.R., 1986. "Summer condensation within dry lined solid walls". *Building Services Engineering Research and Technology*, *7*(3), pp.101-106.

Straube, J.F. and Burnett, E.F., 1998. "Drainage, ventilation drying, and enclosure performance". *Thermal Performance of Exterior Envelopes of Buildings VII*, pp. 189-198.

Straube, J.F. and Burnett, E.F.P., 1998. "Field Measurements of Masonry Veneer Exposure Conditions", *Proc of the Eighth Canadian Masonry Symposium*, Jasper, Alta., Canada, May 31-June 3, pp. 208-221.

Straube, J.F., 2001. "The Influence of Low-permeance Vapor Barriers on Roof and Wall Performance," *Proceedings of Thermal Performance of Building EnvelopesVIII*, Clearwater Beach, Florida, Dec.

Straube, J.F., 2006. *BSD-119: Summer Condensation Problems in Ice Arenas.* Building Science Digest 119. www.BuildingScience.com. October.

Straube, J.F., 2011. BSD-163 *Controlling Cold-Weather Condensation Using Insulation*, www.Buildingscience.com.

Smegal, J., Straube, J.F., Schumacher, C.J. 2018. "Measured Drying Ability of Compact Low-Slope Roofs". *Proc. Canadian Building Envelope Technology Symposium*, Mississauga, ON, Sept. 13-14.

Straube, J.F., deGraauw, J.P., 2001. "Indoor Air Quality and Hygroscopically Active Materials". *ASHRAE Transactions*, Vol 107, Pt 1.

Straube, J.F., Smegal, J., Schumacher, C.J., Jablonka, M., 2009. "Adhered Veneers and Inward Vapor Drives: Significance, Problems, and Solutions", *Proc. of Can. Bldg. Sci. & Tech. Conference*, Montreal, May 6-8, pp. 303-312.

Tamura, G. T., Kuester, G. H., and Handegord, G. O.,1974. "Condensation Problems in Flat Wood-Frame Roofs," *Second International CIB/RILEM Symposium on Moisture Problems in Buildings*, Rotterdam, Sept.

Teesdale, L.V., 1939. "Condensation Problems in Modern Buildings." US Forest Products Lab, FPL Report 1196, Madison, WI.

Timusk, J, Lischkoff, J., 1983. "Moisture and Thermal Aspects of Insulated Sheathings", *Proc. of 2nd Canadian Building Science & Technology Conference*, pp. 72-91.

Timusk, J., Doshi, H.,1986. "Effect of insulating sheathing on heat and moisture flow", *Can. J. of Civ Eng.*, Vol 13, pp. 674-680.

Trainor, T., J. Smegal, J.F. Straube, A. Parekh, 2016. "Measured and Predicted Moisture Performance of High-R Wall Assemblies in Cold Climates", *Proc. Of Buildings XIII*, Clearwater Beach, FL., Dec.

Viljanen, K., Puttonen, J. and Lü, X., 2022. "Hygrothermal performance of highly insulated external walls subjected to indoor air exfiltration". *J of Building Physics*, *46*(2), pp.121-175.

Wilson, A.G. and Garden, G.K.,1965. "Moisture Accumulation in Walls Due to Air Leakage". *RILEM/CIB Symposium Moisture Problems in Buildings*, Helsinki.

Wilson. A.G. 1965. "Condensation in insulated masonry walls in summer" *Proceedings, Rilem/CIB Symposium on Moisture Problems in Buildings.* Helsinki.

Chapter 11

Airflow Control

The physics of airflow within and through enclosures and buildings is covered in Chapter 7. This chapter discusses various practical means of controlling and describing airflow within and through enclosures.

11.1 Air barriers

The primary plane of airflow control in a wall is generally called the air barrier or air control layer. Some define the actual air barrier as the element across which at least half of the total air pressure drop occurs, or as the "pressure boundary" between indoors and outdoors. In actual enclosure assemblies, numerous layers and materials may contribute to the airflow resistance. In design, it is more practical to define a single plane to be the sole, or at least the primary, resistor to airflow. Because in practice such a plane consists of numerous elements, materials, and joints, the term air barrier system (ABS) is often used to emphasize the need to design and build a complete system (Figure 11.1)

11.1.1 Secondary and accidental air barriers

Although this chapter focuses on air barrier systems it must be borne in mind that the layer labelled by the designer or builder as the air barrier may not in fact act as the air barrier in service. In framed, low-rise residential buildings, for example, the primary air barrier system may comprise an inner layer of gypsum wallboard (sealed around the perimeter and at all penetrations) or sealed polyethylene and drywall. Sheathing, (such as gypsum, waferboard, fiberboard, EXPS), cladding, air impermeable insulation, and housewrap or building paper can all provide additional resistance to out-of-plane airflow through the enclosure assembly. Similarly, a housewrap may be sealed and supported to act as the air barrier, but the interior gypsum board may provide a significant amount of airflow resistance. In some cases, unexpected layers may act as the air barrier system. For instance, stucco, roof membranes, and precast panels can often *act* as the air barrier even though other components may be intended by design.

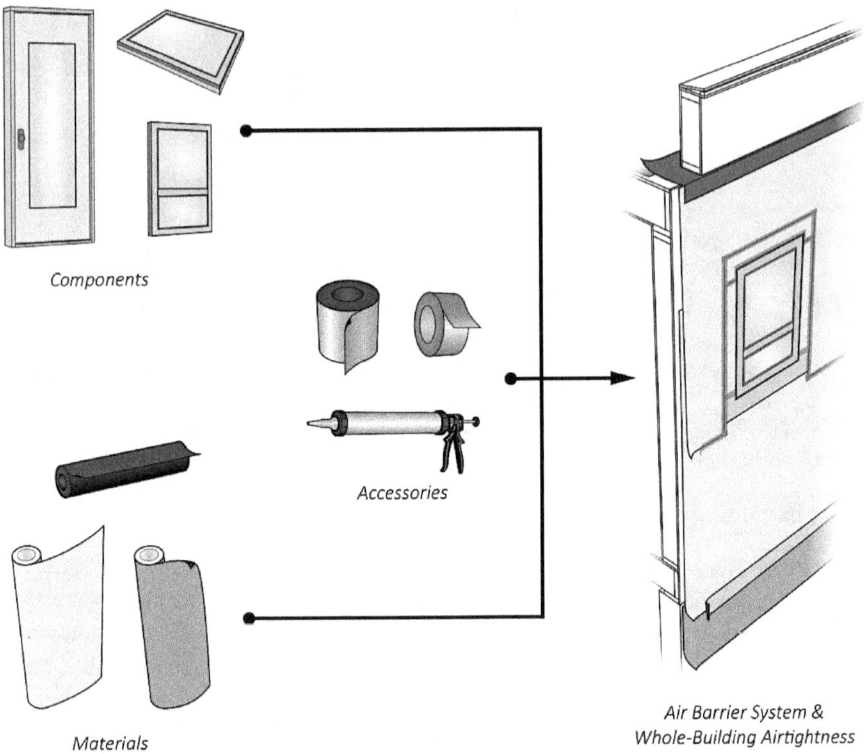

Figure 11.1: Air Barrier Systems (ABS) are comprised of materials assembled into components [Higgins et al 2017]

Secondary planes of airflow resistance almost always improve the performance of an enclosure. Unlike vapor barriers, multiple planes of air tightness are desirable. These secondary air barriers fulfill several functions, either on their own or in conjunction with the other planes of airflow resistance. They not only add marginally to the overall airflow resistance of the assembly, but they also provide a degree of back-up if the primary air barrier is designed, built, or performs imperfectly.

The satisfactory performance of many older wall and roof systems can often be explained by the unintentional, and often synergistic, contribution to airtightness provided by layers such as building paper, plaster, panel sheathing, and brickwork often combined with low interior humidity levels during cold weather.

11.2 Requirements of air barriers

Typically, several different materials, joints and assemblies are combined to provide an uninterrupted plane of primary airflow control. Regardless of how air control is achieved, the following five criteria must be met by the air barrier system (ABS):

1. **Continuity.** This is the most important and most difficult criterion. Continuity through doors, windows, penetrations, around corners, across joints, etc. This can be investigated in the generic case with laboratory testing (e.g. ASTM E2357) but normally requires on-site inspection and whole-building testing (e.g., ASTM E1176 and ASTM E779).

2. **Strength.** If the ABS is in fact much less air permeable than the remainder of the enclosure assembly layers, then it must be designed to transfer the full design wind load (e.g., the 1-in-30 year gust) to the structural system. Fastenings are often critical, especially for flexible, non-adhered membrane systems, and long-term adhesion must be ensured for fully-adhered systems.

3. **Durability.** The ABS must continue to perform for its service life. Consider the ease of repair and replacement, the stresses due to building and wind-induced movement, fatigue, oxidation aging, temperature range, etc.

4. **Stiffness.** The ABS must be stiff enough so that deformations do not change the air permeance, distribute loads through unintentional load paths, or result in fatigue failure of the membrane, fastenings, or adhesion (especially at laps and penetrations).

5. **Air Impermeability.** Naturally, the primary air barrier material must be impermeable to air. A typical required air permeance values for materials that comprise the air barrier system are of less than 0.02 lps/m^2 at a 75 Pa pressure difference (less than 0.004 cfm per square foot at 0.3" w.c.). Although this is an easy property to measure (e.g., using ASTM E2178), it is not as important as might be thought. In practice, the ability to achieve other requirements (especially continuity) are far more important to in-service performance, and the air leakage at joints, cracks, and penetrations (and hence the airtightness of the enclosure assembly and the whole building) outweighs the air permeance of the solid materials that make up most of the area of the ABS.

It is critical to understand that laboratory tests only yield data on the potential airtightness of a material (ASTM E2178) or system (ASTM E2357, ASTM E283, ASTM D8052) under ideal conditions: assembling a building enclosure in the field essentially always results in more air leakage and the performance of a specific assembly or whole building can usually only be tested in the field.

Building on previous recommendations for materials and components, Lstiburek [2005] proposed that the acceptable maximum whole building air leakage rate be ten times the maximum air barrier assembly (or prefabricated component, such as windows, doors) value, that is:

Air Barrier Material 0.02 l/(s m^2) @75 Pa (0.004 cfm/ft^2 @ 0.3" w.c.)

Air Barrier Assembly 0.2 l/(s m²) @75 Pa (0.04 cfm/ft² @ 0.3" w.c.)

Building Enclosure 2.0 l/(s m²) @75 Pa (0.4 cfm/ft² @ 0.3" w.c.)

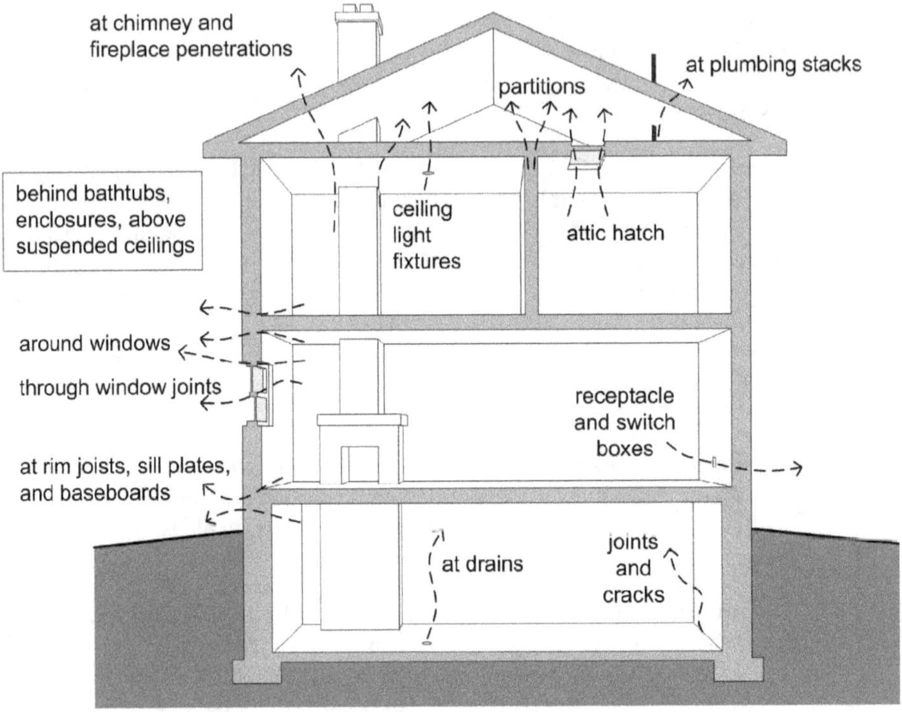

Figure 11.2: Common air leakage sites in buildings

11.1.2 Air barriers versus vapor barriers

Controlling airflow has long been known to be more important than limiting water vapor diffusion [Torp & Graee 1961, Garden 1965]. However, the fact that many vapor diffusion barriers also retard or eliminate airflow sometimes causes confusion. In fact, much of the older literature (and a remarkable proportion of current documents, and most European literature) confuses or combines the function of the air barrier system and vapor barrier. The difference between the two is still one of the most discussed building science issues in design and construction offices. Hence, the distinction, which is well understood and should be clear from the physics presented earlier in this text, is repeated here.

The function of a vapor barrier is simply the control of water vapor diffusion with the intent of reducing the occurrence, intensity, and/or duration of condensation. As such, it has one performance requirement: it must have the specified level of vapor permeance and be installed to cover most of the area of an enclosure. If a small crack or perforation occurs in a vapor barrier, its performance (i.e., its vapor permeance) is not substantially reduced and such imperfections can usually be accepted in practice.

Air barrier systems control airflow and thereby control convective vapor transport. Although air barriers reduce and manage interstitial condensation the control of airflow provides other important benefits such as increased comfort (by reducing drafts), reduced energy consumption, control of odor, pollutant, smoke, and sound transmission. Air barrier systems must meet all five performance criteria listed earlier, only one of which is low air permeance, and none have to do with the vapor permeance of the layer.

Most modern building codes now require an air barrier system in all enclosures. More specifically, in the case of the national Canadian building code an air barrier system is required in enclosures that would be adversely affected by air leakage (generally the adverse impact is condensation). In practice, this means air barriers are required for most practical building enclosures, especially since air barriers do much more than merely control condensation. Codes and building programs are increasingly requiring testing to confirm actual performance delivered in the field.

Vapor diffusion control is also required in all enclosures that would be adversely affected by condensation as a result of poor diffusion control. Building codes once prescribed a low permeance (under 60 ng/Pa/s/m^2 or 1 US perm) layer on the "warm in winter" side of an enclosure[1]. However, a low permeance barrier is rarely required because diffusion transports little moisture in most practical situations [Karagiozis 1993] and depends on other factors such as the interior climate and details of the enclosure assembly [Straube 2001]. Calculations based on the methods presented in Chapter 6 show the relatively small impact on moisture accumulation due to vapor diffusion.

Early building science identified the role of the vapor barrier [Babbit 1939, Rowley 1939, Teesdale 1937, Weber 1942]. This work was published in the 1930's and 1940's. However, it failed to recognize the much greater importance of air barrier systems which only became clear later [e.g., Garden 1965, Quirouette 1985]. The choice of 1 US Perm (57 ng/Pa s m^2) as the definition of a vapor barrier is essentially arbitrary and traditional [Rose 1997], and in many situations such a low level of vapor permeance is not required on the interior. Unfortunately, the important role played by the air barrier and the relatively less significant role of the vapor barrier is still not widely understood. Many codes, standards and building product manufacturers still base their recommendations and requirements on this historical view of the importance of vapor barriers.

The term vapor retarder has become more widely used to describe a layer intentionally provided in the design to control vapor diffusion, but that may have a vapor resistance of well over 1 US perm. Values of 300 to 600

[1] This anachronism of building science still exists in Part 9 of the 2025 Canadian National Building Code.

ng/Pa/s/m² (5 to 10 US perms) are sometimes used as the maximum permeance allowed for a layer described as a vapor retarder. Lstiburek [2002] proposed simple categories that have been adopted by US Codes (Table 11.1).

Table 11.1: Vapor permeance categories

Class	Permeance (US perms)	Permeance (Metric perms)	Description
I	Less than 0.1	Less than 6	Impermeable
II	0.1 to 1.0	6-60	Semi-impermeable
III	1 to 10	60-600	Semi-permeable
none	Over 10	Over 600	Permeable

The vapor permeance of an air barrier system must be considered in the same way the vapor permeance of all other materials in an assembly should be. The vapor permeance of the ABS is no more or less important than the vapor permeance of most other layers in an assembly, such as the cladding, sheathing, insulation, interior finish, etc. For example, in cold climates, a vapor barrier on the exterior is usually not desirable but can be designed for, as it is in an exposed membrane low-slope roof. By contrast, in hot humid climates, locating the vapor barrier on the exterior could be desirable, but a vapor barrier in this location is not necessary if the remainder of the enclosure is designed properly.

The widespread confusion between air barriers and vapor barriers may be due to the fact that some materials (such as glass and sheet metal) can be used as part of an air barrier system while at the same time acting as a vapor barrier. However, materials like unpainted sealed gypsum board, sealed and supported housewraps, and sealed wood panel sheathing can be excellent air barrier materials but provide little vapor diffusion control. To add to the confusion, the vapor permeance of a material is often sufficient to identify a vapor barrier because a modest number of accidental penetrations and un-sealed laps do not drastically change the permeance of a vapor control layer, whereas the same flaws would render an air barrier system mostly ineffective.

11.1.3 Air Barriers vs Water Resistive Barriers

Another more recent phenomenon is the confusion between air barriers and water control layers, or Water Resistive Barriers (WRB) in drained enclosure systems (Chapter 12). This confusion may have arisen because many building products can be used to fulfil both functions. The requirements are quite different however, and so it remains necessary to specifically identify both in design conversations and on design documents. WRBs can be made continuous by simply gravity lapping two components, whereas air barriers must be sealed.

An air barrier must be strong enough to resist peak wind loads, whereas a WRB does not. A WRB must be moisture tolerant whereas an air barrier need not be. Hence a product designed and installed as a water barrier may not be function as an air barrier and vice versa. On the other hand, it has become a common and successful practice to combine the air barrier and WRB in one product.

11.1.4 Air Barrier Location

Theoretically, the air control layer (air barrier) can be located anywhere with within a properly design enclosure assembly: at or near the interior or exterior surface as well as anywhere within. However, there are numerous practical realities that need to be addressed.

When air barriers were first proposed it was common to consider repurposing vapor barrier membranes as the primary component of an air barrier system. As air barrier researchers were predominately in cold climates, vapor barrier membranes were located near the interior surface. As the importance of air barrier continuity to performance became clear, alternate locations of the air barrier began to be move the air barrier plane further to the exterior within enclosure assembly. The benefit of moving the air barrier membrane outward is that far fewer penetrations, particularly due to service penetrations such as electrical receptacles, need to be identified and sealed. For framed enclosures, which are very common in North America and a growing option around the world, installing the air barrier to the exterior of the framing allows the membrane to be continuous past the three-dimensional complexity of floor and roof penetrations (Figure 11.3).

Locating the air control layer on the exterior of the support structure allows it to be inside of exterior insulation layers: as the modern approach to achieving reduced thermal bridging and higher thermal performance involves continuous insulation on the exterior, this is a recommended and widely-applicable solution.

However, locating the primary layer of airtightness outside of most of the insulation (as opposed to outside the support) can create risks of performance problems in cold climates. This has been investigated by numerous European researchers [Langmans et al 2012, 2013, Pihelo and Kalamees 2016] which often confuse air barriers, vapor barriers, wind barriers, and investgate systems with no continuous exterior insulation. If the insulated layers inside of an exterior air barrier have very high air permeance, temperature-induced convection loops could allow condensation and additional heat loss (see Chapter 7). Standard design solutions can and should be used to manage this risk. Air barriers located on the exterior of most of the insulation have been widely used in North America[2] with an excellent track record of success. Nevertheless,

[2] A housewrap used as an air barrier is common in many, perhaps the majority, of wood-frame houses in North America since the 2000's. Standard practise insulation materials and installartion quality has not resulted in performance failures.

convective loop condensation issues are a common failure mode when poor designs are used (for example, deliberate gaps between architectural precast panels and insulation, or between concrete foundation walls and insulation).

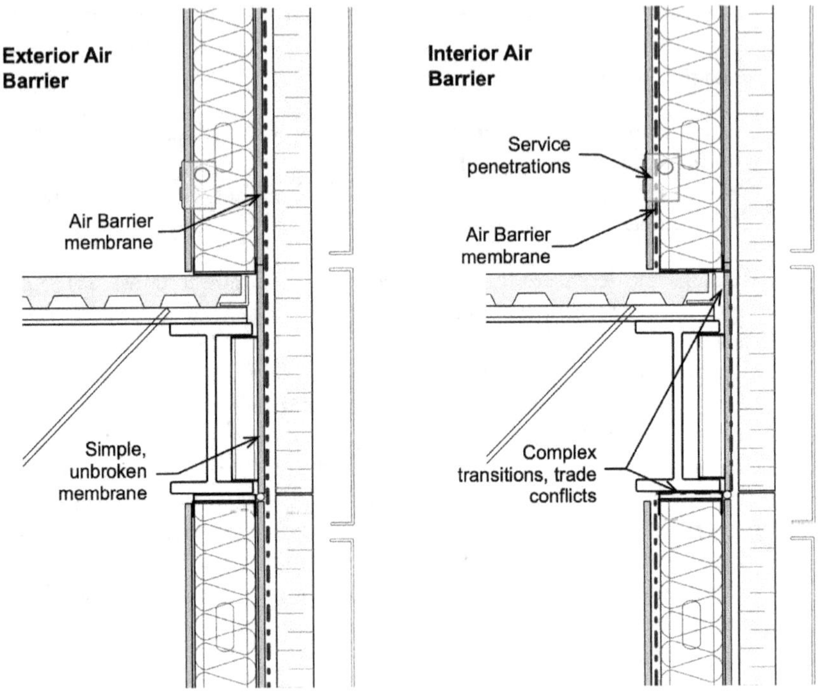

Figure 11.3: Air barrier location within the enclosure

Locating the air control layer near the inside of most or all of the insulation of course avoids any concerns about convection loops from the inside. However, it is also increases the risk of air leakage because the construction difficulty due to geometric complexity is significant (and the resulting poor air tightness has often been documented). Also, wind washing risks increase significantly if all the insulation is outside of the air control layer. Both of these risks can be managed by design of course.

Locating an air control layer on the exterior of all of the insulation has the disadvantage that the materials and joints are exposed to high temperatures in warm weather and cold temperatures in cold weather. These temperatures and variations stress many common air barrier materials (especially polymer-based products) and hence will reduce their durability by mechanisms such as oxidation and fatigue.

11.3 Air barrier systems in practice

Air barriers systems can be categorized by various methods, but the most useful categorizations are based on the location within the assembly, product format (sheet, fluid, rigid panel), and vapor permeance of the air barrier system components.

Since the 1970's, a sheet of 0.15 mm (6 mille) thick polyethylene (poly) has been widely used as a vapor barrier in low-rise rise wood-frame building assemblies in Canada and some northern US states. Poly is inexpensive to buy and install and greatly restricts vapor diffusion. As airtightness was identified as a more important characteristic, it was hoped poly could be repurposed. Although many builders managed to achieve airtightness by sealing polyethylene sheet, the effort required and the uncertainty of the outcome was significant.

Polyethylene sheet it struggles to meet most of the air barrier requirements other than air impermeability. It is difficult and relatively expensive (in labor hours) to achieve continuity, especially since it is pierced by services and enclosure penetrations at many locations. It is likely to fail structurally when exposed to wind gust loads and can fail through fatigue if it flaps because of varying wind pressures. Hence, it may not be very durable. It is so flexible that it can deform and transfer loads through unexpected paths, deform batt insulation, and tear fastenings, etc. Lab testing [Shaw 1985] and field experience highlighted these shortcomings.

Nevertheless, for undemanding applications such as low-rise housing, poly may act as both the vapor barrier and the air barrier, especially if sandwiched between drywall and cavity insulation and sealed (with tape or sealant) at fully supported laps.

The vapor barrier and air barrier functions are often separated in enclosure wall systems employing housewraps (mechanically attached flexible sheet air and water-resistant membrane) as the primary air barrier components. Polyethylene sheet, foil-backed drywall, or low-permeance paint is applied on the interior and acts as the vapor barrier. The housewrap is applied outside of the enclosure structure (not the exterior of the enclosure) so that it spans over all service penetrations, framing intersections, etc.

Since the housewrap is installed on the cold side of the enclosure for most cold-climate applications, it must be reasonably vapor permeable to allow water vapor to move out of the assembly. Unfortunately, housewrap is so flexible it can deforms to such an extent that it or pumps air as it flaps under variable wind pressures (Figure 11.4). Therefore, truly successful applications sandwich the housewrap between two stiff and strong layers of sheet material (e.g., semi-rigid insulation and sheathing), or support it at frequent intervals (e.g., by the use of siding, furring strips, or, unintentionally, by mortar dams). This same mechanism afflicts mechanically attached low slope roof membranes.

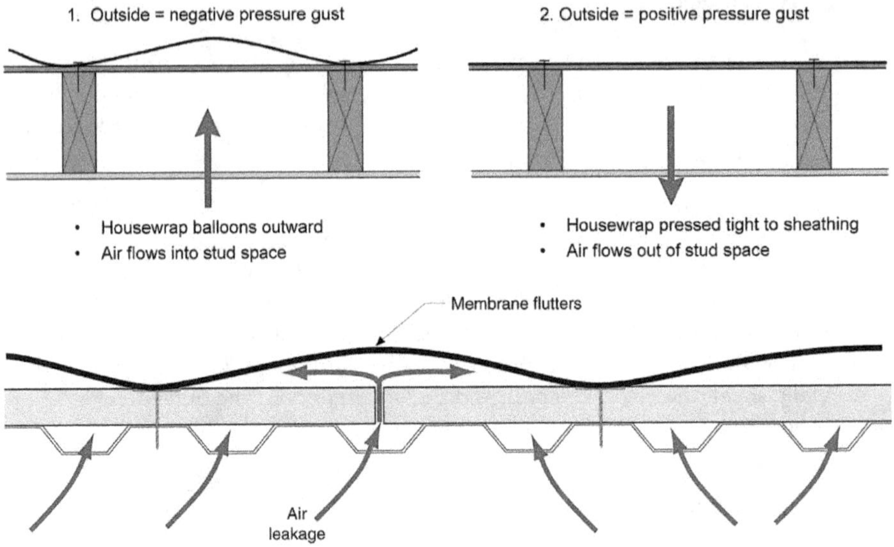

Figure 11.4: Flexible mechanically-attached air barriers can allow air movement into building enclosures

The airtight drywall approach (ADA) was developed by Lstiburek and Lischkoff to address some of the limitations of poly sheet membrane air barriers. The ADA employs the interior layer of painted drywall as the air barrier and poly (or an appropriate type and thickness of paint) as the vapor diffusion barrier. The drywall is stiff and strong enough for most applications and because it is visible, it is easy to ensure continuity and to inspect and repair. However, as in interior air barrier, difficulty in achieving continuity is often encountered at service penetrations, wall-floor interfaces, etc. Despite these difficulties, the airtight drywall approach to air barrier systems is often more successful than a loose-sheet polyethylene system and can be used in a wide-range of steel and wood-framed roof and wall systems.

Another approach to airtightness is the use of air impermeable, usually foam plastic or foil-faced, rigid insulation boards with sheathing tape and/or gaskets at joints. Such systems have the advantage of fewer penetrations but do require special tapes and gaskets to create continuity. The ability of these systems, including their joints, to transfer wind loads through connectors to the structural frame must be investigated and demonstrated.

Reinforced concrete, especially when precast (i.e., no form ties or cold joint penetration and often of higher quality concrete), is often sufficiently airtight to form part of an ABS as long as the concrete is dense and crack size and spacing are controlled by the appropriate use of reinforcing. Blockwork is generally not sufficiently impermeable to act as an air barrier. Shrinkage cracks and porous materials compromise its airtightness, but a wide range of available coatings can produce an airtight assembly

In commercial applications, bitumen-based air barriers, in either fluid-applied or preformed reinforced membrane form, are often adhered to the exterior of blockwork, concrete, or exterior gypsum sheathing. To ensure continuity, compatible membranes are used to bridge cracks and tie the ABS to windows, doors and other penetrations.

These bitumen-based air barriers are vapor-impermeable and thus also act as vapor barriers. Therefore, the location of the insulation must be based on the climate region and interior conditions. In effect, the location of the air barrier system and vapor barrier must be the same.

Vapor permeable self-adhered membranes based on laminated polymer films are also widely available.

Fluid-applied cement-based (non-combustible) air barriers are also available. Basically, thin polymer-modified and fiber-reinforced plasters, these systems require flexible or reinforced joint details around penetrations, but have low flame spread ratings and may be as vapor permeable or impermeable as specified. Most are produced by EIFS manufacturers.

Various latex- and acrylic-based fluid-applied (trowel or spray) membranes are also available. These may be moderately permeable or impermeable depending on the product, but all are combustible and may have a high flame spread or smoke produced rating.

Finally, rigid panel products, based on wood or gypsum, can be air impermeable, rigid air barrier materials. They require special joint detail solutions but are available with a range of water-resistance and vapor permeance.

11.4 Measuring enclosure and whole-building airtightness

Material air permeability, assuming laminar flow, can be described with an air permeability coefficient (see Chapter 7). In recent years, air flow per unit area at 75 Pa (0.3" w.c.) has been chosen, usually tested in a laboratory using ASTM 2178 or similar.

Building assemblies and components, such as a roof, wall, window, door, damper, etc. are tested in the lab or field. Performance is reported as air flow per unit area at a specific pressure difference, often 75 Pa in North American commercial practice, but pressure differences of 50 Pa, 150 Pa, 200 Pa and 300 Pa are also used depending on the standard. Slightly different sample preparation and equipment and reporting requirements are required by different standards. ASTM E2378 is a popular modern standard for enclosure walls incorporating a wide range of penetrations, whereas ASTM E283 was developed many years ago for curtainwalls but was and still is used for other enclosure components with or without penetrations.

The power law is almost exclusively used to describe the airflow through assemblies, parts of enclosures, and entire buildings across a pressure range (see also Chapter 7). The airflow rate, Q, is related to the pressure difference ΔP by fitting a line to a power law with a coefficient, C, and a flow exponent n:

$$Q= C\,(\Delta P)^{n} \qquad \text{(Eq. 11.1)}$$

To provide a simple, one-number rating of airtightness various approaches have been used, the most common of which is to report the flow at an agreed test pressure (usually 50 or 75 Pa) normalized by enclosure area. Results therefore are presented in units such as liters/second per m^2 at 75 Pa, cfm/ft^2 @ 0.3 in w.c. (where pressure is measured in inches of water column), m^3/h per m^2 @75 Pa, or even cfm/sf @ 50 Pa (often abbreviated to cfm50).

The most common reporting metric of whole-building tests in the past was the measured flow rate through the enclosure over the interior building volume (or Air Changes per Hour, symbol ACH) when tested at a pressure difference of 50 Pa. This can be abbreviated at ACH50.

As buildings have a very wide range of enclosure surface area to volume ratios, the normalized air permeance per unit enclosure area has become seen as a preferred metric. In this case the flow measured at a particular pressure is divided by the enclosure area and as such is true measure of the entire enclosure performance. For a whole-building test, the enclosure area includes all six sides, that is, the walls, roof and floors, both above- and below-grade.

11.1.5 Whole-building Airtightness Testing

In residential construction, air leakage tests are often referred to as blower door tests because these tests are most commonly conducted using a piece of equipment called a blower door. A large fan or "blower" is used to extract air from or supply air to the building. Test equipment measures the airflow (the volume of air moved into or out of the building) and the corresponding pressure difference acting across the building enclosure. Such methods were developed initially to research buildings [e.g., Tamura 1975, Kronvall 1978, Orr & Figley 1980] but became deployed more widely after the energy crisis of the 1973.

The flow rate is almost always measured with the aid of a blower door test as prescribed by the CGSB 149.10 Standard, *Determination of the Airtightness of Building Envelopes by Fan Depressurization Method* [CGSB 1986] or the ASTM E779 *Test Method for Determining Air Leakage Rate by Fan Pressurization* [ASTM 2018] (which is essentially the same) or ISO 9972 [ISO 2015]. In large buildings, it is possible, but not common, to use the building's own HVAC system to pressurize or depressurize the building and special instruments to measure the flowrate and resulting pressure difference. CGSB Standard CAN/CGSB-149.15-96 [CGSB 1996] provides a standardized approach for this method. The trend towards tighter buildings and easy-to-use blower door equipment has rendered the use of building mechanical systems a rare option in modern practice.

In all cases, a building must be prepared for testing beforehand by blocking intentional openings such as HVAC intake and exhaust grills, kitchen and bathroom exhaust fans, relief dampers, etc. Numerous excellent resources are now available to guide the test agency and designer [Brennan et al 2013, USACE 2012].

For larger buildings, that is, with a floor area of over 10 000 ft^2 (929m^2), an airtightness test usually requires numerous coordinated blower doors running at the same time. Buildings of nearly 1 000 000 ft^2 (92 900m^2) have been airtightness tested.

11.5 Airtightness Results

A study by Lawrence Berkley Labs [Sherman 1998] summarized the results of a comprehensive survey of American house airtightness. The study concluded that American houses are leakier than previously thought. In fact, based on results from about 400 homes built between 1988 and 1994, the average airtightness was about 7 ACH at 50 Pa. The average air leakage rate of the whole sample set, which included 12946 houses as old as 100 years, was 29.7 ACH at 50 Pa. Canadian houses built in the 1980s and 1990s tended to be tighter, especially in cold climates, and were more likely to have typical test values of 2 to 4 ACH at 50 Pa. More recent homes have continued the trend to increasing airtightness [Chan et al 2013].

Houses are rarely exposed to overall air pressure differentials of 50 Pa — such pressures would often result in operational problems. Pressures of 50 and 75 Pa are used to obtain consistent and comparable test results, and neither number represents a real condition. Typical long-term average pressure differences for low-rise housing are in the range of 4 to 10 Pa. At these lower pressures the actual air change rates are about 0.35 to 0.5 ACH for a well-built 1980's vintage house. More modern Canadian houses are currently being built to routinely achieve less than 1.5 ACH @ 50 Pa, and hence can be expected to have typical in-service air change rates of less than 0.25 ACH.

Airtightness data for commercial buildings are harder to find in older literature and are more variable. However, in the last decade large building airtightness testing has become much more common and mandated for specific building programs and in certain jurisdictions. Hundreds of tests have shown a wide range of airtightness, with one compendium showing values from 0.29 to 19 l/s/m^2 @75 Pa [RDH 2017]

Pressures acting on the air barrier system of commercial buildings are usually higher, especially for larger and taller buildings. HVAC systems and stack effect generate these larger pressures, which may be in the range of 10 to 50 Pa for long periods of time. As a consequence, the actual airflow across the enclosure of a tall building may be much larger than for a smaller single-family home with an enclosure of the same air permeance.

For buildings, including houses, that have achieved good control of air leakage, controlled ventilation is required to ensure air quality and to dilute indoor moisture levels in cold weather. Heat recovery ventilation may be economical but is not required to achieve ventilation.

11.6 Airtightness Targets

Numerous organizations, standards, and codes have set airtightness targets for different building types, climate conditions, parts of the building. For most buildings in North America, different targets are being set for materials, assemblies and whole buildings.

Targets first suggested in 1986 [Quiroette et al 1986] were adopted by the 1995 National Building of Canada which required that the material that provides the principal resistance to air leakage should have a permeance of less than 0.02 $l/s/m^2$ at a pressure difference of 75 Pa.

The current Canadian code references standard ULC-S741 for air barrier materials and limits materials to 0.02 $l/s/m^2$ @75 Pa. Air barrier assemblies must comply with CAN/ULC-S742 "Standard for Air Barrier Assemblies – Specification" and achieve an assembly air leakage rate of less than 0.2 lps/m^2 when tested at 75 Pa (Type A4). ULC-S742 references the same test methods as ASTM E2357.

In the United States, the building industry (e.g., Air Barrier Association of America) has largely adopted the Canadian value and converted them to imperial units: 0.4 cfm/ft^2 at 0.3" H_2O pressure. ASHRAE 90.1 has the same targets for materials and assemblies tested in the same manner, as does the International Energy Conservation Code, and hence, the International Residential code.

More and more owners and code jurisdictions require that all new buildings meet whole building airtightness targets measured by testing: the United States General Services Administration [GSA 2024] and Washington State Building Code are two early examples. The air leakage across the enclosure of commercial buildings is limited to a maximum of 2 lps/m^2 at a pressure difference of 75 Pa (0.4 cfm/ft2 @ 0.3 in w.c.) For higher performance buildings, the U.S. Army Corps of Engineers has a target of 0.25 cfm/ft^2 at 0.3" H_2O, which works out to about 1.3 lps/m^2 at 75 Pa.

High performance buildings sometimes use a target of 1.0 lps/m^2 @75 Pa or even 0.1 cfm/sf (0.5 lps/m^2). Such targets require above average quality control but have been achieved on many projects. The German Passiv Haus Institut requires an airtightness of 0.6 ACH@50 Pa which is easy to achieve on large buildings (because their enclosure area to volume ratio is small) but more challenging for smaller buildings with complex shapes (because the enclosure area to volume ratio is large).

11.7 Complexities

Any airflow control strategy for the building enclosure must recognize the complexity of real buildings. Wind speed and directions change quickly, equipment turns on and off, and occupants open and close openings within the building and across the enclosure. All of this means that the air pressures acting across the enclosure vary in time and across both the inside and the outside face.

Buildings are essentially three-dimensional airflow systems that use the rooms as well as the hollow walls, roofs, plenums, ducts and floors as flow paths [Lstiburek 2000]. Many building performance problems can be understood if one measures the driving pressures and identifies the flow paths. Flow paths are often unexpected and mechanical systems routinely operate in service at different flow rates than expected by the designer or measured by the testing and balancing contractor.

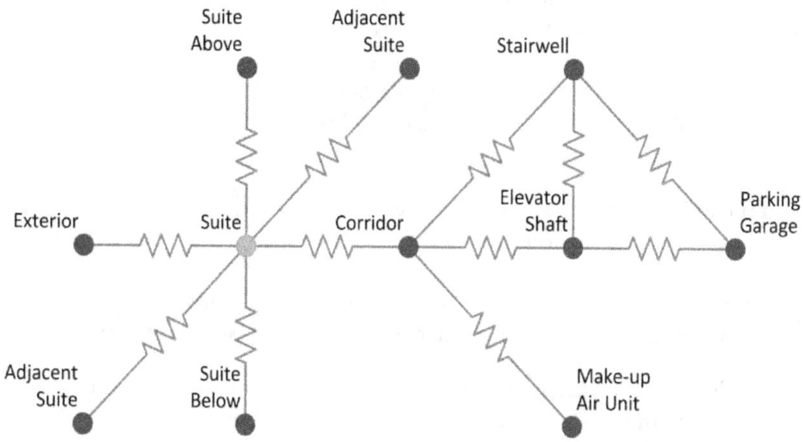

Figure 11.5: Network models can be used to analyse multi-compartment buildings with different pressures in different interior spaces [Ricketts 2014]

Lstiburek has also suggested that the complex pressure distributions can be classified into four distinct pressure zones: the exterior pressure (including below grade), the interior spaces pressure, the interstitial (within wall and floor) pressures, and the conveyance system pressures (within ducts and air handlers). Airflow patterns develop in a complex and changing manner as the pressure fields change. The pressure fields interact through both the intentional and inevitable accidental openings, cracks, penetrations, grilles, and ducts.

Examples of potential problems include airflow from enclosures into partition walls, flow from crawlspaces or attics drawn into building spaces, and many complex HVAC system interactions that draw or force air from unexpected locations to undesirable locations.

Research into the interaction of wind, stack effect, mechanical systems and distributed airflow resistance through complex building systems have been undertaken for decades [Achakji & Tamura 1988, Shaw & Tamura 1977]. Ricketts [2014, Ricketts & Straube 2014] has conducted the most detailed field study to date of pressures and air movement within a building in service. The variability of pressures and surprising connections between seemingly unconnected spaces are highlighted by field measurements.

Network models, where each volume is a node and each air-resistant plane a resistor, have been used in an attempt to model and understand complex pressure fields [Kronvall 1982, Walton 1989]. These can be useful research tools, but have little role in practice.

11.8 Active Pressure Management

Complex buildings, historically significant buildings, or performance-critical buildings (such as some laboratories and special manufacturing facilities) may require exceptionally low leakage rates, or cannot economically be modified or repaired to provide a reasonable airflow resistance. In such cases, mechanical pressurization of rooms, voids, or cavities can be used to control the direction, quality, or volume of airflow through an enclosure.

Active pressure control to manage a lack of a sufficiently tight air barrier is not often the best solution, but the literature is full of examples of special cases dating to the 1960's [Dickens & Hutcheon 1965, Tamura et al 1974]. In some circles an actively pressure-managed cavity in an enclosure was marketed with the appealing "Dynamic Buffer Zone" [Pasqualini 1999] and applications for of the approach continue [O'Brien and Calamari 2012]. See also Chapter 10.

One application for which active pressure control has been found to be very effective, both functionally and economically is for radon control [Leovic & Craig 1994]. Radon gas that enters below-grade spaces from certain soils and rock can be a serious health risk. Active Sub-slab (or Soil) Depressurization (ASD) is used to essentially limit any radon gas from entering by depressurizing the space below the building (crawlspace or slab-on-grade) with a fan (Figure 11.6). ASD has been shown to be very effective [Henschel 1994, Health Canada 2016], can be used for more than radon gas control, and has significant other possible side effects such as the drying of damp basements [Turk & Straube 2017].

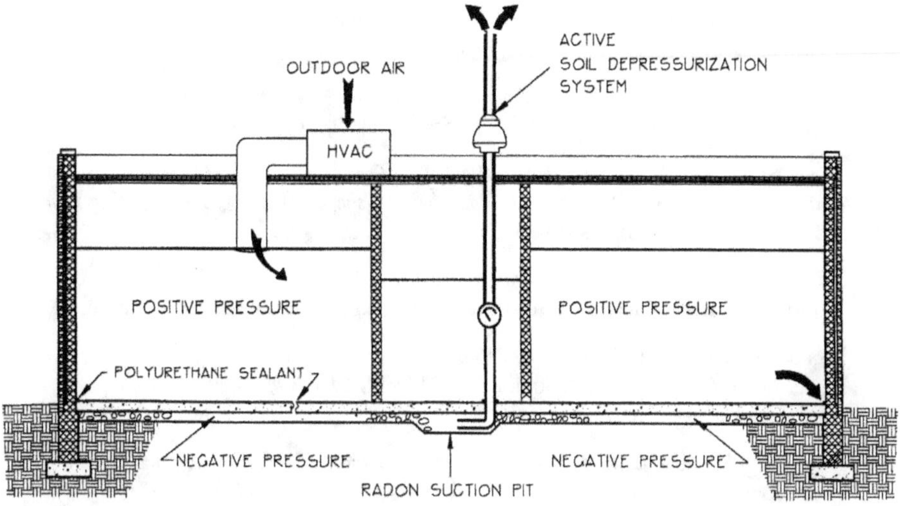

Figure 11.6: Example of an Active Sub-slab Depressurization System from [Leovic & Craig 1994]

11.9 References

Torp, A. and Graee, T., 1961. "The Effect of Air Currents in Moisture Migration and Condensation in Wood Frame Structures". *BYGG*, 9 (1), pp. 1-10. Translated by National Research Council of Canada.

Achakji, G.Y. & G.T. Tamura. 1998. "Pressure Drop Characteristics of Typical Stairshafts in High rise Buildings," *ASHRAE Transactions*, 94(1) pp. 1223-1236.

ASTM D8052-17. *Standard Test Method for Quantification of Air Leakage in Low-Sloped Membrane Roof Assemblies.* ASTM International, West Conshohocken, PA, 2017

ASTM E283-04 (2012). *Standard Test Method for Determining Rate of Air Leakage Through Exterior Windows, Curtain Walls, and Doors Under Specified Pressure Differences Across the Specimen.* ASTM International, West Conshohocken, PA, 2012.

ASTM Standard E779-10 (2018). *Test Method for Determining Air Leakage by Fan Pressurization.* ASTM International, West Conshohocken, PA, 2018.

ASTM E2178-13. *Standard Test Method for Air Permeance of Building Materials*, ASTM International, West Conshohocken, PA, 2013.

ASTM E2357-18 *Standard Test Method for Determining Air Leakage Rate of Air Barrier Assemblies,* ASTM International, West Conshohocken, PA, 2018.

Babbit, J.D., "The Diffusion of Water Vapor Through Various Building Materials," *Canadian Journal of Research*, Vol 17, No. 2, 1939, pp. 15-32.

Brennan T, Anis, W, Nelson G and Olson, C. 2013. "ASHRAE 1478: Measuring Airtightness of Mid- and High-Rise Non-Residential Buildings." Clearwater, FL: *Proceedings of the Thermal Performance of the Exterior Envelopes of Whole Buildings XII International Conference*, Dec. 1-5. pp 1-6.

Chan, Wanyu R, Joh, J. and Sherman, M., 2013. "Analysis of Air Leakage Measurements of US Houses." *Energy and Buildings* Vol. 66 , pp. 616-625.

CGSB. Standard 149.10, 1986. *Determination of the Airtightness of Building Envelopes by Fan Depressurization Method*, Canadian General Standards Board, Ottawa.

CGSB. Standard CAN/CGSB-149.15-96, 1996. *Determination of the Overall Envelope Airtightness of Buildings by the Fan Pressurization Method Using the Building's Air Handling Systems.*: Canadian General Standards Board, Ottawa.

CMHC, 1996. *Controlling Stack Pressure in High-Rise Buildings by Compartmenting the Building.* Report by Morrison-Hershfield for Canada Mortgage and Housing Corporation, Ottawa.

Dickens H.B. and N.B. Hutcheon, 1965. "Moisture accumulation in roof spaces under extreme winter conditions", *RILEM/CIB Symposium on Moisture Problems in Buildings*, Helsinki.

Dumont, R.S, and L.J. Snodgrass, 1990. "Investigation of Chimney Backflow Conditions: A case study in a Well-Sealed House," *ASHRAE Trans* V96(I).

Garden G.K. 1965., *Control of Air Leakage is Important.* CBD 72, DBR, NRC Ottawa.

Health Canada, 2016. *Summary Report on Active Soil Depressurization (ASD) Field Study*. Ottawa.

Henschel, D.B., 1994. "Analysis of Radon Mitigation Techniques Used in Existing US Houses", *Radiation Protection Dosimetry*, Volum 56, Issue 1-4, Pages 21–27.

Higgins, J., Haaland, D., Ricketts, L. 2017. *Illustrated Guide – Achieving Airtight Buildings.* RDH Building Science for BC Housing.

IRC 1997. *Air Barrier Systems for Walls of Low-Rise Buildings: Performance and Assessment* Institute for Research in Construction, National Research Council Canada.

ISO. International Standard ISO 9972 *Thermal performance of buildings - Determination of air permeability of buildings - Fan pressurization method.* International Organization for Standards, Geneva 2006.

Leovic, K.W., and Craig A.B. 1994. *Radon Prevention in the Design and Construction of Schools and Other Large Buildings.* US Environmental Protection Agency, June.

GSA 2024. *PBS-P100 Facilities Standards for the Public Buildings Service. Washington: General Services Administration.* Washington DC.

ISO 2015. International Standard ISO 9972:2015 *Thermal performance of buildings - Determination of air permeability of buildings - Fan pressurization method.* International Organization for Standards, Geneva.

Karagiozis, A.K. and Kumaran, M.K.,1993. "Computer Model Calculation of the Performance of Vapor Barriers in Canadian Residential Buildings," *ASHRAE Transactions*, Volume 99 (2), pp. 991-1003.

Kronvall, J. 1978. "Testing of houses for air leakage using a pressure method". *ASHRAE Transactions* 84(l), pp. 72-79.

Kronvall, J. 1982. *Air flows in building components*. Report TVBH-1002. Division of Building Technology, Lund University of Technology, Sweden.

Langmans, J., Klein, R. & Roels, S. 2012. "Hygrothermal risks of using exterior air barrier systems for highly insulated light weight walls: a laboratory investigation", *Bldg and Environ* 56(10), 192–202.

Langmans, J., Klein, R. & Roels, S. 2013. "Numerical and experimental investigation of the hygrothermal response of timber frame walls with an exterior air barrier", *Journal of Building Physics* 36(4).

Lstiburek, J., 2000. *Toward an Understanding and Prediction of Airflow in Buildings.* Doctoral thesis, University of Toronto, Toronto, Canada.

Lstiburek, J., 2002. Moisture control for buildings. *ASHRAE journal*, *44*(2), pp.36-41.

Lstiburek, J. 2005. "Understanding Air Barriers", *ASHRAE Journal*, Vol. 47, No. 7.

O'Brien, S. and Calameri, V., 2012. "Dynamic Exterior Wall Systems for Solid Masonry Walls in Humidified Buildings", *Proc. of BEST 3: Building Enclosure Science and Technology conference*, Atlanta, GA, April.

Orr, H.W. and D.A. Figley.1980. *An exhaust fan apparatus for assessing the air leakage characteristics of houses.* NRC Building Research Note No. 156. National Research Council of Canada, Ottawa.

Pasqualini, P., 1999. *The control of moisture movement in buildings using the dynamic buffer zone.* M.A.Sc. Thesis, Civil Engineering Dept., University of Toronto, 1999.

Pihelo, P. and Kalamees, T., 2016. "The effect of thermal transmittance of building envelope and material selection of wind barrier on moisture safety of timber frame exterior wall". *Journal of Building Engineering*, *6*, pp.29-38.

Quirouette, R.L., 1985. *The difference between a vapour barrier and an air barrier.* National Research Council Canada, Ottawa.

Quirouette, R.L., Ganguli, U., Lux, M.E., Brown W.C. and Perreault, J.C., 1989. "An Air Barrier for the Building Envelope," *Proceedings of Building Science Insight 1986*, Institute for Research in Construction, National Research Council Canada.

RDH, 2017. *Study of Part 3 Building Airtightness.* Report for National Research Council of Canada

Ricketts, L. and Straube, J. 2014. "A Field Study of Airflow in Mid to High-Rise Multi-Unit Residential Buildings". *Proc. Canadian Conference on Building Science and Technology.* Oct 28 – 30.

Ricketts, L. and Straube J.F., 2014. "Corridor Pressurization System Performance in Multi-Unit Residential Buildings." *ASHRAE Transactions* 120.

Ricketts, L. 2014. *A Field Study of Airflow in a High-rise Multi-unit Residential Building.* MASc Thesis, Civil Engineering Dept., University of Waterloo.

Rose, W.B., 1997. "Control of Moisture in the modern building envelope: The history of the vapor barrier in the United States 1923-1953," *APT Bulletin*, Vol. XVIII, No. 4.

Rowley, F.B, 1939. "A theory covering the transfer of vapour through materials", *ASHVE Transactions*, Vol. 45, p. 545.

Shaw, C, and Tamura, G. 1977. "The Calculation of Air Infiltration Rates Caused by Wind and Stack Action for Tall Buildings". *ASHRAE Transactions* Vol 83, Part 2, pp. 145-158.

Shaw C.Y., 1985. *Air leakage tests on polyethylene membrane installed in a wood frame wall*, Division of Building Research - National Research Council, Ottawa.

Sherman, M.H., Dickerhoff, D.J., 1998. "Airtightness of U.S. Dwellings," *ASHRAE Transactions*, Vol. 104, Pt. 2, Toronto, pp. 1359-1367.

Straube, J.F., 2001. "The Influence of Low-permeance Vapor Barriers on Roof and Wall Performance," *Proceedings of Thermal Performance of Building EnvelopesVIII*, Clearwater Beach, Florida, December.

Straube, J.F. 2001. "Airflow Control in Enclosures," *Proc. of Eight Building Science and Technology Conference*, Toronto, February.

Tamura, G.T., G.H. Kuester and G.O. Handegord, 1974. "Condensation problems in flat wood-frame roofs" *2nd International CIB/RILEM Symposium on Moisture Problems in Buildings*, Rotterdam, September.

Tamura, G.T., 1975. "Measurement of air leakage characteristics of house enclosures". *ASHRAE Transactions*, Vol 81 (1) pp. 202-211.

Teesdale, L.V. 1939. *Condensation in walls and attics.* Forest Products Laboratory, Madison WI.

Turk, B., Straube, J.F., 2007. "Basement Dampness and Ventilation During Radon Mitigation" *ASHRAE IAQ 2007: Healthy and Sustainable Buildings*, Baltimore, Maryland, October 15-17.

U.S. Army Corps of Engineers Air Leakage Test Protocol for Building Envelopes. Version 3, May 2012.

Walton, G.N., 1989. "Airflow Network Models for Element-Based Building Airflow Modeling", *ASHRAE Transactions*, Vol. 95, Pt. 2.

Weber, C.G., Reichel, R. 1942. *Accumulation of moisture in walls of frame construction during winter exposure"*, National Bureau of Standards, Building Materials and Structures Report BMS93, Nov.

Chapter 12

Rainwater Control

12.1 Introduction

Keeping out the rain is one of the oldest and most important functions of building enclosures. In part this is because the amount of water deposited on the above-grade building enclosure by rain, especially wind-driven rain for walls, is generally larger than that from any other moisture source. Despite thousands of years of experience, avoiding rain-related building damage is still one of the most difficult tasks that enclosure designers and builders face. Leon Alberti [1987], in his famous 15th Century treatise on architecture, said

> *"For Rain is always prepared to do mischief, and wherever there is the least crack never fails to get in and do some hurt or other . . ." (pg 16)*

Hence, the choice of rainwater control strategy and the design and quality of construction details are critical. The basic principles for moisture control are outlined in earlier chapters. This chapter examines, in some detail, rain control and enclosure response to rain deposition. A rational and rigorous means of describing rain penetration control is developed. Although rainwater generates surface-water and groundwater flows that must also be controlled, they are not directly considered in this textbook.

Traditionally, rainwater control meant preventing rain from reaching the interior of the building and wetting interior finishes and furnishings. Avoiding rainwater *penetration,* therefore, was critical. It is still a critical criterion, but contemporary enclosures typically must perform at a higher level and control damage *within* the enclosure as well. Rain-related problems can induce damage to finishes or structure (freeze-thaw damage, corrosion, decay); can affect appearance through discoloration, staining, and efflorescence; and can affect human health, for example through mold growth.

Understanding the nature of driving rain and its interaction with buildings is an important, often neglected, part of rainwater control. Chapter 3 considered this in some depth. The quantity of rainwater on a building's surface can vary by as much as two orders of magnitude, a much larger variation in load than designers are accustomed to for structural loads such as snow or wind.

The design and analysis of enclosures for rain penetration control is considered in this chapter in both general and specific terms. The following sections cover wind-driven rain as an enclosure loading; classify the various rain-control strategies; and propose some design strategies for walls.

12.2 Enclosure design strategies

As first identified in Garden's [1963] seminal paper, for water to penetrate the surface of a building enclosure, three conditions must exist:

1. *water* must be present on the surface of the wall,
2. there must be an *opening* on the surface of the enclosure through which the water can pass, and
3. there must be a *force* to drive the water inward through the opening.

If any one condition can be eliminated, rainwater penetration can be eliminated. For rainwater-related damage to occur to a building, a fourth requirement can be imposed:

4. the wetted materials must be *susceptible* to moisture damage.

Because it prevents rain from reaching the exterior surface of a wall, an overhanging eave or balcony can often provide the best form of rainwater protection. In many cases, this solution is either not practical or aesthetically desirable and hence it must be accepted that rainwater will be deposited on the walls of a building. However, rainwater not immediately absorbed upon deposition drains along the enclosure surface and can be shed. Controlling this surface drainage and shedding is an important first step that is often overlooked. Roofs, walls, and even basement designs should all attempt to control surface drainage, both to minimize the amount of water on the surface available to cause further penetration and to control staining. Reducing the rainwater load that the enclosure must deal with by means of these mechanisms is collectively termed deflection and is covered in Section 12.3.

Although deflection will reduce the volume of water on the enclosure, an enclosure rain penetration control strategy must still be developed. For many reasons an opening in the enclosure surface is likely because of the natural porous characteristics of many materials, errors or oversights in construction and maintenance, and/or the natural degradation of materials when exposed to the environment. In some cases, reliably watertight materials such as roof membranes, glass sheets, and precast concrete are constructed with sufficient care and attention that unintentional openings do not occur for many years, often the design life. The three strategies available are described in Section 12.5

A major part of developing a rain-control strategy is the control of rainwater penetration forces. By eliminating, minimizing, or redirecting water penetration forces, the designer can control rainwater penetration into and through the enclosure. Section 12.4 investigates these forces.

Finally, despite our best efforts, some rain is often absorbed into porous materials or penetrates through imperfections, so drying must be provided to remove this incidental moisture. Most enclosures require drying to perform acceptably but some common enclosure components have no drying capacity at all (e.g., insulated glazing units, insulated metal panels).

A multi-pronged approach is usually the most appropriate, reliable, and cost-effective design strategy. Concurrent consideration of all three requirements for rain penetration in the context of the amount of rain deposition expected will result in an effective and redundant enclosure design. This holistic approach to rain control[1], can be summarized and described by the three Ds:

- **Deflection,**
- **Drainage/Storage/Exclusion,** and
- **Drying.**

The following sections of this chapter investigate each in turn.

12.3 Deflection

The climate and the site play a large role in determining the amount of driving rain that a building is exposed to. Many populated parts of the world experience a significant amount of wind-driven rain, and regions exposed to typhoons and hurricanes may suffer extreme exposure.

Although climate may define the *potential* for rain deposition on walls, the actual amount of rainwater that needs to be controlled by the enclosure can be strongly modified (by orders of magnitude) by deflecting wind and rain away from a building or enclosure element. Hence, the first step in rainwater penetration control is the control of rain deposition by the siting, orientation, and massing of the building – exposure to the prevailing driving rains can be defended against by plantings, landscaping, topography, and choosing lower building designs (e.g., bungalows).

[1] Originally developed by Dave Ricketts [Morrison Hershfield 1998] with Drainage as the only enclosure strategy "D", and popularized by CMHC and Hazelden and Morris [1999] who added a fourth D, Durability. We have added Storage and Exclusion to extend the approach to all practical wall, window, and roof systems, not just drained systems.

The shapes of the house roof and overhang also have an important impact that has been well documented if not precisely quantified. Field measurements [Straube 1998, Blocken 2002, Ge 2017] and computer modeling [Blocken 2000] have shown that modest overhangs and peaked roofs reduce rain deposition by approximately 50%. A damage survey of wood frame buildings in British Columbia [Morrison-Hershfield 1997] found that the size of a building's overhang correlated directly with the probability of rain-related damage (Figure 12.1). Peaked roofs and overhangs protect a wall from rain by shadowing and redirecting airflow (Figure 12.2). Hipped roofs create the opportunity to shelter the walls from rain on all four sides of the building.

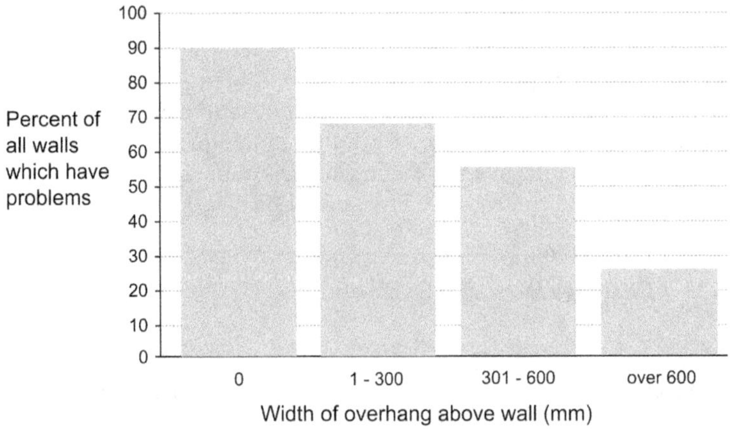

Figure 12.1: Influence of overhang size on low-rise enclosure wall problems in Vancouver, Canada [Morrison-Hershfield 1997]

Field measurements [Lacy 1965, Straube & Burnett 1997, Kuenzel 1994a, Frank 1973, Blocken & Carmeliet 2005, Abuku et al 2009, Ge & Krpan 2007, 2009, Ge et al 2017], computer modeling [Choi 1994; Karagiozis 1995, Blocken 2002], and wind tunnel testing [Inculet & Surry 1994] have provided an indication of the quantity of driving-rain deposition that can be expected on vertical walls. For many low-rise buildings, the amount of rain deposited is in the order of 10 to 20% of the product of wind speed and rainfall intensity. Thus, the amount of rain deposited on the walls of houses erected on exposed sites could be in the order of hundreds of liters per square meter per year (several gallons per square foot). Sheltered locations and single-story houses with wide overhangs are exposed to much smaller quantities of rain.

The design of the enclosure begins to play the dominant role in rainwater penetration control once rain has been deposited on the enclosure surface. However, essentially all enclosures, regardless of the strategy used to control rainwater, can benefit from rain shedding and the control of rainwater on the surface.

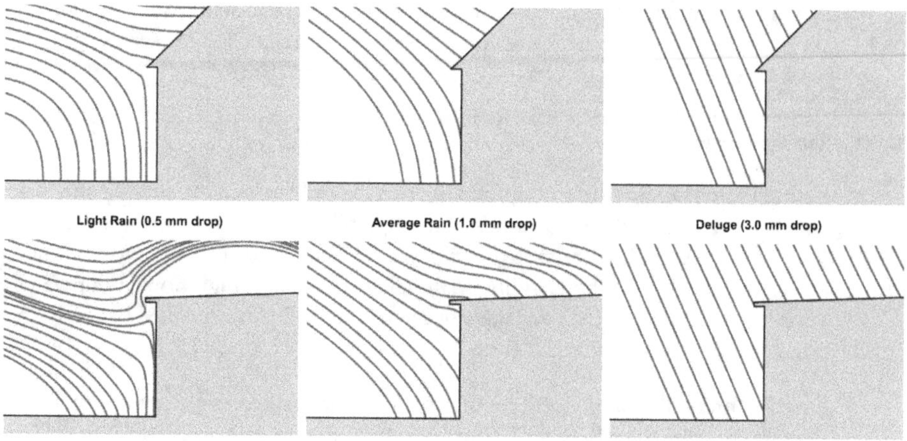

Figure 12.2: Influence of overhangs and pitched roofs on raindrop trajectories [from Blocken 2000]

Shedding is the term commonly used to describe the action of surface rainwater being removed from the surface by drainage. The use of drip edges and slopes ensures that rainwater on the enclosure surface is removed from the building surface as often and as quickly as possible (Figure 12.3).

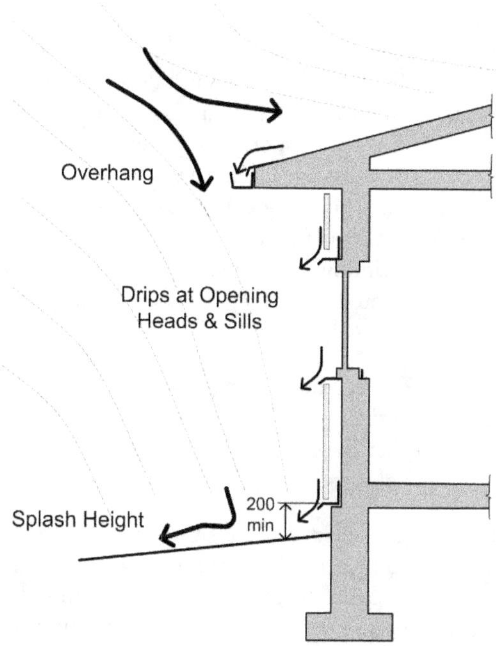

Figure 12.3: Principles of rain deflection

Differences in shedding effectiveness should inform the selection of design alternatives for rain control, the quality of the materials used for water resistance, and the intensity and type of field quality control.

12.3.1 Rainwater on walls

Rainfall intensity and duration, the absorbency and storage capacity of the enclosure material, the smoothness and wettability of the wall surface, the shape of the wall, the effect of the wind and gravity, and the previous wetting and drying history of the cladding material all affect the migration of rainwater once it has been deposited on the surface of the wall [Robinson & Baker 1975, Straube 1998, Blocken et al 2013].

Open-jointed claddings decrease the shedding of surface rainwater and hence tend to increase the amount of rainwater that penetrates deeper into the wall. Pitched roofs with overlapping tiles or shingles and stone-dash stucco finishes rely heavily on shedding to reduce the amount of rain passing through the exterior cladding.

The nature of surfaces above and beside the wall also play a role, since water draining or blown from these surfaces adds to the water on the wall. If water is deposited on a wall at a rate higher than it can be absorbed, a surface film forms, ensuring that any additional deposited water is drained away on the surface.

Once rainwater deposited on a wall forms a film it will begin flowing downward under the force of gravity. Wind flowing over the enclosure surface tends to deflect the flow from this path and, in extreme cases, may even force water upward. Surface features such as trim, texture, and openings can greatly affect the flow paths of this surface drainage, either concentrating or dispersing surface flows. Traditional surface details on old and vernacular buildings (Figure 12.4) often served the function of directing water away from potentially vulnerable areas (e.g., windows) and distributing surface water in such a way as to prevent the concentrated streams that cause staining, freeze-thaw damage, etc.

For walls clad with porous materials, rainwater is absorbed and stored. Hence, a film will only form if the rate of deposition is higher than the rate of absorption or the surface becomes saturated. In practice, many masonry walls will absorb most typical low-intensity rainfalls.

For walls clad with relatively impervious materials (e.g., vinyl siding, painted wood, coated stucco, polished natural stone, metal, glass), there is little or no absorption and a surface film forms very quickly, even in light rainfall. Water flow is encouraged because the surface is generally smooth and, since it is not wettable, it exerts less surface tension on the water.

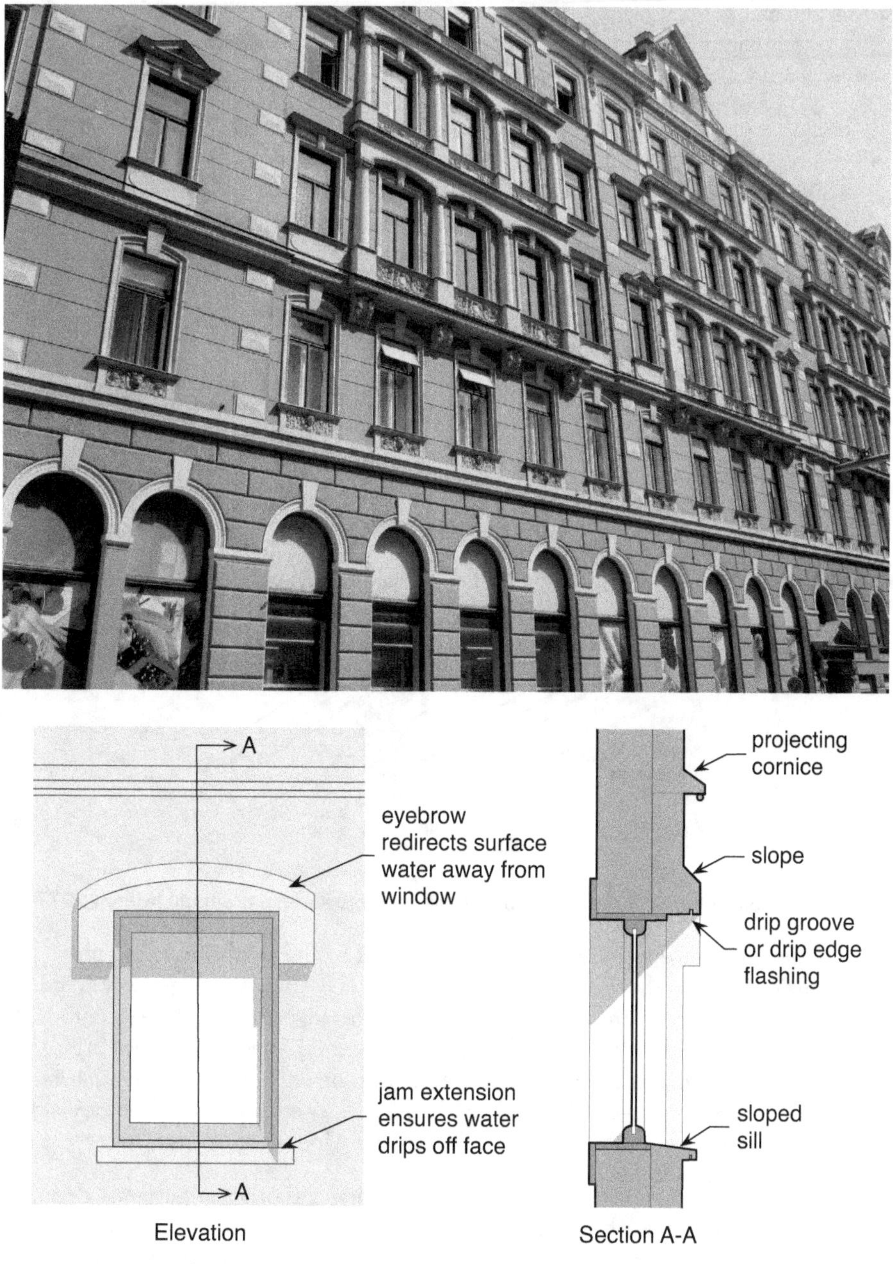

Figure 12.4: Surface rain-shedding features are common in traditional buildings

Surfaces such as dense stucco and smooth, dense concrete absorb a larger amount of water than glass but less than rough brick masonry and therefore act in a manner somewhere between these two examples. The common wall assemblies discussed above are merely two points in a spectrum of wall types which will act relatively more like one or the other type depending on their characteristics. Chapter 8 discusses the water absorption coefficient that can be used to assess the rate of rainwater absorption.

Water flowing on a wall surface tends to concentrate in streams wherever it encounters projections from the flat wall surface. Effective design solutions have been long been developed for common projections, such as window sills (Figure 12.5, Figure 12.8).

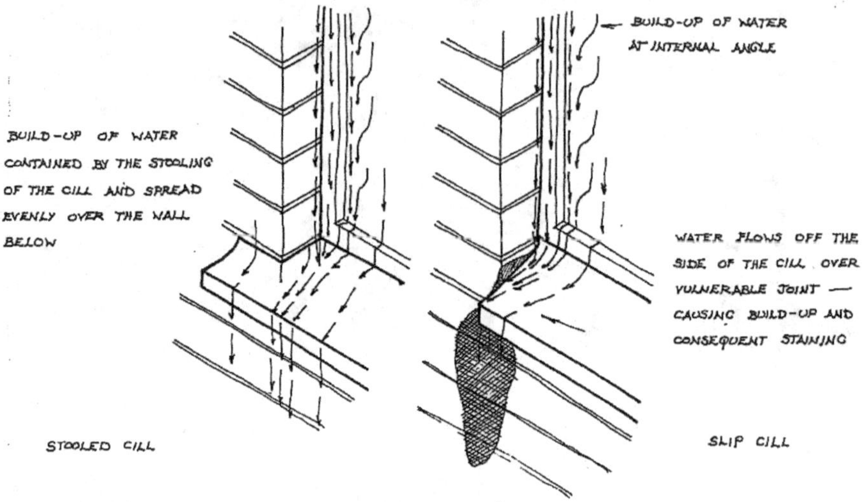

Figure 12.5: Proper window sill-jamb designs manage surface drainage [Marsh 1977]

The non-uniform pressure contours induced by the wind cause secondary airflows perpendicular to the building face. These secondary air flows tend to move from the center of a building face to the edges and act to concentrate the surface rainwater flow towards the outside corners, especially if the surface is smooth and offers little resistance (e.g., glass, metal, polished stone). Surface tension will, however, act to adhere smaller water streams to the leeward side of corners and projections such as flutes and window mullions.

Rainwater surface drainage collected by glazing and other impervious cladding materials can increase the amount of water on a wall below by an order of magnitude. A band of windows on a high-rise building can dramatically increase the amount of rain deposited on the brickwork spandrel below if the driving rain collected by the glazing is not shed and distributed by a proper sill with a projecting drip [Saneinejad & Doshi 2006, Smegal et al 2014].

The concentration of flow can increase the volume of water on small areas of wall to levels many times that predicted by the rain deposition equation. Because of highly irregular concentrations of surface flow, unsightly staining, streaking, and dirt washing can occur, and potentially damaging differential moisture expansion is possible.

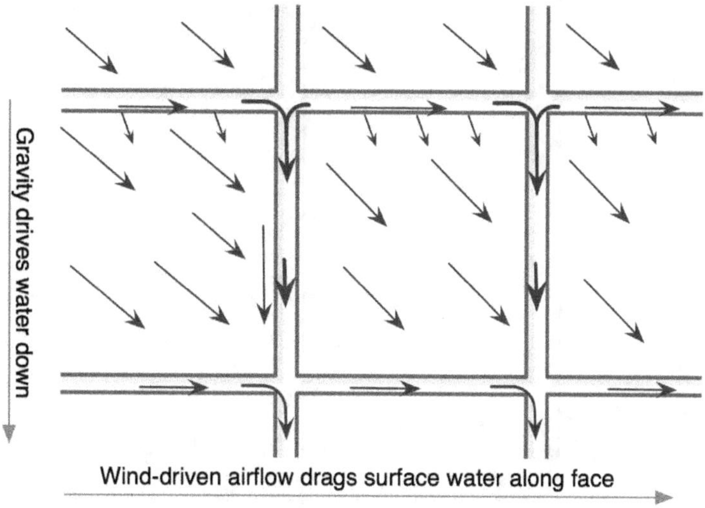

Figure 12.6: Surface features such as joints interact with gravity and surface airflow

As outlined in Chapter 3, an RDF value of 1.0 is probably a reasonable maximum to account for rain deposition. However, poor surface drainage control can easily concentrate the amount of water on a surface by a factor of 10 or more. Such calculations demonstrate why surface drainage and shedding can be so critical to enclosure performance.

Rain deposited on a building face is not the only rain load on a segment of the enclosure: water drained from above and splashed from below can be significant. Different enclosure cladding systems absorb rainwater deposition differently: non-absorptive claddings, such as glass and metal, shed and drain essentially all incident water onto the enclosure below. The shape and surface texture has significant influence, but observation of building wetting patterns leads to some very common enclosure responses, summarized in Figure 12.7.

It is the rain load from the combination of rain deposition and drainage from above that defines the "water load" the enclosure must resist. Standardized tests specify fixed high rates of water impingement. Project-specific design should consider the building exposure, shape, deposition, surface water management and shedding when specifying test rates and interpreting the results of such tests.

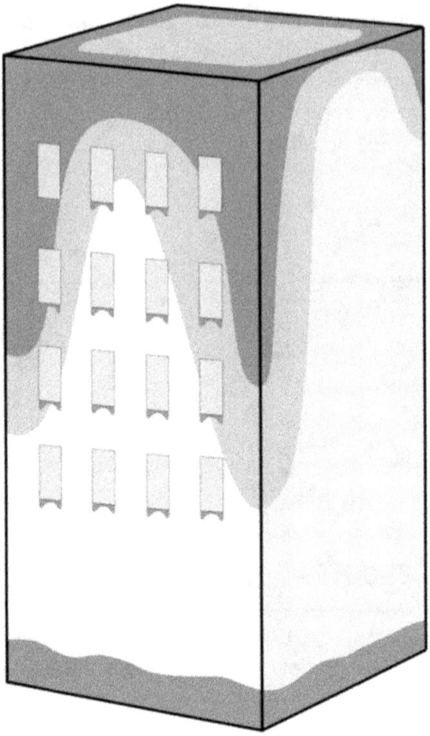

Figure 12.7: Rain deposition, surface drainage, and splashback on a representative building

12.4 Water-penetration forces

Once water has accumulated as a surface film over an opening, whether an intentional joint, unexpected crack, or brick pore, there are essentially five forces that may drive water through: surface tension, capillarity, gravity, air pressure differences, and momentum. Openings include not only designed joints, vents and drains but also the accidental openings left during construction, as well as the cracks and openings that form over time due to rheological forces and exposure to the elements. The relative contribution of each force to rainwater penetration has rarely been measured or studied. The most significant force certainly varies with different enclosure designs, materials, and environmental conditions. For example, wind-induced air pressure differences are not always the most important force, and neither is gravity or capillarity.

12.4.1 Gravity

The force of gravity pulls water down the face of the wall and into openings that lead inwards and downwards. Gravity also causes water to flow horizontally when a small head of water builds on a horizontal surface. Water that penetrates the cladding of a wall runs down the back face of the exterior wythe (adhered by surface tension) until a break in the form of a joint, flashing, mortar bridge, or bulging insulation redirects it. Water penetrating a roof membrane is often transported long distances from the point of penetration by gravity. Gravity is one of the most important water penetration forces in the below-grade enclosure.

The gravity force driving water downward is calculated simply from $f = m \cdot g$. When water flow downward is restricted, so-called *hydrostatic* pressures result. These pressures can be calculated from:

$$p_h = \rho \cdot g \cdot h \qquad \text{(Eq. 12.1)}$$

where p_h is the hydrostatic pressure (Pa),

ρ is the density of water (1000 kg/m^3 at room temperature), and

h is the height of the water (often called pressure head).

Because of the high density of liquid water, standing water of only 10 mm results in hydrostatic pressures of 100 Pascals.

12.4.2 Surface tension

As described in Chapter 8, surface tension is caused by the attraction of polar water molecules to other water molecules and to molecules on the surface of materials. Some materials attract water molecules (e.g., glass) whereas others repel them (e.g., silicone). Many building materials have a strong attraction to water molecules. Rough surfaces, which have a larger effective surface area, exert a larger attractive force than smooth materials. Therefore, a water droplet of smaller size (i.e., acted on by a smaller gravity force) will more readily roll down a smooth surface (e.g., polished granite) than a rough stucco surface.

As surface tension causes water to adhere to and run along surfaces it tends to direct water along surface irregularities (because of the larger effective surface area) and may allow water to adhere to a horizontal surface. Surface tension is the primary force causing deposited rain water to adhere to the building enclosure. However, without the additional force of gravity, capillary suction, or air pressure, surface tension will not force water into or through an opening.

The time-proven approach to dealing with surface tension as a penetration force is to incorporate drip edges (in the form of grooves or bent edges) along the leading edge of horizontal joints and protections. These drips make use of gravity to redirect the water away from a vulnerable opening (such as a vent opening or above a window) or to prevent water concentrations (i.e. below window sills) Figure 12.8. Projections of 12 to 25 mm are often recommended in trade literature. Some research has been conducted [Saneinejad & Doshi

2006, Smegal et al 2014] that also sheds light on the shape of the drip edge as well as the projection.

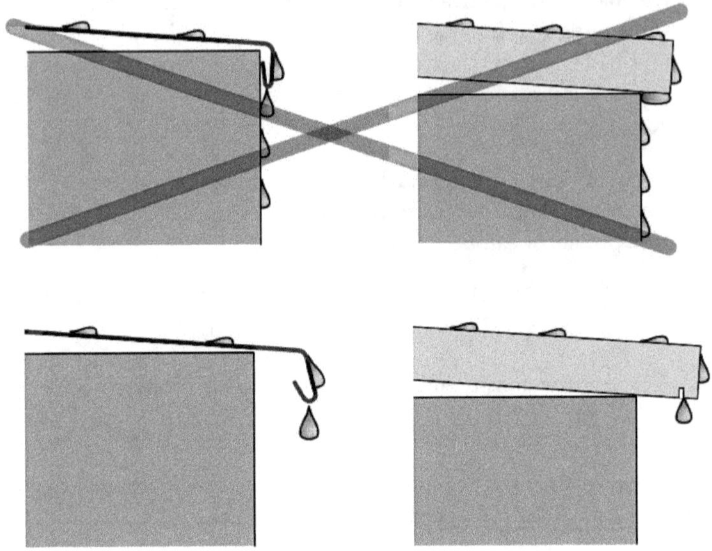

Figure 12.8: Drip edges used to control surface tension at parapets and sills

12.4.3 Capillarity

Capillary suction, a force directly related to surface tension, will draw water into small cracks and voids in a material until the void is full and the perimeter of the void is saturated. The water will not leave the crack or void until another force is applied, such as an air pressure difference or gravity, because surface tension acts over the exit area and resists water penetration once the capillary forces have filled a crack or hole. If this surface tension film is broken, water can flow under a much smaller driving difference (of pressure or gravity). The pressure that induces the capillary rise observed in small tubes can be calculated by the following equation (see Chapter 8 for more details):

$$p_c = 2\frac{\sigma}{r} \qquad \text{(Eq. 12.2)}$$

where p_c is the capillary suction pressure,

σ is the surface tension constant, and

r is the radius of the tube or the width of the crack.

For distilled water exposed to air at 20 °C, the value for surface tension is approximately 0.075 N/m. Lower temperatures increase the surface tension value slightly, and dissolved oxygen or salts may reduce the value by as much as one half.

Capillarity itself does not cause water to leave a material layer (or the assembly of materials that constitute the enclosure). Rather, it draws water into small cracks and openings as well as tiny pores in materials. The equation indicates that for small cracks, significant suction pressures can be developed to draw water into and resist the removal of water from a crack. Capillarity bound water to leave a material only when assisted by gravity pressures (i.e., hydrostatic), air pressure differences, or some other force. Given the limited pressures imposed by air pressures (hundreds of Pascals), water held in small pores and cracks cannot be dislodged by air pressure.

12.4.4 Air pressure differences

If the air pressure on the outside face of a material, wall, or cladding is greater than on its back face, then water from a surface film may be driven through by the unbalanced force. If the flow of air through an opening is of sufficient speed and volume, friction can drag water into and along the air flow path.

Total air pressures induced by the wind and other forces are in the order of 1 kPa (20 psf) for long return periods (e.g., once in 10 years), and short durations (e.g., 3 seconds) for tall buildings. More likely peak gust pressures (return periods of one year) are an order of magnitude less (approximately 200-400 Pa for a few seconds), and *daily* average pressures are in the order of 4 to 20 Pa for low-rise buildings and 10 to 50 Pa for exposed high-rise buildings. Wind pressures coincident with rainfall are usually lower and the 1-in-5 year Driving Rain Wind Pressures from the Canadian National Building Code – a rather rare event that is conservative for most uses -- are below 300 Pa for many locations.

If the air pressure on either side of a layer can be equalized, water on the surface of that layer will not be forced inwards by this force, and the other forces can be dealt with individually. This approach is the theoretical basis of pressure-moderated walls and joints (discussed later).

Simple substitution into Eq. 12.2 shows that cracks (which have an equivalent radius equal to twice their width) of 1 mm width exhibit a capillary suction of about 37 Pa, and cracks of 0.1 mm have a suction force of 375 Pa. Wind pressures and mechanically-induced air pressures are therefore practically unable to drive water from all but the larger cracks. Cracks of 1 mm width and greater are easily visible to the naked eye. Wind pressures can never be large enough to drive water from the microscopic pores of typical building materials (see also Chapter 8).

12.4.5 Momentum or kinetic energy

Droplets of rainwater can be carried through sufficiently large openings (usually intentional openings) by their wind-driven momentum. The conversion of the kinetic energy of a water droplet to a static pressure on impact with a crack may result in a small, short-term increase in water pressure which may overcome the surface tension forces holding water in the crack.

When air currents move through an opening due to a pressure difference, the momentum of a rain drop can be increased (since air may accelerate as it passes through a small opening).

If there is no direct path to the interior, the rain droplets will not move far into the wall and, with proper design, can be directed back out again by gravity. Baffles, splines, catchment trays, and labyrinth joints are all effective methods of countering the penetration of water past the outer surface by kinetic energy [Herbert & Harrison 1974, Isaksen 1974].

Although water droplets can theoretically be carried through openings by this mechanism, the cracks must be much larger than the droplets (typically about 1 mm in diameter) and significant airflows through the opening must occur. Except for open-jointed claddings or very air leaky enclosures, this combination of events is unlikely and hence the amount of water that can penetrate a cladding layer by this mechanism is small.

Open-jointed claddings are likely to allow water past the joints [Herbert 1974], although the amount of surface drainage that can be directed inward by surface tension is larger.

12.4.6 Forces in combination

As noted above, some of the forces that cause rain penetration are significant only when acting together with other forces. Gravity-assisted capillarity and pressure-assisted capillarity cause water to penetrate a material with fine cracks. For some wall claddings (e.g., horizontal lap siding), pressure-assisted capillarity may be a significant water penetration force. Gravity-assisted capillarity is usually the most significant force causing water to penetrate masonry veneers. Surface tension resists water being pulled through a surface opening or crack, but gravity may assist surface tension, and air pressure differences may combine to result in water moving through an opening.

As an example, consider the rain penetration mechanism of single-wythe masonry veneers. Figure 12.9 shows an isometric of a likely situation of debonded mortar at the brick-mortar interface.

Researchers have long reported that water penetrates single-wythe veneers almost exclusively through the interfacial cracks between the mortar and brick at head (vertical) joints [Grimm 1982]. Capillary suction draws the water in (a suction pressure of up to 375 Pa can be calculated for a 0.1 mm wide crack). Gravity head (equal to 10 Pa per mm of height or 500 to 700 Pa for normal head joints) within the head joints drives the water out. If the crack suction pressure is greater than the gravity pressure, no flow will occur. Usually this is not the case (since the cracks are often 0.1 to 0.3 mm wide). As mentioned earlier, air pressure has a relatively small effect since air pressures are almost always much smaller, although they can and do increase flow slightly.

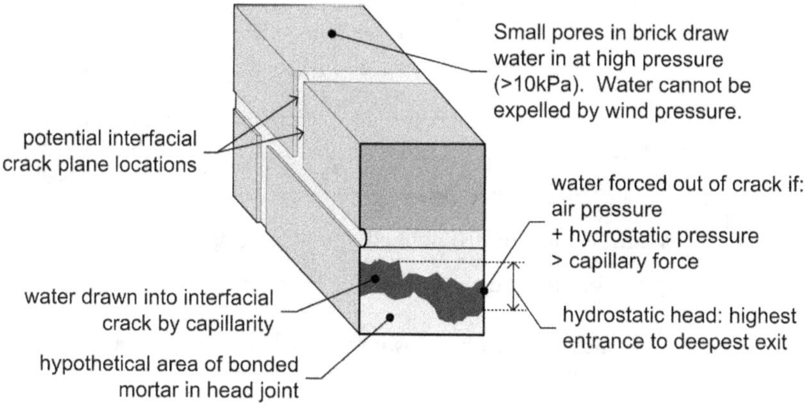

Figure 12.9: Combined water penetration forces in single-wythe masonry

12.5 Rain penetration control strategies

Although there are many forces that drive rain penetration, understanding these can help provide rain penetration resistance. Rain control can be provided by using both traditional details and modern scientific understanding.

The previous sections examined the amount of driving rain that can be expected to be deposited on a vertical above-grade building enclosure and the behavior of rainwater on the surface. However, siting, building shape, and surface rainwater control rarely provide complete rain control[2]: they merely reduce the rainwater load. Hence, some strategy to deal with the rainwater that penetrates the surface of the wall must be employed.

This section examines the response of the building enclosure to the rain penetration loads in more detail. The objective of this section is to develop a rational and rigorous means of describing and classifying rain control for the above-grade enclosure. Such a description and classification system is needed because:

- It assists enclosure designers in understanding the function of each layer and material in an assembly. Such understanding is needed to create better designs that more reliably and economically meet performance requirements.

2 although deep, wraparound porches may eliminate the need for rain control under them

- It allows water penetration tests to be developed and applied that impose realistic loads and more closely match likely service conditions. The results of such testing can also be interpreted to aid in the prediction of actual performance under extreme and typical loadings.
- It provides a means for the accurate assessment of the reasons for enclosure failures and therefore proper repair and retrofit strategies.

There are three fundamental rain control strategies available to the designer [Straube & Burnett 1999]. In order of historical priority, they are:

- storage (also termed reservoir or mass systems),
- drainage (also termed drained, rainscreen, or drainscreen), and
- exclusion (termed perfect barrier, face-sealed, concealed barrier).

These three strategies can be shown to be the result of the rational analysis of rain control described below.

12.5.1 Rainwater control theory

Rain deposited on a wall can either be drained from its face (shed), stored (absorbed by capillarity or attached by surface tension), or transmitted further into the wall. Each layer of a multi-layer wall responds in a similar manner. If water penetrates through the entire wall assembly, the wall is generally considered to have failed. However, transmission of rain water into any sensitive layer of a wall assembly may also cause damage, degrade performance, and affect durability. Only the water that is stored causes wetting, i.e,. it must be subsequently removed by one of the drying mechanisms described in earlier chapters (diffusion, air leakage, ventilation, and surface evaporation).

The above reasoning can be made more rigorous by applying mathematical rules. Consider a one-dimensional enclosure assembly consisting of n layers. At each layer i, the water from the previous layer, i-1, must be either stored (S_i), drained (D_i) or transmitted to the next layer (T_i). Since the enclosure under consideration is one-dimensional, drained water must be considered as water that leaves the assembly. In a 2-D (height and layers) or 3-D (height, width, layers) analysis, drained water ($D_{i,x,y}$) may subsequently become stored ($S_{i,x,y}$) or transmitted water ($T_{i,x,y}$).

The materials and arrangement of most practical assemblies are such that water cannot penetrate further than a specific layer, k, $k \leq n$, or the assembly will be considered to have failed to control rainwater. This is termed the water control layer for both design and analysis. In the case of k=n, every layer of the assembly is presumed to be tolerant of rain wetting, e.g., an all-masonry wall with no finishes in a warm climate. (Note: the precise assessment of layer k is limited by the fact that failure cannot be defined simply as the presence of moisture, the amount of moisture and the length of time and the moisture tolerance of the material are obviously important as well. However, from a design point of view, one can usually choose a layer beyond which rain water should not penetrate, e.g., into the studspace of framed walls.)

To provide a completely general analysis, an enclosure layer must include both materials and interfaces, since interfaces between wall layers often have an important influence on storage, drainage, and transmission characteristics.

From observation of Figure 12.10 and conservation of mass it can be written:

$$\sum_{i=0}^{n} D_i + \sum_{i=0}^{n} S_i + T_n = R \qquad \text{(Eq. 12.3)}$$

for any given period of time, ranging from a few seconds to an entire rain event. R is the amount of water on the surface from both rain deposition (r_{bv}, see Chapter 3) *and* drainage from above.

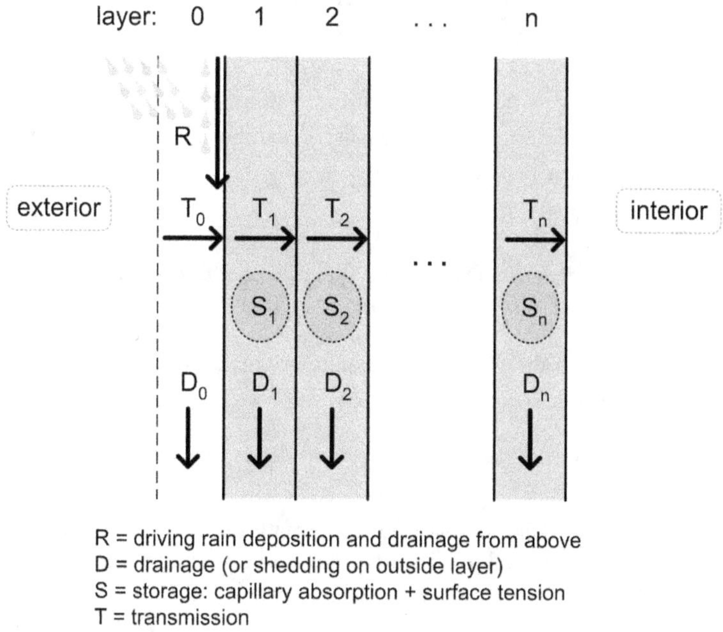

Figure 12.10: Distribution of rain within a multi-layer enclosure assembly

Also, for any given layer,

$$D_i + S_i + T_i = T_{i-1} \text{ for all layers } i = 1 \text{ to } n \qquad \text{(Eq. 12.4)}$$

It is useful to non-dimensionalize values on a per layer basis, thereby providing *fractions* of rainwater. Since, $D_i + S_i + T_i = T_{i-1}$, then

$$\frac{D_i + S_i + T_i}{T_{i-1}} = 1.0 \quad \text{or,} \qquad \text{(Eq. 12.5)}$$

$$d_i + s_i + t_i = 1.0 \qquad \text{(Eq. 12.6)}$$

where the lowercase letters are used to denote fractions.

Analysis usually requires the concept of interfaces in addition to layers. Water does not drain *in* an enclosure layer such as an airspace of over 3 mm thickness: rather, water flows along the interface (a surface) between the cladding and the airspace and the interface between the backup and the airspace. The use of layers and interfaces depends on the need and the analyst.

It is also useful to define an imaginary outer surface layer, layer zero: this is sometimes labelled the *water shedding surface* or layer. This imaginary layer allows one to consider surface shedding as this layer's drainage (D_0), water beading on the surface as storage (S_0), and transmission (T_0) as the water that penetrates the outer surface of the cladding. Therefore,

$$D_0 + S_0 + T_0 = R \qquad \text{(Eq. 12.7)}$$

S_0 is almost always a negligible quantity for walls, although it can be significant for low-slope and flat roofs. For walls, S_0 can be assumed to be zero or nearly zero and Equation 12.7 rewritten as:

$$D_0 + T_0 = R \qquad \text{(Eq. 12.8)}$$

which merely states the obvious fact that rain water deposited on a wall can either penetrate (T_0) or be shed (D_0).

The magnitude of the fraction T_0 is quite important for defining the load on the enclosure system.

The above analysis of rainwater movement leads to a series of related conclusions:

- The relative amount of water dealt with by each fraction, and at which layer(s), is critical to understanding and describing how an enclosure exposed to rain behaves.
- Because the response of materials and interfaces may change due to changes in moisture content, temperature, time, etc., enclosure response may also change with time.
- Based on the previous conclusions, the rate and duration of rain deposition, the time between rain events, and the drying of the enclosure, i.e., the environmental conditions preceding the time in question, may all influence enclosure response to a driving-rain event.

These conclusions have significance for understanding the in-service performance and interpreting the results of rain penetration testing, both in the laboratory and in the field.

12.5.2 Rainwater control strategies

As discussed above, the intended or actual response of an enclosure to driving rain can largely be described by the relative fractions of rainwater dealt with by each mechanism at each wall layer. Regardless of the control strategy, the amount of water shed from the surface is D_0 and the amount that penetrates the outer surface of the enclosure is T_0. Bear in mind that R is the volume of rainwater deposited on the surface plus any drained from above. Several different strategies are used to control rainwater penetration in buildings. These are discussed below in the context of drainage, storage, and transmission.

Perfect barrier systems

A multi-layer enclosure may contain a single layer where the amount of rainwater transmitted inward must be zero (a perfect barrier), at one layer, j:

$$T_j = t_j = 0, \quad j <= k, \qquad \text{(Eq. 12.9)}$$

A rain control strategy with j=1 is often termed "face-sealed" (Figure 12.11) and is just one of numerous perfect barrier approaches.

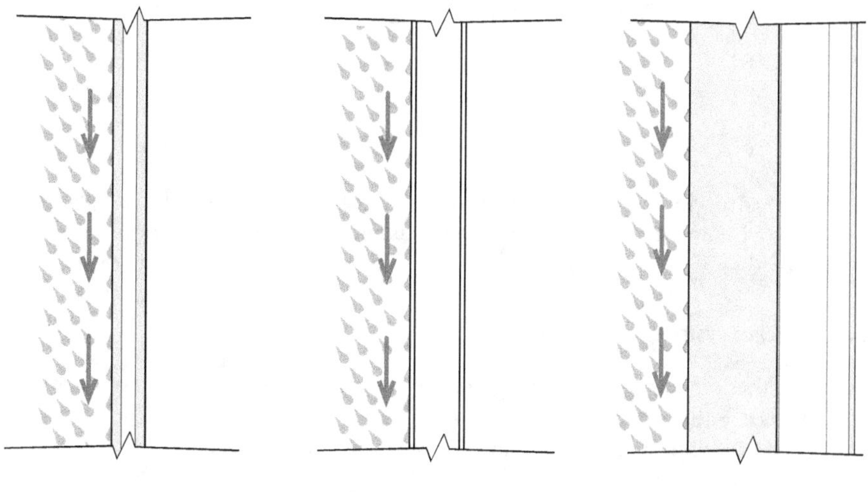

Figure 12.11: Examples of face-sealed perfect barrier wall systems

Although the perfectly water-impermeable layer j is often located on the exterior face of an assembly, i.e., j=1 and t_1=0, it need not be. The layer in the assembly that acts as the perfect barrier is the water control layer for a perfect barrier system. Since the transmission characteristics of the perfect barrier layer are usually independent of moisture content in practice, a perfect barrier approach is independent of previous history (i.e., weather conditions).

A "*concealed barrier*" approach is also used in practice. Examples include stucco directly applied to a single layer of building paper. This system is not designed with any provision for drainage, and the storage capacity of the stucco is rather limited. Hence, the building paper often acts as the perfect barrier in the system (and often fails to perform adequately in this example because building paper is not sufficiently durable or waterproof).

Another concealed barrier enclosure example is the protected membrane roof design used for low-slope applications. As the roof membrane barrier is concealed, or protected, it is likely to perform better than a face-sealed system exposed membrane roof (both because most water is shed by outer layers and because the barrier is protected from solar radiation and temperature swings).

Perfect barriers with j greater than 1 may provide some small amount of drainage or storage in outer layers, but the key differentiator is that perfect barrier systems rely on the water control layer to manage all deposited rainwater.

In the special but common case of a *face-sealed, perfect barrier* enclosure j = 1, and the following characteristics apply:

$$T_1 = t_1 = 0 \qquad \text{(Eq. 12.10)}$$

and so

$$D_0 = R \qquad \text{(Eq. 12.11)}$$

That is, all driving rain deposition and surface drainage from above must be shed. Drainage and storage within the assembly therefore do not play a role in providing rain control in a *functional* face-sealed perfect barrier wall or roof.

Drained-screened systems

Another strategy to controlling rainwater penetration accepts that water will penetrate the outer surface *and* provides drainage to remove it. The drainage need not be provided at one layer only. In fact, if drainage were to be provided at one layer only, the wall would by definition be acting as a perfect barrier, since any layer j where $D_j = T_0$, $t_j = 0$, is the definition of a perfect barrier. Drained walls, instead, control rain penetration by reducing the amount of water transmitted at each of several layers, i.e.:

$$\sum_{i=0}^{\leq k} D_i = T_0 \qquad \text{(Eq. 12.12)}$$

Since the dominant control mechanism is drainage, enclosures using such a strategy are sometimes called "drained" systems. However, the layered and screening nature of the assembly is also important in this approach. The critical characteristic of a drained wall system is that the drainage need not be complete at any one layer for acceptable performance, nor does transmission need to be completely controlled at any given layer, nor is storage required.

Because drained enclosure systems control rain penetration by reducing the successive fractions of water that are transmitted inward, the following relation can be defined:

$$\prod_{i=1}^{\leq k} t_i \approx 0 \qquad \text{(Eq. 12.13)}$$

The implications of Equation 12.13 can be better understood by considering an example assembly consisting of just three layers, each with small penetration fractions of 10%. The amount of rain penetrating the third layer would therefore be $0.1 \cdot 0.1 \cdot 0.1 \cdot (T_0) = 0.001 \cdot (T_0)$.

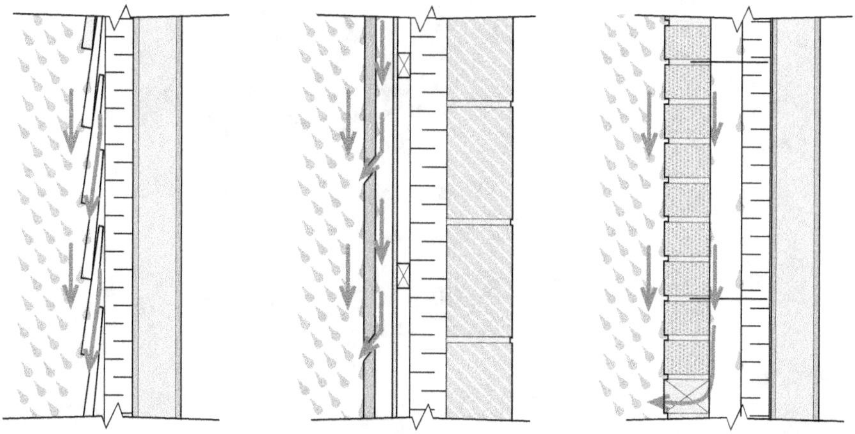

Figure 12.12: Examples of drained-screened wall systems

If all the rain deposited during a particularly intense rain storm on the enclosure penetrates (T0 = R), the amount of water penetrating the third layer will be 0.1% of the total deposited. Assuming a relatively intense deposition rate of 5 l/m²/hr, the total that would penetrate the three layers would 0.001(5) = 5 ml/m²/hr! Providing even a small amount of storage in the third layer will ensure that rain penetration is controlled. In fact, failure to control rainwater in such walls is usually the result of a failure to drain (since the required amount of drainage equals .999· (T1)). This line of reasoning supports the "drained" label applied to this type of system.

The term rainscreen has been rather loosely applied to several special cases of drained walls, especially those that use several layers and several mechanisms to control rain. A rainscreen system typically contains a drainage layer as the second layer of the assembly, (i.e. $d_2 \approx 1.0$). The first drainage layer is actually the shedding on the exterior face of the screen (i.e., the cladding as for all enclosures), and the second drainage layer is often the back side of the screen. Almost without exception, functional rainscreen walls provide several additional drainage layers, for example, the exterior face of continuous insulation, and the sheathing membrane / water resistant barrier (WRB). The screen is, of course, much more than a rainscreen in these wall systems: it is also a screen against the sun, impact, etc.

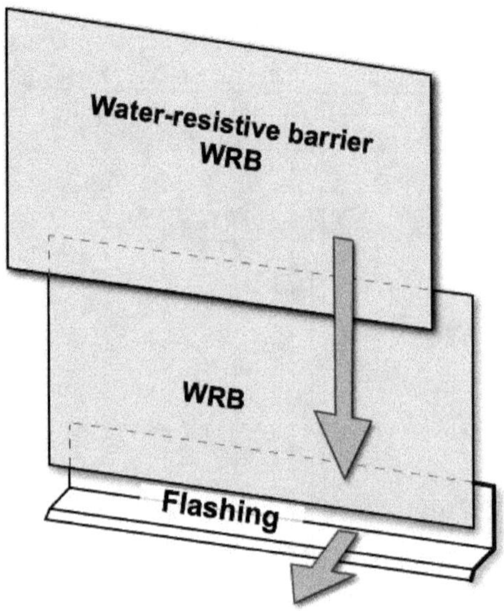

Figure 12.13: Water-resistive barriers integrated with flashing forms a drainage plane

Pressure-equalized rainscreens are a very special case of drained systems that are designed to intentionally moderate wind pressure to reduce the transmission fraction of the screen. These systems still rely on effective drainage because forces other than air pressure drive rain across the screen (especially gravity) and hence do not often provide significant improvements in rain penetration control performance relative to well drained and vented systems.

It is useful to understand the difference between *drainage spaces* and *drainage planes* [Lstiburek 2006]. The former are gaps, airspaces, or highly porous materials that allow unimpeded gravity-driven flow of water. The latter are the hydrophobic and/or waterproof layers that act as capillary breaks and surfaces along which water can flow. Together, the space and the plane (more formally called the WRB or sheathing membrane) form the water control layer of a drained system.

Although large air spaces may be beneficial for other reasons (i.e., managing dimensional variations and providing ventilation drying), drainage spaces often need only be only a fraction of an inch, i.e, less than a millimeter, to allow drainage [Straube et al 2000, Weston et al 2001, Straube & Smegal 2007, VanLinden 2019]..

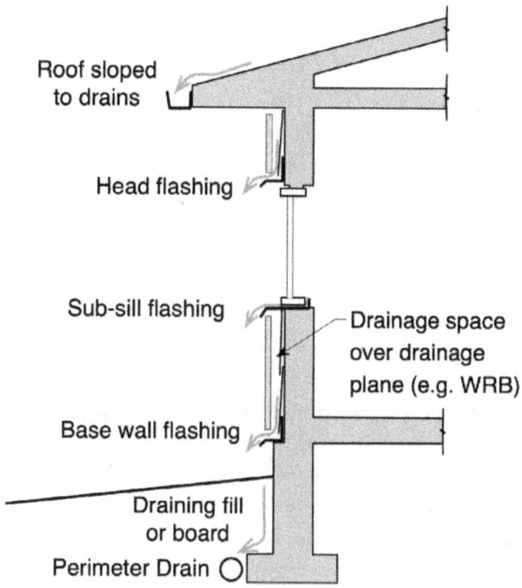

Figure 12.14: Drainage concepts applied to the building enclosure

Mass systems

A third rain control strategy is to provide an enclosure with enough moisture storage to absorb all rainwater that penetrates the outer surface (Figure 12.15). By shedding or absorbing all rain water, such an assembly can prevent inward transmission, i.e.,

$$\sum_{i=0}^{\leq k} S_i = T_0 \qquad \text{(Eq. 12.14)}$$

Enclosures employing this strategy might be termed "storage" systems, they are may also be called "mass" or "reservoir" enclosures since a large quantity of material is usually required to store a meaningfully large quantity of water. Unlike perfect-barrier or drained systems, the maximum quantity of rain that can be controlled is limited by the storage capacity available. Mass systems also do not use drainage within the assembly. Therefore, the performance of mass systems is highly dependent on weather conditions, e.g., if a series of rain events occurs and drying conditions are poor, the storage capacity may be exceeded and failure to control rain will result.

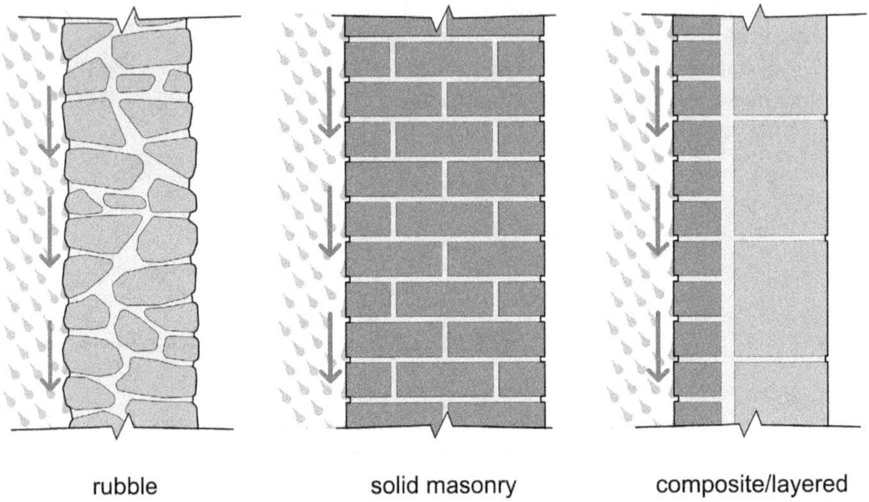

Figure 12.15: Examples of mass or storage wall systems

The storage capacity of the system is not the only system characteristic that is important in rain control. The liquid water permeance of the various layers and interfaces can play a critical role. For example, a 300 mm (12") thick solid multi-wythe brick masonry wall has a certain amount of storage capacity. Applying a thin, 15-mm (5/8") coating of cement-lime based rendering increases the storage capacity only marginally, but the liquid water permeance of this outer layer is so low (even when cracked) that the rain penetration control is vastly improved.

The combination of materials and layers that are intended to, or actually, provide sufficient moisture storage for the rain exposure is the water control layer in a storage system. In many practical scenarios, only some fraction of a layer may be needed (for example, only 50% of the thickness of a solid brick wall may be required).

12.6 Categorization system

Based on the theory presented above, a system for categorizing enclosure wall rain control is presented in Figure 12.16. This categorization system is based only on the method by which the wall system controls rain penetration. The categorization is independent of materials, building function, or design intent.

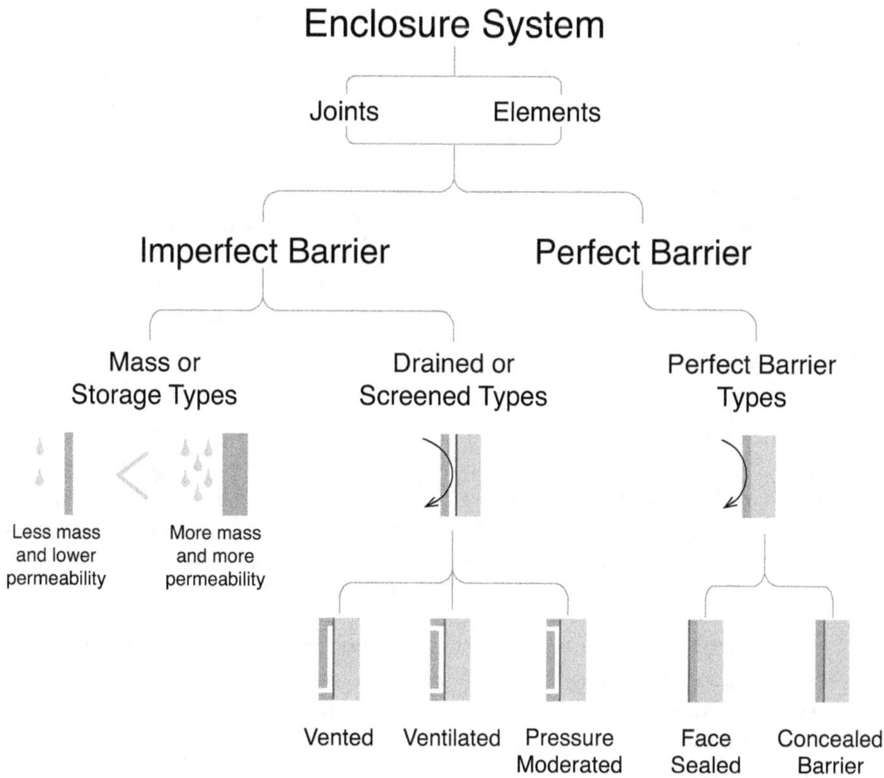

Notes: The categorization is based on either actual behaviour or design intent dependent, which should be declared.
For the purposes of this classification system, the following definitions are necessary:

1. Drained: water that penetrates the exterior surface can be removed by gravity.
2. Cavity: a space that facilitates gravity drainage and possibly air flow and resists the lateral transfer of water (a capillary break).
3. Ventilated: allows sufficient flow of air to promote drying by water vapor movement.
4. Vented: allows some degree of water vapor diffusion through vent and redistribution within a cavity by air mixing and vapor diffusion.
5. Pressure-moderated: an approach that moderates air pressure differences across the cladding. "Pressure equalized" describes perfect moderation, a theoretical condition.

Figure 12.16: Wall categorization system (by rain penetration control)

Walls can be considered to be composed of elements and the joints between these elements. Both the elements and joints may be classified in the same manner, but the classification of the joints in any given system may be quite different from that for the element. The primary and initial classification is whether a wall element is a perfect barrier or an imperfect barrier. If an imperfect barrier approach is selected, the choice becomes between mass storage and drained systems.

Perfect barriers stop all water penetration at a single plane, much as a perfect air barrier would prevent air flow. Examples of perfect barrier walls are some window frames, EIFS, and metal and glass curtain wall systems. Because it is very difficult to build and maintain a perfect barrier wall, most walls are designed as, or perform as, imperfect barrier wall systems of either the mass type or the screened type.

Mass systems control rain penetration by absorbing and storing rain water which penetrates the exterior surface. In a functional mass wall, this moisture is eventually removed by evaporative drying before it reaches the inner surface of the wall. Some examples of mass walls include adobe, solid multi-wythe brick masonry, and single-wythe block masonry. Undrained masonry cavity walls (i.e., those with blocked airspaces or weep holes) often behave as multi-layer mass walls.

Drained walls control rain penetration by draining any water that is not shed from the front face. Drained walls are also imperfect-barrier type wall systems in that this approach acknowledges that some rain water will penetrate the outer surface. Supplementary mechanisms, such as capillary breaks, storage, and multiple drainage planes are usually employed to resist further inward movement of the water that penetrates the cladding ("screen").

Some examples of drained wall systems include masonry cavity walls, brick and stone anchored veneers, rainscreen panel claddings, vinyl siding, two-stage joints, and drained EIFS. Note that the outer layer of a multi-layer drained wall is much more than a rain screen: it must also resist wind, snow, solar radiation, impact, etc.

A small gap or cavity behind the screen provides a capillary break (to prevent liquid transport from the cladding to the layer behind the cladding), as well as a clear path for gravity drainage and a path for air flow. A cavity is defined for our purposes as any space behind the screen, which may be filled with a porous material, that is sufficiently continuous and permeable to fulfill these functions. An air space is the most common choice. (Note: from a structural point of view, a cavity in a masonry wall is the space between two wythes. It may be empty or partially filled by insulation).

Given a cavity behind the screen, four major sub-classifications relating to air movement and vents are possible: vented, pressure-moderated, ventilated, or pressure moderated and drained.

A *vented wall* system allows some degree of water vapor diffusion and air mixing between the cavity and the exterior. Venting (and, to a greater degree, ventilation) provides a mechanism for the removal of water that does not drain from behind the cladding. Venting, or better still, ventilation, may also remove water vapor that may have diffused outward from the inner wythe. By increasing the flow of air into and through the cavity, a relatively large volume of water vapor can be transported from the cavity. Such a *ventilated* wall will assist the drying of both the inner wythe and the cladding.

A *pressure-moderated* wall system promotes the moderation of the pressure difference across the cladding. The proper choice of venting, i.e., size, number, and location, and the division of the cavity into stiff, airtight compartments are necessary requirements. Although such walls are sometimes described as *pressure equalized rainscreens* (PER) in Canada, instantaneous pressure equalization rarely occurs in reality, and the screen deals with more than rain. Hence the more realistic term, *pressure-moderated drained* wall, is preferred.

12.6.1 Actual behavior

Many functional enclosure systems may perform quite differently from the way in which they were designed to perform during most rain events. Many enclosure walls may have such effective rain shedding that water does not penetrate at all during most rain events over most of the area of the enclosure. Of course, the success of an enclosure design is usually based on its ability to manage rain penetration during all but perhaps extremely rare rain events and over all of the enclosure area. Understanding how enclosure actually perform, rather than how they were intended to perform, is critical for assessments of existing enclosures.

If drainage, storage, or both are provided in an enclosure designed and/or built as a perfect barrier system, the perfect barrier need not be perfect. For example, despite the many cracked and imperfect Exterior Insulation and Finish Systems (EIFS), there are millions of square meters of EIFS in North America that continue to control rain. This is not because the lamina allows absolutely no transmission, but because the substrate has sufficient storage to store the very small fraction t_1, that penetrates. In other cases, this storage capacity may act in conjunction with drainage that is inadvertently provided by interface layers and large pored or fibrous insulation materials. Modern EIFS usually include full drainage behind the insulation layer, mostly to deal with rain penetration that occurs at penetrations such as windows.

Storage may also play an important role in drained wall systems. For example, in most masonry veneers, $S_1 = R$ during many rain events and therefore no water is available to be shed or transmitted inward. However, during extended rainy periods and intense rain events, the veneer may become saturated, and some water will penetrate the permeable veneer (since masonry veneers are remarkably water-permeable).

The layers behind the veneer need to perform their drainage function only during extreme events, and it is therefore only during these events that drainage becomes critical. Inspection of many functional veneer walls shows that drainage is often blocked by mortar bridges and dams. The exposure conditions and storage capacity of such walls suggest that they are actually behaving as mass walls, with several low transmittance layers (e.g., the building paper or housewrap) behind the storage. These walls may be functional, but if they are used in an area with more driving rain deposition or less drying potential, the storage capacity is likely to be exceeded and rain-control failure will result. This has commonly been observed to occur in practice when brick veneer wall designs with a record of good performance in single-family housing fail when used in high-rise construction or custom homes with tall walls unprotected by overhangs.

As another example, consider common residential systems with cladding fasteners that penetrate building papers and housewraps: these fastener penetrations may allow very small amounts of water to pass through, but because of the storage capacity of the sheathing and drying capacity such systems historically have performed very well.

Low-slope roof assemblies provide many examples of the perfect barrier approach. Most low-slope roofs are designed as face-sealed, perfect barrier systems. Because of the often shorter-than-desired life-span of these roofs, a different roof design philosophy emerged. This approach, labeled the "self-drying roof" by Desjarlais [1995] accepts the fact that the exposed membrane will not be perfect for the design life of the roof. Numerous layers of storage are provided, typically in the form of wood fiberboard, which are sized to allow for the safe storage of moisture equal to $T_1 \neq 0$. Solar heating and appropriately vapor-permeable inner layers assist in removing this water by evaporation and diffusion to the interior. This type of roof therefore becomes a mass system when the membrane fails, with a very low transmittance outer layer and drying potential to the interior.

A *protected membrane roof* (PMR, sometimes inaccurately called an "inverted" roof) is a classic example of a concealed barrier strategy. Such roofs have a waterproof membrane ($t = 0$) located below several layers of moisture-tolerant materials such as crushed stone ($i=1$) , a geotextile ($i=2$), and extruded polystyrene insulation ($i=3$). In this assembly, $T_4 = 0$, and usually $k=4$. Observations have shown that a large fraction of water is drained along the top of the insulation (typically, d_2 is in the order of 70%). Nevertheless, the success of the design rests on the perfection of the protected membrane layer. Modern PMR designs may use a water-repellent layer above the insulation [e.g., Merkel 2001] to reduce the quantity of water reaching the concealed perfect barrier roof membrane to a minimum, and thereby also protect the insulation from excessive wetting.

Sloped (usually over 3:12 or 14 degrees), shingled roofs are drained systems with the special characteristic that the majority of drainage occurs on the outer surface (i.e., shedding). The overlapping shingles provide a drainage layer below the exposed shingle layer. If the roof deck is covered with a water-resistant layer ("roofing paper", "underlay", "sarking") this provides the final layer of drainage. For lower-slope applications (typically between 1:12 and 3:12), manufacturers of asphalt shingles generally recommend doubling the layers of shingles, i.e., providing more drainage layers.

Drainage is widely seen as an effective rain control strategy. However, no system drains perfectly. Water is absorbed into materials and retained on surfaces. Only a small gap is required to allow drainage since hydrostatic head pressure builds up and drives water through small gaps. Hydrostatic pressures at the drainage plane make it possible for water to penetrate further into the wall through small openings and flaws. Research has shown that small gaps, such as those between two sheets of building paper, can allow rapid and high rates of drainage but may slightly increase the water pressure on the drainage plane membrane [Straube et al 2000, Weston et al 2001, Smegal 2006, Smegal & Straube 2007]. Surprisingly, small gaps (less than around 1 mm) formed with hydrophobic sides *store* less water because large droplets (often 2-3 mm in dimension) cannot form [Smegal 2006, van Linden et al 2018].

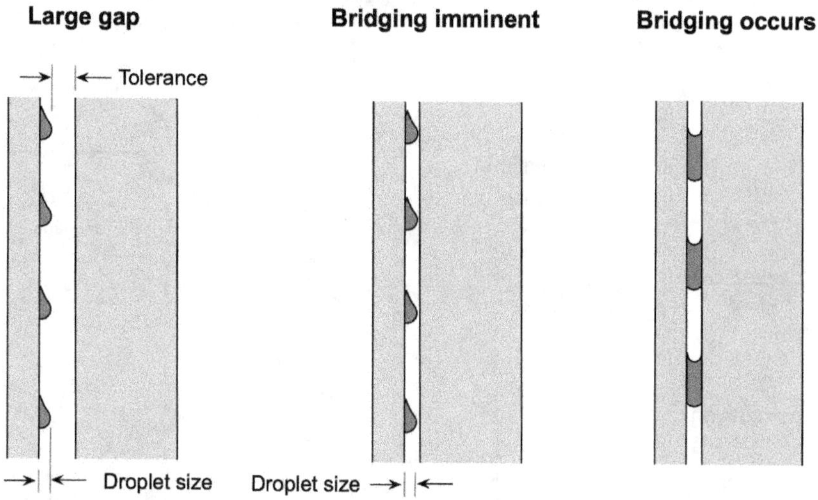

Figure 12.17: Gaps larger than the size of a droplet prevent moisture bridging

Larger gaps (e.g., over about 3 mm) *may* be desirable as they avoid moisture bridging across the gap (Figure 12.17) as well as to essentially eliminate any build-up of even very small hydrostatic pressures on the water-resistant layer. As rainwater passes through the cladding primarily at penetrations (windows, doors, balconies), and hence often directly onto the water-resistant layer, the ability to resist moisture bridging is not always that important in practise. Even controlled laboratory tests by Lacassse [2003] showed:

> *" … test results also indicated that even with the presence of this clear drained cavity behind the cladding, a small quantity of water could still find its way into the stud cavity."*

The use of large gaps in practice (e.g., ½" to 1") is justified to accommodate dimensional variations or to encourage ventilation drying: none of the research has shown that large gaps provide any measurable benefits to drainage. Research into the nature of small drainage gaps continues to the current time.

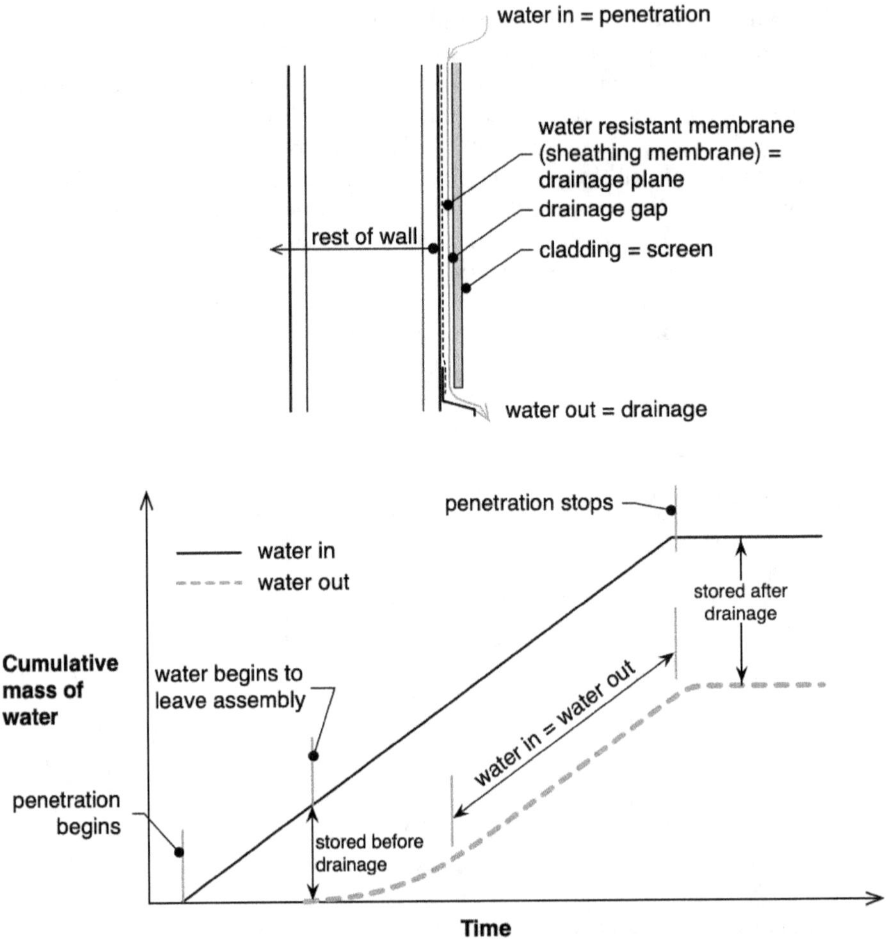

Figure 12.18: Response of a drainage space

Drainage spaces tend to respond to rain penetration as shown in Figure 12.18. Water enters the space through some flaw or imperfection. Any water not absorbed into materials lining the gap will be stored in gaps, held by surface tension. When the amount of water generates gravity forces that exceed surface tension forces, drainage begins. When the penetration of rainwater stops, drainage also ceases, leaving water in the gap. The amount of this stored water will depend on the assembly and materials used but is typically in the range of 100 to 300 g/m^2 (0.3 to 1.0 ounces/square foot), greater if the cladding is highly absorbent and thick [Smegal & Straube 2007]. This remaining moisture must be removed by drying mechanisms.

12.6.2 Summary

It should be clear from the preceding discussion that the rain control of a perfect barrier enclosure is dependent on a single layer resisting all inward water transmission, a mass system relies on sufficient storage capacity, and a drained system depends on drainage. As pointed out above, despite the fraction that dominates, in practice, all fractions play a role during field performance in almost all wall systems. If some storage and drainage is possible in many perfect-barrier walls, these walls may not fail if and when the perfect barrier is less than perfect. In the same way, many drained walls may, in fact, have poor drainage locally and therefore can avoid damaging rain penetration only if sufficient storage is available. However, the designer is often concerned with the behavior of the wall under extreme events and should choose a rain control strategy from one of the three.

12.7 Pressure Equalized Rainscreens

Pressure Equalized Rainscreens (PER) are one special type of enclosure wall system that has received a lot of attention. Pressure equalization can be defined as a performance state where the wind-induced pressures on the face of a cladding are exactly equalized by air pressures behind the cladding. This state is desirable both because a) it reduces the structural loads the cladding needs to be designed for and b) because it eliminates air pressure as one of the forces causing rain penetration of the cladding.

A pressure-equalized system comprises the rainscreen cladding, a vent, and an airtight chamber (Figure 12.19). The vent in the cladding is needed for pressure equalization to allow air to enter or leave with the changing wind pressure. The theory assumes that by reducing air pressure differences across the cladding, rain will not be forced across openings in the cladding by this force, while the features of a drained system, i.e., drainage, WRB, and flashing deal with the other rain penetration forces. To achieve this an interior air barrier, an air chamber behind the cladding, and vent holes in the cladding are required.

Garden, in his influential Canadian Building Digest *Control of Rain Penetration* [Garden 1963-2] described how a vented cladding installed over a wall can control rain penetration:

> *"In essence the outer layer is then an "open rain screen" that prevents wetting of the actual wall or air barrier of the building."*

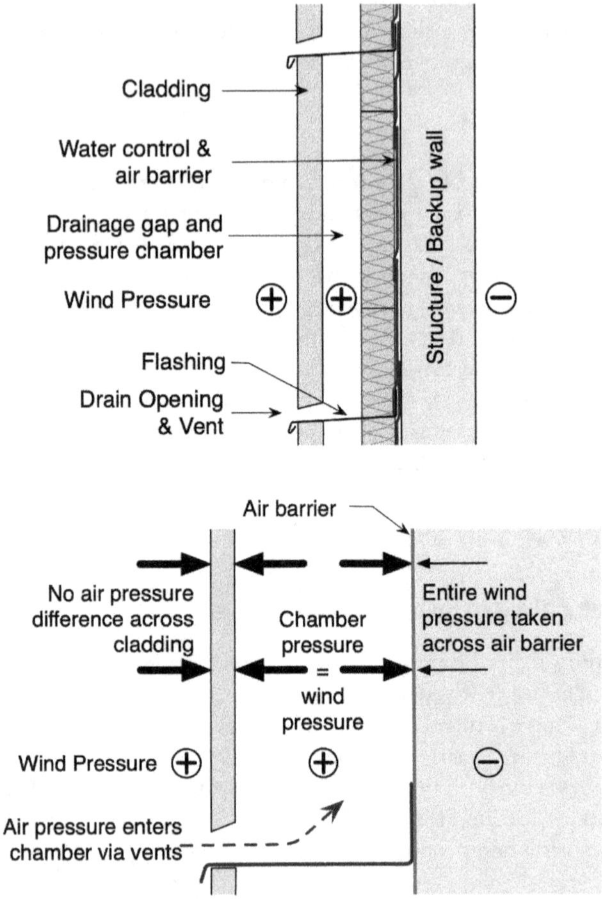

Figure 12.19: Pressure Equalized Rainscreen (PER) concept

He began the serious discussion of air pressures and pressure equalization, a topic investigated by many building research establishments in the 1960's (although there is little to no mention of either drainage or WRBs).

The curtainwall industry grasped the rainscreen principle and pressure equalization early on, perhaps because they could benefit the most. The AAMA publication *The Rain Screen Principle and Pressure-Equalized Wall Design* [AAMA 1971] was a seminal document that informed many manufacturers and practitioners. In this the authors defined rainscreen "the *component*" as distinct from rainscreen "the *principle*".

Although not always clearly stated, pressure equalization cannot exist on its own as a rain control strategy as it can neither manage all rain penetration mechanisms nor does it perform perfectly all the time. Pressure equalization is always a complementary addition to the fundamental drainage approach to rain penetration control.

Field measurements have shown that pressure equalization of dynamic wind gusts is rarely achieved in practice. For example, Brown et al [1991] field tested an average quality (for the time) brick veneer steel stud system designed to be a rainscreen. Measured performance ranged from 35 to 65% pressure moderation. Field measurements of well-vented and compartmentalized brick veneers at the University of Waterloo studied performance in the frequency domain [Straube & Burnett 2000]. They found good equalization at average pressures (over 95%) but poor (less than 50%) moderation of wind gusts. Kumar et al [2003] and van Bentum & Geurts [2015] measured the performance on a high-rise building and found similar results.

The National Research Council of Canada [e.g., Poirier et al 1992, Brown et al 1995] and others [Alqhoury 1990, Ganguli & Quirouette 1987] tested assemblies with spatially uniform and carefully controlled time-variant pressure waves. These tests showed much better pressure moderation than field tests of similar systems. The primary reason equalization cannot be achieved in the field is because of the spatially variant dynamic nature of the wind [Inculet & Surry 1996, Straube 1998, Inculet et al 1997, van Straaten et al 2010, van Straaten 2017], not the temporal wind pressure variations or wall characteristics. Test methods, such as AAMA 509 [AAMA 2022] do not impose spatial variations, and hence cannot capture the actual field performance [Dernayka 2011]. Researchers at least as far back as Dalgliesh and Garden [1968] described this challenge:

> "*... pressure equalization is not always achieved in a cavity which is open at two or more locations. The cavity must be closed at strategic locations to prevent air flow behind the outer screen. Knowledge of both the overall and local pressure variations is necessary for full exploitation of the principle of pressure equalization as a means of controlling rain penetration*".

However, high levels of static pressure moderation (i.e., pressure moderation of wind pressures averaged over ten to thirty seconds and longer) can be achieved by modest cladding venting area combined with effective air barriers. Effective air barriers are now part of most modern enclosure assemblies.

Laboratory, simulation and field studies can be summarized to conclude that the degree of equalization (or pressure moderation) depend on two general classes of factors [Straube 1998, Van Den Bossche 2013]:

A. the enclosure system characteristics, and

B. the nature of the wind loading.

The enclosure system characteristics (A) can be further sub-divided into categories of characteristics that:

I. minimise the volume of flow required to achieve pressure equalisation, and

II. maximise the ease of flow of air into the chamber.

Factors included under the first category (I) of enclosure characteristics are:

1. air barrier leakage, or rather, the nature and magnitude of leakage;

2. the volume of air in the chamber;

3. the spatial extent of the chamber, i.e. compartmentalisation; and

4. chamber deformability (or stiffness, including the flexibility of both the cladding and the air barrier);

The second category of enclosure characteristics (II) includes factors such as:

5. venting, or more precisely the nature, magnitude, and spatial distribution of cladding air permeance; and

6. the air flow characteristics within the chamber.

The wind load characteristics (B) that affect pressure moderation are:

1. the mean pressure across the enclosure system;

2. the mean gradients (in two dimensions) across the exterior of each compartment;

3. the time-varying pressures across the system (i.e., speed of gusts); and

4. the instantaneous pressure field acting across the exterior of each compartment (i.e., the severity and duration of spatio-temporal pressure variations).

All four load characteristics are strongly dependent on the building shape, size, and orientation and the nature of the upwind topography.

Pressure equalization is more effective for joints than walls (primarily because of the limited spatial extent of joints) [Killip & Cheetham 1984]. Early research on joints [Isaksen 1965, 1972, Platts & Sasaki 1965, Garden 1963] that penetrated through the entire enclosure showed the importance of airtightness. Bear in mind that an interior plane of water tightness (a "WRB") was not explicitly called out in the designs of the 1960s and 70s (but was implicitly provided in the aluminum joinery that formed curtainwalls).

Joints between windows and walls are a common detail whose water penetration resistance can be significantly improved by draining and pressure equalization. Although drainage is fundamental, the addition of pressure equalization to window frames and window-wall joints greatly improves performance [Basset et al 2003, Van Den Bossche et al 2008]. Using a sloped sill or backdam works together with pressure-equalization by ensuring that water does not contact the interior air seal (Figure 12.20).

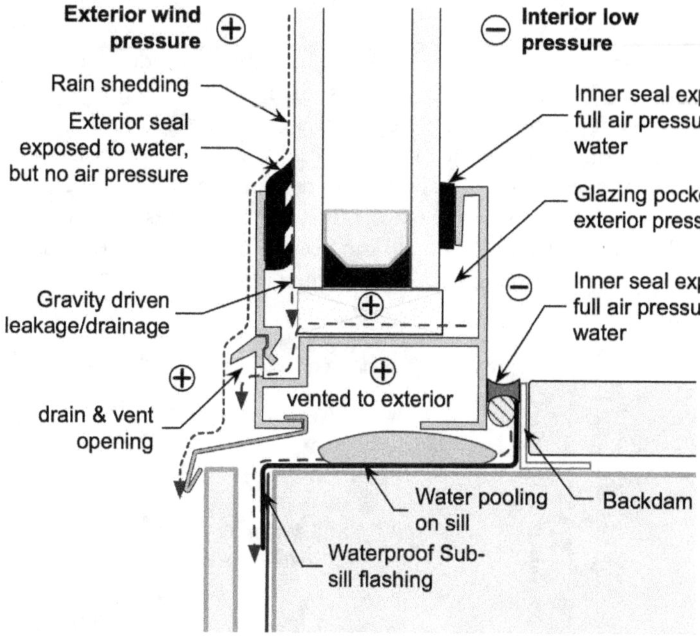

Figure 12.20: Pressure equalization of window joints can be effective

The current state of knowledge accepts that some pressure equalization (pressure moderation) occurs, not perfect pressure equalization, and that a protected water control layer (drainage plane or water resistant barrier) with drainage is both far more important and a *prerequisite* for pressure moderated systems in any event.

The rain penetration resistance of joints made of non-absorbent materials likely benefit the most from pressure moderation [Lstiburek 2008]. Drained walls with leaky claddings or leaky penetrations (e.g., windows) but good water control layers and flashings benefit the least (if at all) from pressure equalization.

In short, it is critical to understand that without drainage and a water control layer, pressure-moderation will not deliver good or even acceptable rain control performance. However, pressure moderation offers only a modest benefit to most walls with good drainage but can be important for joints with unwetted inner air barrier seals.

Today, industry in North America[3] has come full circle, with a focus on providing very high levels of airtightness provided by a complete uninterrupted air- and water-resistant barrier (AB & WRB) membrane inside of a drainage gap (often much larger than needed for just drainage). While this approach can result in excellent performance, designs that do not manage the amount of driving rain passing through the rainscreen cladding (or penetrations of it) risk significant quantities of water draining over the combined AB & WRB membrane, which now usually has essentially all the air pressure difference across it (Figure 12.21). This scenario can occur and cause failures in sloped walls and, to a much lesser extent, open-jointed claddings.

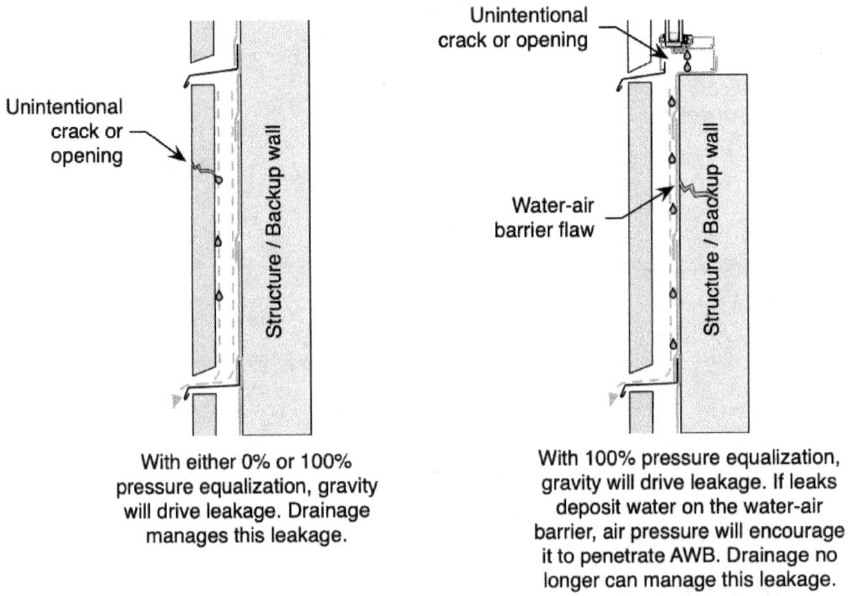

Figure 12.21: Pressure equalization can exacerbate water leakage through penetrations

Using baffled joints that appear open, cladding designs with better rain shedding and/or complete additional rain shedding layers behind the cladding (e.g., in the case of sloped walls) are some design solutions. Figure 12.22 is a conceptual design that acknowledges the cladding in a sloped wall will provide little rain shedding (the extreme case of open joints is shown) and thus provides a secondary, non-airtight, WRB as a rain shedding layer. This acts to greatly reduce the amount of water on the inner combined AB & WRB which will be exposed to the full air pressure difference imposed by the wind.

---

[3] Even into the 2020's, European and Scandinavan literature does not place much emphasis on water control layers, and relies heavily on the presumption that water will not cross a 40-50 mm gap. Water control layers, however, are increasingly being used by practioners even in these countries.

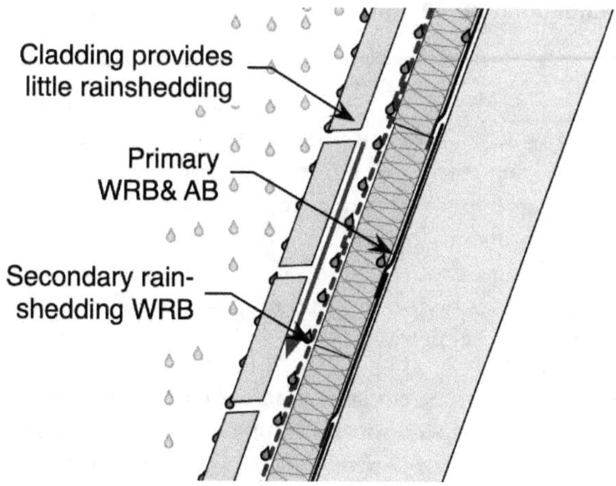

Figure 12.22: Cladding which provide little screening will benefit from an additional rain-shedding layer

A drained and vented wall that is both pressure-moderated and ventilated is not only feasible but may be considered the preferred solution for many applications (Figure 12.23). However, there should be no doubt about the primacy of drainage as the most important feature, whereas ventilation, while often helpful, is often not needed. Pressure moderation is also sometimes beneficial, but rarely critically important to successful performance.

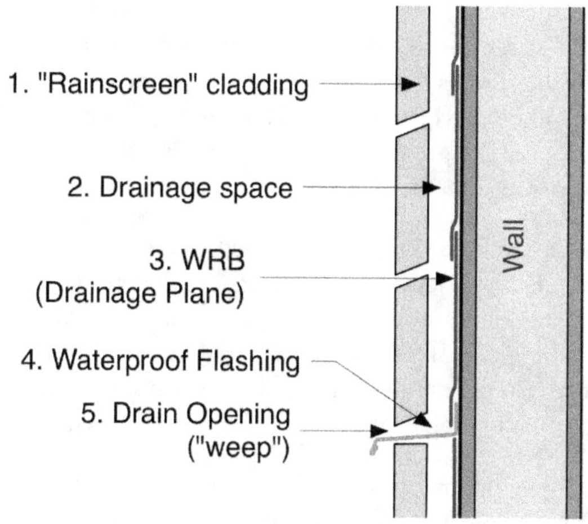

Figure 12.23: The components of a common drained (aka "rainscreen") wall system

12.8 Recommendations for design

12.8.1 Exposure

Regardless of which strategy or wall assembly is considered, reducing the amount of rain deposited (i.e., controlling exposure by the judicious choice of site and building shape) and increasing the fraction shed from the exterior face will reduce the moisture load on the assembly. Controlling surface water using generous overhangs, proper windowsills, and other surface drainage features should almost always be the first and most important step in designing for the control of rainwater penetration.

Exposure must always be considered in the choice of strategy, materials, penetration details, inspection, and testing. The exposure of the enclosure is more than just a factor of climate, height, and rain shedding, tilt is in some ways the most critical factor and explains why roofing is a separate trade that uses different materials and methods than walls.

A wall tilted off-vertical by only a few degrees requires much better rain control performance, not only because there is no longer any deflection possible (drip edges protect only small areas, and overhangs are mostly ineffective), but because gravity will drive rainwater more deeply into the assembly. Any water that passes the cladding will be directed across the drainage cavity (regardless of width) then past any insulation (at joints) and onto the water control layer. Hence, drained systems with a single water control layer will have little benefit over concealed barrier systems in such applications.

Any holes in a water control layer of an off-vertical enclosure will now be challenged by gravity-driven water. In a vertical enclosure, water at holes and penetrations will be driven downward by gravity, not inward. The more an assembly is sloped, the greater the increase in gravity force and duration of rainwater exposure (Figure 12.24).

As the slope passes some culturally defined value the enclosure is called a roof, and tradition (i.e., experience of performance) demands better materials, more effective rain shedding, more focus on gravity laps, avoidance of penetrations, etc. Asphalt shingle roofs, for example, are limited to use at 3:12 pitch (14-degree slope, or 76 degree off vertical) by most manufacturers. For slopes as low as 2:12, asphalt shingles may be used, but they must used in two layers.

Commercial low-slope roofs generally require slopes of 1% (1:100) or more with a recommendation of 2%. The materials used as low-slope roof water control layers are much thicker and more durable and use better laps and seals than vertical wall water control layers of course. However, warranties are generally void if ponding occurs on such low-slope roofs, because this means hydrostatic pressure exists for longer periods of time than the even such systems can resist. This should underline the fact that these membranes and their details of installation, while far more robust than control layers in walls protected by cladding or overlapping shingles, are still not sufficient to resist standing water pressure of even an inch (or tens of mm) for more than a few hours.

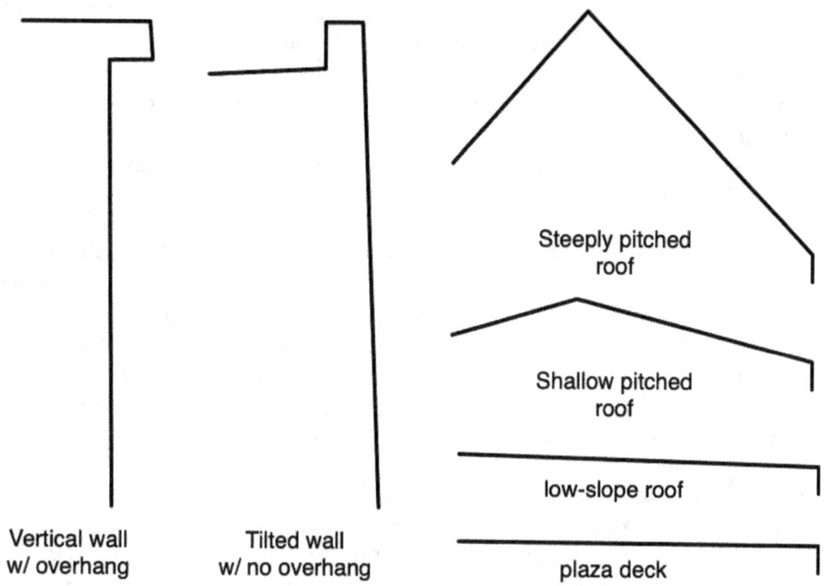

Figure 12.24: Enclosure slope has a profound effect on rain control performance needed.

For a roof which is expected to experience ponded water for some time, manufacturers offer a range of higher quality membrane products, with premium materials, greater thickness, more demanding installation requirements, and more inspections. These are described as plaza deck waterproofing systems.

12.8.2 Rain penetration control strategy choices

There is a vast range of rain penetration control performance between different systems, materials, and details. Dogma should be avoided. For example, all drained systems are not more effective than all perfect barrier systems and mass systems may be a better choice than a concealed barrier system in some cases.

Drained wall systems are inherently more forgiving than either mass or perfect barrier systems for walls. Properly designed and built drained wall systems will provide economical and durable rain penetration control. Failures in such systems tend to occur because drainage was not provided (either through a design or construction failure) or because flashing (which is routinely exposed to standing water) fails. So long as sufficient drainage is provided, the combination of the number of layers, their absorbencies, and their penetration fraction will determine the ability of a screened system to control rainwater.

Pressure-equalized systems offer only marginal benefits in most walls. So long as drainage and capillary breaks are provided, the reduced water penetration that may (and in many cases may not) result from pressure equalization does not aid rain control -- failure will still be due to a failure to drain. Designing enclosure walls with water control layers that are protected (screened) from most direct rain contact and excessive quantities of draining water while at the same time reducing the air pressure across them is, however, universally beneficial.

Storage walls are usually prohibitively expensive (in materials, structural weight, or floor space) in modern construction. Failure of joints (designed as perfect barriers) between wall elements or the interface with windows, doors, balconies, etc. often causes rain penetration, even when the wall element itself is functional. If sufficient storage is provided for the exposure conditions and absorbency of the material, performance of a mass wall system will in practice depend on the performance of joints and interfaces.

Many vernacular and traditional mass walls (e.g., adobe, multi-wythe masonry) rely on storage to control rain penetration. Provided the absorbency and safe storage capacity are well matched with the exposure conditions, mass walls may be a reasonable choice. Mass storage walls are still used in some low-rise commercial buildings (in the form of single-wythe concrete block), and some high-rise residential buildings (exposed concrete shear walls). These can perform, but it is often simple and inexpensive to design some of these systems as drained systems, with the benefit of significant storage in the "rainscreen".

Experience has shown that many perfect barrier systems, especially face-sealed versions, often fail to perform as designed, if not initially, then after many years of exposure. Some perfect barrier systems that appear to function in the field may lend their performance to the fact that other layers are providing drainage and/or storage. In effect these "face-sealed perfect barrier systems" are acting as drained systems or mass systems with small quantities of penetration and small moisture storage.

Perfect barrier enclosure wall components built with tight quality control in factory conditions and connected in the field with drained-screened types of joints have been shown to provide good performance for a long service life. Successful examples from the field include precast concrete, insulated glazing units, insulated metal panel systems and even some EIFS – but note that all these systems require drained joints (or frames) for acceptable levels of performance when exposed to driving rain [Williams & Williams 1998]. Extensive years of experience with rain control forensics has shown that the failures that occur in perfect barrier wall elements are almost always a result of either a failure to design and/or construct joints that will drain water that penetrates or a failed attempt to design and construct perfect joints.

The classification system presented earlier is based on how an enclosure system manages water deposited by rain. Water that is absorbed by building materials must also be dealt with. Small leaks, imperfections in materials, construction moisture, absorption of water by cladding materials and drainage planes all result in water stored in an enclosure. Therefore, drying must be considered as the final part of any rain control strategy.

For mass or storage walls, drying is a critical variable that defines how often a wall can be wetted before it fails. For drained walls, drying removes the sometimes-considerable amounts of moisture that is absorbed by the porous materials (like wood and paper products) often used to construct walls. Drainage is also never perfect, and mortar dams, grooves in siding products and window extrusions (especially vinyl and aluminum), etc., will store water. Finally, perfect barrier walls that can store sufficient volumes of water safely and dry quickly in the climate to which they are exposed can better tolerate a small amount of rainwater that may penetrate.

12.9 Drying

Despite all attempts to resist and drain water, field experience has shown that some water may still penetrate or be built in during construction. Drying of this moisture must be provided for. How much drying is needed depends on the tolerance of wetted materials, and the amount and frequency of wetting. Drying, of course must be provided to deal with other sources of moisture, such as construction moisture built-in and occasional condensation events.

Moisture can be removed from an enclosure wall by (Figure 12.25):

1. evaporation from the inside or outside surfaces (with water wicked to either surface through capillary active materials);
2. vapor transport by diffusion, air leakage, or both, either outward or inward;
3. drainage, driven by gravity; and
4. ventilation, if provided for.

Drainage is capable of removing the greatest volume of water in the shortest period of time. Hence, as described above, it is a very important mechanism for rain penetration control. (Note that drainage may also be useful for removing cold weather, air-leakage condensation that can form on the back side of cladding). Provided that a drainage path exists (e.g., cavities, slopes, drainage openings), a large proportion of rainwater penetration can flow out of a wall.

A small but significant amount of water usually remains attached to surfaces by surface tension and is absorbed into materials by capillary forces, even in walls with excellent drainage. Absorptive claddings such as wood and cement-bonded sidings, stucco, and masonry also absorb and store moisture. This moisture can be removed from a wall system only by diffusive drying or by air movement.

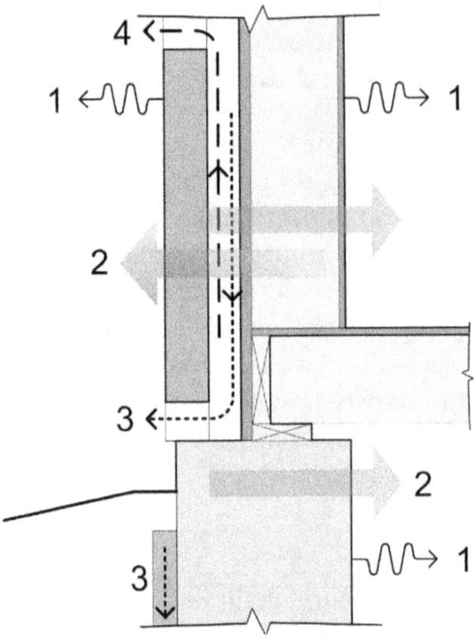

Figure 12.25: Drying mechanism in walls

Diffusive drying can transport moisture in either direction, depending on the wall system and the weather. In colder weather the vapor flow is outwards. During warm weather and especially when the sun heats the surfaces of the walls, inward drying can be an important mechanism.

Air movement (or leakage) through the enclosure can, under the proper conditions, move a large quantity of moisture. While both cold weather air leakage outward and warm weather air leakage inward can cause condensation and hence potentially damaging wetting, the opposite conditions (e.g., inward flow in cold weather) will provide some drying. In some cases, this drying can be significant. Because air leakage through the enclosure is difficult to control, expensive in terms of energy, and potentially dangerous for indoor air quality and durability, the modern approach is to limit airflow through the wall to nearly zero. This also eliminates the potential for airflow drying, so drying must be achieved by other means.

Ventilation, or exterior air flow behind the cladding driven by wind pressure differences on the face of the building or solar heated air rising, is useful since it accelerates drying. Ventilation bypasses the high vapor resistance of claddings such as vinyl siding, metal panels, and dense natural stone, thereby allowing outward drying. A seminal Canadian study [Straube and Burnett 1995] and early European research [Popp et al 1980] has shown that ventilation drying occurs and may be important. A more extensive ASHRAE research project [Burnett et al 2004] solidified the understanding and confirmed that meaningful rates of drying could be achieved. All of this research shows that large clear spaces (at least 19 mm) should be provided to encourage ventilation, and that clear large vents (of the order of 0.4 to 2% of wall area) should be placed at the top and bottom of the cavity for the most effective ventilation (Figure 12.26). Smaller vent spaces, such as those behind vinyl siding, also appear to be effective.

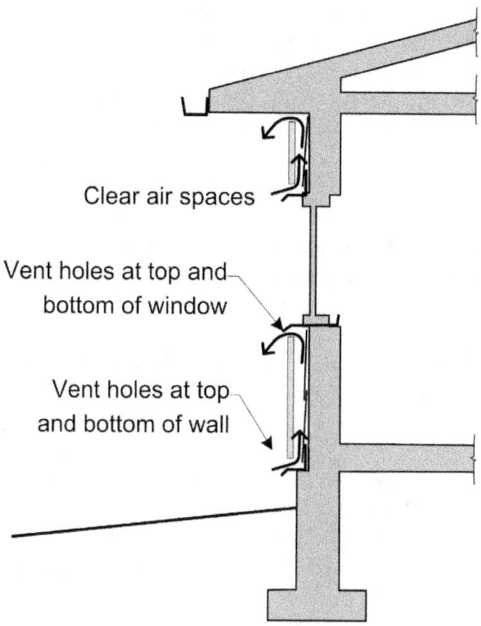

Figure 12.26: Ventilation drying concepts

To minimize resistance to air flow, ventilation gaps need to be large enough. Chapter 7 introduced the equations needed to predict ventilation flow. As described earlier, drainage gaps need only be small. Hence, the size of the drainage space can be defined on the basis of drainage capacity as a lower limit and ventilation drying capacity as an upper limit. Ventilation, not drainage, is a reason to use larger 19 or 25 mm (¾" to 1") gaps behind cladding. There is little documented hygrothermal benefit to even larger gaps, but there are real drawbacks such as fire spread and animal infestation.

Claddings such as stucco, brick, wood, and cement-board absorb water. When these materials are heated by the sun during warm weather, very large inward vapor drives develop (which accelerate drying of the cladding). These inward drives can cause dangerous summertime condensation within some types of walls, especially if a very permeable exterior assembly and a low-permeance interior vapor barrier (e.g., polyethylene or vinyl wallpaper) are used.

Experience suggests that this mechanism may be a concern in both hot-humid and cold climates [Wilson 1965, Anderson 1988, Sandin 1991, Straube and Burnett 1998]. Ventilation is often able to reduce or control these inward vapor drives by safely redirecting the vapor to the exterior even in hot-humid conditions.

12.10 Conclusion

Rain control is an important function of the building enclosure and a serious challenge to achieve in practice. Although little quantitative research has been conducted, the principles laid out in this chapter should provide a rational foundation on which to make design and analysis decisions.

12.11 References

AAMA. 1971. *The Rain Screen Principle and Pressure-Equalized Design Details of Three Recent Buildings.* CW-2-71 American Architectural Manufacturer's Association, Schaumburg, Illinois.

AAMA 509-22. 2022. *Voluntary Test and Classification Method for Drained and Back Ventilated Rainscreen Wall Cladding Systems.*

Abuku M, Blocken B, Roels S. 2009. "Moisture response of building facades to wind-driven rain: field measurements compared with numerical simulations". *J Wind Eng Ind Aerodyn* 97(5-6), pp. 197-207.

Alberti, Leon. *The Ten Books of Architecture.* New ed of 1755 edition, Dover Publications, 1987.

Alkhoury, E. 1990. "Window Rain Screen Evaluation". *Proceedings of 5th Canadian Building Science Tech.*, Toronto, Canada, March 1-2, pp. 240-257.

Andersen, N.E., 1988. "Summer Condensation in an Unheated Building," *Proc. of Symposium and Day of Building Physics,* Lund University, August 24-27, pp. 164-165.

Bassett, M. R., Burgess, J. C., & Camilleri, M. J. 2003. "The weathertightness of window-to-wall joints–dependency on installation details". *Proc of IRHACE Annual Conference*, Hamilton, New Zealand.

Blocken, B., and Carmeliet, J.,2000. "Driving Rain on Building Envelopes—I. Numerical Estimation and Full-Scale Experimental Verification," *J. of Thermal Insulation and Bldg Envelopes*, Vol 24, No 4, pp. 61-85.

Blocken, B., and Carmeliet, J., 2000. "Driving Rain on Building Envelopes— II. Representative Experimental Data for Driving Rain Estimation," *J. of Thermal Insulation and Bldg Envelopes*, Vol 24, No 4, pp. 89-110.

Blocken B, and Carmeliet J. 2002. "Spatial and temporal distribution of driving rain on a low-rise building." *Wind and Structures*. No. 4, 2002, pp. 441–462.

Blocken, B. and Carmeliet, J. 2005. "High-resolution wind-driven rain measurements on a low-rise building - Experimental data for model development and model validation". *J. of Wind Engng and Indust. Aerodynamics* 93(12), pp. 905-928.

Blocken, B., D. Derome, J. Carmeliet, 2013. "Rainwater runoff from building facades: A review". *Bldg and Environ*, Vol 60, pp. 339-361.

Brown, W. C., Rousseau, M. Z., and Dalgliesh, W. A. 1991. "Field Testing of Pressure-Equalized Rain Screen Walls," *Exterior Wall Systems: Glass and Concrete Technology, Design, and Construction*, ASTM STP 1034, B. Donaldson, Ed., American Society for Testing and Materials, Philadelphia, pp. 59-69.

Brown, W.C., Ullett, J.M, Dalgliesh, W.A. 1995. *Measured Pressure Equalized Performance of a Brick Veneer/steel Stud Assembly: Performance of Pressure Equalized Rainscreen Walls*. CMHC, June. and *Measured Pressure Equalized Performance Two Precast Concrete Panels: Performance of Pressure Equalized Rainscreen Walls, a Collaborative Research and Development Project.* CMHC, May.

Burnett, E.F., Straube, J.F., Karagiozis, A.K., 2004. *ASHRAE 1091 Development of Design Strategies for Rainscreen Sheathing Membrane Performance in Wood Frame Walls– Synthesis Report and Guidelines*, ASHRAE, Atlanta.

Choi, E.C.C., 1994. "Determination of the wind-driven-rain intensity on building faces". *J. of Wind Engng and Indust. Aerodynamics*, Vol 51, pp. 55-69.

CMHC. 1996. *Survey of Building Envelope Failures in the Coastal Climate of BC.* Report by Morrison-Hershfield for CMHC, Ottawa, Nov. 1996.

Cornick S.M.and M. A. Lacasse, 2004. "A Review of Climate Loads Relevant to Assessing the Watertightness Performance of Walls, Windows and Wall- Window Interfaces," Performance and Durability of the Wall-Window Interface: ASTM STP 1484, B. G. Hardman, C. R. Wagus and T. A. Weston, Eds., ASTM International, West Conshohocken, PA.

Cornick, S. and W. A. Dalgliesh, 2009. "Adapting rain data for hygrothermal models". *Building and Environment* 44(5), pp. 987-996.

Cornick, S.M. and M. A. Lacasse., "An Investigation of Climate Loads on Building Façades for Selected Locations in the United States." *Journal of ASTM International* Vol. 6, No. 2, pp. 1-33.

Dernayka, S.D., 2011. *Pressure-Equalized Rainscreen Wall System: A Full-Scale Experiment.* M.Eng.Sc. Thesis, Faculty of Engineering, Western University, London, ON.

Desjarlais, A.,1995. "Self-Drying Roofs: What?! No Dripping!" *Proc. Of Thermal Performance of Building Envelopes VI*, Clearwater, Florida, December 4-8, pp. 763-773.

Dalgliesh,W.A. and Garden,G.K. 1968. "Influence of wind pressures on joint performance". *Proceedings of the International Symposium Weathertight Joints for Wall*, held in Oslo, Norway, September 25-28, 1967. CIB Report No. 11, NBRI Report 51C, January, pp. 329-331.

Frank, W., 1973. "Entwicklung von Regen and Wind auf Gebaeudefassaden," Verlag Ernst & Sohn, *Bertichte aus der Bauforschung*, Vol 86, pp. 17-40.

Ganguli, U., & Quirouette, R. L. 1987. "Pressure equalization performance of a metal and glass curtain wall". *CSCE Centennial Conference Montreal*, Quebec, May 19- 22, Vol. 1, pp. 127-144.

Garden, G.K., 1963. *Joints between prefabricated components*. Building Note 40, NRCC, Div of Bldg Research, Ottawa, February.

Garden, G.K. 1963. *Rain penetration and its control*. Canadian Building Digest 40, pp. 4, April.

Ge, H., Chiu, V., Stathopoulos, T., " Effect of overhang on wind-driven rain wetting of facades on a mid-rise building: Field measurements". *Building and Environment*, Vol 118, pp. 234-250, 2017.

Ge, H., R. Krpan, et. al. 2009. "Field measurements of wind-driven rain: a study of seven buildings in Metro Vancouver". *Proc. of the 12th Canadian Conference on Building Science and Technology* Montreal, Quebec.

Ge, H. and Krpan, R., 2007. "Field measurement of wind-driven rain on a low-rise building in the coastal climate of British Columbia". *Pro. of the 11th Canadian conference on building science and technology*, March, Banff, Alberta.

Grimm, C.T., "Water Permeance of Masonry Walls: A Review of the Literature", Masonry: Materials, Properties, and Performance, ASTM STP 778, J.G. Borchelt, Ed., American Society for Testing and Materials, Philadelphia, 1982, pp. 178-199.

Hazleden, D.G. and Morris, P.I. 1999. "Designing for durable wood construction: The 4D's". *Proc of Durability of Building Components 8*. Research Press, Ottawa.

Herbert, M.R.M., and Harrison, H.W. 1974. *New ways with weatherproof joints*. Current Paper 90174. Garston, England: Building Research Establishment, 12 pp.

Herbert, M.R.M. 1974. *Open-jointed rain screen claddings*. Current Paper CP89174. Garston, England: Building Research Establishment.

Inculet, D., Surry, D., Davenport, A.G., "Unsteady Pressure Gradients and Their Implications for Pressure-Equalized Rainscreens", *Proc. of ICBEST '97*, Bath, U.K., pp. 457-463, 1997.

Inculet, D., Surry, D., *The Influence of Unsteady Pressure Gradients on Compartmentalization Requirements for Pressure-Equalized Rainscreens*, Report by BLWTL, The University of

Western Ontario for CMHC, Ottawa, June, 1996.

Inculet, D.R., Surry, D., "Simulation of Wind-Driven Rain and Wetting Patterns on Buildings," *Report BLWT-SS30-1994*, U. of West. Ontario, London, Nov, 1994.

Isaksen, T. "Rain Penetration in Joints: Influence of Dimensions and Shape of Joints on Rain Penetration". *Proc of the RILEM/CIB Symposium on Moisture Problems in Buildings*, Helsinki, August 16th-19th, 1965.

Isaksen T. 1972. *Driving rain and joints: testing of model joints between elements.* Norwegian Building Research Institute.

Kala, S., T. Stathopoulos, et al. . 2008. "Wind loads on rainscreen walls: Boundary-layer wind tunnel experiments". *Journal of Wind Engineering and Industrial Aerodynamics* 96(6-7): 1058-1073

Killip, I.R., D.W. Cheetham, 1984. "The prevention of rain penetration through external walls and joints by means of pressure equalization", *Bldg & Environ*, Vol. 19, No. 2, pp. 81-91.

Kumar, K. S., Stathopoulos, T., & Wisse, J. A. 2003. "Field measurement data of wind loads on rainscreen walls". *J. of Wind Engineering and Industrial Aerodynamics*, 91(11), 1401-1417.

Künzel, H.M., *Bestimmung der Schlagregenbelastung von Fassadenflaechen.* Fraunhofer Institut für Bauphysik, Mitteilung 263, 1994.

Künzel, H.M., *Regendaten für Berechnung des Feuchtetransports.* Fraunhofer Institut für Bauphysik, Mitteilung 265, 1994.

Lacy, R.E., *Driving-Rain Maps and the Onslaught of Rain on Buildings.* Building Research Station Current Paper 54, HMSO Garston, U.K., 1965.

Lstiburek, J., 2006. "Understanding Drainage Planes", *ASHRAE Journal*, Feb 2006, pp. 30-35.

Lstiburek, Joseph W. "A cup in the rain." *ASHRAE Journal*, vol. 50, no. 4, 2008.

Merkel, Holger, 2001. "Inverted roofs with reduced heat losses due to a water–repellent separating layer." *Proc. of Thermal Performance of the Exterior Envelopes of Buildings VIII*, Clearwater Beach, FL.

Marsh P. 1977. *Air and rain penetration of buildings.* The Construction Press Ltd., Lancaster, England.

Morris, S., 1975. *A wall penetration map of Canada.* Atmos. Environ. Service Downsview, Ontario, Canada

Morrison Hershfield / RDH Building Engineering, 1998. *Best Practice Guide for Wood*

Frame Construction in the Coastal Climate of BC. for CMHC Ottawa.

Platts, R.E. and Sasaki, J.R.,1965. *Rain Leakage Tests on vertical Through Joints.* Internal Report 323. Ottawa: National Research Council Canada, Division of Building Research, Ottawa.

Poirier, G.F., Brown, W.C., Baskaran, A., "Pressure Equalization and the Control of Rainwater Penetration", *Proc. Sixth Conference on Building Science & Technology*, Toronto, March 5-6, 1992, pp.45-64.

Robinson G, Baker MC. *Wind-driven rain and buildings.* National Research Council of Canada, Division of Building Research, Technical Paper No. 445, Ottawa, July 1975.

Sandin, K., 1991. *Skalmurskonstruktionens fukt- och temperaturbetingelser.* Rapport R43:1991 Byggforskningsrådet, Stockholm, Sweden.

Saneinejad, S. and Doshi, H., 2006. "Testing of metal flashing for water-shedding effectiveness". *RCI Interface Journal.*

Smegal, J., A. Lukachko, J. Straube & T. Trainor, 2014. "Quantitatively Evaluating the Effectiveness of Different Drip Edge Profiles", *Proc. 14th Can Bldg Sci & Tech Conf.* Toronto, pp. 89-99.

Smegal, J., 2006. *Drainage and Drying of Small Gaps in Wall Systems.* MASc Dissertation, Civil Engineering Department, University of Waterloo, 2006.

Straube, J.F. and Burnett, E.F.P., 1998. "Drainage, Ventilation Drying, and Enclosure Performance," *Proceedings of Thermal Performance of Building Envelopes VII*, Clearwater Beach Florida, December 4-7, pp 189-198.

Straube, J.F. Burnett, E.F.P., 1995. *Vents, Ventilation Drying, and Pressure Moderation.* Building Engineering Group report for CMHC, Ottawa.

Straube, J.F., and Burnett, E.F.P., 1997. "Driving Rain and Masonry Veneer," *Water Leakage Through Building Facades*, ASTM STP 1314, R. Kudder & J.L. Erdly, Eds., American Society for Testing and Materials, Philadelphia, pp. 73-87.

Straube, J.F., 1998. *Moisture Control and Enclosure Wall Systems.* Ph.D. Thesis, Civil Engineering Department, University of Waterloo, April.

Straube, J.F. and Burnett, E.F.P., 1999. "Rain Control and Design Strategies," *J. Of Thermal Insulation and Building Envelopes*, July 1999, pp. 41-56.

Straube, J.F. and Burnett, E.F.P., 2000. "Pressure Moderation and Rain Control for Multi-Wythe Masonry Walls". *Proc of International Building Physics Conference*, Eindhoven, September 18-21, pp. 179-186.

Straube, J. and Smegal, J. 2007. "The Role of Small Gaps Behind Wall Claddings on Drainage and Drying", *Proc. 11th Can. Building Sci. and Technology Conf.*, Banff.

Straube, J.F., J. Lstiburek, A. Karagiozis and C. Schumacher "Appendix D: Preliminary Investigation of Drainage in Full-Scale Walls Clad with Stucco and Horizontal Vinyl Siding", *Oak Ridge National Laboratory Building Enclosure Hygrothermal Study*, Oak Ridge, Tennessee, 2000.

van Bentum, C. and Geurts, C., 2015. "Full scale measurements of pressure equalization on air permeable façade elements". *Proc 14th Internation Conference on Wind Engineering*, ICWE14, Brazil.

Van Den Bossche, N., Janssens, A. Moens, J., 2008. "Pressure equalisation as design strategy for watertight windows". *Proc. Nordic Building Symposium*, Copenhagen, Denmark, pp. 769-776.

Van Den Bossche, N., *Watertightness of Building Components: Principles, Testing and Design Guidelines*. Doctoral thesis, University Ghent, Belgium, 2013.

Van Linden, S., Lacasse, M., & Van Den Bossche, N., 2018. "Drainage and retention of water in small drainage cavities: Experimental assessment." *7th International Building Physics Conference*, Syracuse, NY. Z2

Van Linden, S., & Van Den Bossche, N. 2019. "Water Management in Cladding Systems with Small Drainage Cavities: A Review". *Thermal Performance of the Exterior Envelopes of Whole Buildings XIV International Conference*, pp. 867-874.

Van Straaten, R. A., Kopp, G. A., & Straube, J. F. 2010. "Testing water penetration resistance of window systems exposed to "realistic" dynamic air pressures". In *Proceedings of International Conference of Building Envelope Systems and Technology* (ICBEST), Vancouver.

Van Straaten, R. A. 2017. *Pressure Equalization of Wind-Induced Pressures on Residential Vinyl Siding Cladding in Full-Scale.* PhD Thesis, Western University, London, Canada.

Weston, T. A., Stachnik, M., & Waggoner, J., 2001. "Development of a textured spun-bonded polyolefin water barrier for stucco and EIFS". ASHRAE Transactions, 107, p. 433.

Williams, M. F. & Williams, B. L., 1998. "An Overview of Water Leakage Problems in Single-Family Residences Clad with Exterior Insulation and Finish Systems (EIFS)," *Water Leakage Through Building Facades*, ASTM STP 1314, R. J. Kudder and J. L. Erdly, Eds.

Wilson, A.G., "Condensation in Insulated Masonry Walls in the Summer," *Proc. Of RILEM/CIB Symposium*, Helsinki, 1965, pp. 2-7.

12.12 Notes

Chapter 13

Movements, Tolerances, & Joints

13.1 Introduction

Buildings are large products and this fundamental characteristic results in the need for joints: joints that allow movement and joints that allow for assembly from smaller components.

Buildings move. Members and materials expand and contract. Foundations settle, floors sag, beams deflect, and windows bow. This movement, static or dynamic, short-term or long-term, must be accommodated to avoid performance problems. Building elements that have not been designed to accommodate this movement often crack, crush or bulge. Such responses could result in one or more performance problems, and local or even complete, failure.

Cracks, usually tension failures, are a common building problem. They may be unsightly, reduce structural integrity, and allow the entry of water, air, insects, etc. Cracking and crushing, i.e., compression failure, and excessive deformation (a serviceability failure) can be avoided either by allowing movement to occur unrestricted (by providing joints to prevent or limit restraint) or by using members and materials with the appropriate strength, stiffness, and range of movement.

Buildings are assembled from many premanufactured components. Joints are needed to allow offsite manufacture of components. These joints must accommodate the dimensional variations inherent in the manufacturing process, the expansion and contraction of materials, and the assembly process. The trend to prefabrication of entire enclosure elements (e.g., windows, walls) means more joints that pass through many layers of the enclosure at one location: this renders these joints critical to performance.

In this chapter movements, tolerances, and joints are considered from a completely general perspective starting at the level of the building and then its major component parts. The focus is joints that occur in the building enclosure, but roofing is not discussed.

13.2 Categories of joints

In any built facility there are likely to be many joints and many different types of joint. A joint in a building may generally be defined as the junction between two or more elements, or components, of the building. The various professional groups involved in the design and construction of buildings have, over the years, developed numerous terms, definitions, and typical details. Across professions, however, there is less consistency and even within professions—such as structural engineering—there is often confusion about what, for example, constitutes an expansion joint or a control joint. Hence it is necessary to identify and define each of the various types of joint, especially those in the enclosure.

With one possible exception, joints are provided to fulfill one or more of the following functions:

- **constructability**: to facilitate construction, assembly or manufacture,
- **movement**: to accommodate movement,
- **control**: to promote the formation and control the location of shrinkage cracks, and
- **aesthetics**: to provide a visible architectural pattern.

Although the interfaces between the layers of an assembly (for instance, the interface between the masonry wythe and the thermal insulation in a wall system) could be considered a joint, by convention such transitions are not considered a joint. Nor does every connection (e.g., cladding attachments, light framing fasteners) constitute a joint. However, joints between sheets of water-resistant barrier, OSB, or gypsum wallboard are considered joints.

Constructability (construction) joints are formed during the construction process primarily to facilitate construction. They may be a temporary joint that, with further construction, they have served their purpose. Construction joints are also often are required to accommodate dimensional variations between parts, to accommodate the offsite manufacture, etc,.

Typical joints include those that occur when casting large concrete subsystems such as a floor or a long wall. Other examples include the joint between two sheets of plywood in a shear wall.

Movement joints: Joints that accommodate movement may variously be categorized into different types (Figure 13.1), for instance separation joints (e.g., between buildings), expansion[1] and contraction joints. They could, and usually should, also be called movement joints if they are intended to accommodate movement in either direction.

---

[1] It is problematic that in practice the term "expansion joint" is widely used imprecisely: joints that only allow expansion are rare, whereas movement joints are common.

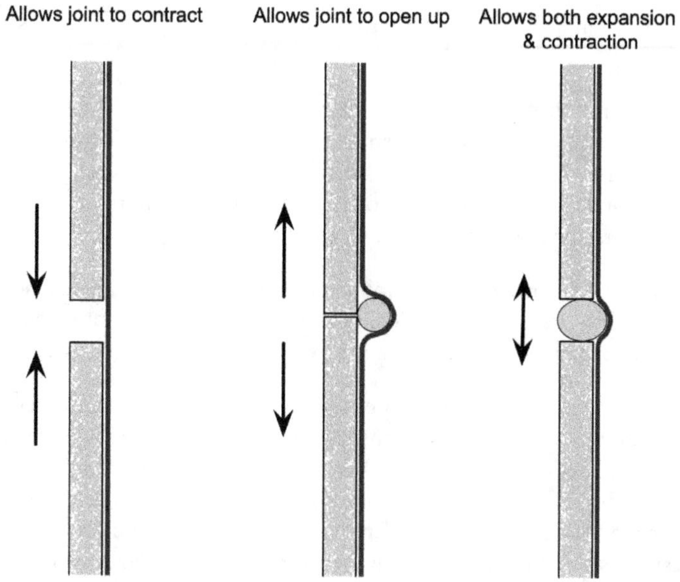

Figure 13.1: Joints can be designed to allow for contraction, expansion or both

Control Joints, or more properly, *crack* control joints, are joints designed to encourage cracking that is expected to occur because of shrinkage to occur where the designer intends. These are commonly formed by saw-cutting green concrete (Figure 13.2) or adding specially designed trim to exterior stucco cladding or interior drywall finishes.

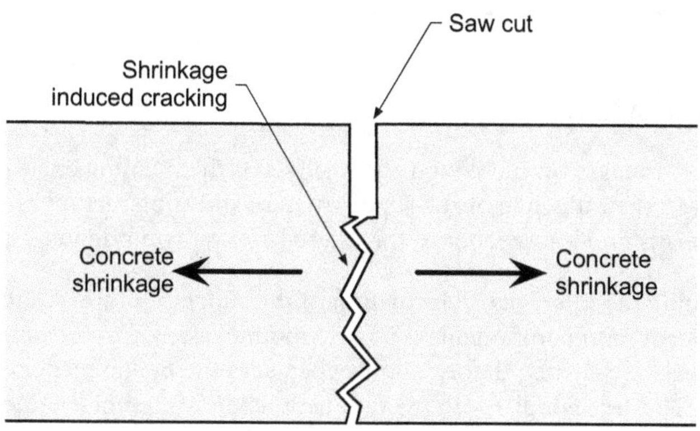

Figure 13.2: A common crack control joint used in concrete slabs on grade

Aesthetic Joints are provided to produce a desired appearance. A significant effort is often invested in aligning the other types of joints with aesthetic desires.

As previously discussed, any building can be considered to consist of four major physical parts: the superstructure, the building enclosure, the service systems, and the fabric. It follows that joints may also be categorized as being joints in one of these four component parts of the building: structural joints, finish joints, air-control layer joints, etc.

There are many other types of joints in building systems that are not directly connected to the building enclosure. Although these will not be considered in this text, two types are identified below.

Service system joints are joints between elements of the various service systems in a building. The service systems also have joints where they interface with the superstructure or enclosure. Joints in wiring, piping, conduit, and ducts are often required for the same reasons as superstructure joints: constructability and accommodation of movement. However, limiting the distribution of vibration is a not uncommon joint requirement for services which is rare for superstructures.

The types of service system joint are many and varied, and they are outside the scope of this text.

Fabric joints[1] are very common in those elements that constitute the building fabric (partitions, dropped ceilings, millwork) and interior finishes. They can range from joints in non-structural partition walls to joints in a floor tile system. Most of these joints exist for reasons of manufacture and assembly, i.e., constructability, but aesthetics is often of critical importance. These joints are not discussed any further in this text.

13.3 Enclosure movements

This text focuses on the design and analysis of the building enclosure. However, the influence of the superstructure and substructure on enclosure movements and joints requires that these building components be considered.

Evaluating the in-service deformation of the superstructure and its structural sub-systems and components is a fairly routine task for the structural engineer of record. Evaluating deformational change in the building enclosure is less routine and less familiar. In many situations the structural engineer of record may not be involved in the design of the enclosure.

A thorough and comprehensive discussion of the in-service deformation of the enclosure assembly and its component parts is not possible here. The following

[1] This is an unsatisfactory label, but "finish joint" is incorrect, and we have not found an accurate or widely used label.

is merely an attempt to identify and discuss some of the more important issues, among them some of relevant loadings, causes of movement and the methods of quantifying the movement.

All parts of a building move: each layer, each assembly, the enclosure, the superstructure, the foundation, etc. However, depending on the scale of the movement and the restraint of the surroundings the movement may need to be accommodated or can be ignored.

Movements are quite important for enclosure design, both because there are significant differential movements between the enclosure and superstructure and because both differential and absolute material movements within the enclosure are large due to the nature of temperature and moisture conditions.

Movements are primarily caused by temperature changes, moisture content changes, and extension and compression due to mechanical forces (Brooks 2014, PCA 1982, Baker 1964). These movements can also be classified in numerous ways, for example:

1. short-term or long-term,
2. reversible or irreversible,
3. internal (i.e., material changes) or external forces

When designing movement joints, with or without sealant, ASTM Standard C1472 *Standard Guide for Calculating Movement and Other Effects When Establishing Sealant Joint Width,* [ASTM 2022] is a useful and authoritative reference.

13.3.1 Movement of the super-structure

Almost all historical buildings (i.e., made of solid masonry, logs, adobe) combined enclosure support and superstructure functions into a single system. As described in Chapter 2, the enclosure support function is explicitly separate from the superstructure in many modern buildings. The difference in movement between the enclosure support and building superstructure systems must be accommodated by joints (Figure 13.3). In some common current building systems, such as load-bearing light wood framing, load-bearing (low-rise) masonry, and load-bearing precast concrete, the support and superstructure systems are the same and such joints are not needed and in fact cannot function with joints.

Modern enclosure designs with light-gauge steel framing and aluminum curtainwalls are almost always designed to separate the enclosure support (the steel studs/ aluminum tubes) and the superstructure (often of reinforced concrete or hot-rolled steel). Essentially all high-rise enclosures (including those with masonry enclosure support) are designed in this way.

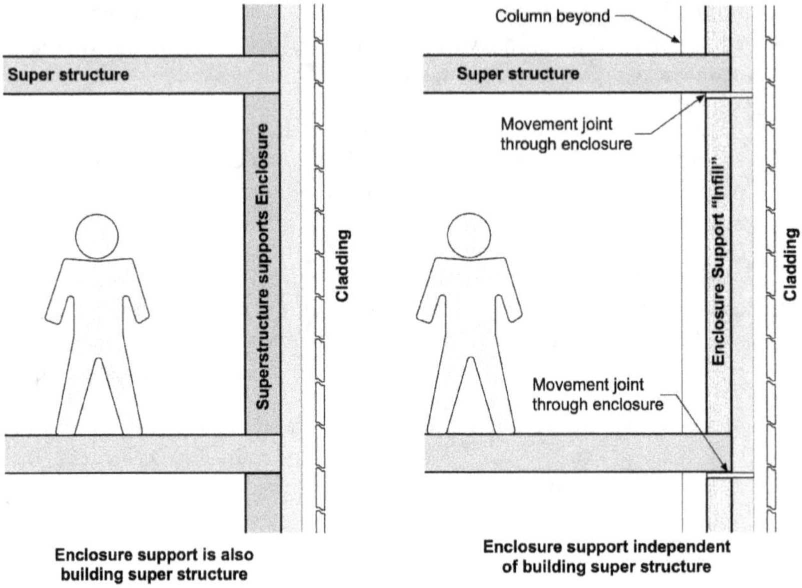

Figure 13.3: Movement joints are required between support and superstructure

Building superstructure movement can be categorized as

- short-term elastic movements such as deflections from gravity loads (Figure 13.4)
- side sway due to wind and seismic forces (Figure 13.5), and
- long-term irreversible movements, such as creep, drying shrinkage, and foundation settlement.

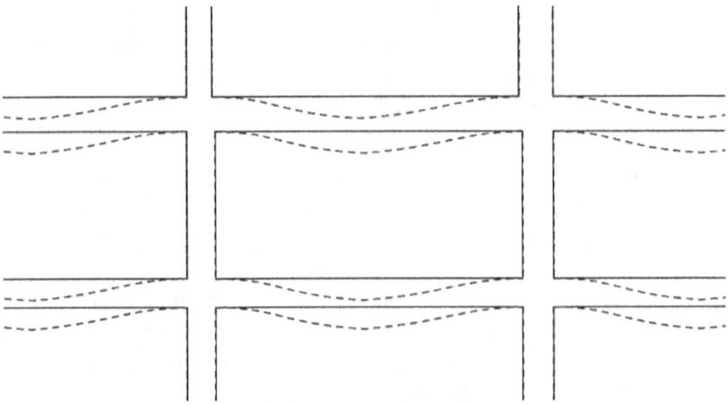

Figure 13.4: Schematic deflected superstructure shape due to imposed gravity loads

Live-load deflection of floor beams is often limited during design by the structural engineer. A common limit for deflection of live loads in North America is 1/30th of span: a 10-meter span beam would be designed to limit deflection to 27.8 mm under live loads. This movement will not be resisted by the enclosure support structure (as it is not designed to be loaded vertically) and hence must be allowed for by joints. Such joints must be designed to resist out-of-plane forces (in particular wind forces but also seismic loads) while allowing vertical movement.

The side sway of buildings must also be accommodated. The amount of movement is often described as the difference in horizontal movement between adjacent floors. The lateral movement of the superstructure will often be designed to limits of 1/300 to 1/600 of storey-height, although much larger movements can occasionally be accepted.

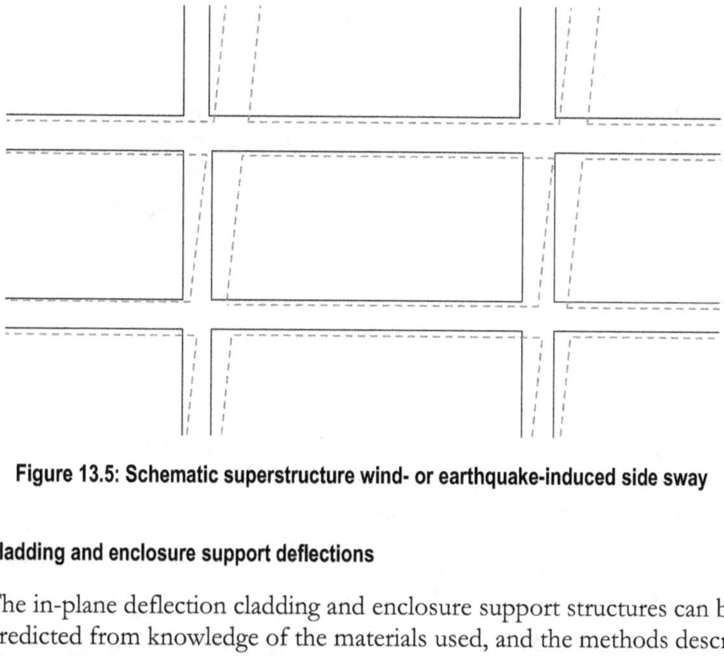

Figure 13.5: Schematic superstructure wind- or earthquake-induced side sway

13.3.2 Cladding and enclosure support deflections

The in-plane deflection cladding and enclosure support structures can be predicted from knowledge of the materials used, and the methods described above. However, out-of-plane deflections due to wind (and seismic) loads are calculated using traditional solid mechanics and knowledge of the material properties.

These out-of-plane deflections are more likely to be specified and constrained than temperature or moisture movements. Field experience with different claddings and enclosure support structures has resulted in the development of different allowable design deflections. For example, a vertical curtainwall mullion may be designed to allow deflections of span/90 under wind loads, whereas a steel stud wall might be designed for span/600 when supporting a masonry veneer and span/120 when supporting corrugated steel sheet cladding. These allowable out-of-plane deflections, and any differences between them, need to be accommodated by the cladding, the enclosure support system, and the building superstructure.

13.3.3 Reversible movement due to changes in temperature

A linear relationship between a change in temperature and the change in volume or dimension is assumed to exist. This relationship involves a constant of proportionality commonly referred to as the coefficient of thermal expansion (sometimes shortened to CTE) and contraction (α). This is also assumed to have constant value for a given, isotropic material. Values for selected materials are provided in Table 13.1. Dimensional changes in a given unrestrained member can be determined using the following relationship:

$$\Delta L = \alpha \cdot \Delta T \cdot L \qquad \text{(Eq. 13.1)}$$

where L represents the initial length of the component.

Provided that a temperature change does not induce a change in the phase of the material and the material is uniform throughout (i.e., isotropic), the above relationship should provide a reasonably accurate prediction of the movement due to thermal variation. Provided there is no restraint, a positive change in temperature would induce a uniform increase in size (expansion) of the component.

13.3.4 Reversible movement due to a change in moisture content

Changes in moisture content cause materials to undergo movement in a manner similar to that resulting from thermally induced changes. Linear proportionality is assumed but may not always be the case. Moisture content changes are usually difficult to predict, but they tend to occur more slowly than temperature changes. Values for the coefficient of expansion and contraction due to moisture change (μ), can be found in Table 13.1.

The following relationship can be used to determine the dimensional change in an unrestrained member due to a change in moisture content:

$$\Delta L = \mu \cdot \Delta M \cdot L \qquad \text{(Eq. 13.2)}$$

where L represents the initial length of the component,

ΔM is the change in moisture content, and

μ is the moisture change coefficient (often as a percentage).

As with thermal movement, an increase in moisture content causes expansion of the material and a decrease in moisture content causes contraction.

Moisture movements in wood have great practical importance. Wood is used for finishes, structure, sheathing and cladding and hence is widely encountered in enclosure design. The moisture-induced movement of wood varies significantly with grain direction, and somewhat with species, wood source, etc. Figure 13.6 provides a graphical summary, but much more detail can be found in US Forest Products Labs publications.

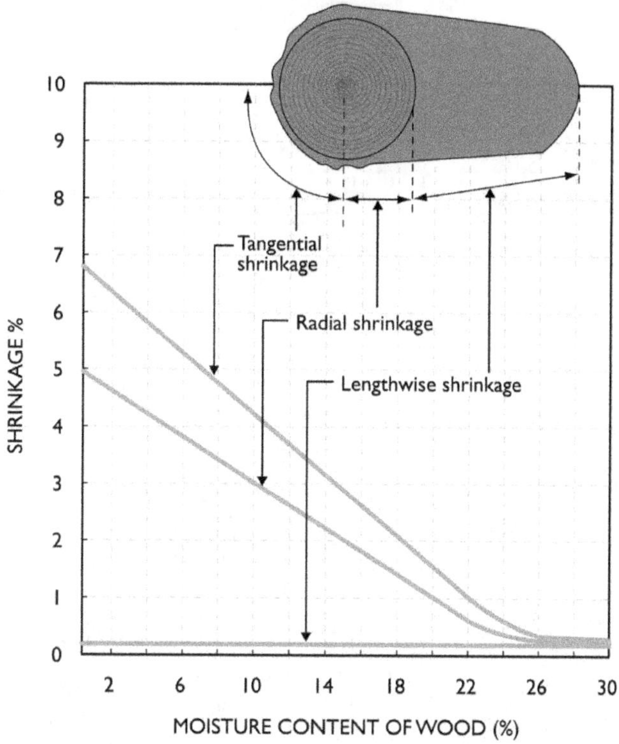

Figure 13.6: Reversible moisture movement of wood varies with direction (RDH Building Science)

13.3.5 Irreversible, moisture-related volume change

The initial drying (oven-baked brick) or wetting (wet cast concrete) of a material is often the cause of initial shrinkage or expansion in materials such as concrete or masonry. This initial change in volume is not reversible. Values for this irreversible dimensional change are provided as a percentage in Table 13.1. To evaluate the extent of any unrestrained dimensional change, the designer may use the same approach as for reversible moisture-related change.

Table 13.1: Properties needed to assess change in dimension of materials

Note: Convert thermal coefficients to °F by multiplying by 1.8. Convert MPa to psi by multiplying by 145.

	Material[1]	Coefficient of Thermal Expansion/ Contraction (α) °C x10^{-6}	Coefficient of Moisture Expansion and Contraction (μ) %	Irreversible Moisture Movement (+) Expansion (-) Shrinkage %	Modulus of Elasticity (E) MPa
Natural Stone (dry to saturated)	Granite	8-10			20-60
	Limestone	3-5	0.01		10-80
	Marble	4-6			35
	Sandstone	7-12	0.07		3-80
	Slate	9-11			10-35
Cement-Based Composite (Concrete values for plain concrete)	Mortar and Fine Concrete	10-13	0.02-0.08	0.04-0.10(-)	20.35
	Dense Aggregate Concrete				
	Gravel Aggregate	12-14	0.02-0.06	0.03-0.08(-)	15-36
	Crushed Rock (Except Limestone)	10-13	0.03-0.10	0.03-0.08(-)	15-36
	Limestone	7-8	0.02-0.03	0.03-0.04(-)	20-36
	Steel-Fiber-Reinforced Concrete	5-14	0.02-0.06	0.03-0.06(-)	20-41
	Aerated Concrete	8	0.02-0.03	0.07-0.09(-)	1.4-3.2
	Lightweight Aggregate Concrete				
	Medium Lightweight	8-12	0.03-0.06	0.03-0.09 (-)	8
	Ultra Lightweight (Perlite)	6-8	0.10-0.20	0.20-0.40(-)	(6)
	Glass-Reinforced Cement	7-12	0.15-0.25	0.07(-)	20-34
Gypsum Based Composite	Gypsum wallboard	18-21	(gypsum materials are highly susceptible to moisture damage. Moisture changes are not reversible)		16
	Sanded Plasters	12-15			8.5-16
	Lightweight Plasters	16-18			1.5-4
	Glass-Reinforced Gypsum	17-20			16-20
Brickwork and Blockwork	Concrete Brickwork and Blockwork				
	Dense Aggregate	6-12	0.02-0.04	0.02-0.06(-)	10-25
	Lightweight Aggregate (Autoclave)	8-12	0.03-0.06	0.02-0.06(-)	4-16
	Aerated (Autoclave)	8	0.02-0.03	0.05-0.09(-)	3-8
	Calcium Silicate Brickwork	8-14	0.01-0.05	0.01-0.04(-)	14-18
	Clay or Shale Brickwork or Blockwork	5-8	0.02	0.02-0.07(-)	4-26
Metals	Cast Iron	10	(Metals are not subject to expansion or contraction due to moisture changes. Corrosion of some metals can result in significant irreversible expansion.)		80-120
	Plain Carbon Steel	12			210
	Stainless Steel				
	Austenitic	18			200
	Ferritic	10			200
	Aluminum and Alloys	24			70
	Copper	17			95-130
	Bronze	20			100
	Aluminum Bronze	18			120
	Brass	21			100
	Zinc				
	Parallel to Rolling	33			140
	Perpendicular to Rolling	23			220
	Lead	30			14

Table 13.1 (cont.): Properties needed to assess change in dimension of materials

	Material[1]	Coefficient of Thermal Expansion and Contraction (α) °Cx10^{-6}	Coefficient of Moisture Expansion and Contraction (μ) %	Modulus of Elasticity (E) MPa
Wood and Plywood [(5)] (G is specific Gravity)	Softwoods:			
	Along Grain (Longitudinal)	3.1 - 4.5		
	With Grain (Tangential)	32.4*G* + 18.4	0.01(0.6-2.6)[(1)]	5.5-12.5[(4)]
	Across Grain (Radial)	32.4*G* + 9.9	(0.45-2.0)[(1)]	
	Hardwoods:			
	Along Grain (Longitudinal)	3.1 - 4.5		
	With Grain (Tangential)	4-6	(0.8-4.0)[(1)]	7-21[(4)]
	Across Grain (Radial)	30-70	(0.5-2.5)[(1)]	
	Plywood:			
	With Grain	(5)	0.15-0.20[(2)]	6-12[(4)]
	Across Grain		0.20-0.30[(2)]	
Wood-based Material	Hardboard		0.30-0.35[(2)]	3.0-6.0
	Medium Density Board		0.30-0.40[(2)]	1.7-3.3
	Chipboard		0.35[(3)]	2.0-2.8
	Wood-Wool-Cement		0.15-0.40	0.6-0.7
Rubber and Plastics Etc.	Asphalt	30-80		
	Thermoplastics		(While most plastics do not move with moisture, some plastics may be liable to irreversible progressive shrinkage due to loss of volatiles and related causes.)	
	PVC, UPVC, CPVC	40-70		2.1-3.5
	Polyethylene (Low Density)	160-200		0.1-0.25
	Polyethylene (High Density)	110-140		0.5-1.0
	Polypropylene	80-110		0.9-1.6
	Polycarbonate	60-70		2.2-2.5
	Polystyrene	60-80		1.7-3.1
	Acrylic	50-90		2.5-3.3
	Acetal	80		2.8-3.7
	Polyamide	80-130		1.0-2.7
	ABS	60-100		0.9-2.8
	Thermosets (laminates)			
	Phenol and Melamine Formaldehyde	30-45		5.5-8.5
	Urea Formaldehyde	27		10
	Cellular (expanded):			
	PVC	35-50		
	Phenolic	30-90		
	Urea-Formaldehyde	20-40		
	Polyurethane	20-70		
	Polystyrene	15-45		
	Reinforced:			
	GRP (Chopped Strand)	20-35		6-12
	Carbon-Fiber (Parallel to Reinforcement)	0-0.05		180-220
	Carbon-Fiber (Perpendicular to Reinforcement)	30-70		180-220
Glass	Plain, Tinted and Opaque	9-11	(Glass is not impacted by moisture)	70
	Foamed (Cellular)	8.5		5-8

Notes:
(1) Based on 60% and 90% relative humidities.
(2) Based on 33% and 90% relative humidities.
(3) Based on 65% and 90% relative humidities.
(4) At 12% moisture content; values reduce at higher moisture contents.
(5) Specific data can be found in the US Forest Products Laboratory *Wood Handbook*.

13.3.6 Reversible Out-of-plane movements - Bowing

In many situations, the thermal and moisture conditions are not uniform across a layer. For example, there may be a thermal gradient across a layer, or the exterior may have much higher moisture content than interior. In the case of a single unrestrained layer of material the extent of the bowing, due to a differential temperature of ΔT, can be evaluated from the following expression:

$$\Delta_{bow} = \frac{\alpha \cdot \Delta T \cdot L^2}{8 \cdot w} \qquad \text{(Eq. 13.3)}$$

where L is the length of the member and

w its thickness.

In many instances some degree of restraint or tieback may be present, and these effects must be accounted for.

13.4 Construction and dimensional tolerances

Building products and systems are not built to perfect dimensions nor are they located precisely where intended with respect to adjoining objects: almost nothing is, whether a computer monitor or warship. Acceptable variations are termed "tolerances", and the precise definition involves careful statistical descriptions. The detailed investigation of this important topic is the subject of entire textbooks [e.g., Ballast 2007], and a study more detailed than the brief introduction provided here is recommended.

In general, the larger the size of the component or product, the larger deviations from the intended dimension should be expected. A car door will often deviate from its nominal dimension by a millimeter (1/16") or more. A precast concrete panel, with ten times the dimension of a car door, is likely to have dimensional variation of as much as 10 mm. Hence, the tolerance of the joint between the car door and the chassis might be specified as 5 mm, and the joint between large precast panels specified as 25 mm.

Depending on the system or materials used, construction tolerances can vary greatly. Different standards for design and products often include dimensional variations. For example, the PCI Handbook [Section 4.6 of PCI 2007] is a useful source of information on recommended tolerances for precast concrete panels, ACI 117-90 from the ACI is a proposed specification for concrete, and the AISC *Code of Standard Practise* [ACI 2022] and associated ASTM product standards provide standard tolerances for steel construction.

In the building industry two approaches are used to manage dimensional variation:

1. allow for on-site variation during construction (such as cutting or pouring materials to fit the space available), or
2. use interchangeable components that are designed to fit together with joints having sufficient tolerance to accommodate the dimensional variations.

Although reducing variations of product dimensions (and thus reducing the required tolerance) is a viable design and production goal, the cost of designing joints to accommodate tolerances is often far less than the effort required to consistently produce large building components with small dimensional variations. As joints inevitably need to accommodate significant amounts of movement, increases in joint size also accommodate tolerances is neither difficult nor onerous and is the preferred design approach.

13.5 Joints in the building enclosure

Joints within the enclosure system can be both within and between enclosure elements. A typical vertical enclosure (Figure 13.7) often consists of several different components: opaque wall, windows, and a door, each of which is a distinct enclosure element. There are joints at the perimeter of the windows and door where the enclosure elements interface with each other. In addition, there may be service system penetrations, such as a duct or an air-conditioning unit that penetrates through the enclosure element. The interface between each of these components requires a joint, one which fulfills the full function of the enclosure components it joins (albeit usually with some lower performance).

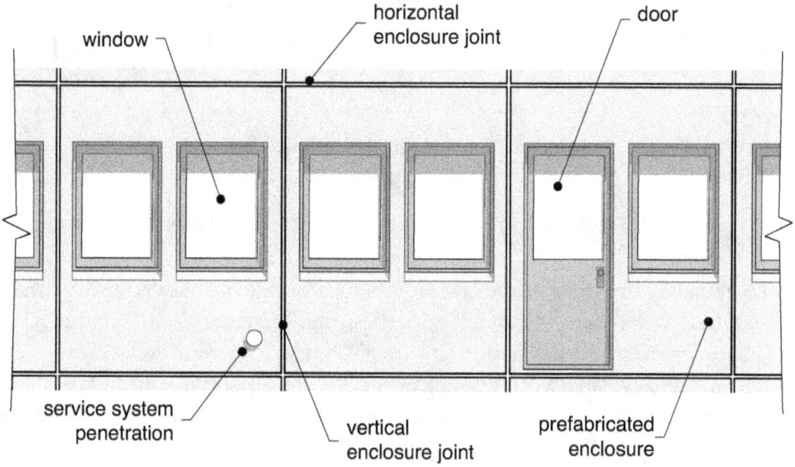

Figure 13.7: Enclosure joints between building enclosure elements

Most modern enclosures are multi-layered assemblies, comprised of contiguous layers that provide the control functions (Chapter 2). Although the function of some layers (i.e., non control layers) may not require continuity it is critical that control layers be continuous across joints. Some of these layers are thin and flexible, e.g., membranes, such as water-resistant or air barriers. One or more layers may be an air space and layers are often thick (such as insulation or support systems). Note that a layer may be something other than one continuous thickness of a specific material; for example, a wall of stud framing filled with insulation is an assembly layer comprised of two materials in a three-dimensional form.

The interface between different layers in the assembly does not constitute a joint. However, joints can exist within the plane of a layer, and these are considered to be layer joints.

13.5.1 Layer joints

Layer joints are those occurring at the perimeters of the in-plane elements making up a single layer. Depending on enclosure design, layer joints may require continuity of layer function, that is water, air, vapor, fire, smoke control functions or mechanical load resistance. For example, the joints of an air-water barrier membrane need to be airtight and watertight, or the joints may need to block view and UV-radiation for a rainscreen cladding. Layer joints can be made with a wide range of geometries. Some common layer joint geometries are shown in Figure 13.8. These joints also often facilitate alignment, accommodate dimensional tolerances, and may need to transfer mechanical forces.

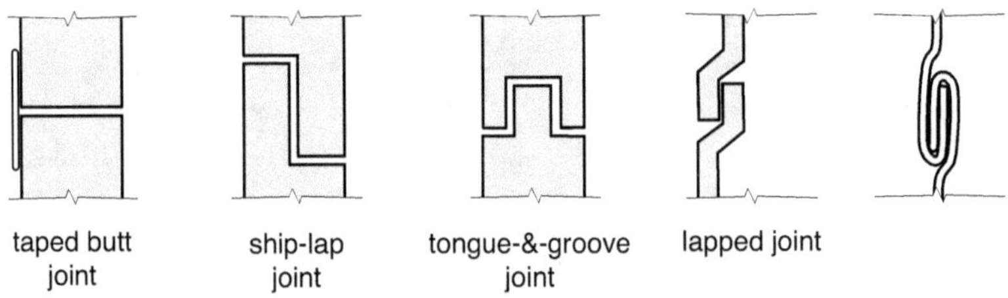

Figure 13.8: Pre-formed, intra-layer continuity joints

To achieve continuity some layer joints may be filled or sealed using other materials, often with different properties than the layer itself (Figure 13.9). Examples of layer joint continuity are gaskets within windows, mortar between bricks, adhesives between air barrier membrane laps, non-shrink grout between load-bearing precast concrete panels, and sealant between precast concrete panels.

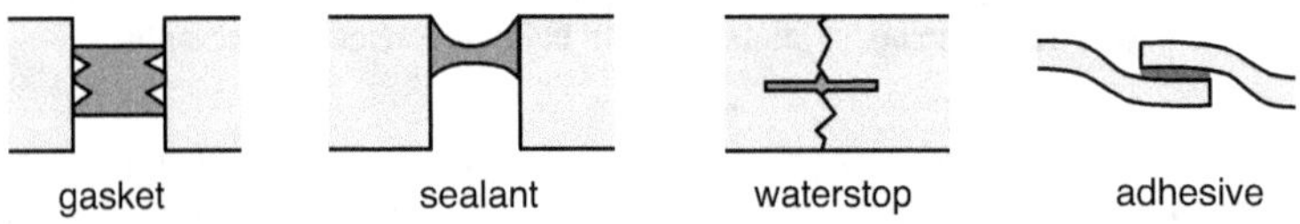

Figure 13.9: Layer joints using filler materials to provide continuity

Figure 13.10 provides a sample of some of the many ways that an air or water control layer membrane can be made continuous. The design of these joints is often proposed by the manufacturers of the products used as control layers, but must be understood by the enclosure designer to ensure they are appropriate for the project. Job-site supervision and review also requires an understanding the joint design intent.

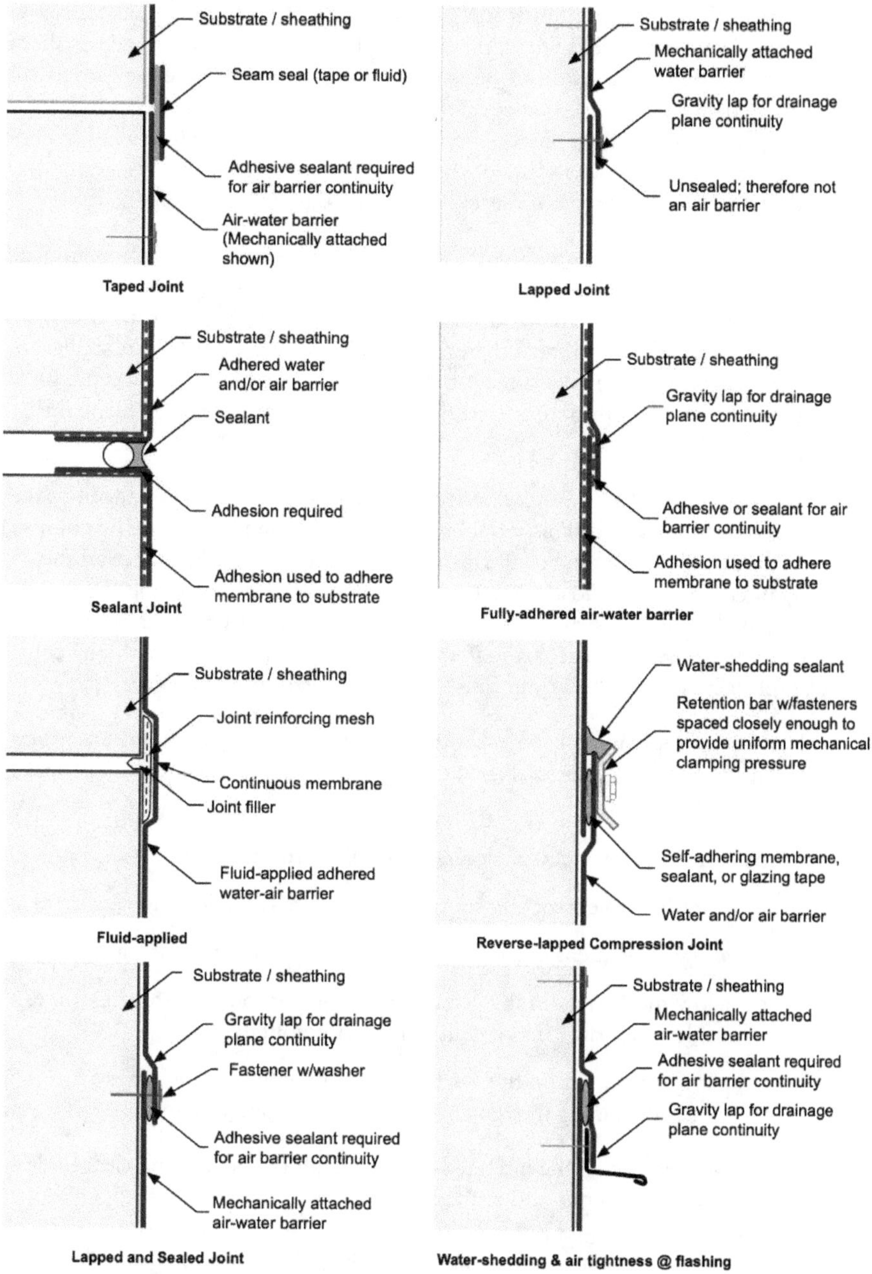

Figure 13.10: Different approaches to providing air and/or water continuity at a layer joint

13.5.2 Joints between enclosure elements

Enclosure joints[1] are out-of-plane joints between adjacent elements of the building enclosure (e.g., window to wall, wall to cantilever balcony slab). At these joints all control, finish and support layers of an element are interrupted but must be designed, built, and maintained to ensure continuity across the joint. The aesthetics of these joints are also important because they are usually visible (inside and out). Inadequate performance of these enclosure joints is far too common, and the consequences have been costly. The following is a brief attempt introduction to some of the more important concerns regarding the design and performance of these joints in the building enclosure.

Three major criteria, namely, constructability, movement accommodation, and esthetics must be met. These joints are also exposed to the same loadings (e.g., rain, temperature difference, air pressures) as the adjacent elements of the building enclosure and must therefore perform in a manner that is similar to the enclosure elements (although reduced performance is often accepted). These joints are common, and with the increase in prefabricated elements, increasingly so.

There are several steps to the design of an enclosure joint. First the joint must be located and sized. This requires that the movements involved be computed or estimated using some of the information presented earlier. Second the required mechanical load transfers, both direction and magnitude must be designed. This is a structural or façade engineering activity. Finally, the joint must be designed to act as an effective part of the enclosure, that is, all control (especially water and air) and finish functions need to be satisfied.

It is not a simple matter to design these joints. It is difficult to ensure proper installation and even more difficult to ensure their maintenance and proper repair when needed.

From a structural perspective joints in practice are designed to:

- restrain movement and transfer forces across the joint in all axes,
- allow movement in all axes without transferring mechanical forces,
- allow movement in the vertical direction without transferring forces, while restraining movement in both horizontal axes, or
- allow movement in the vertical direction without transferring forces, while restraining movement in the out-of-plane horizontal axis.

It is fundamental that the intended force transfer of the joint be clear in design, on contract documents, and in the field.

[1] Although the term "through-wall joint" may be applied, some joints labelled this way do not penetrate through all of enclosure, e.g. if flashing only penetrates through an assembly's air, water, and thermal control layers, it can be call a "through-wall flashing".

Common joint designs use bolts, shear blocks, and made-for purpose products to transfer mechanical forces. Figure 13.11 is an example joint that exemplifies modern design. It utilizes a steel C-clip to allow differential vertical movement between a CMU support structure and a reinforced concrete superstructure while transferring all out-of-plane loads to the superstructure. Of course, this joint design also allows the predicted thermal movement without damage, accommodates expected tolerances, and provides continuity of the control functions. The interior and exterior finishes (e.g., gypsum board and cladding respectively) would require their own joint to allow movement in all axes but would not need to provide any control functions.

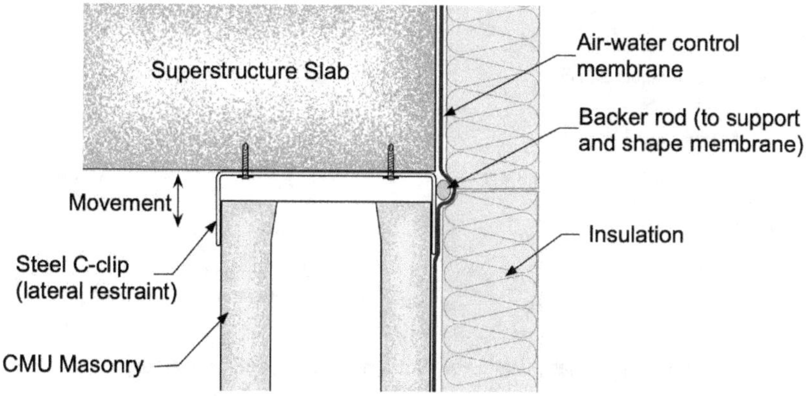

Figure 13.11: Construction and movement joint between CMU support and superstructure

13.5.3 Enclosure joint rain penetration control

The control of rain penetration is a fundamental requirement of the building enclosure. Enclosure joints are often a source of rain penetration (see the window joint section above), and proper rain penetration control must be a primary consideration in their design. A significant of research in the past [Platts and Sasaki 1965, Isaksen 1965, Killip & Cheetham 1984] produced fundamental understanding of joint performance that remains highly relevant over 50 years later.

Even in light rain, joints between panels of low absorption materials such as metal and glass will be exposed to large quantities of water drained from the panels. The shape, size, and orientation of a joint may also tend to collect and trap water. These combined effects result in a high potential for rainwater penetration through joints.

As for the building enclosure element, the three approaches to controlling rain penetration: perfect barrier, mass, and drained, apply equally well to the design of enclosure joints. Figure 13.12 outlines a classification system for types of enclosure control joints based solely on the means of dealing with liquid water (see also Chapter 12).

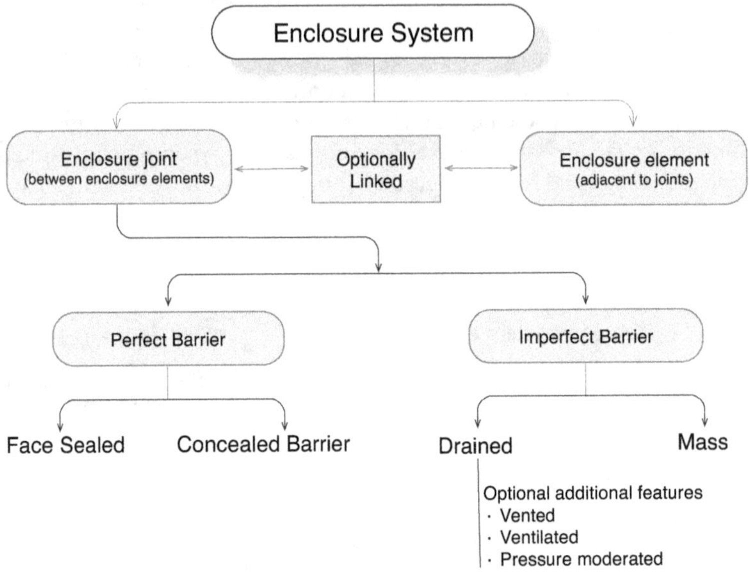

Figure 13.12: Classification of joints based on rain control strategy

Figure 13.13 provides a simple graphical representation of the three rain control strategies applied to common joints.

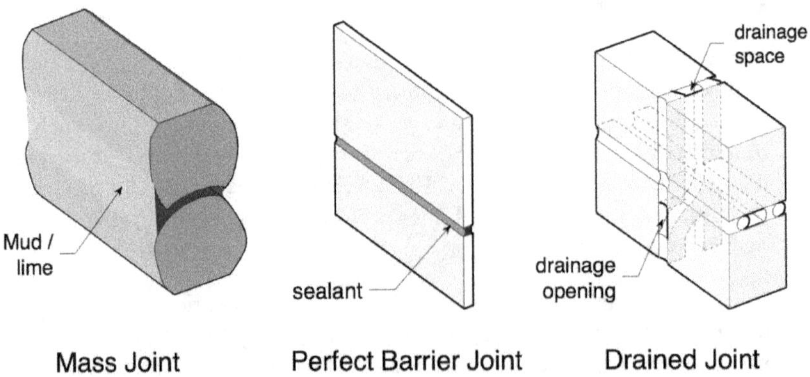

Figure 13.13: Examples of each of the three rain control joint types

Note that the joint and the adjacent enclosure element may or may not function completely independent of each other. In some systems the joints are *linked* in such a manner that draining water can move from the joint to the adjacent enclosure and vice versa. In such cases it is essential that the enclosure as a whole be considered to be a drained system even if the cladding itself is ostensibly a perfect barrier to wind-driven rainwater. Exterior Insulated Finish Systems (EIFS) from the early-1990s are an example where the linkage between joints and enclosure element resulted in rain penetration control failure, costly repairs, and subsequent litigation.

13.5.4 Enclosure joint rain control design approaches

Consider the series of idealized joints illustrated in Figure 13.14. The joint could be left open, as shown in Figure 13.14(i), but for many applications this may not be a viable or permissible option due to precipitation, ultraviolet radiation, insects, birds, small animals, and appearance.

However, a layer joint, such as a joint between cladding panels, may be open if the enclosure control functions are provided by other layers nearer the interior. This has long been accepted as desirable approach for cladding [Herbert 1974], as it protects the materials used for control continuity from direct exposure to rain and thc environment and yet allows for movements and tolerances.

Unlike layer joints enclosure joints require some degree of closure. Providing a seal at a single line close to the outside face is one easy approach. This single-stage approach (shown in Figure 13.14(ii)) was once common in early panelized enclosure systems such as curtainwall and precast concrete.

The seal shown uses a flexible polymeric sealant and a backer rod, but other forms of sealing could be used. Nonetheless, irrespective of the type of seal, this still constitutes a single-stage joint, and the basic design assumption is that the single seal, with appropriate maintenance and repair, will always provide a perfect seal to precipitation, especially wind-driven rain, and air. Face-sealed fluid sealant joints are inherently sensitive to workmanship as well as deterioration due to UV radiation exposure and oxidation. Unfortunately, the risk of failure is great with only one seal, especially if site installed and located at the face, and thus this joint design is considered risky, especially regarding rain penetration control. Such joint designs are still used as a repair strategy (where the more effort in installation quality and maintenance required can often be less expensive than other repair strategies).

Such single-stage enclosure joints are, by definition, face-sealed systems in that a single perfect barrier to water penetration is placed at the exterior surface. Many serious cladding and wall problems particularly with older undrained EIFS designs, non-drained metal and glass curtain walls, and stucco direct over single paper on wood frame walls can be blamed on the choice of a perfect barrier system for the joint as well as the opaque wall. If the joint and the adjacent enclosure element are not independent of each other in relation to water entry, then water will travel internally from the failed joint to the wall element; this would constitute a linked system. As shown in Figure 13.5, linked enclosure systems should be designed using the drained approach to rain control.

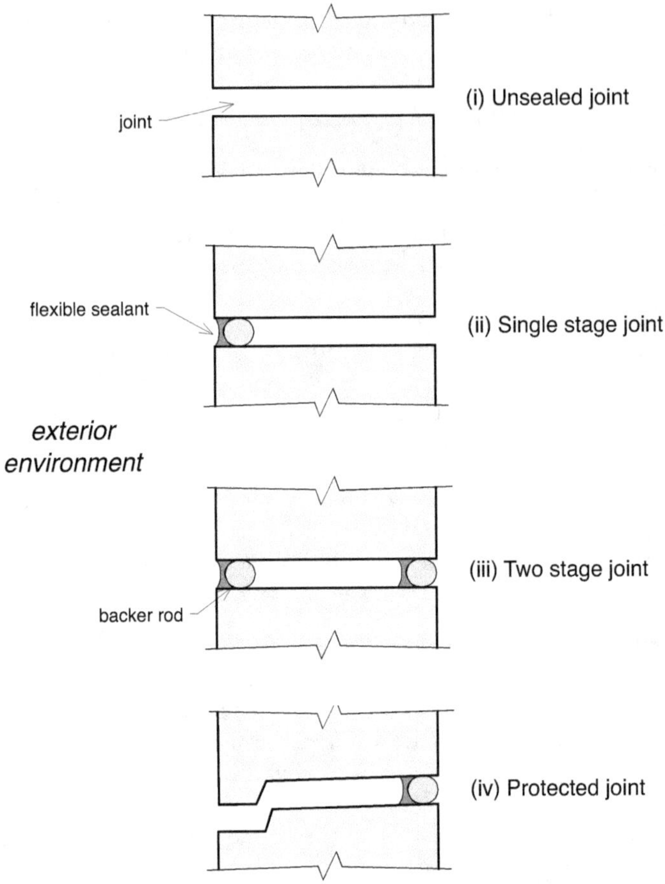

Figure 13.14: Conceptual sections of an idealized horizontal joint

Proper joint design should, in general, utilize the screened and drained approach to rain control as listed in Figure 13.12 and discussed in Chapter 12. This leads to the basic two-stage design approach in which the outer seal is a screen to rain and an interior seal is provided that is a barrier to the air and the penetrant rainwater loading at this location. The concept has long been studied and implemented [Svendsen 1967]. Care is needed to ensure that the two-stage joint actually constitutes a properly drained system. For example in Figure 13.14(iii) there are two lines of seals but there is no indication of any provision for drainage (usually provided by discrete drain openings).

To reduce the required drainage capacity and the water load on the inner seal, it is highly desirable to stop or manage both wind-driven rain and accumulated surface drainage water at or close to the outside face. To meet this requirement joints are usually sealed at or close to the outside.

In a drained system the outer seal does not have to be watertight, and in fact must have drain openings. Exterior seals that are not perfect barriers to rain and air are desirable, for example, compressed EPDM or silicone gaskets, sprung polymer or metal inserts, or, most commonly, with a flexible, wet-applied sealant over a preformed foam backer rod. Maintenance and repair are not as critical for this seal because provision has been made for a small amount of rain penetration past the outer exposed layer. Dry gaskets are often difficult to rely on for perfect water tightness (because of the imperfectly smooth substrate against which they are compressed) but can be a good choice for exterior "rain screen" seals where they are only relied upon to shed water, block direct raindrop entry, and provide shelter from UV (i.e., they act as a rainscreen).

Figure 13.14 (iv) shows an example of a horizontal joint designed to use geometry to act as the outer "stage" against rain (and sun). This is strictly speaking a two-stage joint and can be very effective, but as shown is only applicable to a horizontal joint.

Well-designed protected outer stages have been used in joints for many years in the harsh climates of coastal Norway and United Kingdom [Herbert & Harrison 1974, Isaksen 1972]. Research has confirmed the design principles of many practical baffled and protected joints, e.g., [Platts & Sasaki 1965].

Labyrinth joints have a complex geometry (Figure 13.15) intended to reduce rain penetration (that is, increase the protection offered by the geometry) but still rely on an air-water seal as the second stage (not shown in the figure). They are more appropriately described as joints for a rainscreen layer, not an enclosure joint. Research in both laboratories and, importantly, under natural exposure conditions suggest that if a labyrinth joint is used, the momentum of the water droplets travelling in the airstream through the joint could be used to deposit them in areas of the joint away from the fast airstream, where vertical drainage can then take place. Alternatively drop penetration can be reduced or prevented by designing the joint so that the air velocity through it was kept below that which would transport water droplets.

Two-stage joints produced using sealants are a very common enclosure joint design for large enclosure elements such as between precast concrete panels. The precast industry has promoted and deployed such joints since the 1970's [Schutz 1973]. Figure 13.16 is an example of a vertical and horizontal joint designed to as a two-stage joint: an outer screen to rain, a gap to allow drainage downwards and out, and an inner seal that provides continuity of the air- and water-barrier.

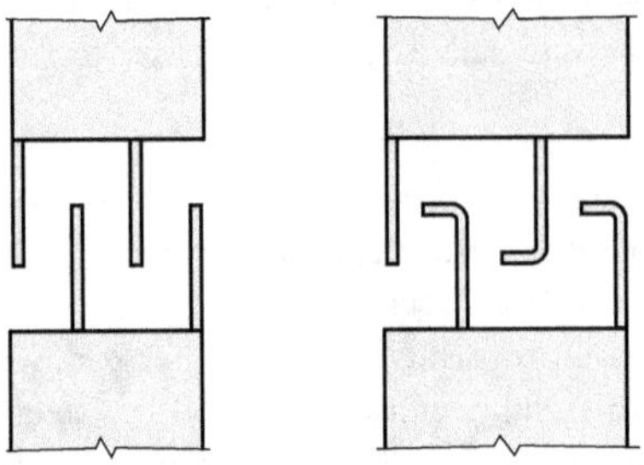

Figure 13.15: Example labyrinth joints designed to reduce cladding rain penetration

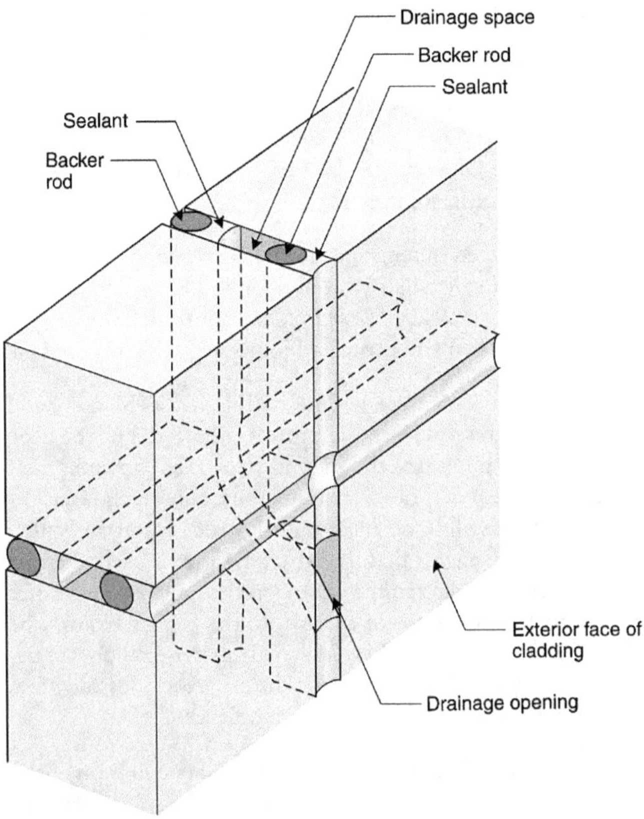

Figure 13.16: A two-stage joint formed from sealant

One benefit of the two-stage approach is that the interior seal can be expected to be more durable it is in a more thermally stable location protected from wind, rain, and, critically for sealant, ultraviolet radiation. Water that penetrates the first seal can drain between the two layers, and the interior air barrier will prevent any flow of water or air through the joint. The interior second stage of a two-stage joint does not need to be made of sealant but could be made from a self-adhering air- and waterproof membrane, tight conforming gasket, or any other reliably water and air-tight system.

Two-stage joints require that drain openings, usually every 3 to 6 m, but there is little science to guide this decision. Drain and vent openings should have a minimum dimension of 9 to 12 mm (3/8" to 1/2") to avoid blockage with a water film. Tubes may be inserted into wet sealant, but the internal diameter of the tubes should also be at least 9 mm (3/8") and such tubes can often restrict drainage.

Both vertical and horizontal joints follow the same basic principles. However, it is desirable to slope horizontal joints outward to encourage water that penetrates to flow laterally against the outer seal until reaching a drain opening. In practice, field-applied vertical two-stage joints constructed of sealant will also slope outward just above the drain opening (Figure 13.17, Figure 13.16).

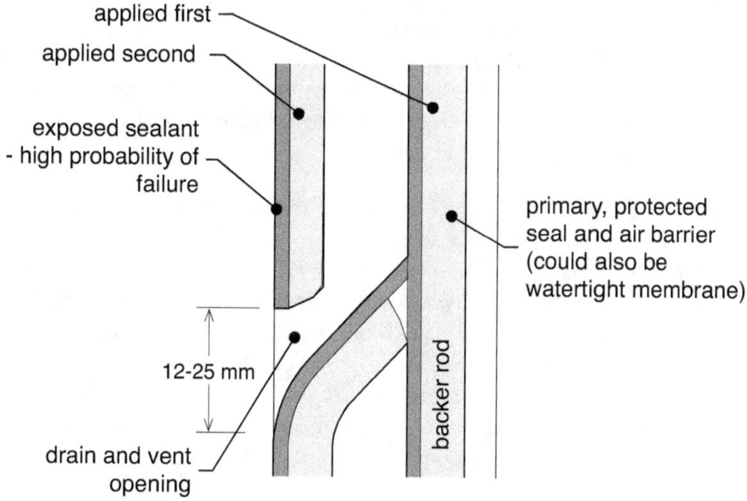

Figure 13.17: Vertical section through a two-stage joint

The control of air flow across the enclosure joint is of course also an important design consideration. It is almost always the case that the interior water control seal also acts as the air control.

Two-stage enclosure joints will, even without deliberate design intent, benefit from pressure equalization. The nature of most enclosure joints: airtight, large vent area to volume ratio, limited spatial extent, are ideal for achieving high levels of pressure moderation (see Chapter 12, section 7). In fact, joints are the part of the enclosure likely to benefit the most from pressure moderation.

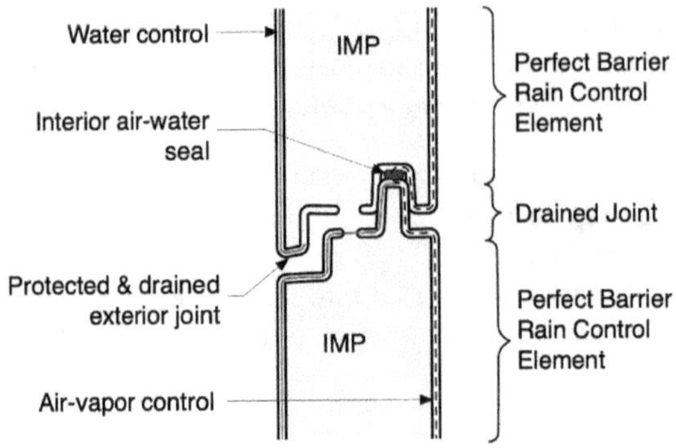

Figure 13.18: Insulated metal panel systems (IMP) use two-stage drained joints

13.5.5 Window and door enclosure joints

All windows and doors require enclosure joints between these prefabricated enclosure elements and the opaque wall. This common joint is one of the most significant sources of performance problems in practise: the separation of responsibility between trades, the lack of design detail often provided, the potential of link water in the fenestration product with the joint and wall, and inaccurate understanding all conspire to result in failure.

The poor control of rain penetration at windows has been well documented over the last 50 years, with surprisingly modern descriptions provided by Herbert [1974], Scott [1984] and Lies & Faith [1991]. Modern research has increased over the last 25 years with dozens of published papers. Such as RDH [2002], Basset 2003, Hardman [2005], Lacasse [2007], and entire symposia subsequently devoted to the topic (ASTM 2004, 2007). The introduction to the ASTM E2112-18 *Standard Practice for Installation of Exterior Windows, Doors and Skylights* states "Rainwater leakage has been the leading reason for dissatisfaction of building owners with performance of fenestration installations".

The wide range of different types of walls, cladding, and types of windows result in a vast range of possible combinations. As a result, the design of the window-wall interface detail is commonly unique to each project. Adding to the design and construction challenge are the changing requirements for better insulation values and controlling (limiting) thermal bridging and the need to consider trade sequencing during the construction process.

Clarity of understanding the control layers is fundamental to the design of enclosure joints, and this is exemplified by window enclosure joints. Figure 13.19 conceptually depicts how a joint around a window can allow for dimensional tolerances, differential movement while also ensuring there is continuity of the air and water control. The approach also ensures pressure moderation of air pressures of the joint around the window (see Section 12.7).

Underlying this design is the assumption that window frames may leak water into the joint, and that airtightness can best be achieved by sealing to the inside face of the window.

Not shown for clarity are the structural connections (which allow vertical and lateral movement), and trims (continuity of the exterior and interior finish) such as window sills (which also act as a water shedding rainscreen). Some designs will also increase the insulation value of the joint by adding water tolerant flexible insulation. This conceptual joint design allows for some thermal continuity, but improvements can be made.

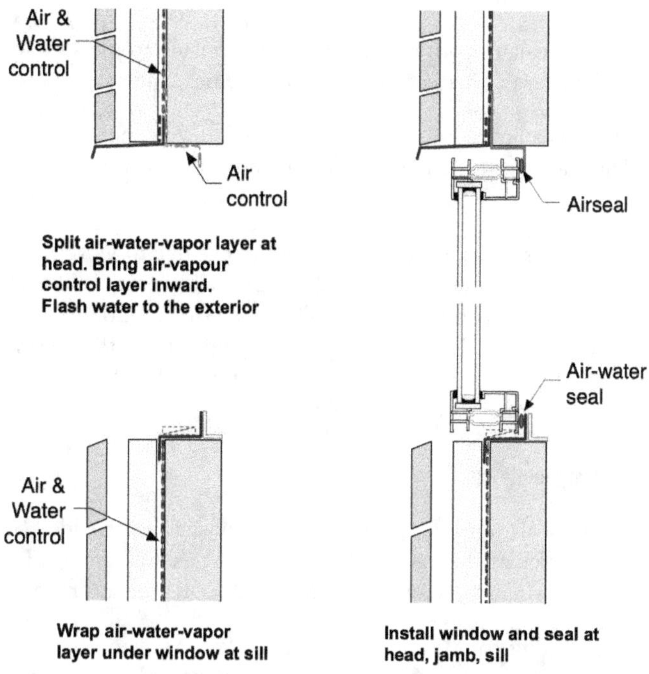

Figure 13.19: Enclosure joint at window penetration

Newer window installation standards, such as the CSA A440.6:20 *High exposure fenestration installation*, are more explicit in their requirements and reviewing this standard is highly recommended.

13.6 Sealants for enclosure joints

The flexible sealant is very common in modern buildings and is used for both rain and air control. These sealants were originally made of such materials as tar-impregnated sisal rope but today are made of sophisticated chemical compounds with remarkable properties of durability, extensibility and adhesion. However, most installed sealants have a limited service life which may be as little as five years but as much as over 30. Failure of the sealant, and thus of the sealed joint, is a major problem as the resulting water penetration can result in hidden damage. Poor joint design can result in premature failure, and poor workmanship resulting in failure is common. It is estimated that each year many miles of sealed enclosure joints fail due to inadequate provision for movement, improper selection of materials, poor workmanship, poor detailing, and poor design, or combinations thereof.

Cladding is often designed for a 50- to 150-year service life. Current polymeric sealants often need to be replaced at 15- to 30-year intervals. Maintenance and repair are therefore a continuing expense. Proper design and the appropriate choice and placement of materials is essential.

Protecting the sealant from the extremes of exposure to the sun, wind, rain, and temperature can greatly extend the replacement interval.

What follows is a limited discussion of, and some guidance on, sealed enclosure joints. This, however, is a bare minimum and is not considered sufficient for the design of joints.

As noted earlier, sealant joints are inherently sensitive to workmanship. However, the use of fluid-applied sealants on building site is likely to continue because these have the best ability to allow for construction tolerances, accommodate rough substrates, and compensate for errors in fabrication and installation. The successful response to the limitations of sealants as enclosure joints has been to use a two-stage joint.

13.6.1 Design of sealant joints

The design of sealant joints is described in publications provided by organizations such as the Sealant, Waterproofing, & Restoration Institute [SWRI 2013], standards (such as ASTM C1193), and sealant manufacturers.

Figure 13.20 shows four common failure modes of sealant joints. The sealant can fail to adhesively bond (termed an adhesive failure) or tear apart (termed cohesive failure). The substrate may also fail if it is weaker than the sealant.

Figure 13.21 shows details of a well-designed, flexible sealant-on-backer rod joint seal. This seal comprises the following:

- A backer rod, often of closed-cell polyethylene foam (other products may not stop sealant from adhering or absorbing water). The backer rod should be sufficiently compressed in the joint to ensure that it will remain in place during application of the sealant. The sealant is to be shaped by, but not bonded to, the backer rod.
- Surfaces or substrates to which the sealant must be firmly attached. The surfaces must be clean and capable of accepting sealant adhesion.

Sealants do not bond well to many surfaces, especially synthetic waterproof coatings because these coatings are designed to avoid dirt collection, water absorption, graffiti, etc.

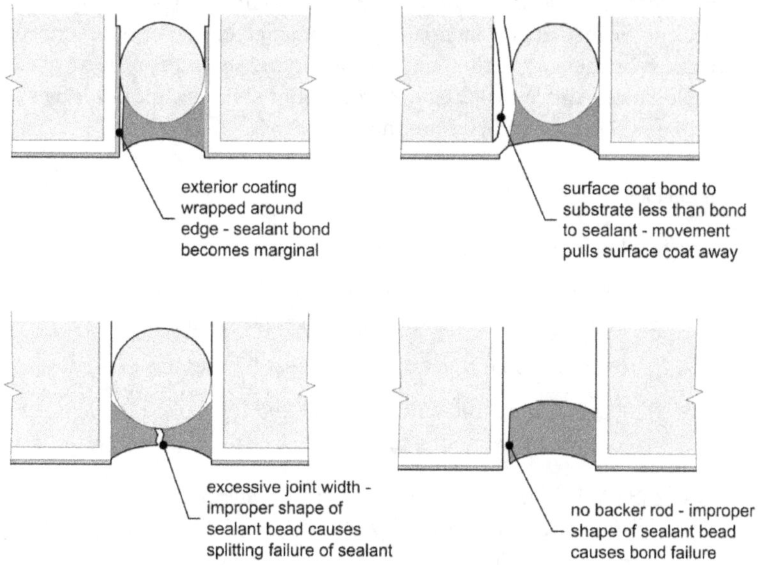

Figure 13.20: Four common types of sealant failure

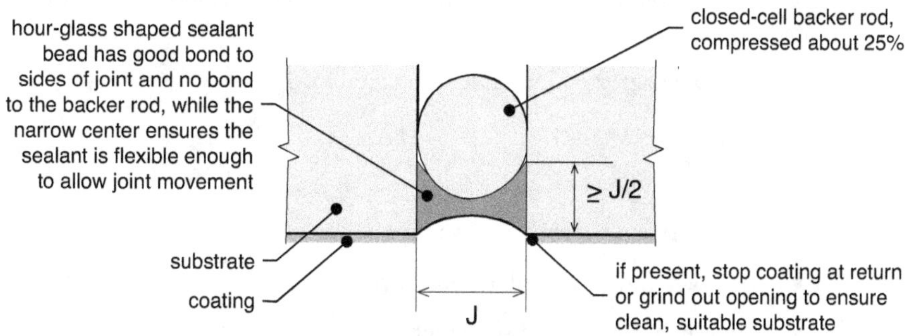

Figure 13.21: Sealant design features for enclosure joints exposed to movement

- The sealant, which should be applied and shaped into an hour-glass section. The hour-glass shape helps ensure that the stress in the sealant is greatest in the middle (and thus this is where most tensile elongation will occur) and least at the bonded ends (because the higher contact area reduces the stress). The sealant should be tooled to ensure compaction and thus ensure bubble- and void-free surface. The sealant material should be sufficiently flexible, strong, adhesive, and durable.

A common recommendation is that the bonded contact length be no less than half the width of the joint to be sealed. In other words, if J is the design joint width, then the length of adhesion should be greater than or equal to J/2.

The selection of sealants and the design of joints is a specialization and a field with significant science, standards, and practice itself. An excellent fundamental specialist reference for fluid-applied polymeric sealant is *Sealants in Construction*, by Klosowski and Wolf [2015]. An older but still relevant book by Panek and Cook [1991] can also be recommended.

13.6.2 Sealant joint size

The size, that is, its width, is dependent on three factors:

1. The size of the movements expected (movement of the superstructure as well as of the enclosure layers being joined).
2. The dimensional tolerances that must be accommodated
3. The capacity of the joint geometry and joint filled (sealant, gasket, etc.) to tolerate movement and cycling
4. ASTM C1472 *Standard Guide for Calculating Movement and Other Effects When Establishing Sealant Joint Width,* [ASTM 2022] is specifically intended to help designers predict the amount of movement that a sealant will be exposed to.

Joints intended to allow unrestrained movement must be designed with all three factors in mind. The contributors to movement include but are not limited to:

- in-service deformation under structural loadings
- temperature change
- moisture content change
- creep (inelastic deformations under sustained load)
- shrinkage (both reversible and irreversible)
- differential foundation settlement
- dynamic loadings such as impact, vibration, etc.

In general, the accommodation of construction and dimensional tolerances has a significant influence on the sizing of most enclosure joints.

13.6.3 Sizing enclosure movement joints

In order to develop a reasonable joint dimension, all the variables discussed to this point must be incorporated. It is customary to evaluate the width of the joint using the following joint size equation:

$$J \geq \frac{A}{X} + B + C \qquad \text{(Eq. 13.4)}$$

A is the total reversible movement (the summation of dimensional movements) in inches or mm.

B is the total irreversible movement (the summation of rheological and inelastic deformations) in inches or mm.

C is an allowance for construction and manufacturer tolerances (in inches or mm). A seismic provision can be incorporated if necessary, and

X is the capacity for movement (in %) of the sealant material, gasket, or geometry.

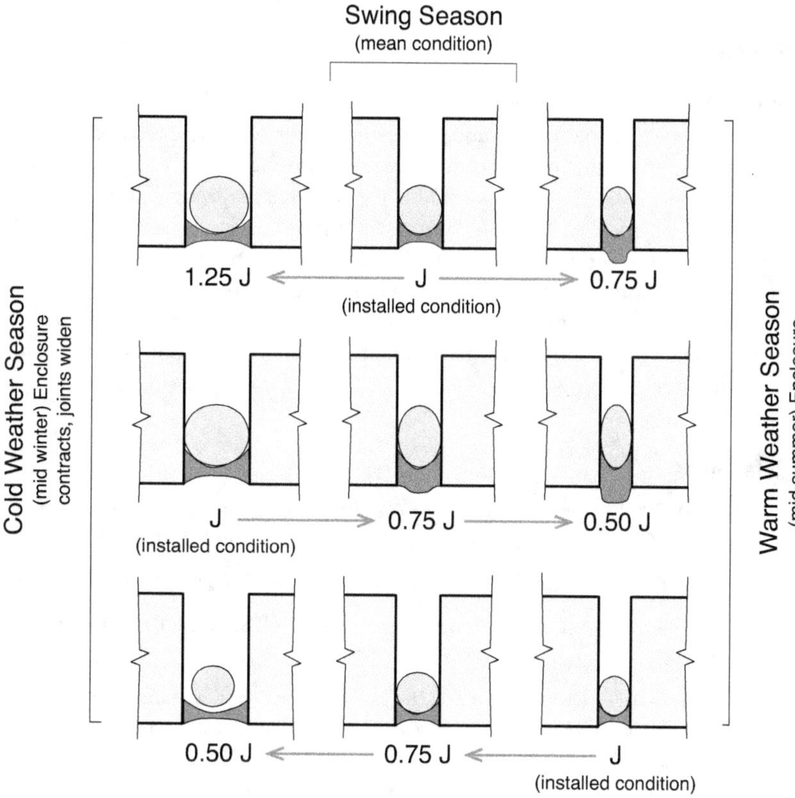

Figure 13.22: Temperature of installation impacts

The capacity for movement of the jointing material is important in the design of enclosure joints. Modern sealants should be chosen that have a stated movement rating. Some are rated for movements of +/-25%, others may be +25/-50%, of +/-50%, etc.

When sizing joints for it is also important to take into account the environmental conditions during installation. The joint size J should be determined considering the temperature conditions that will exist during installation.

Figure 13.22 demonstrates the importance of proper consideration of climatic conditions during installation to the size of enclosure joints. If the designer has considered the movement of the enclosure under mean conditions and has specified a certain dimension for the joint, installation under different climatic conditions will result either in the backer material being forced into too narrow a joint, or into one that is too large. In either case, the resulting sealant joint will probably fail prematurely. Larger joint sizes and large sealant movement capacity (up to 50% are readily available) can overcome these concerns.

13.7 Summary

In the preceding sections an attempt has been made to document some fundamental information on enclosure joints within a single chapter. This chapter is neither sufficiently detailed nor comprehensive enough to be the sole basis for design. As many have said, "*The devil is in the details,*" and the proper design and performance of the joints is as important as the performance of whatever is between the joints. Of all the joints in a built facility, it is usually the enclosure joints that are the most vulnerable and the most likely to develop a performance problem. Hence careful consideration and analysis of both control layer joints and enclosure joints must be an important part of building enclosure design.

13.8 References

ACI 117-90, 2002. *Standard Tolerance for Concrete Construction and Materials and Commentary*, ACI 117R-90 Reapproved 2002, American Concrete Institute

AISC, 2022. *Code of Standard Practice for Steel Buildings and Bridges.* ANSI/AISC 303-22,

ASTM C1193 *Standard Guide for Use of Joint Sealants.* ASTM, West Conshohocken, Pennsylvania.

ASTM C1472-2016 (Reapproved 2022). *Standard Guide for Calculating Movement and Other Effects When Establishing Sealant Joint Width.* ASTM, West Conshohocken, Pennsylvania.

ASTM E2112 *Standard Practice for Installation of Exterior Windows, Doors and Skylights.* ASTM, West Conshohocken, Pennsylvania.

ASTM, 2004. *Proc of Performance and Durability of the Window-Wall Interface Symposium*, Salt Lake City, Utah, April 18. STP 1484.

ASTM, 2007. STP 1509 *Proc. of Up Against the Wall: An Examination of Building Envelope Interface Techniques and Systems Symposium.* Tampa, FL, November.

Ballast, D.K., 2007. *Handbook of Construction Tolerances*, John Wiley & Sons.

Bassett MR, Burgess JC, Camilleri MJ. 2003. "The Weathertightness of Window-to-Wall Joints—Dependency on Installation Details". In *Proceedings of the IRHACE Annual Conference*, Hamilton, NZ.

Bishop D, Webster CJD, Herbert MRM. 1968. *The performance of drained joints.* Paper No. 64c, CIB Report No. 11, Jan.

Brooks, J., 2014. *Concrete and masonry movements.* Butterworth-Heinemann.

CSA A440.6, 2020. *High exposure fenestration installation.* Canadian Standards Association.

Hardman, B.G. and Katsaros, J.D. 2005 "A Rediscovery for Fenestration Installation; Correcting the Mistakes of the Past Century". *Proc. 10DBMC International Conference On Durability of Building Materials and Components,* Lyon, France, 17-20 April.

Herbert, M.R.M. 1974. *Window to wall joints.* Building Research Establishment Current Paper CP86/74. Garston, England.

Herbert, M.R.M. 1974. *Open-jointed rainscreen claddings.* Building Research Establishment Current Paper CP89/74. Garston, England

Isaksen T. 1972. *Driving rain and joints: testing of model joints between elements.* Norwegian Building Research Institute, 30 p.

Isaksen, T. 1965. "Rain Penetration in Joints: Influence of Dimensions and Shape of Joints on Rain Penetration." *Proc. of the RILEM/CIB Symposium on Moisture Problems in Buildings*, Helsinki, August 16th-19th.

Klosowski, J., Wolf, A.T., *Sealants in Construction*, Second Edition, CRC Press, 2015.

Killip, I.R., D.W. Cheetham, 1984. "The prevention of rain penetration through external walls and joints by means of pressure equalization", Building and Environment, Vol. 19, No. 2, pp. 81-91.

Lacasse, M.A., M. Manning, M. Rousseau, S. M. Cornick, S. Plescia, M. Nicholls, S. Nunes, 2007. "Results on Assessing the Effectiveness of Wall-Window Interface Details to Manage Rainwater", *Proc of 11th Canadian Conference on Building Science and Technology* Banff, Alberta.

Lies, K. M. and Faith, B. A., 1998. "Window Detailing Considerations for Leakage Prevention," *Water Leakage Through Building Facades*, ASTM STP 1314, R. J. Kudder and J. L. Erdly, Eds., American Society for Testing and Materials.

M. C. Baker, 1964. *Thermal and Moisture Deformation in Building Enclosures.* Canadian Building Digest, 56, August, 4 pp.

Panek, J. R. and Cook, J.P. 1991. *Construction Sealants and Adhesives*, 3rd Edition, John Wiley, New York.

PCA, 1982. *Building Movements and Joints.* Portland Cement Association, Chicago, IL.

PCI, 2007. *Architectural Precast Concrete*, 3rd ed. Precast/Prestressed Concrete Institute, Chicago, IL.

Platts, R.E. and Sasaki, J.R. 1965. *Rain Leakage on Vertical Through Joints*. Internal Report 323. Ottawa: National Research Council Canada, Division of Building Research, Ottawa, October.

RDH 2002. *Water Penetration Resistance of Windows*. Report for CMHC, Ottawa.

Scott, D.L. 1984, "Rain Leakage in Wood Frame Walls: Two Case Histories", *Building Research Note No. 210*, National Research Council of Canada.

Schutz, R.J.1973. "Architectural Precast Joint Details". Report of PCI Committee on Architectural Precast Joint Details *PCI Journal*, V. 18, No. 2, March-April, pp. 10-37.

ASTM STP 1422. *Performance of Exterior Building Walls*, Proceedings of an ASTM Symposium held in Phoenix, AZ on March 31-April 1, 2001. 282 p.

Svendsen, S., 1967, "The principles of one-stage and two-stage seals". *Weathertight Joints for Walls: Proceedings of the International CIB Symposium*, pp. 25-28.

SWRI, 2013. *Sealants: The Professionals' Guide to Sealant.* Sealant Waterproofing, & Restoration Institute, Kansas City.

Chapter 14

Hygrothermal Modeling

14.1 Introduction

Previous chapters have presented techniques that can be used to analyze the hygrothermal performancc of building enclosures. Most of these techniques have, of necessity, been steady-state, one-dimensional, and highly simplified. More detailed—and often more accurate—computer-based methods of analysis are also available. In practice it is these computer models that will often be used, but it is critical that users of such software be highly knowledgeable of building science and practices. This chapter discusses the nature of computer modeling and its use in the design of building enclosures. A range of computerized tools are also introduced and briefly reviewed.

14.2 Computer-based hygrothermal tools

In the design of modern engineered buildings it is customary to use a variety of mathematical models to simulate the performance of the primary structural system and the service (i.e., HVAC) systems. Structural, mechanical and electrical engineers use different mathematical models to analyze the response of the modeled system or subsystem and then improve, adjust, or revise the system as needed until a final design is arrived at. Analysis, even repeated analysis, and often with more than one model, is necessary to design a new facility, system, or subsystem as well as to assess an existing building or part thereof.

The building industry is moving towards a similar situation with building enclosures. However, we still have some way to go with regard to developing a professional consensus on which models are to be preferred, when they should be used, what analysis procedures are cost and qualitatively effective, and how users will develop the necessary experience to use these models properly. Rapidly changing technologies, e.g., materials and interior building environments, combined with higher expectations of performance for both the enclosure and the building, have created a very real need for the development and use of practical methods of hygrothermal analysis.

Coupled heat and moisture transport through building enclosures can now be calculated for each hour of several years' worth of exposure in a few minutes using currently available, one-dimensional, hygrothermal computer models. This capability can be very useful for the quantitative and qualitative analysis of building enclosure performance. Heat, air, and moisture analysis (HAM) is, of course, only one small part of the enclosure design process, but it is an emerging, important, and poorly executed part deserving more attention.

The general goal of hygrothermal analysis is the evaluation of the temperature and moisture conditions that might prevail across and within a portion of any building enclosure over time. Different individuals or groups will have different needs. Three general analysis needs for can be defined: design, assessment, and study (Figure 14.1). Design professionals such as architects and engineers generate the first two needs. Researchers and students study enclosure performance and hygrothermal behavior.

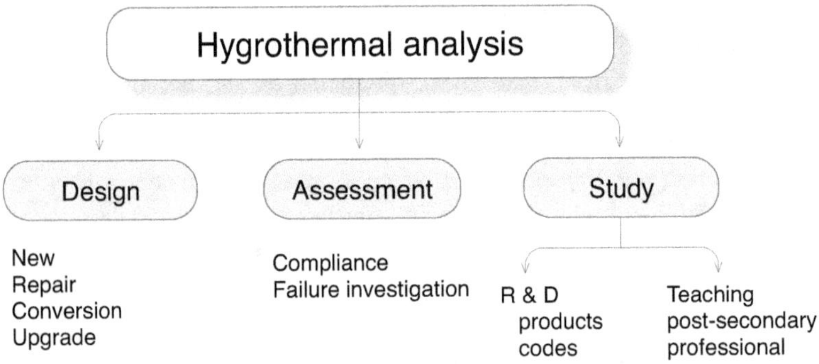

Figure 14.1: Hygrothermal analysis: needs and user groups

The process of designing a new enclosure or demonstrating regulatory compliance involves choices of assemblies and materials, many of which can be guided by hygrothermal analysis. Of course, enclosure design choices rely on much more than analysis.

The need to assess an enclosure—whether for the purpose of a condition assessment, forensic investigation, conversion, or space-conditioning energy calculation—usually involves an existing building.

The process of learning to conduct a HAM analysis, even at a relatively simple level, is a very effective means of deeply engraining knowledge of the basic physics governing hygrothermal performance, material properties, boundary conditions, and material properties. The results provide good training for the relevant value of different variables, that is, the relative important of driving rain, vapor permeance, sorption isotherms for different assemblies in different applications. Hence modeling can be a helpful tool to aid study of enclosure performance provided the learning is guided by knowledgeable teachers with extensive field performance experience.

Hygrothermal analysis for the purposes of research, development, or policy development is also a form of study. Often trends in performance, gaps in material properties, the role of boundary conditions, and other factors not specific to a single building are investigated. More accurate models with more complex physics and properties may be necessary for some research, whereas quite simplified models may be helpful for studying thousands of variations.

14.3 Hygrothermal modeling process

At the highest level of abstraction most hygrothermal analysis is intended to provide the required information of sufficient quality needed for decision-making. Hence, if an analysis is deemed necessary, the next step is to further refine the intent of the activity, or put more simply, what question should be answered by conducting the analysis and how the results of the analysis will be used to make a decision.

The process of a typical hygrothermal modeling activity is outlined schematically in Figure 14.2.

Given a clear question, the next step is to choose which methodology will be used to answer it. This may involve numerous models (with a range of simplifications and limiting assumptions), lab testing, field investigations, etc. For the modeling part of the process, the chosen model will need to be provided with the geometry of the enclosure, material properties and boundary conditions. These will be discussed in more detail later. The model is then run and the results are assessed and compared to the performance thresholds decided upon. If results are not satisfactory, the design of the assembly, the modeling assumptions or even performance thresholds can be changed and the process repeated.

The primary goal of most hygrothermal analysis is often the identification of a performance problem, e.g., excessive condensation; mould growth; differential movement-induced bowing, freeze-thaw damage. It can be appreciated that such analysis can (or should) be part of an enclosure design cycle that begins with the proposed enclosure design (or proposed change to existing building), analysis to identify problems, modifications, and analysis in an iterative manner. In this type of normal design process, executing the calculations is rarely the technically challenging or time consuming activity: selecting the proper boundary conditions, estimating the material properties, defining the definition of failure, and assessing the likely failure modes (performance criteria) of the as-built assembly require experience and knowledge. That "run analysis" is such a small part of Figure 14.2 graphically highlights this reality.

The choice of appropriate analysis technique should be based on the need to be fulfilled. The quality and quantity of information available and the resources available must be consistent with the analysis technique chosen. In practice, the financial resources available often limit options.

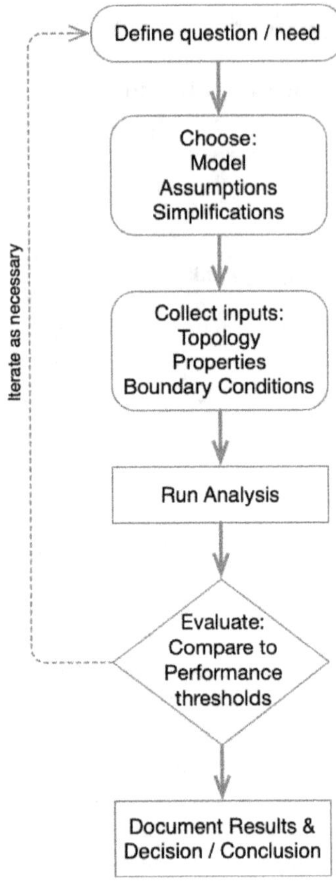

Figure 14.2: General hygrothermal modeling process

For example, consider the case where the designer seeks to avoid a specific enclosure problem. This problem may require only a one-dimensional, steady-state analysis of one extreme set of climatic data and material properties. Although a simple analysis technique may provide neither absolutely correct nor accurate results, so long as a satisfactory decision can be made (i.e., a safe design) on the basis of this information, the technique fills the need. Consider also the situation when conducting a parametric analysis where the accuracy of the difference between results (i.e., relative results) may be much greater (and much more relevant) than the absolute value of any particular result. Indeed, in many buildings no analysis is required because of long and successful experience with that specific assembly in that specific climate.

Depending on the need, an appropriate analysis technique should be chosen. The quality and quantity of information required must be consistent with the analysis technique chosen. For example, consider the case where the need is simply to avoid a specific enclosure problem at a schematic stage of design.

This problem may be simple enough to require only a one-dimensional, steady-state analysis at one extreme set of climatic data and material properties. Although a simple analysis technique may provide neither absolutely correct or accurate results, so long as a satisfactory decision can be made (i.e., a safe design) with this information, the technique fills the need. In general, there is a benefit to choosing the simplest method that can meet the need, as more complex models can often increase the risk of difficult-to-detect modeling errors and certainly require more time and information. Complex and highly accurate analysis methods are also rarely required when conducting a parametric analysis because the accuracy of the difference between results (relative results) is what is important. Many analysis techniques may be much more accurate in their ability to rank relative performance than in their absolute accuracy of any particular result. Again, it should not be noted that in many buildings no quantitative analysis is required because of long and successful experience with that specific assembly in that specific climate.

More detailed, and more accurate, analysis is often required when the potential cost of a problem is high, a new and untried product is to be used, to demonstrate conformance to regulatory bodies, or when it is worth effort to simplify the enclosure design. A detailed analysis, however, requires a much higher level of experience on the part of the analyst, more and more detailed material and boundary condition information, clearer performance criteria, and of course more time. For example, if asked to define the risk of failure mode such as "mold growth" an analyst must first define what level of mold growth is excessive (as some mold exists at the microscopic level on many materials as delivered), identify which layers in the proposed assembly are susceptible to mold growth, what mold growth model will be selected, and what dynamic (it must be dynamic to reflect the dynamic nature of mold growth) hygrothermal model will be used, all before choices about appropriate boundary conditions and material properties are made.

The proliferation of hygrothermal modeling has also resulted in a rise of problems, mostly because of the poor choices and limited knowledge of the analyst. Little guidance and essentially no widespread training (other than how to use the software) is available. Guidelines [Lstiburek et al 2016, Lacasse et al 2018, Holcroft & Wang 2024] and standards, such as ASHRAE 160 *Standard Guide for Characterization and Use of Hygrothermal Models for Moisture Control Design in Building Envelopes* [2021], EN 15026 *Hygrothermal performance of building components and building elements - Assessment of moisture transfer by numerical simulation* [2023], and ASTM 3054 [2016] *Standard Guide for Characterization and Use of Hygrothermal Models for Moisture Control Design in Building Envelopes* have begun to be emerge.

One of the key problems often encountered in practice is a failure to explicitly define a precise question that modeling can help answer. A second common deficiency is the lack of clear and measurable targets. Without a well-defined question a modeling exercise may generate a large volume of results, but these results cannot aid a design decision. Sometimes the chosen modeling approach was inappropriate to answer the question at hand. Even with a precise question and proper model, if the results cannot be compared to a performance target, the results are not useful. For example, predicting the moisture content and temperature of a surface for every hour of a typical year is not useful without understanding what combination of such conditions will result in performance problems.

Regardless of the quality of the model chosen, or the depth of analysis, it is critical that engineers and analysts recognize that the models they build and use are simplifications of the real world. George Box, a British statistician, is generally credited with the quote:

> *"All models are wrong, but some are useful".*

Even so-called "complex" hygrothermal models make a large number of significant simplifications, such as assuming that

- layers are flat,
- adjoining materials are in contact with each other,
- materials have consistent properties throughout a layer and from building to building,
- material properties don't change with age,
- interior building conditions are well controlled,
- exterior conditions are predictable in the future.

Most analysts realize that workmanship variations, changing material properties, and simplified physics and boundary conditions are simplifications that can have a profound impact on the performance of some enclosures in some situations. The role of experience and judgement is to know which simplification matter and when.

Although the physics of moisture storage and transport are reasonably well understood, predicting the moisture and temperature conditions inside building enclosures is seldom a simple task (if reasonably accurate and useful results are desired). The prediction of the hygrothermal performance of the building enclosure requires several categories of information that are usually not known with precision or are even unknowable. The categories of required information to even create a model include geometry, boundary conditions, and material properties.

1. **Geometry** of the enclosure— including all macro building details (e.g., building shape and height), enclosure assembly details, and micro-details (e.g., cracks, holes). As shown in Figure 14.3, the building enclosure is usually discretized into smaller representative elements, such as walls, roofs, windows, etc., for more detailed study.

One-dimensional models cannot of course directly include features such as cracks, holes, and wrinkles. However, effective material properties can be used that modify actual properties in such a manner that behavior of a layer approximates that of a real 3-D material layer. Just as importantly, thc analyst almost never has detailed information about cracks, holes, and wrinkles, and so a statistical average impact on material performance can in fact be more useful for the purpose of design (although maybe not for research or study purposes).

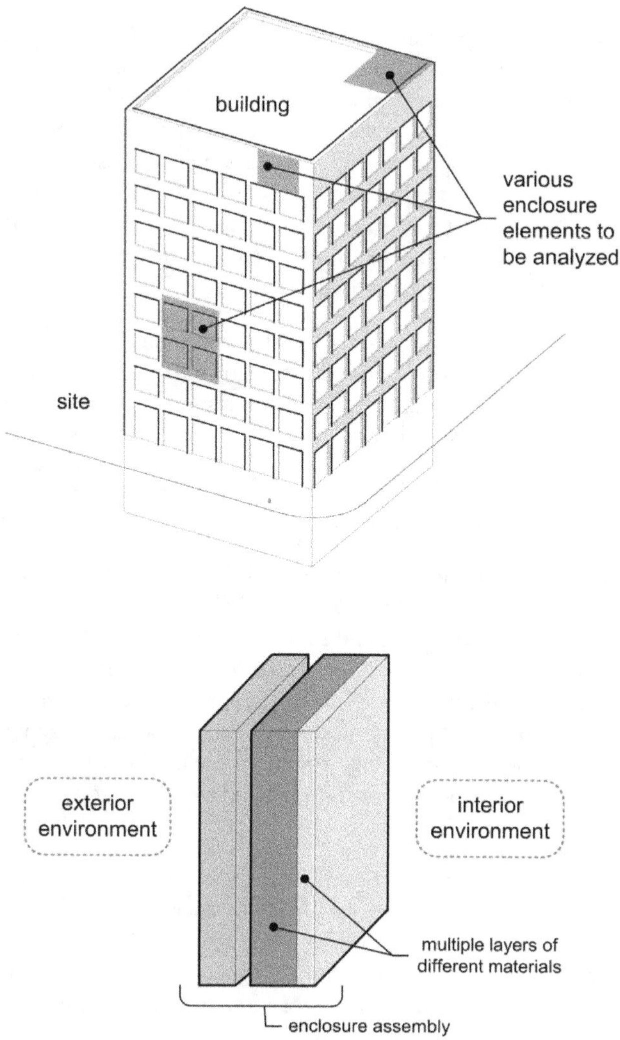

Figure 14.3: The building enclosure as part of the building, site and its boundary conditions

2. **Boundary conditions** around the enclosure element, including:

    i) interior environment, including the interaction of the enclosure with the interior environment;

    ii) exterior environment, including the interaction of the building with the exterior environment; and

    iii) if the enclosure is broken into discrete elements, the boundary conditions along the edges of the elements.

3. **Material properties** and their variation with temperature and moisture content. Of course, other factors such as swelling/shrinkage, aging, salt content, production variations, and the chemical interaction with other materials may be important but are not included in most HAM models. Accommodating these impacts is difficult, and usually rendered unnecessary by making different design choices. Like unintended geometry changes caused by construction variations, the lack of actual information about material property variations limits the practical value of having models that could include this.

In most cases today, analysts are choosing computer packages that include some of the above information in libraries of information. It is the analysts' responsibility to ensure that the default values in a purchased model are appropriate for the case at hand.

4. **Physics**, chemistry, thermodynamics, and mathematics of combined heat, air, and moisture transport.

    Pre-packaged models include fixed, almost always simplified, approaches to the physics, chemistry, thermodynamics and mathematics of combined heat, air, and moisture transport. If these scientific underpinnings are not appropriate for the question or need, it does not mean that the model cannot be used, rather it means the analyst must come up with alternate means and simplifications to achieve their ends. Sometimes call modeling "tricks", these methods allow models to be helpful for situations that are beyond the standard implementation of the computer package.

These four categories are sufficient to conduct an analysis. However, analysis cannot, or should not, be done in a vacuum; there must be both context and limitations. In any enclosure problem, one must know the general performance conditions as well as the important performance thresholds. Thus, a fifth category of information is required for effective applied hygrothermal analysis:

5. **Interpretation & evaluation**, based on the material and system performance threshold, i.e., the conditions under which a material or assembly will cease to perform as intended, are used to evaluate the results of an analysis and aid in making decisions.

Five major categories of necessary knowledge and information are listed above. Each of the five involves mathematical representation, requiring assumptions and approximations. Analysts in professional practice are often forced to make gross assumptions because of a lack of information and knowledge. In practical situations, such as a design problem, the constraints of time and money also have an impact on which approximations and assumptions are made.

Most champions of complex models emphasize the accuracy of the modeling of the building physics, the correctness of their material property approach, or the number of dimensions, etc. For example, the EN 15026 [2023] standard requires a model to predict the results of several benchmark problems. The ability of a model to match real performance, however, depends on the collective, usually cumulative, influence of all other aspects as well, and in practice the predictive power of hygrothermal models are limited primarily by the data input by the analyst. Because of the many unknowns, the most reliable models for the most important questions are calibrated with at least one year of field data measured throughout representative enclosure assemblies. Given a calibrated model the analyst can then change materials, exposure, climate, and dimensions and confidently predict performance.

The information required to model hygrothermal performance can be further subdivided by:

a) Dimension—one, two, or three dimensional;

b) Time—steady-state, quasi-static, or dynamic;

c) Quality and availability of information; and

d) Stochastic nature of each data set (e.g., material properties, weather, construction quality).

The degree to which these factors are taken into account is usually considered to be a measure of the sophistication of a model and is certainly a measure of its complexity. For example, a three-dimensional, dynamic model that uses measured material and boundary-condition data and accounts for their variation with time could be considered to be a reasonably comprehensive, and therefore sophisticated, model. Nevertheless, the accuracy of other input data (boundary conditions, material properties, and geometry) and the performance thresholds will limit the accuracy and utility of the results. Furthermore, from a practical point of view, the value of the results should be consistent with the effort, time, computational resources, and cost entailed.

14.4 Information required for hygrothermal analysis

Each of the five categories of information required for a HAM analysis is covered below in more detail. It should be emphasized that the study of each of these topics is a significant undertaking in itself, and only the most important points can be discussed here.

14.4.1 Enclosure geometry

The actual enclosure geometry must be modeled before any hygrothermal analysis can begin. In simple methods the geometry is almost always reduced to a series of one-dimensional layers. Detailed three-dimensional analysis of the enclosure is actually quite uncommon, partly because of the difficulty of accurately representing enclosures in full detail.

One-dimensional models cannot of course directly include features such as cracks, holes, and wrinkles. The shape of the enclosure may also change with time, due, for example, to wind pressures, shrinkage, settlement, etc. However, even in 1-D analysis the impact of gaps and discontinuities may need to be included, usually if experience or testing shows that they create a contact resistance or break for capillary flow, allow airflow, drainage, etc. However, effective material properties can be used that modify actual properties in such a manner that behavior of a layer approximates that of a real 3-D material layer.

An understanding of the realities of building construction and tolerances by the analyst can be very important. For example, although construction documents may show, for example, a panel cladding applied directly to a framing system, the natural variations in dimensions of framing (and the changes with moisture content of wood framing) result in most of the panel not being in direct contact. For most applications this is unimportant, but if capillary transport between the panel and the framing is of interest, even a 0.1 mm air-filled crack is a capillary break. Similarly, even a 0.5 mm gap can provide a functional drainage gap. However, for assessing the risk of condensation and the rate of drying of rain-wetted materials such small gaps are unimportant.

Cracks, and punctures allow airflow, but how much is unknown. Often one assumes a rate of air leakage through an assembly and makes assumptions about what proportion condenses. Field studies and detailed research modeling has shown that rarely does all leaking air condense [Ojanen & Kumaran 1996]. The analyst must make assumptions. Small amounts of air flow between spaces in the enclosure and the outdoors or indoors can result in the vapor resistance of entire layers being bypassed such as is commonly the case for low permeance cladding [Straube & Burnett 1995] or interior vapor barrier sheets [Ten Wolde et al 1998]. Calculation of the equivalent vapor permeance is explained in detail in Straube & Burnett [1995, 2005].

It also important to note that the analyst almost never has detailed information about cracks, holes, and wrinkles, and so a statistical average impact on material performance can actually be more useful for the purpose of design (although not for research or study purposes)

In practise most analyses are conducted on ideal walls. Unfortunately, the reason for most performance problems is unknown or unpredictable imperfections in the enclosure. The ability to model the actual enclosure geometry, including the inevitable imperfections, can often be the most important factor for the accurate prediction of true hygrothermal enclosure field performance. Until this challenge can be overcome, the value of three-dimensional enclosure analysis will remain limited, and idealized one- and two-dimensional analysis will continue to be used.

14.4.2 Boundary conditions

The boundary conditions imposed on a mathematical model are often as critical to its accuracy as the proper modeling of the moisture physics. Chapter 3 of this text provides information and methods of predicting and defining boundary conditions, so this does not need to be repeated here.

The boundary conditions that require the most focus are tend to be driving rain deposition, solar radiation, and airflow (both through the enclosure and ventilation flow).

The time domain is also important. Almost all computer models employ hourly time steps, since most weather data are available in this form. Simple analysis methods employ monthly averages, binned data, or even seasonal averages. The choice of time step is not critical for most models: a 15-minute time step provides no increase in true accuracy (although it may provide an increase in precision) over a one-hour time step that is not overwhelmed by the uncertainties of the input data. Modeling active enclosures or mechanical systems (such as heating and cooling systems) will usually require a minute-by-minute analysis, but even in these cases employing boundary conditions at hourly intervals is often sufficient.

14.4.3 Material properties

The material properties required for hygrothermal analysis depend on the type of problem that needs to be solved and the analysis tool chosen to assist in the solution. Simple models often require only a single value for the vapor permeability and thermal conductivity. Such data are tabulated in various references for many materials, but it is sometimes difficult to find and is often inaccurate and out of date.

Detailed models will require air and vapor permeability, moisture diffusivity, heat capacity, and thermal conductivity values, all as a function of temperature and moisture content (RH). Such complete detailed material property data sets are still rare but becoming more available, with the notable contribution of the data set provided in the newest Chapter 26 of the ASHRAE *Handbook of Fundamentals* [ASHRAE 2021].

While detailed material property data exists for a limited number of materials few studies have been conducted to quantify the variability of ostensibly similar materials, e.g. pine, or exterior gypsum board. It is known that some materials (e.g., wood, concrete) can exhibit very wide variations in properties depending on source, manufacturing technique, etc. Significant variations between testing methods and testing labs are also common. Round robin studies have recently begun [Feng et al 2020] to help understand variations between labs.

Rarely does an analyst have access to the material properties of the specific assembly being analyzed. Although modern HAM models often have a library of materials it is quite common that the library properties provided differ from those in the subject assembly. A library material with a label of "brick, red" can be wildly different than the red brick being proposed and may or may not refer to the masonry assembly of brick and mortar. The analyst must use care and judgement when selecting material properties and this requires an understanding of which material properties are critical for the analysis at hand, which might vary from library values, and how to estimate properties for those materials that are not documented.

14.4.4 Modeling the physics

Chapter 8 provides a deep review of hygrothermal physics in porous media. Other chapters provide information about understanding air leakage, ventilation, and drainage.

Early research-based hygrothermal models improved upon the more limited models of Philip and de Vries [1957] and Luikov [1965] from the soil sciences and chemical engineering. Building science applications are however more dynamic than soil problems; boundary conditions are less accurately known than in chemical process engineering; and, unlike most other disciplines, multi-layer three-dimensional assemblies must be dealt with. Heat flow modeling is well developed, but moisture and airflow modeling is far more difficult. Despite these difficulties, great strides have been made and many models have been developed, ranging from the very simple to the most complex that are practical given the computer resources and knowledge available.

Each detailed model is based on a particular means of modeling the moisture physics. One approach is to choose a driving potential and lump all mechanisms into one total moisture-diffusivity function. Another approach is to separate vapor diffusion from liquid transport. In the latter case, one can model the flow as either a parallel process (vapor diffusion and capillary transport) or a series (i.e., vapor diffusion functions to a certain moisture content, then capillary conduction takes over). In reality the flow is parallel, although the series approach may be sufficiently accurate in some cases.

Almost all models use an average moisture-storage function that does not exhibit hysteresis.

There is a range of possible moisture driving potentials: vapor pressure, relative humidity, capillary suction stress, or moisture content. (Chemical potential is another little used potential). Vapor pressure drives only vapor diffusion, and hence is not typically used alone. The disadvantage of using moisture content as a driving force, while physically valid, is that it is discontinuous at material interfaces and hence its use adds mathematical difficulties to the calculations. Capillary suction is likewise a discontinuous function. Relative humidity does not actually drive liquid or vapor flow directly but is continuous across an assembly. All of the potentials can be related to one another and can be used with the proper transformations (i.e., via Kelvin's equation and the sorption isotherm / moisture storage function).

Vapor diffusion is supposedly a well understood transport mechanism, although more work is necessary on the measurement and understanding of different mechanisms that enhance vapor flow to permit a consensus to be reached. Knudsen diffusion (effusion) is explicitly ignored by all building models but is usually implicitly included in the vapor permeability. Few computer models account for the different temperature dependencies of Fickian and Knudsen diffusion, probably because the differences are small in comparison to the variability of the measured vapor permeability.

Surface diffusion is discussed as a transport mechanism in much of the work behind the development of models. However, few models explicitly deal with the fact that the adsorbed moisture-density gradient is the driving force. Surface diffusion may be implicitly included in models that use measured total moisture diffusivities, but temperature effects must be accounted for, and many models use material properties that include only capillary flow driven by suction. In fact, it is important to understand that moisture flow cannot simply be driven by vapor diffusion or capillary suction but that surface diffusion also acts, and all three mechanisms may be acting at certain times.

Liquid conductivity is included in most of the detailed models described later, although some of the earlier and simple models use constant diffusivity (even though it usually varies by several orders of magnitude with changing moisture content). This is worrying, since Karagiozis and Salonvaara [1995] have shown, through parametric modeling, that the use of the proper liquid diffusivity is very important for accurate predictions in some applications. One model includes different functions for liquid wetting, drying, and redistribution. If water content is used as a driving potential it must be coupled to the suction curve to avoid the erroneous calculation of liquid flow in the super-saturated region (a fictitious liquid diffusivity might also be used).

Gravity-driven liquid flow (i.e., drainage) may be important for the accurate modeling of some types of walls and some conditions (rain penetration and condensation run down). Liquid water not absorbed in the pores of capillary active materials will cling to surfaces until gravity forces overcome surface tension and drainage flow begins. The amount of moisture that clings is a function of the surface on which it is deposited [Kuenzel 2023]. This surface water can be modeled by assuming a surface material layer with certain moisture storage properties [Straube 2007]. Most of the models that consider drainage assume perfect drainage (e.g., the water is removed from the enclosure) after a certain amount of moisture is deposited on a surface. In addition, rainwater penetration is similar to air leakage in that both occur at unknown locations with unknown intensity and rely on unintended but unavoidable defects in the geometry. Nevertheless, the information and modeling techniques cited above have been developed for both design and assessment that allow for equivalent layer properties that can closely represent actual rain penetration and drainage.

Convective vapor transport, i.e., air leakage and ventilation, is accounted for in some of the most comprehensive models. The proper modeling of convective airflow and its moisture transport is important to some types of buildings (especially lightweight, framed enclosures with incorrectly installed or low-density insulation). Unfortunately, convection is even more difficult to model than diffusive and capillary moisture transport. Models exist, but both laboratory and field research continue to show that predicting natural convection in framed wall assemblies often require more information than available in a design environment.

14.4.5 Modeling and analyzing flaws

The challenge of modeling enclosure geometry, when small variations are inevitable, has been noted above. Similarly, the behavior or drainage and liquid water storage requires modeling techniques or approach to enable realistic results.

Small and/or localized flaws in the control of air flow and rainwater penetration are both common place and complex. Most designers assume that no wall is perfectly airtight (and testing in the lab and field repeatedly confirms this) or allows no rainwater penetration at all (drained "rainscreen" systems assumes that some liquid water will penetrate). In real building enclosures these flaws are more stochastic than deterministic: if one knew where the flaw would occur, it would be easy to avoid in construction.

One design method of accommodating the imprecise but likely nature of flaws, either in the control air flow or rain penetration, is to define a "tolerable flaw" in terms of its moisture influence. Modeling packages such as WUFI have the ability to inject water, air, or heat into specific layers [Kuenzel 2013, 2014] which allow the impact of flaws to be investigated. This is a powerful capability but using it places much more emphasis on the knowledge and experience of the analyst. Numerous technical and research papers are available that explore how the modeling of flaws can be used in practice, eg. [Lstiburek et al 2016, Moore & Lacasse 2020, Van Den Bossche & Van Linden 2023]

An example of this approach is the ASHRAE 160 standard, which assumes that a specific percentage of driving rain (1% of that impinging the surface of the cladding) penetrates the cladding and is then deposited on the face of the Water Resistant Barrier. This guidance, like most of the standard, is one reasonable estimate, but in fact the quantity of penetration varies tremendously with cladding types, joints, penetrations, etc. As the Standard recommends, realistic rain penetration testing of actual enclosures should be used to define this quantity, and how penetrating water is distributed through different layers of the enclosure.

Since almost all building enclosures are assumed to have a nearly perfect air barrier, simulating the effects of air leakage due the flaws in the "perfect" design is useful. There is not a lot of information about how estimate the amount of airflow because field test methods like the blower door test provide only an average air permeance result, and it is well known that most of the air leakage is through discrete openings and cracks. Any models that do include air-leakage effects must deal with the fact that the results are only as accurate as the estimate of the flaw (its size, nature, and location) in the air barrier. With these limitations in mind, models such as WUFI that do include air leakage have been shown to be quite useful as research tools and to quantify the tolerance of proposed designs in practice.

When air leakage is being considered, a flow equation coefficient (in the form $C\ \Delta P^n$) is defined over a unit area of enclosure. A reasonable coefficient can then be developed by calibrating the models moisture and temperature results with those measured in the field. Defining the geometry of the air leakage pathway, often three-dimensional and changing rapidly is much more difficult, and often impossible.

14.4.6 Performance thresholds

The temperature and moisture conditions at which performance is impacted are covered in much greater depth in Chapter 8. The threshold moisture-content level that corresponds to most moisture-related damage mechanisms is often equivalent to that material's moisture content when that material is in equilibrium with an environment of approximately 80% RH [Ashton 1970, Baker 1969, Sereda 1975]. At this relative humidity, both fungal growth and corrosion can be sustained, provided temperatures are favorable. This is a first-order estimate, since wood may require higher RH levels for decay fungi to act, and steel may corrode at lower RH levels.

Although it may be reasonable and conservative to use the moisture content of a material at 80% RH as a threshold level for performance problems, the actual performance threshold varies with time, temperature, type of deterioration, etc. Most materials used as cladding, roofing, or on the exterior of the enclosure more generally will always experience large amount of time above 80%RH: the proven field performance of these materials shows that different performance thresholds must be used to more accurately predict if failures can occur.

Significant progress has been made in developing predictions of mold growth (Ojanen 2010) which have been incorporated in ASHRAE Standard 160.

However, consensus has not yet been reached on the thresholds of other common failure mechanisms such as corrosion and freeze-thaw. Much more work is required to define the conditions under which most materials will deteriorate.

14.5 Available analysis tools

Since all models are simplifications of real behavior, it is difficult to define a demarcation point between simple and detailed models based on their modeling parameters alone. It may, instead, be more useful to differentiate between models based on the need they are intended to meet. This chapter assumes that the differentiation is based on the intent of the model: detailed models aim to predict actual performance, while simplified models are intended primarily to provide sufficient information to allow designers and analysts to make decisions.

In many design or assessment situations, the results of an analysis must provide sufficient information to enable a designer to decide whether to accept or reject a particular assembly or material. The relative performance of several assemblies is far more important to a designer with a choice to make than the actual performance of each assembly. A great deal can be learned from "what-if" analysis, especially when tracking the influence of a single variable. In any case the designer often lacks the resources (time, knowledge, material properties, etc.) to conduct a more detailed analysis. Simple models have been developed to fill this need.

In summary, simple models are not necessarily intended to predict performance accurately but are intended to provide predictions of sufficient accuracy for the purposes of decision making. Such models must include information from all five of the basic data sets, but they simplify the data significantly. For example, monthly average conditions can be used to represent boundary conditions; three-dimensional airflow can be simplified to one-dimensional steady state, material properties are assumed to be constant, etc.

It is often useful or necessary to conduct a detailed analysis for research, product development, litigation, and historical renovation work. Detailed models have undergone dramatic development in recent years. They are briefly reviewed below.

14.5.1 Heat flow models

Heat and moisture flow through building enclosures are inextricably coupled. However, knowledge of only the temperature conditions in an enclosure can still be very useful to the analyst. For example, knowledge of the temperature conditions can be used to predict surface condensation, the rate of heat flow, or the amount of differential movement.

Numerous computer models exist for the prediction of heat flow through buildings. These programs can be differentiated by the number of dimensions that can be modeled, whether dynamic analysis is possible, and on how they handle radiation and convection at surfaces and in cavities. The most widely used programs in North America, such as DOE's Therm, are two-dimensional steady-state models that are especially useful for assessing the thermal performance of windows and other lightweight assemblies. The temperature predictions can be used to estimate the occurrence of condensation by comparing surface temperatures to interior dewpoint temperature. This analysis is quite sensitive to surface transfer coefficients, and these must be carefully considered: it is rare that standard window values can be directly applied.

Dynamic and steady-state one-, two-, and three-dimensional heat flow models are commercially available. For example, Swedish programs HEAT2 and HEAT3 allowing for the dynamic analysis of two- and three-dimensional structures and the Belgian firm Physibel offers a range of software for for calculating two- and three-dimensional heat flow analysis either as steady-state or dynamic models, and able to model both typical rectangular shapes and free form shapes. These models all focus on solid materials with simplified models of convection heat transfer and a range of solutions for radiation transfer. Very powerful Computational Fluid Dynamics models from the industrial design and manufacturing sector, such as Nx from Siemens and Solidworks Flow Simulation provide full dynamic simulation of all heat flow modes in gas, liquids, and solids, but require at least an order of magnitude more investment and time to use. Hence, these models are used for specific research problems. All are commercially available.

14.5.2 Highly-simplified HAM models

Although detailed finite-element based models are available and user friendly, there are still important roles for simplified methods. Simplified methods are very useful for education. Detailed models are "black boxes" for those who do not understand the building physics that underpin them.

One of the first, and most widely referenced, simple models is Glaser's method [1958, 1959], originally published in 1958-59 as a graphical method. This model assumes that the building enclosure is one-dimensional and that all moisture transport is driven by vapor diffusion. The ASHRAE *Handbook of Fundamentals* has included a cursory example of this method since the 1981 edition. Typically, Glaser analysis assumes steady-state boundary conditions for periods ranging from a few days to a few months and fixed material properties.

Several European codes accept the use of Glaser's method for supporting an enclosure design. The German moisture standard, DIN 4108 [2024], for example, provides the thermal conductivity and vapor permeance of a range of materials, defines the boundary conditions and period of time to be used for both wetting and drying analysis, and even recommends acceptable performance thresholds (e.g., by giving maximum safe moisture contents for various materials). Most North American publications are far more limited. They describe Glaser's method assuming only one set of boundary conditions (wetting) and consider any condensation at all as failure. Glaser analysis, although a vast simplification, can be useful if the user understands and includes modifications such as assessing the wetting/drying balance and damage thresholds, use of effective material properties, realistic boundary conditions, etc.

While diffusion may be an important moisture-transport mechanism in enclosures made of solid, capillary active materials (such as the plaster-finished masonry walls often used in Europe), exfiltration condensation is more important for both energy consumption and moisture tolerance of the lightweight, framed assemblies widely used in North America. A simple extension of Glaser's diffusion method can be made which adds simple convection in parallel with diffusion (see Chapter 6). Such a model considers air leakage to be a diffusive process, uncoupled from heat flow, with no accounting for latent heat effects. By the further expedient of ignoring hygroscopic adsorption and convective heat flow, several simple computerized and spreadsheet models have been developed.

Chapters 5 and 6 present physically sound but highly simplified models for calculating heat flow and vapor diffusion, as well as methods to estimate air leakage condensation. These models can be applied at one point in time, once per season, monthly, etc. Their primary purpose is pedagogical: applying them will help users to develop an understanding of the primary design variables. Computer models have become available that can be used to conduct a large number of simulations that include more effects, more accurately. Once the simple methods are understood, most users will progress to computerized models for more challenging analysis problems.

Handegord [Handegord 1985, Reginato & Handegord 1990] developed EMPTIED, Envelope Moisture Performance Through Infiltration Exfiltration and Diffusion for Canada Mortgage and Housing Corporation. This early computer-based model incorporated essentially all of the simplification approaches mentioned above. It used monthly bin temperature data (e.g., it does not consider heat storage), assumed parallel vapor diffusion and diffusive air leakage and output plots of the monthly amount of condensation, drainage, and evaporation. EMPTIED was very easy to operate, and provided fast, generally conservative, results. Even today such models can be useful in the schematic phase of enclosure design.

14.5.3 Historical review of detailed hygrothermal computer models

In 1996, Hens published a comprehensive review of heat, air, and moisture (HAM) models as part of the International Energy Agency Annex 24 project, providing a snapshot of the state of the art at the time [Hens, 1996]. This excellent review describes the state of the art at a time when hygrothermal models was growing rapidly but had not yet matured. Most of these models are no longer available

TRATMO (Transient Analysis Code for Thermal and Moisture Physical Behaviours of Constructions), developed by Kohonen [1984] at the Finnish VTT, was one of the first relatively complete, detailed, and useful computerized building-enclosure models. It employed moisture content and temperature as driving potentials. The sorption isotherm, moisture diffusivity, and vapor permeability were considered as functions of temperature and moisture content. A total moisture diffusivity was calculated by summing vapor, liquid, and surface diffusion and including the proper temperature effects for vapor flow.

WALLDRY [Scanada 1986, Schuyler 1989] was a simple Canadian model that attempted to model the drying of wood framed wall assemblies common in Canadian housing. Although highly simplified, it tackled the complexity of 3-D wood framing and ventilated cladding. The model considered moisture transport to be exclusively via vapor diffusion, since capillary transport in wood is a rather slow process. Simplified handling of cladding ventilation drying was included, as this was found to be important for the type of North American walls of interest. In field validation trials, the model was unable to capture finer details of the drying process, although it was able to model some features of the moisture transport process and broadly match trends of wall drying from realistic field test hut studies.

Carsten Rode (formerly Pedersen) [1989, 1992] used both the sorption and suction curves to define the moisture storage function in his one-dimensional model, MATCH. In the hygroscopic regime the sorption isotherm (defined by an equation which allowed hysteresis) was used and moisture transport was by vapor flow only, driven by vapor pressure differences and defined by the vapor permeability of the material. In the capillary regime the suction curve was used together with the hydraulic conductivity to model moisture transport.

The more recent research-only version accounts for diffuse air leakage, enthalpy flow, and latent heat. Some validation was carried out [Korsgaard and Pedersen 1995, Rode et al 1995] laboratory results, although none of the work involved driving-rain deposition or air leakage, both subsequently shown to be of significant interest.

Burch's MOIST model [Burch et al 1989] was similar in many respects to Rode's earliest seminal MATCH model. It was important as it was released freely to the American market by NIST Moisture transport was modeled as vapor flow, driven by vapor pressure gradients and capillary transport driven by capillary pressure gradients. The vapor permeability and hydraulic conductivity were both given as functions of moisture content. The model accounted for the latent heat of phase changes and the increased heat capacity provided by wet materials. Fibrous insulations were assumed to have no moisture storage capacity, which can be a problematic assumption for some assemblies.

Simple laboratory validation tests of MOIST were conducted in the hygroscopic range [Burch et al 1995], with good results. Field comparisons [Ten Wolde and Carll 1995] (without solar exposure, although it could calculate solar effects) have shown reasonable agreement, so long as the moisture content remained in the hygroscopic range and rain deposition or air leakage was not involved.

Ojanen et al built on Kohonen's work to produce TCCD2, (Transient Coupled Convection and Diffusion 2 Dimensional) [Ojanen and Kohonen 1989, Ojanen et al 1989, Ojanen and Simonson 1995], a two-dimensional program developed primarily for the analysis of framed building walls. The model used the same basic physics and mathematical formulation as used in Kohonen's model. A major improvement over Kohonen's model was the use of moisture-content-dependent diffusivity. An important addition was that convection airflow was accounted for as well as condensation (and frost formation) and evaporation. Befitting its focus on framed walls, capillary transport and surface diffusion were simplified by being lumped into the vapor-diffusion process. It has been validated with laboratory experiments and has been shown to provide useful information regarding the impact of convective flows on hygrothermal performance [Kohonen 1995].

14.5.4 Current Detailed Hygrothermal Models: WUFI and Delphin

The Windows-based WUFI [Kuenzel 1997] was developed by Hartwig Kuenzel supported by the comprehensive work of Kiessl, Krus and other workers at the Fraunhofer Institut fuer Bauphysik. This model uses a full moisture retention function, from the sorption isotherm and suction curve. Surface diffusion and liquid transport are driven by RH (and capillary suction via Kelvin's equation) and governed by a combined moisture diffusivity. Vapor diffusion is considered separately. All necessary material properties can be defined arbitrarily as a function of moisture content (or RH) by entering a series of points (from, for example, measurements), or they can be approximated from several important behavioral markers, such as the absorption coefficient, capillary saturation, and dry-cup vapor permeance.

Important features of this model are its ability to incorporate driving-rain deposition as part of its boundary conditions, the use of different moisture diffusivities for wetting and drying/redistribution processes, the ease of use, the stability of the calculations, and the degree of field validation. The close fit between model predictions and many full-scale field validation exercises on a variety of walls and roofs over several years demonstrates the quality and robustness of this model [Straube & Schumacher 2002, Petersen 2010, Stoekl et al 2014].

WUFI has become the most common hygrothermal computer model in North American practice [Karagiozis et al, 2001] because of its combined ease of use, effective graphical communication, and very good agreement with field measurements by many organizations.

WUFI's major limitations in its original form was its inability to directly handle air flow and draining water. For many lightweight framed assemblies used in North America, including the popular drained and ventilated "rainscreen" enclosure designs, this was a serious limitation. Current versions of WUFI use simplifications of airflow and drainage by allowing user defined, dynamically varying additions or removals of energy or moisture from arbitrary layers of an enclosure assembly [Kuenzel 2014]. Techniques for managing drainage were noted earlier.

A two-dimensional version is also available with many of the same features. More user effort is required and significantly more (at least an order of magnitude) time and computing resources are also necessary.

DELPHIN is a HAM simulation program broadly similar to WUFI developed initially by John Grunewald and supported by the Technical University of Dresden [Haupl et al 1997, Nicolai et al 2007]. It allows for coupled heat and moisture transport in one- or two-dimensions, allows user-adjustable material properties, accommodates means of simulating leaks or air water, and includes climate and material data bases. The underlying approach to moisture transport physics is somewhat different than WUFI but this is rarely visible to the user.

The model appeals more to the research community because it allows for more control of advanced features such as laminar airflow, the possibility of modeling salt movement [Nicolai et al 2007], and custom coding of input scripts.

The results provided by WUFI and DELPHIN have been compared to the EN 15026 Standard and laboratory results [Hejazi et al 2019, Deng et al 2023]. The results are generally very similar, but there can be unique aspects of each that result in deviations for some types of materials or some transport modes. In the cited literature the choice of how material property data and boundary conditions are entered in the two software packages can result in differences. Again, the analyst cannot rely on a model to provide reliable results without understanding the specifics of not only the data entered but how the model treats these entries.

14.6 Conclusions

Computer-based performance simulation has become a powerful tool for improving design in many industries, among them the aerospace and automotive industries. Simulation is used in designing electronic components, developing new pharmaceuticals, understanding material science, optimizing supply chains, and predicting the future of the global climate. The building industry, despite its economic significance, has yet to fully embrace simulation as a tool. There are numerous reasons for this situation, but lack of software is not necessarily one of them. For example, software exists for mechanical engineers to undertake hourly simulations of the energy flows in buildings and for lighting engineers to consider the influence of orientation and weather on daylighting. A significant proportion, if not all, buildings could reduce their energy consumption and improve the quality of the indoor environment with this type of modeling.

Although simulations could be used, full-scale mockups of curtainwall enclosures for major buildings are regularly built to develop an appreciation of what they will look like. Such mockups are tested for rain penetration and strength, both in laboratories and on site. Structural systems and components, which may be even more important with regard to life safety and often more expensive than a curtainwall, are not tested in this way; instead, detailed calculations are performed and trusted. These structural calculations are essentially simulations of performance under a range of loadings.

One result of this lack of predictive simulation is that a building or its parts do not work as their designers intended. For instance, many opportunities provided by site and climate are not exploited, service systems are often vastly oversized, and energy consumption is higher and comfort lower than desired.

Simulations have an almost untapped potential to improve the performance of buildings and dramatically improve reliability. In general simulations can allow innovative enclosures to be designed that are economical, perform better, and provide more reliable behavior. For example:

- Simulation of air movement driven by buoyancy and wind permit naturally ventilated buildings to be designed that would otherwise not be possible (due to a lack of reliable ventilation);
- Simulation of heating, cooling, and lighting systems allow designers to optimize for better comfort and energy efficiency,
- Simulation of how buildings will look allow one to predict, for example, a view of a future lobby or entrance in all kinds of weather, at different times of the day, and to different people (e.g., children, wheelchair-bound, etc.);
- Simulation of the acoustics of a space, enable, for example, the noise from a cafeteria to be reduced to a comfortable level.

In short, simulations will enable innovative enclosures to be designed that are more robust, more economical, and perform better. Buildings will reliably and inexpensively achieve a higher level of performance in the future only with the aid of simulations, analysis, and the application of building science.

14.7 References

ASHRAE Standard 160-2021, 2021. *Criteria for Moisture-Control Design Analysis in Buildings*. Atlanta.

Ashton, H. E., CBD 124, 1970. *Biological Attack on Organic Materials*, NRCC, Ottawa.

ASTM 3054-16, 2016. *Standard Guide for Characterization and Use of Hygrothermal Models for Moisture Control Design in Building Envelopes*. ASTM International.

Baker, M. C., 1969. CBD 111: *Decay of Wood*, NRCC, Ottawa, March.

Burch, D.M., Thomas, W.C., Mathena, L.R., Licitra, B.A., and Ward, D.B., 1989. "Transient Heat and Moisture transfer in Multi-Layer, Non-Isothermal Walls - Comparison of Predicted and Measured Results," *Proc. of Thermal Performance of the Exterior Envelopes of Buildings IV*, Clearwater Beach, Fla, Dec 4-7, pp. 513-531.

Burch, D.M., Zarr, R.R., Fanney, A.H., 1995. "Experimental Validation of a Moisture and Heat Transfer Model in the Hygroscopic Regime," *Proc. of Thermal Performance of the Exterior Envelopes of Buildings VI*, Clearwater Beach, Florida, Dec 4-8, pp. 273-281.

Chen, P., and Pei, D.,1989. "A Mathematical Model of Drying Processes," *Int. J. Of Heat Mass Transfer*, Vol. 32, No. 2, pp. 297-310.

Dang, X., Janssen, H. and Roels, S., 2023, "Hygrothermal Modelling of one-dimensional Wall Assemblies: inter-model Validation between WUFI and DELPHIN". *Journal of Physics: Conference Series* Vol. 2654, No. 1, p. 012040.

DIN 4108-*3, 2024. Thermal protection and energy economy in buildings - Part 3: Protection against moisture subject to climate conditions - Requirements, calculation methods and directions for planning and construction.*

DIN 4108-3, 2020. *Thermal protection and energy economy in buildings Part 4: Thermal and moisture protection design values.*

EN 15026-2023, 2023. *Hygrothermal performance of building components and building elements - Assessment of moisture transfer by numerical simulation.*

Fortes, M. And Okos, M, 1980. "Drying Theories: Their Bases and Limitations as Applied to Foods and Grains," *in Advances in Drying*, vol. 1., ed. A. Mujumdar, Hemisphere Publishing, Washington, D.C., pp. 119-154.

Fortes, M., and Okos, M., 1981. "Heat and Mass Transfer in Hygroscopic Capillary Extruded Products," *AIChE Journal*, Vol. 27, No. 2., March, pp. 255-262.

Glaser, H. 1958. "Waermeleitung und Feuchtigkeitsdurchgang durch Kuehlraumisolierungen," *Kaltetecknik*, Vol. 3, pp. 86-91.

Glaser, H., 1959. "Graphisches Verfuhren zur Untersuchung von Diffusionsvorgangen," *Kaltetecknik*, Vol. 11, pp. 345-355.

Handegord, G.O., 1985 "Prediction of Moisture Performance of Walls," *ASHRAE Transactions* Vol 91, Part 2.

Hartley, J.G., 1987. "Coupled Heat and Moisture Transfer in Soils: A Review," in *Advances in Drying*, vol. 4., ed. A. Mujumdar, Hemisphere Publishing, Washington, D.C., pp. 199-248.

Häupl, P., Grunewald, J., Fechner, H. Stopp, H., 1997. "Coupled heat air and moisture transfer in building structures". *International Journal of Heat and Mass Transfer, 40*(7), pp.1633-1642.

Hejazi, B., Rodrigues Marques Sakiyama, N., Frick, J. Garrecht, H., 2019. "Hygrothermal simulations comparative study: assessment of different materials using WUFI and DELPHIN software". In *Building Simulation 2019* (Vol. 16, pp. 4674-4681).

Hens, H., 1996. *Enquiry on HACat Codes. Heat, Air and Moisture Transfer in Highly Insulated Building Envelopes*, Annex 24 Final Report Task 1.

Holcroft, N., Wang J. 2024. *Hygrothermal Modelling Guide for Wood Buildings* SP-548. FP Innovations.

Imakoma, H., Okazaki, M., Toei, R., 1992. "Drying Mechanism of Adsorptive Porous Solids," in *Advances in Drying*, vol. 5., ed. Arun Mujumdar, Hemisphere Publishing, Washington, D.C., pp. 77-107.

Karagiozis, A. and Salonvaara, M., 1995. "Influence of Material Properties on the Hygrothermal Performance of a High-Rise Residential Wall," *ASHRAE Transactions:* Symposia, CH-95-3-5, pp. 647-655.

Karagiozis, A., Kunzel, H.; Holm, A. 2001. WUFI-ORNL/IBP—A North American hygrothermal model. In *Proc of the Performance of Exterior Envelopes of Whole Buildings VIII*, ASHRAE, Clearwater Beach, FL, USA, pp. 2–7 Dec..

Kohonen, R., 1985. "Thermal Effects of Airflows and Moisture on Exterior Wall Structures," *Proc. of Thermal Performance of the Exterior Envelopes of Buildings III*, Clearwater Beach, Florida, Dec 2-5 , pp. 583 - 605.

Kohonen, R., 1984. "Transient Analysis of the Thermal and Moisture Physical Behaviour of Building Constructions," *Building and Environment*, Vol. 19, No. 1, pp. 1 - 11.

Kuenzel, H.M., Kiessl, K., 1997. "Calculation of Heat and Moisture Transfer in Exposed Building Components," *Int. J. Heat Mass Transfer*, Vol. 40, No. 1, pp. 159-167.

Kuenzel, Hartwig, M.,1995. *Simultaneous Heat and Moisture Transfer in Building Components-One- and Two-Dimensional Calculation Using Simple Parameters*, IRB Verlag, Stuttgart, Germany.

Künzel, H., Tanaka, E. Zirkelbach, D., 2023. "Tipping point for condensation water drainage on surfaces and interfaces of insulated wall assemblies-experimental method to define water content limits for hygrothermal simulation models." *Proc of 56th Intl Conf of Arch Sci Assoc.,* Tasmania, Australia, 2023.

Künzel, H.M., 2014. "Accounting for unintended moisture sources in hygrothermal building analysis". *Proceedings of the 10th Nordic Symposium of Building Physics*, pp.947-953.

Künzel, H.M. Zirkelback, D. 2013. "Advances in hygrothermal building component simulation: modelling moisture sources likely to occur due to rainwater leakage". *J Building Performance Simulation*, V 6, p. 346.

Lacasse, M.A., Ge, H., Hegel, M., Jutras, R., Laouadi, A., Sturgeon, G. and Wells, J., 2018. *Guideline on design for durability of building envelopes.* National Research Council (NRC) of Canada, Ottawa, ON, Canada

Lstiburek, J., Ueno, K., & Musunuru, S., 2016. *Strategy Guideline. Modeling Enclosure Design in Above-Grade Walls.* Building Technologies Office, US Dept of Energy, Washington, DC.

Luikov, A.V., 1965. *Heat and Mass Transfer in Capillary Porous Bodies.* Pergamon Press, Oxford.

Moore, T. Lacasse, M., 2020. "Approach to incorporating water entry and water loads to wall assemblies when completing hygrothermal modelling". In *Symposium on Building Science and the Physics of Building Enclosure Performance* (pp. 157-176). ASTM International.

Nicolai, A., Grunewald, J. and Zhang, J.J., 2007. "Recent improvements in HAM simulation tools: Delphin 5/CHAMPS-BES". In *Proc of 12th Symposium of Building Physics.* pp. 866-876. Dresden, Germany: Technische Universität Dresden.

Nicolai, A., Grunewald, J. and Zhang, J.S., 2007. "Salztransport und Phasenumwandlung–Modellierung und numerische Lösung im Simulationsprogramm Delphin 5". *Bauphysik, 29*(3), pp. 231-239.

Ojanen, T. and Simonson, C., 1995. "Convective Moisture Accumulation in Structures with Additional Inside Insulation," *Proc. of Thermal Performance of the Exterior Envelopes of Buildings VI*, Clearwater Beach, Fla., Dec 4-7, pp. 745-752.

Ojanen, T., and Kohonen, R., 1989. "Hygrothermal Influence of Air Convection in Wall Structures," *Proc. of Thermal Performance of the Exterior Envelopes of Buildings IV*, Clearwater Beach, Fla., Dec 4-7, pp. 234-242.

Ojanen, T., Kohonen, R., Kumaran, M.K., 1994. "Modelling heat, air and moisture transport through building materials and components," Chpt. 2 in *Moisture Control in Buildings*, ed. H. Trechsel, ASTM Manual Series MNL 18, Philadelphia.

Pedersen, C.R., 1992. "Prediction of Moisture Transfer in Building Constructions," *Building and Environment*, Vol. 27, No. 3, pp. 387-397.

Phillip, J.R., and deVries, D.A., 1957. "Moisture Movement in Porous Media Under Temperature Gradients," *Trans. of the Amer. Geophys. Union*, Vol. 38, pp. 222-232.

Reginato, L., Handegord, G.,1990. "A Procedure for Estimating the Moisture Performance of Building Envelopes," *Proc. of Fifth Conference on Building Science and Technology*, Toronto, March, pp. 123-132.

Rode, C., Burch, D.M.,1995. "Empirical Validation of a Transient Computer Computer Model for Combined Heat and Moisture Transfer," *Proc. of Thermal Performance of the Exterior Envelopes of Buildings VI*, Clearwater Beach, Fla, Dec 4-8, pp. 283-295.

Salonvaara, M., 2011. *RP-1325 - Environmental Weather Loads for Hygrothermal Analysis and Design of Buildings, ASHRAE*, Atlanta.

Scanada Consultants, 1986. *Computer Model of the Drying of the Exterior Portion of Framed Walls - Updated Version*, Prepared for Morrison Hershfield Ltd.

Schuyler, G.D., Swinton, M., Lankin, J., 1989. "Walldry - A Computer Model that Simulates Moisture Migration Through Wood Frame Walls - Comparison to Field

Data," *Proc.of Thermal Performance of the Exterior Envelopes of Buildings IV*, Clearwater Beach, Fla, Dec 4-7, pp. 492-505.

Sereda, P. J., 1975. CBD 170: *Atmospheric Corrosion of Metals*, NRCC, Ottawa.

Stöckl, B., Zirkelbach, D. and Künzel, H.M., 2014. "Hygrothermal simulation of green roofs–new models and practical application". In *Proceedings of the 10th Nordic Symposium on Building Physics—NSB.*

Straube, J.F. and Burnett, E.F.P., 2000. "Simplified Prediction of Driving Rain Deposition". *Proc of Int Bldg Physics Conf*, Eindhoven, September 18-21, pp. 375-382.

Straube, J.F., and Schumacher C.J., 2006. "Assessing the Durability Impacts of Energy Efficient Enclosure Upgrades using Hygrothermal Modeling", *Journal of the International Association for Science and Technology of Building Maintenance and Monuments Preservation*, Vol 2, 2006.

Straube, J.F., Schumacher, C.J, 2003. "Hygrothermal Enclosure Models: A Comparison with Field Data", *Proc. of 2nd Int. Conf. on Building Physics*, Leuven, Belgium, Sept. 14-18, pp. 319-326.

Straube, J.F., Schumacher, C.J., 2002. "The Role of Hygrothermal Modeling in Practical Building Design: Case Studies", *Proc. Of e-Sim 2002*, Montreal, Quebec, Canada, Sept 11-13, pp. 185-192.

TenWolde, A., Carll, C., 1995. "Moisture Accumulation in Walls: Comparison of Field and Computer-Predicted Data," *Proc. of Thermal Performance of the Exterior Envelopes of Buildings VI*, Clearwater Beach, Fla., Dec 4-8, pp. 297-305.

Van Den Bossche, N., & Van Linden, S. 2023. "Modelling rain water ingress in hygrothermal simulations". *13th Nordic Symposium on Building Physics.* Aalborg University.

Chapter 15

Design and Future Directions

15.1 Introduction

This text provides the intellectual framework to study building enclosure performance along with some of the science and mechanics. It is, however, neither a design guide nor a manual for the design of building enclosures. Understanding and predicting enclosure performance, and designing building enclosures fit for their purpose, requires a synthesis of science, mechanics, and technology. To properly apply the information set out in this text, an enclosure specialist needs extensive practical experience in construction technology and processes (to understand what can be built and how) as well as knowledge of the building materials, systems, and techniques available. No textbook can provide all this essential knowledge; it is best gained by practice in the field.

The preceding chapters present a coherent and rigorously defined view of what a building enclosure is, what its functions are, and how it fits into the context of a built facility. The loads that enclosures are expected to resist and accommodate and the performance thresholds of various materials have been covered. The remainder of the text is devoted to developing the science to understand and predict why and how enclosures behave as they do with respect to heat, air, and moisture.

This concluding chapter provides some general design recommendations and then identifies some of the trends that are expected to influence the future of building enclosure design and analysis.

15.2 Design approach

Design, including the design of building enclosures, is an iterative process. Ideally, the process starts with the identification of a problem and the constraints, and it proceeds with the choice of a possible solution and the assessment of the suitability of this solution. On the basis of this assessment, the designer accepts the solution for further development or chooses a different solution and largely repeats the process. The more experienced the designer and the better defined and the fewer the constraints on the problem, the fewer the number of iterations required to arrive at an acceptable solution. In some cases, this process is formal and explicit, but in others it may occur in seconds in the mind of the designer.

In many disciplines, the process of assessment has been codified and the analysis is often quantitative. The hygrothermal assessment, quantitative analysis, and design of building enclosures have rarely followed this process. In enclosure design, choices are often made without any design computation or even rigorous assessment. Much depends on the experience of the designer and builder. A formal and comprehensive assessment of the building enclosure is rarely conducted because of a lack of awareness of the potential methods to do so, or a lack of financial and technical resources. However, the knowledge and methods of detailed hygrothermal analysis are increasingly applied — especially for forensic purposes and high-profile or mission-critical buildings.

Unlike, say, structural engineering, there is no well-defined professional discipline known as building enclosure engineering; nor is building science, especially the hygrothermal aspects, even taught in most engineering schools (this situation is slowly changing). The superstructure of a building is designed subject to published codes of practice, supported by materials standards and specified standardized minimum loads, and assured by quality control procedures. The enclosure is not designed in this manner, partly because the supporting professional infrastructure has still to develop; that is, few engineers and fewer architects are educated and trained to design enclosures using quantitative, predictive methods.

In life safety terms the consequences of enclosure performance failure is often (not always) viewed as less significant than structural collapse. Avoiding a failure in enclosure performance is, however, often desirable for reasons of cost, comfort, health, and convenience, and the level of design rigor should be appropriate. Since comparisons of performance (relative rather than absolute results) are often considered in enclosure design, the accuracy of an analysis is not as critical as is sometimes thought. Much like mechanical building systems, performance failures of enclosures are rarely dramatic, abrupt, or fatal (but they are often expensive and may have serious health and safety impacts). Hence, relative comparison may suffice and there may be less need for accurate and precise performance assessment. Even if precise assessment were desirable, the required load data, performance thresholds, materials properties, and analysis techniques are not yet widely available. Since the first edition of this textbook there have been notable gains in all of these areas, but massive development is still possible.

In practice, enclosure design is often reduced to the selection of an assembly from among a few limited options (i.e., precast sandwich panel vs insulated CMU cavity wall). This choice would, it is to be hoped, be based on some experience. A rational and more reliable approach is to make a choice and subsequently conduct an analysis to verify compliance or to make comparisons to known successful designs. Quantitative analysis by itself is, of course, not sufficient and can never replace professional judgment, experience, and testing. For design purposes, the role of analysis is that of a tool in support of decision-making. Analysis can often be used to distinguish between several competing alternatives, not to quantify absolute performance. In some cases, the alternatives being compared are different enclosure assemblies, (e.g., a wall system with or without a vapor barrier, or a cathedral ceiling with or without ventilation). In other cases, a known, successful enclosure design is analyzed to assess the influence of different boundary conditions (e.g., the effect of moving a wall system used in Denver to Montreal, or converting a roof system used for office space to a roof for a swimming pool).

Analysis techniques available range from simple calculations of R-value to more detailed hourly assessments of wetting and drying using temperature- and moisture-dependent material properties. This type of analysis have been earlier in this text. Computer-based tools have been developed and widely deployed to assist designers; they are described in Chapter 14.

Enclosure design cannot be properly conducted in isolation from the design of other aspects of the building. Although the relationship of the enclosure to the mechanical conditioning systems should be clear, the structural system, the finishes chosen, and even life safety systems are all inter-related even if not fully integrated. Hence, a whole building or an overall systems approach should be used in the building design process, with enclosure design being one aspect of the whole. The building enclosure designer, whether an architect, an engineer, or a specialist, should therefore work closely with the architect and the structural and mechanical engineers. A truly well-designed building is the result of a comprehensive design process that incorporates the input and meets the needs of building code officials, bankers, insurance providers, owners, occupants, and many others, including the general public.

15.3 Summary principles of building enclosure design

The starting point in the design of a building enclosure element should be a consideration of what environments are to be separated (exterior environment: for example, the climatic region—cold-dry, hot-wet, arid, etc.; and interior environment: e.g., museum, apartment, factory etc.) and what micro-climatic conditions are expected (due to site, orientation, height, etc.). Given these loadings, and knowing the client's expectations and resources, the designer can specify the required level of enclosure performance. The designer must then balance the design, including all the attributes identified earlier, with the available resources, including money, labor, materials, technology, and time.

Regardless of the performance requirements, or of the available resources, the designer should follow certain principles. The list of recommendations that follows is roughly in order of relative importance based on the authors' experience gained from years of design and forensic consulting.

1. **Establish a complete load transfer path,** including all connections, to the ground for all structural loads from the enclosure elements down to the ground. This includes the load paths (i.e., connections) of roofs to foundations, floors to walls to foundation, and from windows or canopy to walls. A very common practical problem occurs as a result of the poor connection of secondary components to the superstructure (e.g., windows to wall). Another far-too-common structural performance problem is that of too much flexibility (resulting in deformation and vibration) caused by excessive slenderness or flexibility.

2. **Control rain penetration** by proper siting, building shape, surface features, and enclosure design strategy. Deflection, drainage, and drying are the proven approaches to rain control. Orient and site the building to deflect rain if possible, and use surface features, drip edges, and flashing to redirect water off the building and away from critical joints.

   Where possible, use a drained-screened system to control rain penetration. The ability of the design to drain water in the likely as-built condition should be a high priority. Remember that the enclosure system includes joints as well as elements.

   - *Drawings should show and label all flashing, including the laps and seals, identify the water control layer (i.e., its function and the specific materials), and make the rain control strategy clear. Slopes and the size and drip edges should be both identified on the drawings and dimensioned.*

   If a mass/storage system is chosen, ensure that the amount of mass (storage and drying capacity) and its rain absorptance is matched to its exposure and the climate zone. Deflect as much rain from the building as possible by means of good details, and beware of joints and openings.

   If a perfect barrier system must be used, pay extra attention to surface drainage, joints, and buildable details. Protect critical seals from direct outdoor exposure. Quality control is essential on site, and maintenance may be important.

3. **Control air flow** by using an effective air barrier system. A strong, stiff, durable and, above all, continuous air barrier can be placed anywhere within the assembly. Consider inspection, maintenance, repair, and the possibility of replacement. The buildability of the air barrier is critical; all joints and interfaces in three dimensions must be tight and must remain so.

   - *The air barrier system should be labeled (i.e., its function and the specific materials) on drawings, complete with details, method of installation, and the materials required to ensure continuity at joints, penetrations, and transitions.*

4. **Insulate the structural components of the assembly**. Place as much thermal mass, superstructure, and support structure inside the thermal insulation as is practical without violating other principles. Always consider the size and effect of thermal bridges. The durability of most materials (e.g., water and air control layer membranes) will benefit from being placed inside the insulation. Thermal mass is useful in all permanently occupied buildings.

    - *A continuous layer of insulation, including thermal breaks at windows, should be provided around, and preferably external to, the structure from the foundation walls to the center of the roof.*

5. **Design the enclosure to accommodate movements and construction tolerances**. Movement joints should be designed and provided wherever movements can be expected. Use dry (not caulked) joints when possible but wet seals when needed to ensure continuity. Provide a design that allows for realistic construction tolerances and installation methods.

    - *All joints must be indicated and dimensioned on drawings. Construction joints must be anticipated, realistic construction tolerances provided, and requirements for sealant performance specified.*

6. **Control diffusion** by providing appropriate layers of vapor resistance throughout the assembly. A specific vapor-barrier plane is frequently not required. Vapor barriers rarely solve problems on their own, and can cause problems. Reduce the risk of condensation within walls by making it difficult for vapor to enter the assembly and relatively easy for it to escape. A layer with the appropriate vapor resistance should be located close to the annual average warm side of the wall. Beware of inward vapor drives from rain-wetted claddings in air-conditioned buildings during warm weather: some vapor diffusion resistance inside such claddings is beneficial.

7. **Provide a moisture-tolerant design** by balancing the potential for wetting, drying, and storage. Despite the other measures in this list, always assume that some wetting will occur. Assess the susceptibility of each material to moisture damage in light of its vulnerability (i.e., its potential for getting wet). Provide as much drying capacity as is practical. Consider the benefits of providing moisture storage.

8. **Consider the enclosure as part of the building system**. Neither the enclosure, the superstructure, nor the HVAC systems should be considered in isolation. Resources expended on the enclosure will often result in savings elsewhere, e.g., the resources expended on HVAC systems. Control interior conditions to levels that the chosen enclosure can tolerate (e.g., ventilate to control RH in cold weather).

- *The specifications should indicate what temperature and humidity conditions are to be maintained for each season for the successful operation of the building enclosure.*

9. **Control unwanted solar gains** by careful orientation and site plantings, and by providing exterior shade and solar control glazing to control glare and excessive solar heating. Interior shading does not control solar heat gain well. Glass facing the sunset causes the worst complaints and the most overheating. Face windows toward the equator—almost twice as much energy falls on a south-facing wall in the northern hemisphere during the winter than during the summer.

- *Specifications should list the overall window U-value, SHGC, VT, and CRI.*

10. **Consider the future**. Try to determine what the life cycle of the building is likely to be including how it will be demolished, retrofitted, or recycled. Thus, the enclosure design should allow for the commissioning, operation, maintenance, repair, improvement, changes in use, additions, and demolition.

15.4 Trends

Several important trends in the building industry will surely influence the need for an improved understanding of building science. Although their effect can only be speculated upon, some of these trends are discussed in this section.

Consumers are more demanding in that they expect warm, dry, healthy, comfortable, durable and energy-efficient buildings that are delivered more quickly and economically. The increased likelihood of legal claims (partly due to collective action by condominium owners and class-action law suits) and rising insurance premiums have increased the sensitivity of designers and builders to the risks associated with poor performance.

Some consumers and local municipalities are also beginning to demand "green" or more sustainable built facilities. This trend is most evident in Europe, but it is becoming more common in North America. Building codes are becoming more stringent, and new building programs (such as Passive House, Living Building Challenge, and Architecture2030) are setting quantitative targets for operational energy and carbon, as well as embodied carbon. States and countries are setting policies that will in the near future require new buildings that consume no more energy in a year than they consume (Net Zero Energy buildings). These trends will promote a demand for more resource-, energy-efficient, and durable buildings and building enclosures, and will require better integrated systems engineering.

To reduce human-induced climate, buildings that reduce carbon emissions (not just by lower energy use, but by deliberate energy strategies) and trap (embody, such as high-rise and mass timber structures and wood fibre insulation) carbon in their materials of construction are becoming more common. These changes will require new understanding to use successfully.

Internationally, the developing world is growing at a much faster rate than the developed world. The earth's population will grow by about three billion people (a 50% increase), in the next 50 years, one billion of whom will be in China and India. The citizens of these countries want more and better buildings (primarily housing), and the needs are tremendous. The cost to the world's economy and to the environment of poor choices in building design could be staggering because of the massive scale of the endeavor. Again, this trend will drive the demand for new types of high-performance building designs supported by rational and quantitative predictive tools.

Building materials, elements and systems continue to develop and change. They are likely to become more sophisticated and will be designed for specific uses. In fact, smart building materials are already available, such as a vapor control layer that changes its permeance with relative humidity. Advanced materials such as electrochromatic films offer the potential to control solar radiation with unprecedented ease and efficiency, while offering significant energy savings and comfort. Nanoporous and aerogel insulations promise very high insulation values in much thinner layers than before—flexible, nanoporous foam insulations with a thermal resistance of R20 per inch are already possible on a commercial (albeit expensive) scale. An increasing proportion of building materials will be manufactured from rapidly renewable plant material with complex moisture and mechanical properties. Designers will require a better understanding of materials if they are to be used to their best advantage.

Mechanical systems are already available with controlled output (such as variable speed fans and modulating boilers) and enhanced sensing (particularly light and human occupancy levels) and control systems that quickly respond to changes in the weather and how buildings are used. Although many modern building automation systems are complex, difficult to design, commission and operate, simple, more robust versions of such systems can dramatically reduce the cost of both installation and operation while improving comfort. HVAC systems that are networked to other systems and to service providers also promise better comfort at lower capital and operating costs.

The importance of information technology as a tool to assist the design and delivery process as well as the measurement of building performance in service is likely to grow. Temperature and humidity are routinely logged in buildings (including in many millions of residential thermostats), and energy consumption and air quality are increasingly being measured. The increase in the quality and quantity of in-service performance data is likely to stoke the desire for buildings with improved and more reliable performance.

The manner in which power is produced, distributed, and consumed could be revolutionized by linking building-integrated power production (provided by, for example, photovoltaics, combined heat and power units, or mini wind turbines) to building operation and power consumption in a dynamic, two-way manner. Using this technology together with network communications, buildings and building enclosures will become active parts of the energy network that communicate with each other and with the community to produce power at the least economic and environmental cost. At the same time, demands on the traditional power production and distribution systems will be rationalized.

As a result of these trends, there will be a need for a better understanding of the technical performance of buildings and their components, so that improved performance, or better defined and more reliable performance, can be delivered more economically. This trends also leads to an increase off on-site testing of the as-installed enclosure: testing the water tightness of windows, structural adhesion of glazing, and airtightness of the entire enclosure is becoming commonplace. Hence, the design of enclosure is being modified to ensure it is inspectable and testable, in the same manner as many traditional mass-produced products.

15.5 Closure

As noted in the previous chapter, the potential for effective performance simulation remains largely untapped. In addition, there is likely to be more real-time monitoring of buildings which would complement the advances in simulation by providing feedback to improve predictions. Simulation will be needed to allow for high-performance, integrated designs that make the best use of materials and design time while resulting in the most reliable enclosure assemblies. However, simulation remains a tool. It must be used in the context of practical experience, a sound theoretical understanding, and awareness of the cultural, historical, and social forces that drive building designs.

It is not an exaggeration to say that built facilities, their buildings, and building enclosures are vital to the well-being of humanity. They can and should perform much better than they now do, and they can and should make more efficient use of the world's resources. Buildings can achieve improved and more reliable performance in the future through a better understanding of building science and building enclosures and improved teamwork among designers, builders, owners, and occupants. The signs of momentum for change are evident, not only from the builder/developer but also from the insurance industry, the legal profession, and the educational institutions. It is hoped this textbook can provide some support for this progress.

Appendix A

Control Layers & Continuity: The "pen test"

An increasingly common and useful method of analyzing building enclosure design and construction in practice is the so-called pen test. Variations of this approach have been used for decades by some but Brennan's work in a US government document [EPA 2013] has done much to popularize and document the concept.

Experience has shown that the most significant and often easiest to solve building enclosure design and construction problems revolve around a lack of control layer continuity, that is, holes, cracks, gaps, and penetrations through the control layers, or their complete absence. An important focus of enclosure detailing must therefore be to ensure this continuity. This not a hygrothermal analysis or science investigation, it is an intensely practical activity that requires a good understanding of building materials and technology, construction processes, and the ability of workers to build what is drawn.

The general process is best learned by doing, but in general one proceeds as follows:

1. The first step is to identify and select the pertinent building enclosure sections or details to investigate (consider Figure A1 as an example). Vertical sections at different parts of the building with different shapes (e.g., at canopies, terraces, overhanging floors), or different cladding should each be investigated; plan sections are equally helpful for certain details and assemblies. A full 3D image of the building can be very helpful in identifying how many sections might be needed, and which details can/should be assessed as their own case. Figure A2 provides some guidance.

2. Next, the building materials (s) that providing each of the control functions must be identified for each of the enclosure assemblies in the section. It is helpful to list the control function(s) provided by the designated material(s) beside the section/detail drawing, in addition to providing the appropriate colour-coded legend for that control function. It is traditional to use blue lines for water, green for air, and a yellow highlighter for the thicker thermal control layer; solid lines are used for primary/continuous layers, and dashed lines may be used for secondary layers (e.g. secondary air barrier, or water shedding surfaces). In

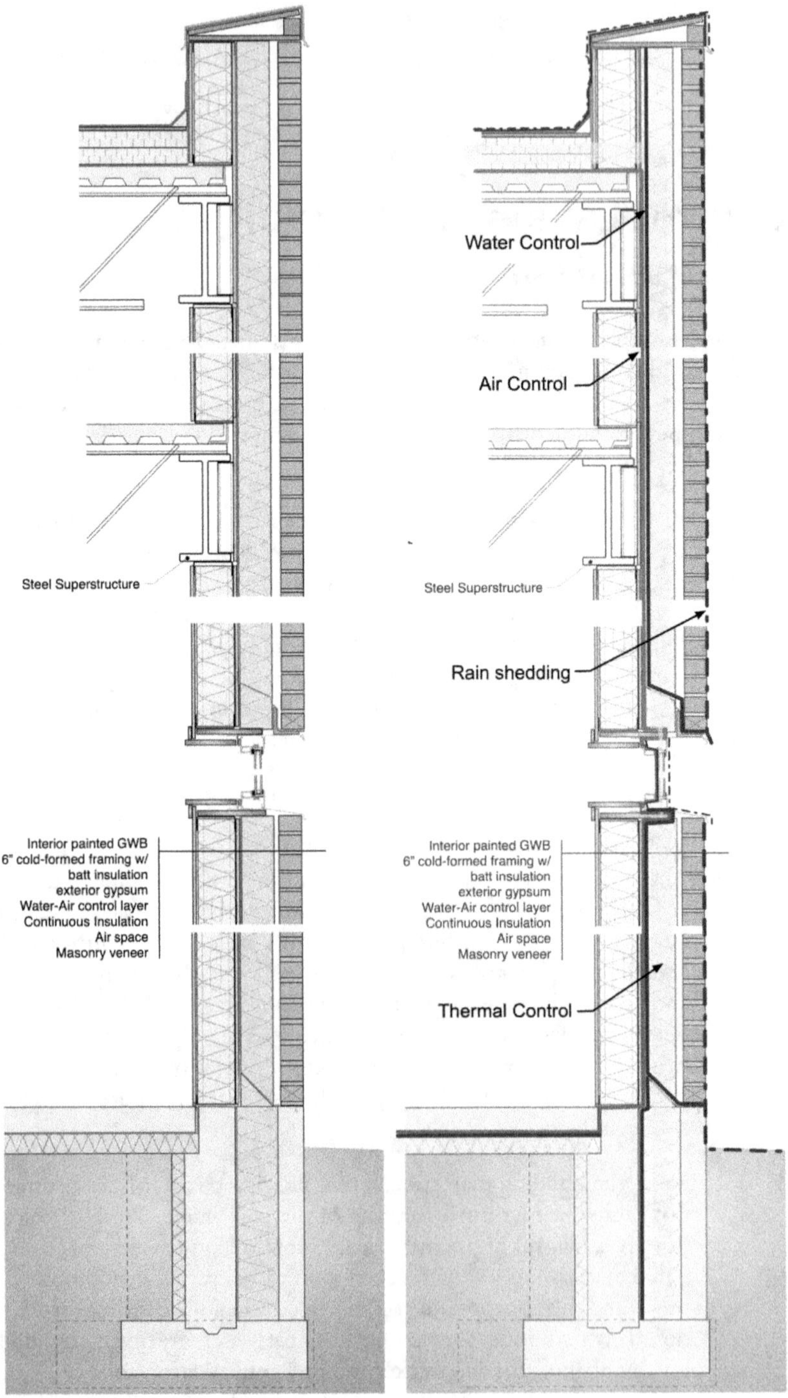

Figure A2: Example Vertical Enclosure Section

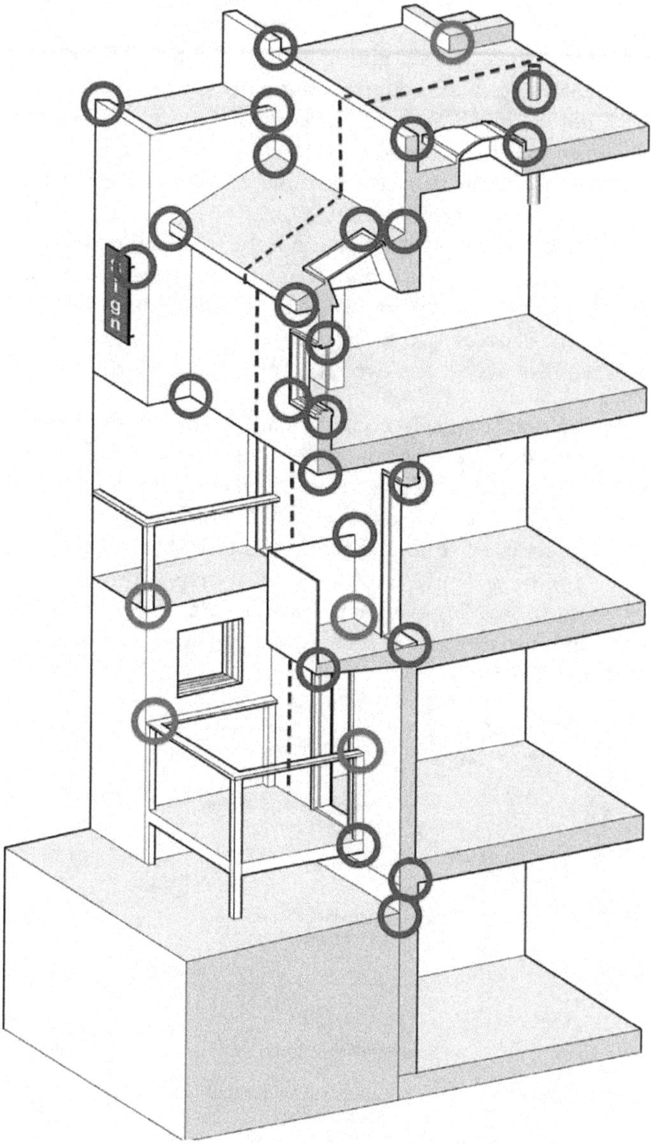

Figure A2: Common details required continuity inspections

some cases, a number of different materials combine to fulfill the function, and in most cases the control layer is formed of different materials at different parts of the enclosure (e.g. the vapor control layer below grade is rarely the same as the vapor control layer in the roof, and the air control layer in the wall is different than the air control in the window).

3. The third step is to draw a notional (or real if working with a paper copy) line, without raising your pen, through the entire section you are analyzing. Vapor control continuity is rarely very important and so may not be drawn. Any gaps in control layers identified must be addressed with a design change or a stated acceptance of the "leak" that will result. It is common, for example, to tolerate leaks in the thermal control layer: a Z-bar or a masonry anchor passing through the insulation would be a heat leak that might be tolerated by some. It should still be identified. It is also important to recognize that not all holes on a drawing are interruptions in control continuity… a vertical gravity lap joint between two water control layers can be continuous rain water control even if the two layers are spaced apart.

4. The final step is to confirm that the line you have drawn is actually buildable at each of the transitions, e.g., one may draw a green line to follow along an air barrier defined as a sheet membrane of polyethylene, but if that layer spans over a 3" (75 mm) wide movement joint without the required structural support to resist peak wind loads, it fails the test. Another example my draw a housewrap as the air control, but if laps are not shown (impossible to build) or are simply a membrane overlaying another, then this is actually a hole (because air pressures will force these apart).

5. 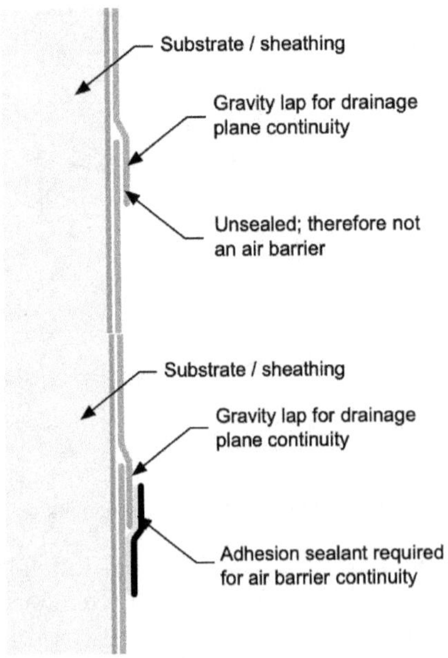

Figure A3: A water-control layer (left) may look a lot like an air and water control layer (right) unless detailed inspections of transitions are undertaken

The pen test seems quite simple but doing it properly requires a lot more skill and knowledge than simply drawing lines. Informed choices need to be made. For example, in the detail at the base of the wall (Figure A4), the thermal control layers have a large discontinuity: the stem wall foundation. The vertical insulation on either side of the stem wall resists heat flow in the horizontal direction, but the concrete in the stem wall has little insulation value (thermal conductivity of about 1.8 W/mK or R0.08/inch). The significance of this break, however, is small, as there is 1.2 m (4') of concrete (U=1.5 W/m^{2}K, R-4) separating the footing and soil from the slab-on-grade edge. The soil at this depth will not be very cold (relative to the air temperature above-grade) nor ever hot. Hence the driving force for heat flow is also small.

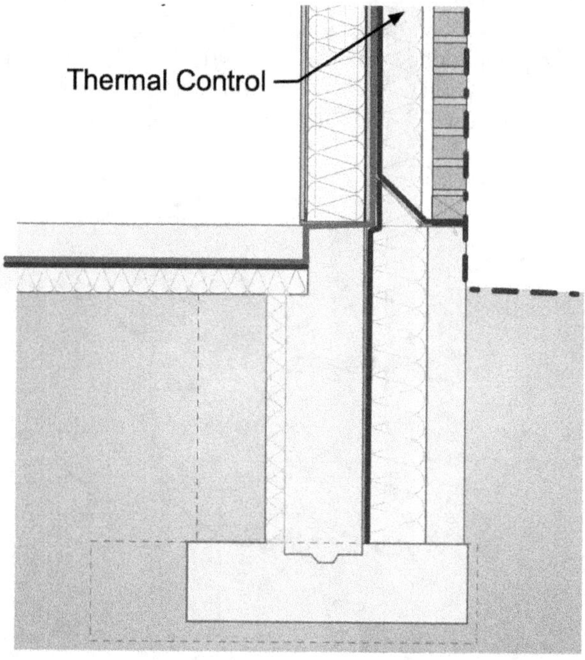

Figure A4: Grade Detail

Note also that the top of the stem wall does not have a water control layer in the example. While an air, vapor, and water barrier is installed below the slab on grade (above the insulation), and a water barrier is installed on the exterior of the sheathing and stem wall, there is no water barrier on top of the stem wall connecting the two water barriers. The purpose of such a barrier would be to stop liquid water wicking upward through the concrete stem wall and diffusing into the building. In some cases, a water barrier (capillary break & vapor barrier) should be installed in this location: wet soil or high water table, poor quality concrete, and moisture sensitive framing (e.g., wood). In this case, well drained soil and site, high quality structural grade concrete, and moisture tolerant steel framing mean the risk is so low as to be ignored.

However, one should not ignore the air barrier continuity at this detail: soil gas, especially radon, can be a concern in most buildings, and hence the slab is sealed to the stem wall with sealant and the air barrier of the wall is sealed to the stem wall to create continuity. The analyst must recognize that concrete can perform as an adequate air barrier for the low pressure differences involved and this can act as an effective transition from the exterior air barrier on the wall sheathing and the slab-membrane assembly that acts as an air barrier in the floor.

Another example of a design judgement can be seen at the parapet (Figure A5). The rain shedding layer is the metal cap, including drip edge. The rain shedding function is taken over by the brick veneer rainscreen below. The water control layer on the roof (the roof membrane) carries up and over the top of the parapet under the metal. It ends at the lower end of the face of the parapet wood framing. The water control layer in the wall is the fully-adhered air-water barrier, which ends at the top of the gypsum wall sheathing.

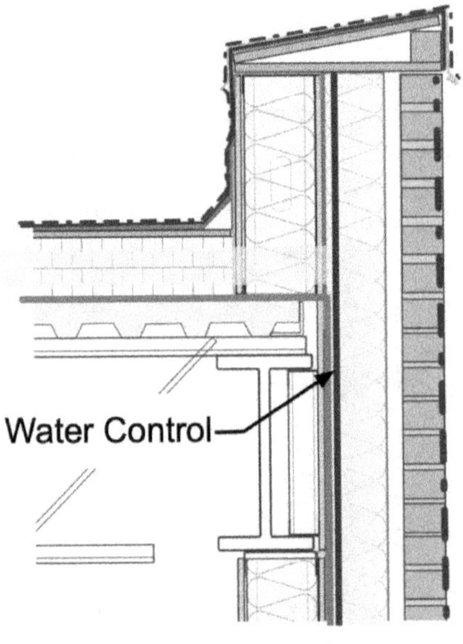

Figure A5: Parapet Detail

The water control is clearly discontinuous along the underside of the wood parapet framing. However, the overlapping between the roof membrane above and the wall air-water membrane below is very generous, and the likelihood of any rainwater penetrating the brick veneer and angling upwards is essentially zero. A designer may choose to fill this discontinuity by specifying a water barrier be applied to the underside of the plywood framing (annoying for trades, but often done) but it is rare this would ever help wall performance.

The deck-level air control layer on the roof is continuous with the wall membrane with the help of a transition membrane that runs under the short steel stud parapet wall. This requires coordination and careful sequencing on the job site. While tempting to carry the air control layer upward along the interior face of the parapet backup from the roof deck, over to the face of the exterior sheathing to join with the wall air barrier membrane, this would risk natural convection loop air flows causing condensation in that void, as well as adding complexity, and cost, due to the extra transitions and joints.

A.1.1 Limitations and Complexities

Only a few of the intricacies and subtleties of a proper enclosure continuity analysis have been described above. The pen test is intended as a simple high-level assessment that precedes a complete design review.

Designers need to make decisions about what assumptions they should make about what the function of each layer is and what risks are tolerable. Some materials that are not designated or expected to act as control layers may nonetheless perform as such in-situ.

Forensic investigators must assess how the enclosure actually behaves in service: what layer(s) is actually providing air flow resistance, rainwater drainage, and vapor control can often be quite different than labels on the construction drawings.

A deeper review must also consider how the plane sections transition between each other in 3-dimensions. Achieving continuity at three-dimensional transitions (e.g., T- and X shape joints) is often challenging. It also illustrates potential vulnerabilities in the control of heat, air, and moisture flows of the assembly at complex interfaces (e.g. jamb/sill or balcony saddle details, etc.).

The pen test cannot identify the ability of a control layer to perform as required, for example, important questions that still will need to be answered include:

- Are the vapor control layers in the proper location to allow drying and avoid condensation accumulation?
- Does the air barrier have sufficient strength and stiffness to withstand applied wind pressures?
- Will the insulation resist the moisture it is likely to be exposed to?
- Can any of the control layers be built given the construction or dimensions shown?
- Are the dimensional tolerances (both installation and material dimensions) between components realistic and achievable?

- How will the superstructure, support, and control layers perform given building movement or settlement?
- Are there any compatibility issues between materials in contact?

A.2 Alternate enclosure systems

The example shown in the pen test example above, and most in the literature, consider the perfect enclosure arrangement. Of course, many enclosure roofs and walls use alternative arrangements, often for very good reasons. Enclosures of the past often used different sets of materials and systems to provide control functions. Hence a series of example enclosure wall examples are shown below with control layers identified. This is in aid of step 2 of the pen test.

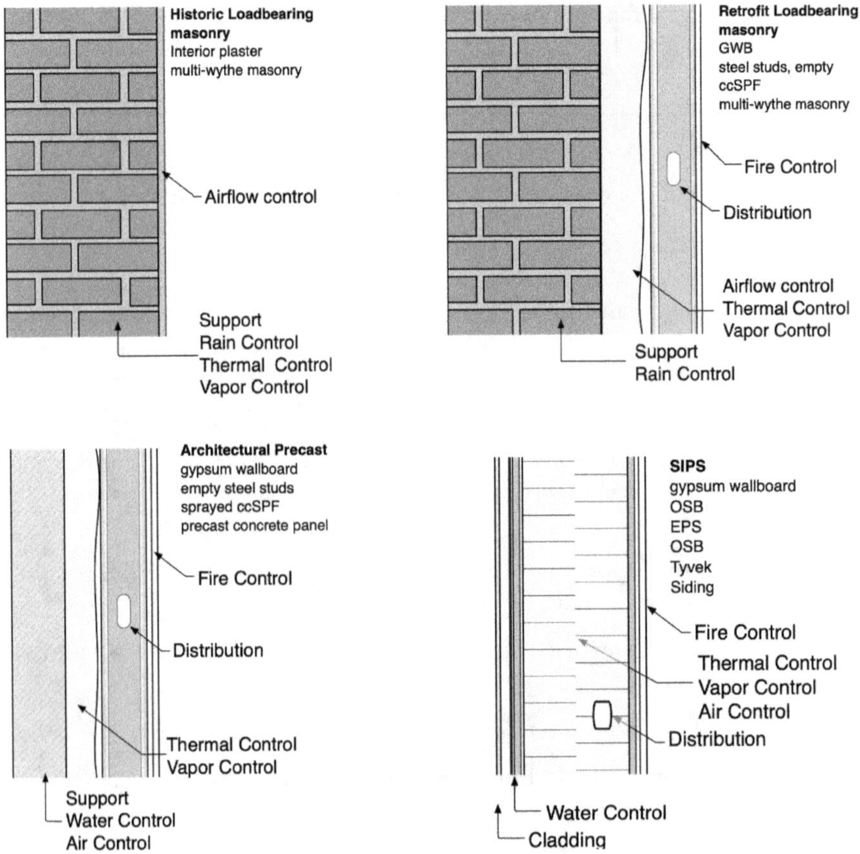

Figure A6: Example control layer definitions in alternative enclosures

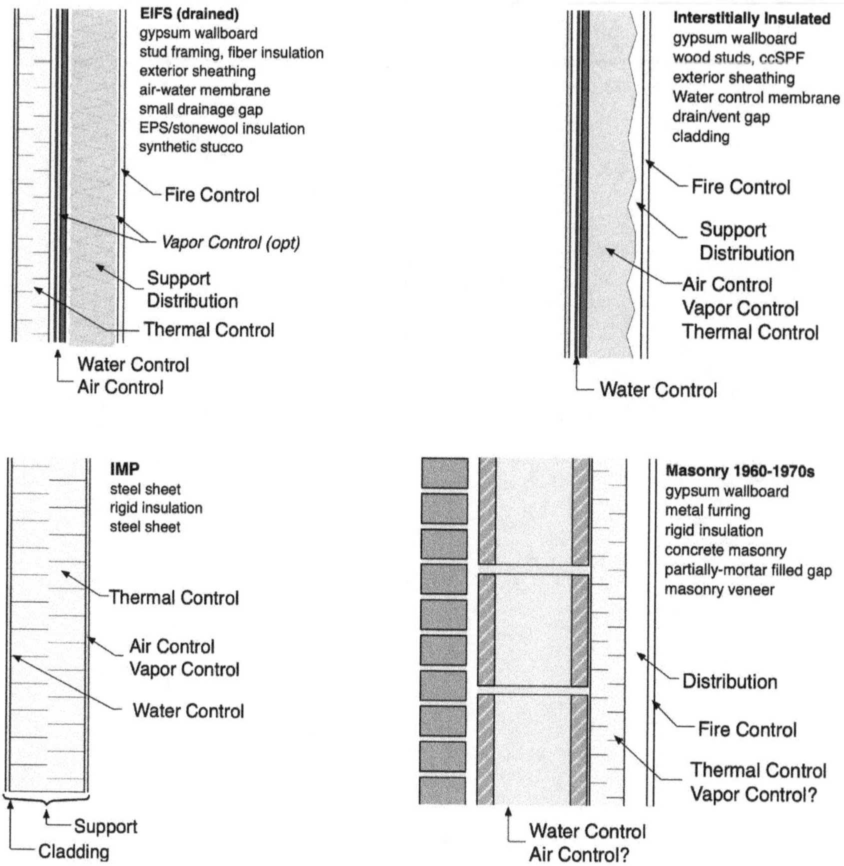

Figure A7: Example control layer definitions in alternative enclosures

A.3 References

EPA. *Moisture Control Guidance for Building Design, Construction and Maintenance.* U.S. Environmental Protection Agency, December 2013.

A.4 Notes

Appendix B:
Climate profiles for representative cities

Climate Summary for Building Design — Atlanta, GA USA

Location	
City	Atlanta
State/Prov	GA
Country	USA
Lat (°)	33.7
Long (°)	-84.4
Elev (m)	315
Time Zone (hrs)	-5

Design Temperatures	
Heating (°C)	-4.9
Cooling (°C)	32.6

Degree Days	
HDD (18°C)	1571
CDD (10°C)	2909

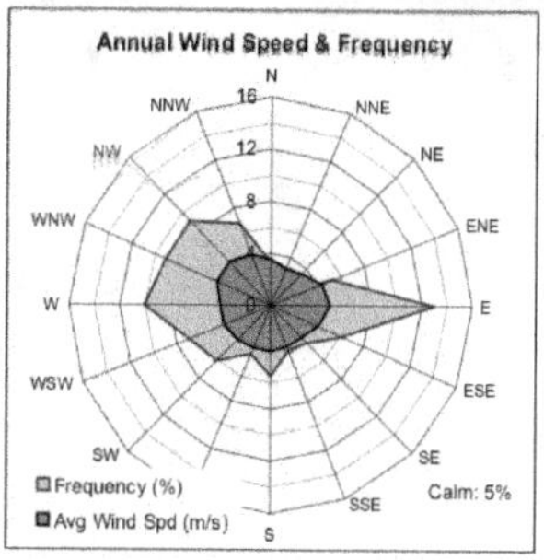

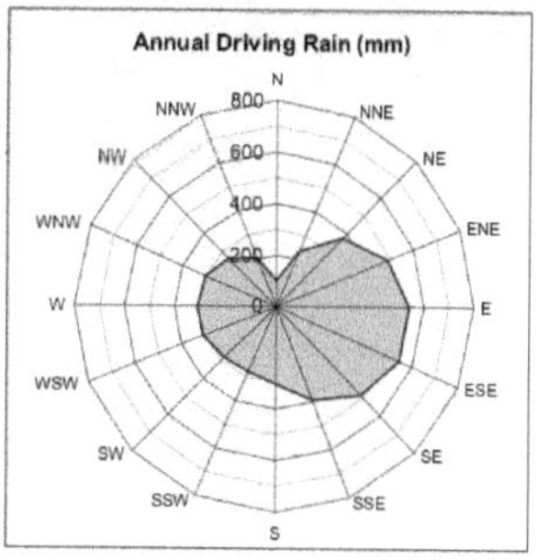

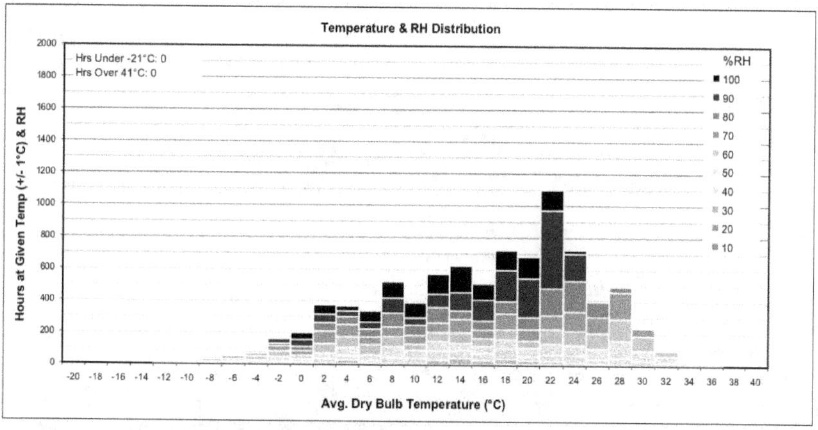

Averages	Jan	Feb	Mar	Apr	May	Jun	Jul	Aug	Sep	Oct	Nov	Dec	Year
Temperature (°C)	4.7	6.6	11.6	16.1	21.0	24.1	25.2	24.5	22.6	16.5	10.5	6.6	15.8
RH (%)	69	61	62	62	64	64	74	75	79	70	69	68	68
Humidity Ratio (g/kg)	3.88	4.13	5.28	7.23	9.85	11.86	14.54	14.28	13.39	8.34	5.73	4.53	8.59
Wind Speed - All Hours (m/s)	3.5	3.9	4.9	4.6	4.3	4.0	3.1	2.8	3.7	4.1	3.5	5.0	3.9
Wind Speed - During Rain (m/s)	4.6	4.3	6.2	5.8	5.0	3.2	3.9	3.6	4.7	4.4	4.2	5.9	4.6

Totals		Jan	Feb	Mar	Apr	May	Jun	Jul	Aug	Sep	Oct	Nov	Dec	Year
Snowfall (cm)		2.5	1.3	1.0	0.0	0.0	0.0	0.0	0.0	0.0	0.0	0.0	0.5	5.3
Rainfall (mm)		125	118	136	92	100	92	130	93	104	79	104	97	1270
Driving Rain (mm/m²)		113	97	151	90	92	56	102	56	84	59	83	110	1092
Hours with Rainfall (hrs)		84	64	58	20	45	16	56	19	31	23	47	55	518
Solar Radiation (kWh/m²)	N	18	19	30	35	48	53	50	43	35	23	19	17	389
	E	43	50	66	75	87	84	93	89	63	60	47	39	795
	S	92	90	90	75	62	55	58	70	77	108	99	92	967
	W	41	48	66	86	83	84	74	73	68	66	45	40	775
	Horiz	80	96	137	178	196	194	196	180	141	129	89	75	1692

Design Temperatures: ASHRAE HOF 2001
Degree Days: NCDC Climate Normals
Rainfall: NCDC Climate Normals, TMY2
T, RH, Wind, Sun: TMY2

2005.06.07
V 1.0

Climate Summary for Building Design — Boston, MA USA

Location	
City	Boston
State/Prov	MA
Country	USA
Lat (°)	42.4
Long (°)	-71.0
Elev (m)	5
Time Zone (hrs)	-5

Design Temperatures	
Heating (°C)	-11.3
Cooling (°C)	30.7

Degree Days	
HDD (18°C)	3128
CDD (10°C)	1656

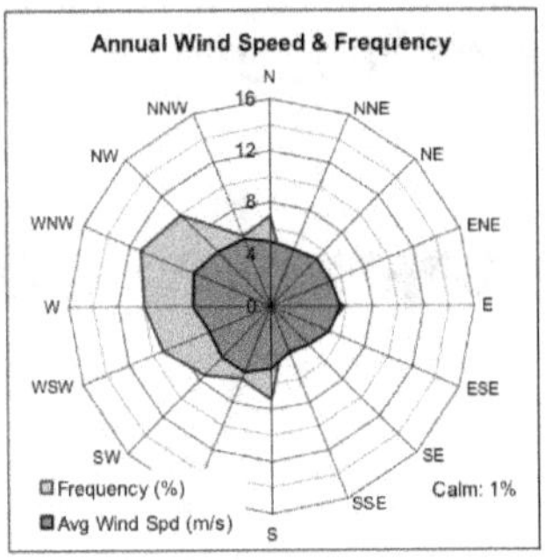

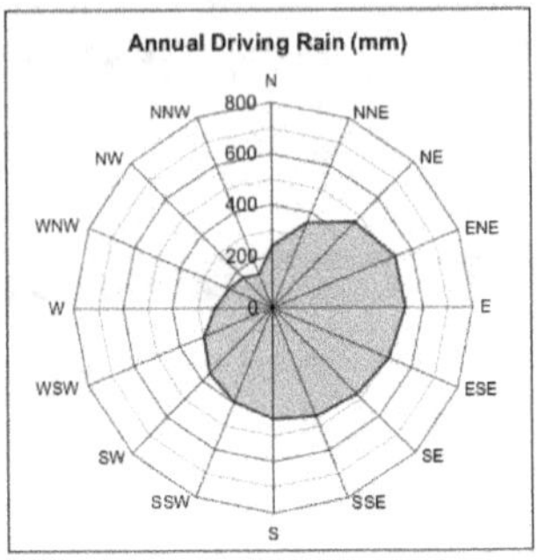

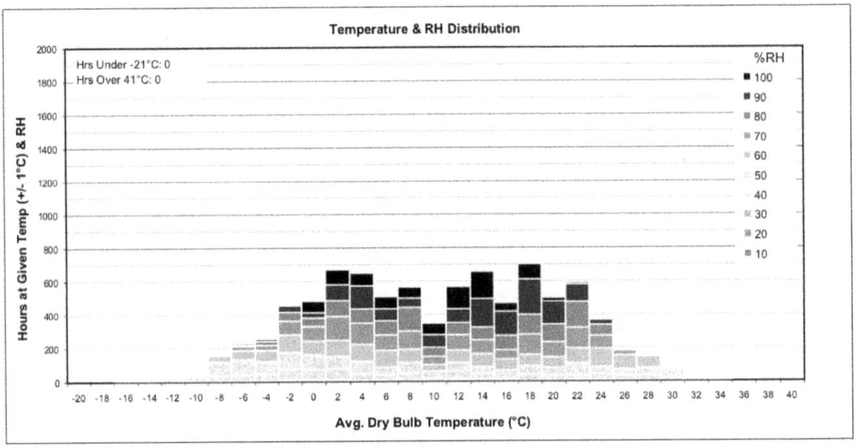

Averages		Jan	Feb	Mar	Apr	May	Jun	Jul	Aug	Sep	Oct	Nov	Dec	Year
Temperature (°C)		-1.5	-0.5	2.5	7.4	14.9	19.3	22.6	21.5	18.1	12.3	5.7	1.2	10.3
RH (%)		56	58	63	70	69	70	68	66	69	69	68	62	66
Humidity Ratio (g/kg)		2.07	2.24	2.86	4.58	7.36	9.65	11.51	10.50	9.00	6.26	4.14	2.93	6.09
Wind Speed - All Hours (m/s)		6.4	5.9	6.1	5.5	5.5	4.2	4.9	4.5	4.6	5.7	6.0	5.7	5.4
Wind Speed - During Rain (m/s)		8.1	5.0	6.3	5.2	7.7	3.4	4.9	4.2	5.0	7.7	8.0	6.2	6.0

Totals		Jan	Feb	Mar	Apr	May	Jun	Jul	Aug	Sep	Oct	Nov	Dec	Year
Snowfall (cm)		32.0	30.2	20.6	2.3	0.0	0.0	0.0	0.0	0.0	0.0	3.3	19.8	108.2
Rainfall (mm)		68	54	77	89	82	82	78	86	88	96	98	75	972
Driving Rain (mm/m²)		96	54	103	98	141	59	78	68	83	137	158	90	1165
Hours with Rainfall (hrs)		22	28	65	115	100	73	50	36	72	44	66	62	733
Solar Radiation (kWh/m²)	N	14	19	24	34	42	49	47	39	28	20	15	13	345
	E	33	41	61	65	79	83	88	79	64	52	30	29	705
	S	83	84	89	74	70	64	69	81	87	99	71	74	945
	W	32	42	56	67	76	81	79	79	60	49	29	27	679
	Horiz	57	77	116	137	173	181	187	168	127	96	56	49	1424

Design Temperatures: ASHRAE HOF 2001
Degree Days: NCDC Climate Normals
Rainfall: NCDC Climate Normals, TMY2
T, RH, Wind, Sun: TMY2

2005.06.07
V 1.0

Climate Summary for Building Design — Houston, TX USA

Location	
City	Houston
State/Prov	TX
Country	USA
Lat (°)	30.0
Long (°)	-95.4
Elev (m)	33
Time Zone (hrs)	-6

Design Temperatures	
Heating (°C)	1
Cooling (°C)	34

Degree Days	
HDD (18°C)	652
CDD (10°C)	4255

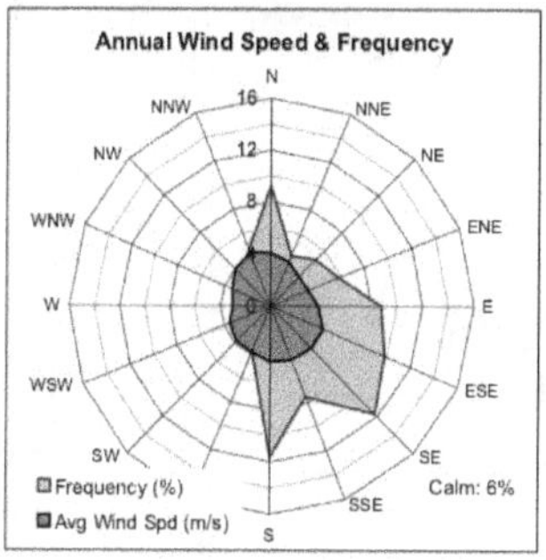

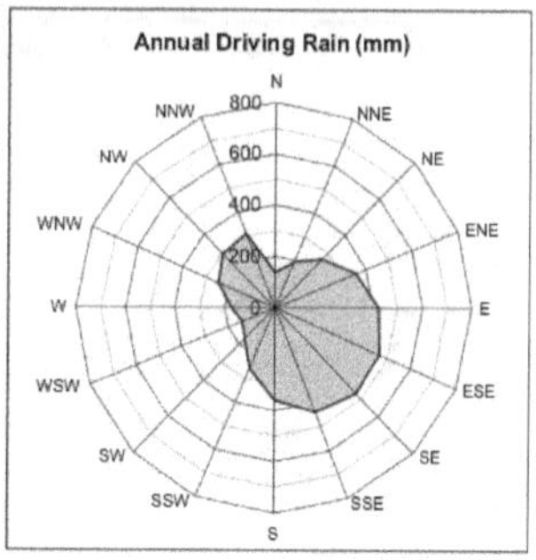

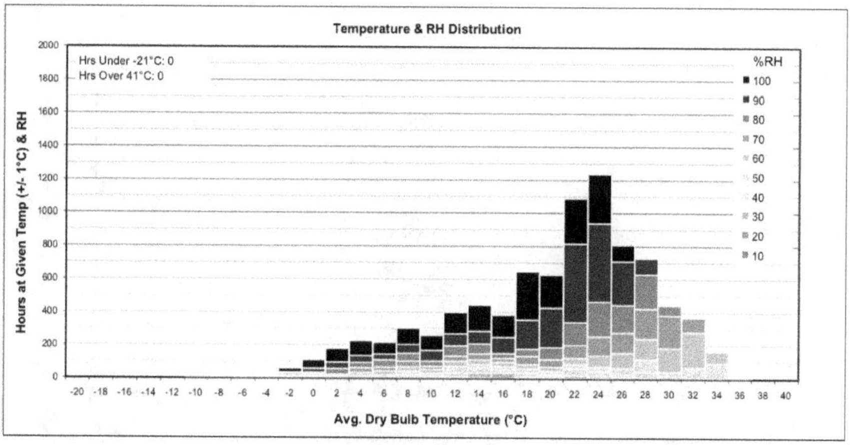

Averages		Jan	Feb	Mar	Apr	May	Jun	Jul	Aug	Sep	Oct	Nov	Dec	Year
Temperature (°C)		11.9	10.9	16.2	20.5	23.9	26.5	28.0	27.3	25.3	20.9	17.1	11.4	20.0
RH (%)		73	79	69	75	74	74	75	75	73	81	76	79	75
Humidity Ratio (g/kg)		7.36	6.76	8.33	11.41	13.83	15.83	17.56	16.86	14.69	12.51	9.64	7.35	11.84
Wind Speed - All Hours (m/s)		4.5	3.7	4.6	3.9	4.1	3.2	3.5	3.2	3.6	3.5	3.7	3.8	3.8
Wind Speed - During Rain (m/s)		5.1	4.8	5.8	4.4	4.3	2.9	4.7	7.4	3.5	5.8	6.1	4.8	5.0

Totals		Jan	Feb	Mar	Apr	May	Jun	Jul	Aug	Sep	Oct	Nov	Dec	Year
Snowfall (cm)		0.5	0.5	0.0	0.0	0.0	0.0	0.0	0.0	0.0	0.0	0.0	0.0	1.0
Rainfall (mm)		93	75	85	91	131	136	81	97	110	114	106	94	1214
Driving Rain (mm/m²)		81	55	91	66	96	51	68	91	63	96	111	77	945
Hours with Rainfall (hrs)		41	51	55	33	50	5	24	32	19	31	31	49	421
Solar Radiation (kWh/m²)	N	20	23	29	38	44	53	51	42	32	25	20	19	396
	E	41	41	58	63	72	76	81	79	65	62	45	38	721
	S	83	72	79	60	50	49	53	60	77	95	91	77	847
	W	43	48	64	66	77	83	77	77	67	62	46	38	747
	Horiz	84	93	131	147	174	184	182	171	146	129	94	75	1611

Design Temperatures: ASHRAE HOF 2001
Degree Days: NCDC Climate Normals
Rainfall: NCDC Climate Normals, TMY2
T, RH, Wind, Sun: TMY2

2005.06.07
V 1.0

Climate Summary for Building Design — Los Angeles, CA USA

Location	
City	Los Angeles
State/Prov	CA
Country	USA
Lat (°)	33.9
Long (°)	-118.4
Elev (m)	32
Time Zone (hrs)	-8

Design Temperatures	
Heating (°C)	7.4
Cooling (°C)	27

Degree Days	
HDD (18°C)	708
CDD (10°C)	2709

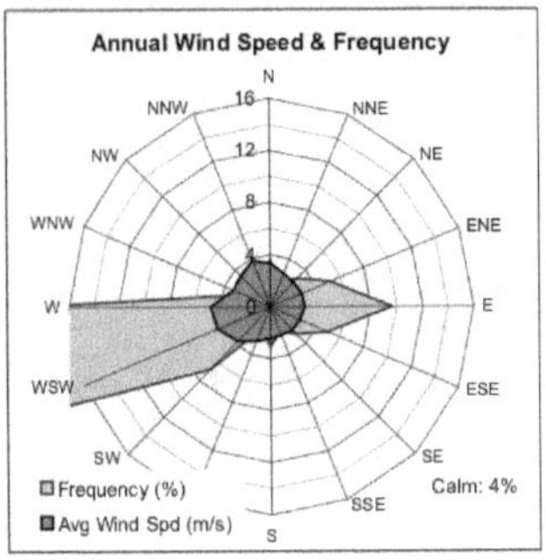

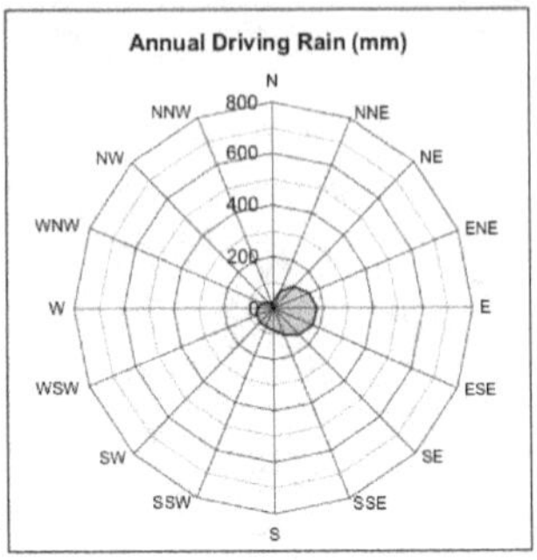

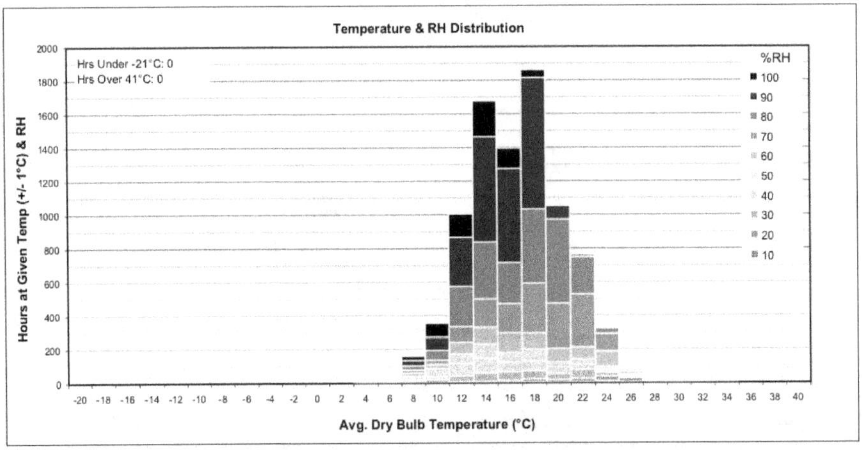

Averages		Jan	Feb	Mar	Apr	May	Jun	Jul	Aug	Sep	Oct	Nov	Dec	Year
Temperature (°C)		13.4	13.3	14.6	15.5	17.0	17.6	19.8	20.1	19.9	18.1	16.0	14.4	16.6
RH (%)		59	67	67	70	75	79	75	77	72	72	68	60	70
Humidity Ratio (g/kg)		5.41	6.19	6.71	7.49	8.99	9.85	10.77	11.28	10.23	9.31	7.58	5.92	8.31
Wind Speed - All Hours (m/s)		3.3	3.8	4.1	3.8	3.8	3.4	3.9	3.8	3.4	3.5	3.2	2.8	3.6
Wind Speed - During Rain (m/s)		6.2	5.3	4.0	4.4	2.2	0.0	3.5	0.0	0.0	3.9	4.2	4.5	3.2

Totals		Jan	Feb	Mar	Apr	May	Jun	Jul	Aug	Sep	Oct	Nov	Dec	Year
Snowfall (cm)		0.0	0.0	0.0	0.0	0.0	0.0	0.0	0.0	0.0	0.0	0.0	0.0	0.0
Rainfall (mm)		76	79	61	16	6	2	1	4	7	9	29	45	334
Driving Rain (mm/m²)		85	63	47	13	3	0	1	0	0	9	24	38	283
Hours with Rainfall (hrs)		26	10	32	10	6	0	9	0	0	17	23	26	159
Solar Radiation (kWh/m²)	N	18	19	29	32	45	46	46	36	29	24	19	16	359
	E	44	49	66	72	73	64	82	83	62	54	46	43	738
	S	104	101	93	77	63	55	59	76	86	106	107	111	1038
	W	43	56	69	83	96	93	94	85	73	63	48	42	846
	Horiz	89	107	147	172	203	204	217	204	153	127	95	84	1802

Design Temperatures: ASHRAE HOF 2001
Degree Days: NCDC Climate Normals
Rainfall: NCDC Climate Normals, TMY2
T, RH, Wind, Sun: TMY2

www.balancedsolutions.com
2005.06.07
V 1.0

Climate Summary for Building Design — Kansas City, MO USA

Location	
City	Kansas City
State/Prov	MO
Country	USA
Lat (°)	39.3
Long (°)	-94.7
Elev (m)	315
Time Zone (hrs)	-6

Design Temperatures	
Heating (°C)	-15.4
Cooling (°C)	33.8

Degree Days	
HDD (18°C)	2916
CDD (10°C)	2188

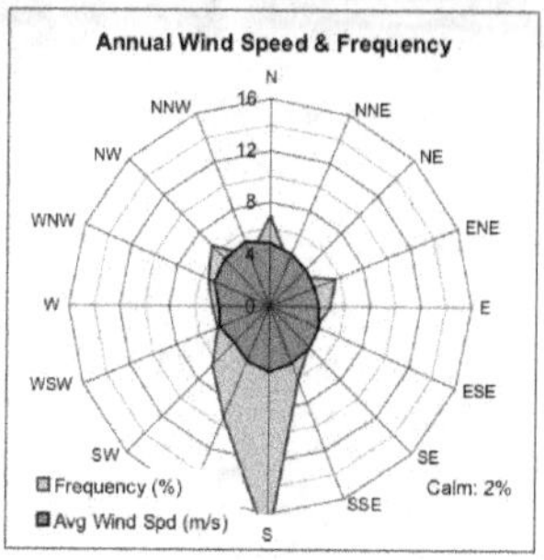

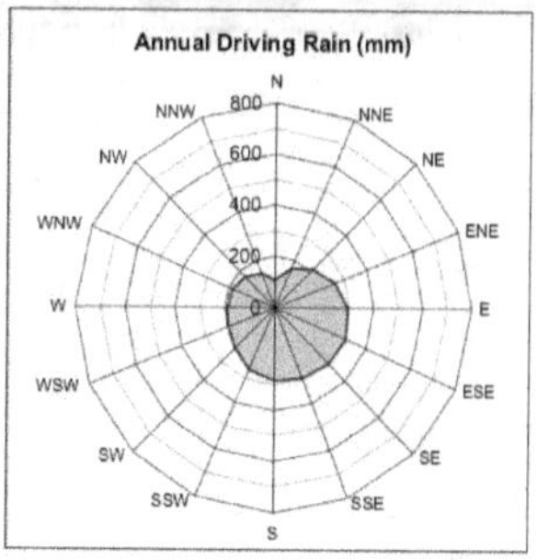

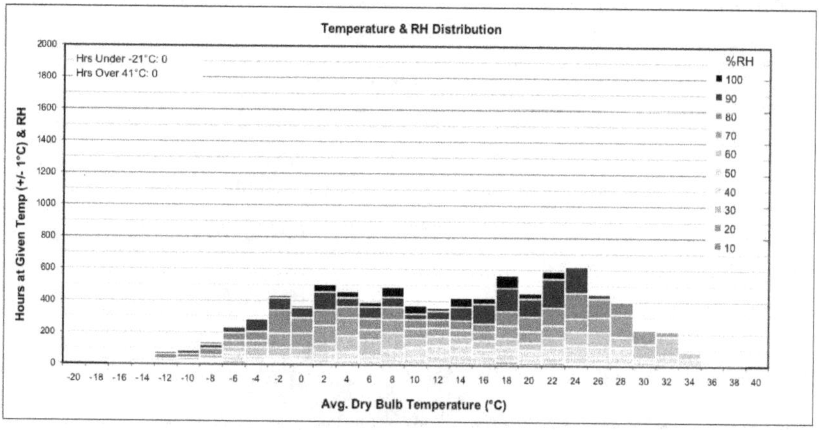

Averages		Jan	Feb	Mar	Apr	May	Jun	Jul	Aug	Sep	Oct	Nov	Dec	Year
Temperature (°C)		-2.6	0.1	5.1	14.6	17.9	23.3	27.6	24.2	19.8	15.2	5.7	0.4	12.6
RH (%)		69	55	59	57	64	74	62	74	69	65	72	65	65
Humidity Ratio (g/kg)		2.22	2.22	3.16	6.06	8.09	13.30	14.21	13.95	9.89	7.26	4.29	2.73	7.28
Wind Speed - All Hours (m/s)		4.5	4.2	5.0	5.2	4.7	3.9	4.1	4.7	3.4	4.4	4.9	4.7	4.5
Wind Speed - During Rain (m/s)		5.2	4.1	5.5	6.5	4.9	4.5	3.9	5.1	3.0	5.1	4.6	5.5	4.8

Totals		Jan	Feb	Mar	Apr	May	Jun	Jul	Aug	Sep	Oct	Nov	Dec	Year
Snowfall (cm)		14.0	11.4	8.6	2.0	0.0	0.0	0.0	0.0	0.0	0.3	3.3	10.9	50.5
Rainfall (mm)		15	22	53	84	137	113	112	90	118	84	55	31	914
Driving Rain (mm/m²)		13	18	55	91	103	77	75	81	53	82	44	36	729
Hours with Rainfall (hrs)		3	14	48	19	19	34	29	26	26	31	27	23	299
Solar Radiation (kWh/m²)	N	15	19	27	33	43	48	46	35	28	20	15	14	343
	E	40	41	61	74	82	90	85	83	65	55	36	31	742
	S	97	83	91	77	68	60	65	77	88	108	85	84	983
	W	38	41	63	74	85	81	96	82	67	57	36	32	751
	Horiz	68	80	126	155	189	194	203	181	135	107	67	58	1564

Design Temperatures: ASHRAE HOF 2001
Degree Days: NCDC Climate Normals
Rainfall: NCDC Climate Normals, TMY2
T, RH, Wind, Sun: TMY2

2005.06.07
V 1.0

Climate Summary for Building Design — Miami, FL USA

Location	
City	Miami
State/Prov	FL
Country	USA
Lat (°)	25.8
Long (°)	-80.3
Elev (m)	2
Time Zone (hrs)	-5

Design Temperatures	
Heating (°C)	9.8
Cooling (°C)	32.2

Degree Days	
HDD (18°C)	83
CDD (10°C)	5394

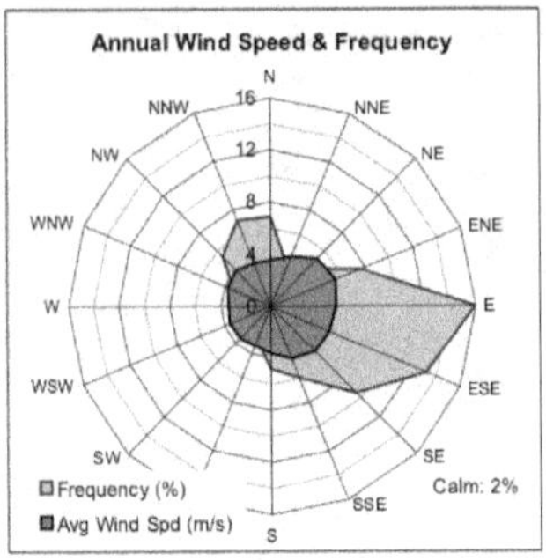

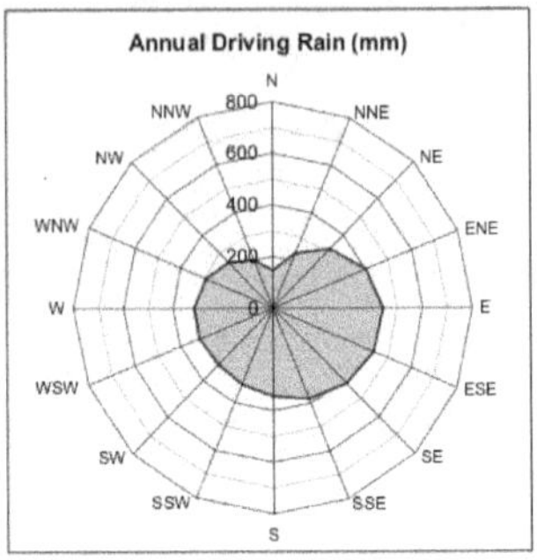

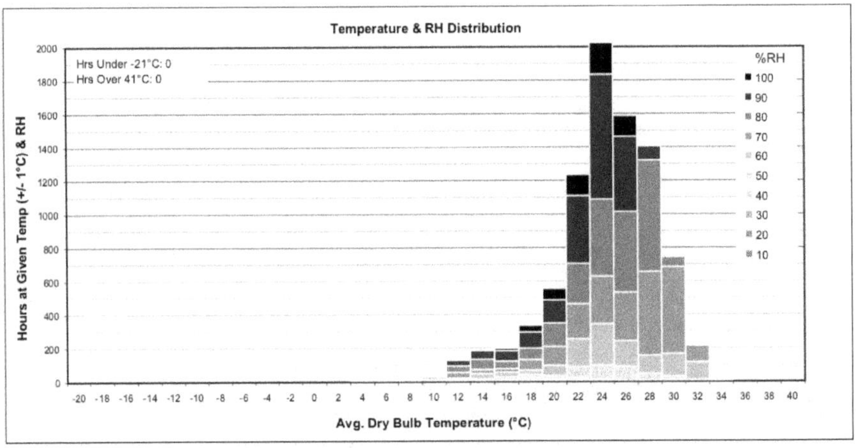

Averages	Jan	Feb	Mar	Apr	May	Jun	Jul	Aug	Sep	Oct	Nov	Dec	Year
Temperature (°C)	20.0	20.8	21.6	24.5	25.8	27.3	28.0	27.9	26.9	25.1	23.2	20.6	24.3
RH (%)	75	71	68	63	76	72	76	74	78	77	70	70	73
Humidity Ratio (g/kg)	11.34	11.22	11.42	12.29	15.83	16.33	17.92	17.35	17.28	15.32	12.53	10.67	14.12
Wind Speed - All Hours (m/s)	4.3	4.8	5.6	5.6	4.5	3.6	3.9	4.0	3.0	3.5	4.8	4.4	4.3
Wind Speed - During Rain (m/s)	4.9	4.1	6.0	5.1	4.1	3.7	4.5	4.5	3.6	4.1	4.8	6.4	4.6

Totals		Jan	Feb	Mar	Apr	May	Jun	Jul	Aug	Sep	Oct	Nov	Dec	Year
Snowfall (cm)		0.0	0.0	0.0	0.0	0.0	0.0	0.0	0.0	0.0	0.0	0.0	0.0	0.0
Rainfall (mm)		48	53	65	85	140	217	147	219	213	157	87	55	1487
Driving Rain (mm/m²)		47	42	66	64	85	133	128	147	142	85	71	46	1057
Hours with Rainfall (hrs)		32	26	14	8	35	45	38	30	70	22	22	2	344
Solar Radiation (kWh/m²)	N	22	23	32	36	48	54	55	49	36	31	24	22	434
	E	54	57	76	78	85	80	84	77	66	64	53	48	821
	S	101	92	84	62	48	47	50	60	67	87	92	102	892
	W	51	55	75	79	76	69	75	79	61	61	48	47	776
	Horiz	107	122	159	184	186	173	185	175	146	133	105	103	1780

Design Temperatures: ASHRAE HOF 2001
Degree Days: NCDC Climate Normals
Rainfall: NCDC Climate Normals, TMY2
T, RH, Wind, Sun: TMY2

2005.06.07
V 1.0

Climate Summary for Building Design — Minneapolis, MN USA

Location	
City	Minneapolis
State/Prov	MN
Country	USA
Lat (°)	48.6
Long (°)	-93.4
Elev (m)	255
Time Zone (hrs)	-6

Design Temperatures	
Heating (°C)	-23.7
Cooling (°C)	31.1

Degree Days	
HDD (18°C)	4376
CDD (10°C)	1510

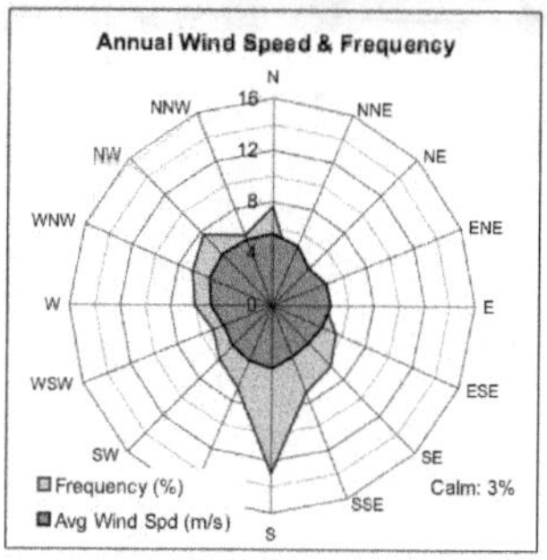

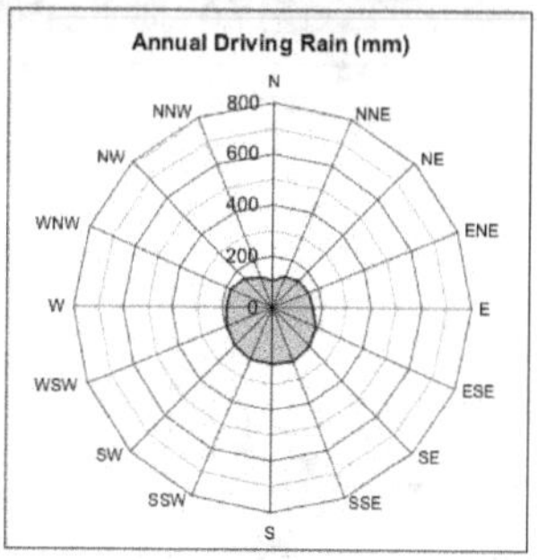

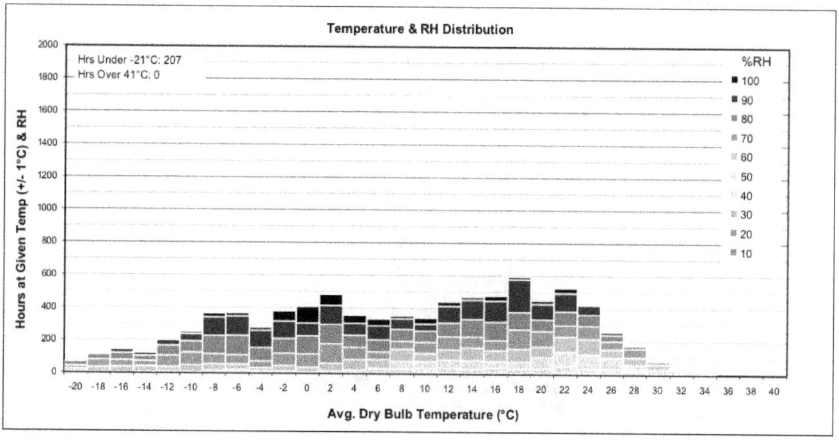

Averages	Jan	Feb	Mar	Apr	May	Jun	Jul	Aug	Sep	Oct	Nov	Dec	Year
Temperature (°C)	-11.8	-8.1	0.3	8.7	16.0	20.5	22.0	21.6	16.1	8.7	0.7	-8.2	7.2
RH (%)	76	69	67	62	61	65	70	72	73	65	78	75	69
Humidity Ratio (g/kg)	1.38	1.54	2.83	4.43	7.05	9.85	11.70	11.54	8.34	4.63	3.20	1.70	5.68
Wind Speed - All Hours (m/s)	4.8	4.4	5.4	5.2	4.7	4.2	4.4	4.2	4.1	4.9	4.5	4.6	4.6
Wind Speed - During Rain (m/s)	0.0	0.0	6.8	6.3	5.0	4.9	5.6	4.7	4.9	7.3	5.0	7.7	4.9

Totals		Jan	Feb	Mar	Apr	May	Jun	Jul	Aug	Sep	Oct	Nov	Dec	Year
Snowfall (cm)		27.2	20.6	26.7	7.1	0.3	0.0	0.0	0.0	0.0	1.3	19.8	23.9	126.7
Rainfall (mm)		0	0	21	52	82	110	103	103	68	52	29	2	622
Driving Rain (mm/m²)		0	0	28	77	70	99	91	77	60	68	28	3	599
Hours with Rainfall (hrs)		0	0	55	80	17	11	15	15	27	16	14	2	252
Solar Radiation (kWh/m²)	N	14	19	27	32	43	47	47	37	26	20	14	13	339
	E	34	44	59	70	87	90	92	83	61	48	30	26	722
	S	95	95	97	80	83	77	81	89	96	96	67	70	1028
	W	35	44	58	66	83	83	91	81	66	47	28	27	708
	Horiz	53	74	110	133	182	193	195	165	121	82	48	41	1396

Design Temperatures: ASHRAE HOF 2001
Degree Days: NCDC Climate Normals
Rainfall: NCDC Climate Normals, TMY2
T, RH, Wind, Sun: TMY2

2005.06.07
V 1.0

Climate Summary for Building Design — Seattle, WA USA

Location	
City	Seattle
State/Prov	WA
Country	USA
Lat (°)	48.0
Long (°)	-124.6
Elev (m)	122
Time Zone (hrs)	-8

Design Temperatures	
Heating (°C)	-2.2
Cooling (°C)	27.4

Degree Days	
HDD (18°C)	2665
CDD (10°C)	1113

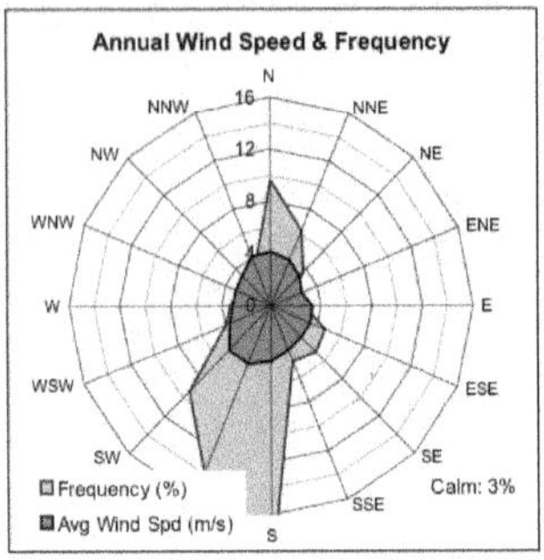

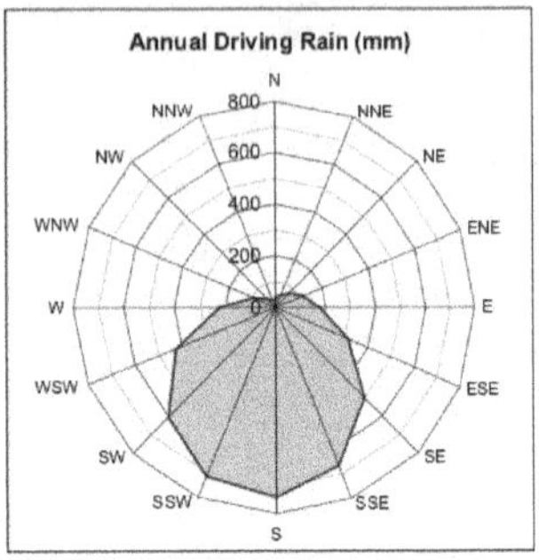

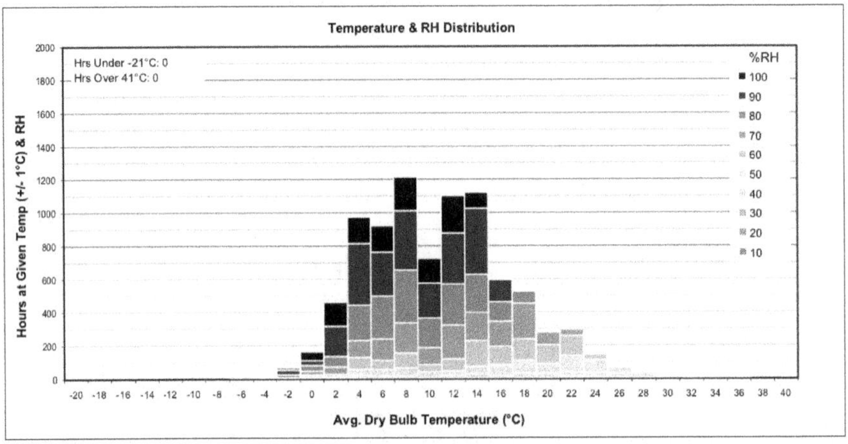

Averages		Jan	Feb	Mar	Apr	May	Jun	Jul	Aug	Sep	Oct	Nov	Dec	Year
Temperature (°C)		4.5	6.7	7.0	9.5	13.0	15.2	17.7	18.0	15.4	11.4	7.1	4.8	10.9
RH (%)		77	70	72	71	72	67	65	70	73	80	76	85	73
Humidity Ratio (g/kg)		4.07	4.20	4.52	5.21	6.54	7.03	7.96	8.75	7.62	6.67	4.94	4.61	6.01
Wind Speed - All Hours (m/s)		4.1	3.5	3.5	4.9	3.9	3.7	3.8	3.7	3.8	3.7	4.9	3.3	3.9
Wind Speed - During Rain (m/s)		5.1	4.3	4.6	5.6	3.8	3.7	4.3	4.9	4.0	4.7	5.6	3.8	4.5

Totals		Jan	Feb	Mar	Apr	May	Jun	Jul	Aug	Sep	Oct	Nov	Dec	Year
Snowfall (cm)		12.4	4.1	3.3	0.3	0.0	0.0	0.0	0.0	0.0	0.0	2.8	6.1	29.0
Rainfall (mm)		118	102	92	66	45	38	20	26	41	81	147	137	913
Driving Rain (mm/m²)		126	98	101	90	41	29	19	20	36	84	181	122	948
Hours with Rainfall (hrs)		110	126	150	129	81	40	25	10	52	96	170	192	1181
Solar Radiation (kWh/m²)	N	10	14	23	33	46	47	42	39	20	18	13	9	312
	E	18	26	48	60	75	80	80	64	55	31	20	13	569
	S	37	49	71	71	79	73	81	82	93	64	37	29	765
	W	14	24	43	57	81	83	92	79	65	38	18	12	606
	Horiz	28	46	88	121	170	182	192	154	114	63	34	23	1213

Design Temperatures: ASHRAE HOF 2001
Degree Days: NCDC Climate Normals
Rainfall: NCDC Climate Normals, TMY2
T, RH, Wind, Sun: TMY2

2005.06.07
V 1.0

Climate Summary for Building Design — Toronto, ON Canada

Location	
City	Toronto
State/Prov	ON
Country	Canada
Lat (°)	43.7
Long (°)	-79.4
Elev (m)	173
Time Zone (hrs)	-5

Design Temperatures	
Heating (°C)	-17.2
Cooling (°C)	28.7

Degree Days	
HDD (18°C)	3570
CDD (10°C)	1457

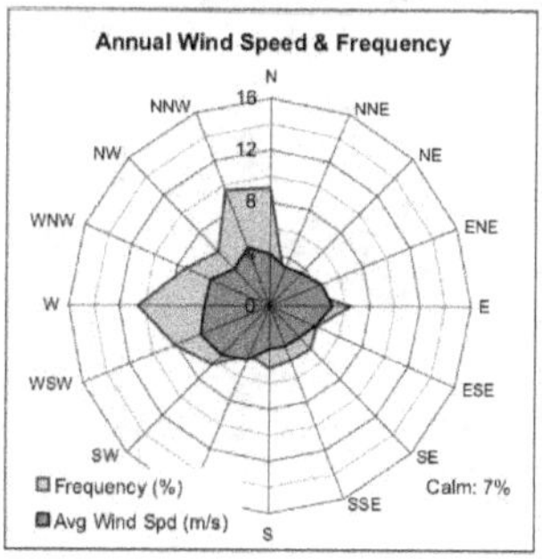

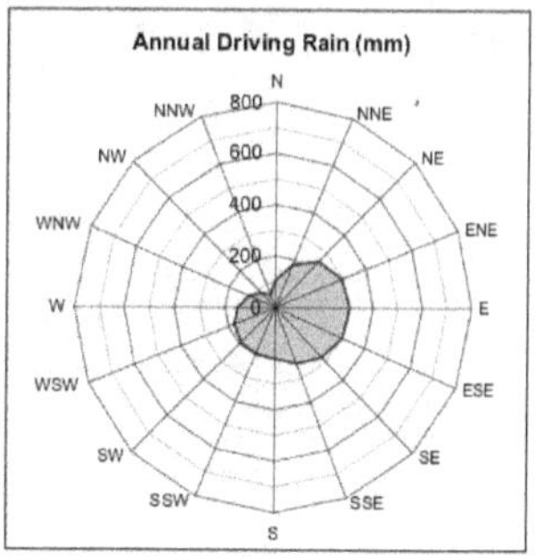

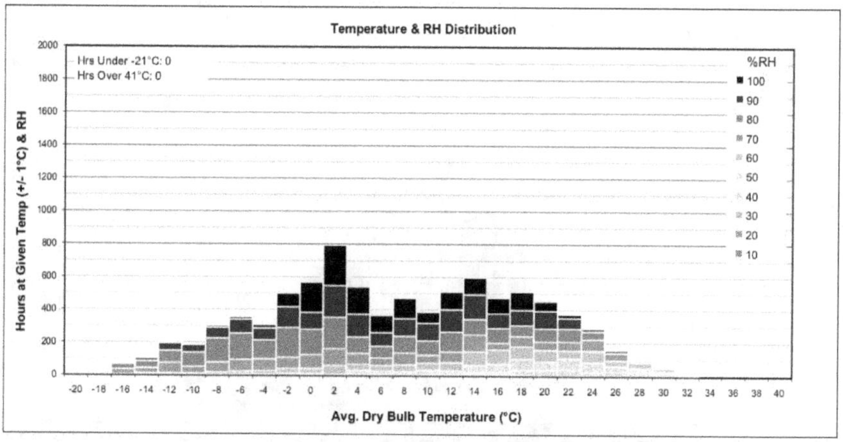

Averages	Jan	Feb	Mar	Apr	May	Jun	Jul	Aug	Sep	Oct	Nov	Dec	Year
Temperature (°C)	-5.8	-5.7	-0.7	5.7	12.0	17.7	20.8	19.8	15.0	8.5	3.5	-2.5	7.4
RH (%)	79	75	75	70	62	68	71	71	75	77	83	80	74
Humidity Ratio (g/kg)	2.11	2.02	2.80	4.18	5.50	8.64	10.75	10.31	8.01	5.58	4.22	2.70	5.57
Wind Speed - All Hours (m/s)	4.5	5.3	5.5	4.8	4.3	3.8	3.2	2.1	3.3	4.3	4.4	5.7	4.3
Wind Speed - During Rain (m/s)	4.5	5.8	6.2	5.2	4.8	4.7	3.4	2.3	3.8	4.7	5.3	6.7	4.8

Totals		Jan	Feb	Mar	Apr	May	Jun	Jul	Aug	Sep	Oct	Nov	Dec	Year
Snowfall (cm)		31.1	22.1	19.2	5.7	0.1	0.0	0.0	0.0	0.0	0.5	7.6	29.2	115.5
Rainfall (mm)		25	22	37	62	72	74	74	80	78	63	62	35	685
Driving Rain (mm/m²)		27	31	50	71	71	67	51	33	59	62	75	49	646
Hours with Rainfall (hrs)		44	38	39	70	54	39	46	26	71	51	95	32	605
Solar Radiation (kWh/m²)	N	12	16	23	29	45	51	47	40	26	21	12	9	333
	E	24	32	55	64	94	88	92	80	67	41	19	18	675
	S	65	69	76	73	71	67	69	78	87	75	42	49	820
	W	26	36	46	62	71	81	80	76	58	43	21	18	618
	Horiz	45	63	98	130	174	184	187	159	121	78	39	33	1311

Design Temperatures: ASHRAE HOF 2001
Degree Days: Cdn Climate Normals 1971-2000
Rainfall: Cdn Climate Normals 1971-2000, CWEC
T, RH, Wind, Sun: CWEC

Climate Summary for Building Design — Vancouver, BC Canada

Location	
City	Vancouver
State/Prov	BC
Country	Canada
Lat (°)	49.3
Long (°)	-123.4
Elev (m)	4
Time Zone (hrs)	-8

Design Temperatures	
Heating (°C)	-4.7
Cooling (°C)	23.2

Degree Days	
HDD (18°C)	2927
CDD (10°C)	901

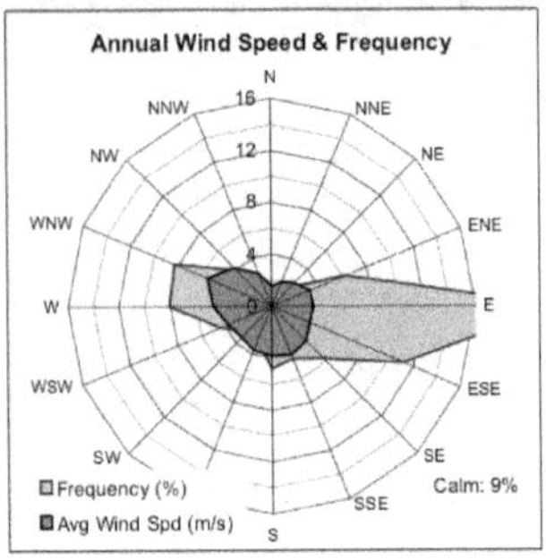

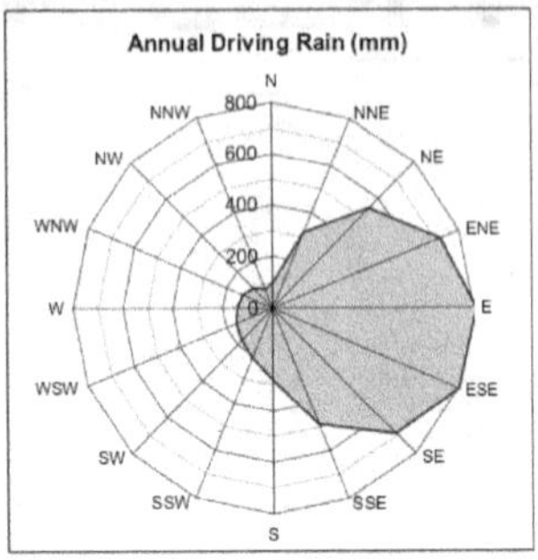

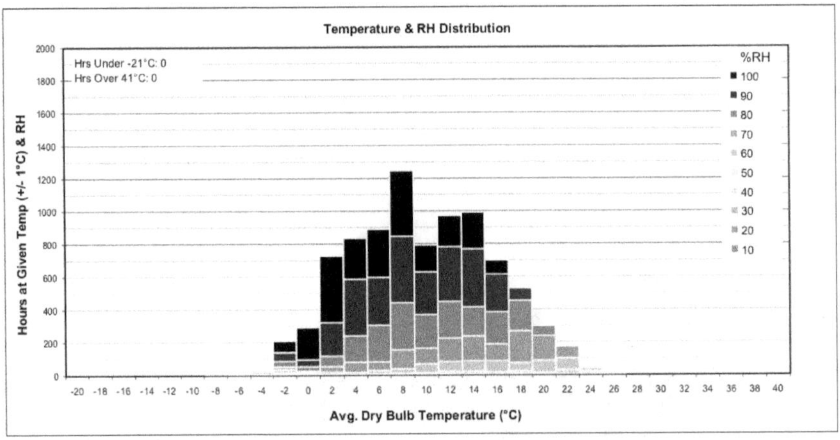

Averages	Jan	Feb	Mar	Apr	May	Jun	Jul	Aug	Sep	Oct	Nov	Dec	Year
Temperature (°C)	3.2	5.1	6.1	8.7	11.8	15.1	17.0	17.1	13.8	9.8	5.3	3.6	9.7
RH (%)	87	86	81	73	73	73	76	79	80	82	82	88	80
Humidity Ratio (g/kg)	4.20	4.71	4.77	5.06	6.28	7.77	9.02	9.52	7.86	6.23	4.72	4.40	6.21
Wind Speed - All Hours (m/s)	3.2	2.7	3.4	3.2	3.4	3.0	2.9	3.3	3.8	3.5	3.4	4.3	3.3
Wind Speed - During Rain (m/s)	4.0	4.2	5.0	3.9	3.0	3.2	2.2	3.1	5.4	3.9	4.4	5.3	4.0

Totals		Jan	Feb	Mar	Apr	May	Jun	Jul	Aug	Sep	Oct	Nov	Dec	Year
Snowfall (cm)		16.3	10.4	2.8	0.5	0.0	0.0	0.0	0.0	0.0	0.1	3.2	13.3	46.6
Rainfall (mm)		137	131	115	89	68	58	41	42	70	122	182	183	1239
Driving Rain (mm/m²)		120	118	129	76	45	39	20	28	82	115	165	217	1156
Hours with Rainfall (hrs)		136	115	146	99	73	50	44	41	60	235	140	210	1349
Solar Radiation (kWh/m²)	N	9	11	22	28	43	51	47	35	22	14	9	6	295
	E	13	23	40	64	80	82	89	75	59	31	15	12	583
	S	33	57	69	86	80	72	83	90	95	60	38	33	798
	W	15	26	47	73	91	93	111	82	63	26	16	12	656
	Horiz	25	44	82	136	174	179	203	164	116	54	28	20	1223

Design Temperatures: ASHRAE HOF 2001
Degree Days: Cdn Climate Normals 1971-2000
Rainfall: Cdn Climate Normals 1971-2000, CWEC
T, RH, Wind, Sun: CWEC

2005.06.07
V 1.0

www.ingramcontent.com/pod-product-compliance
Lightning Source LLC
LaVergne TN
LVHW052014200926
848481LV00005B/1487